Introduction to Remote Sensing

Introduction to Remote Sensing

FIFTH EDITION

JAMES B. CAMPBELL
RANDOLPH H. WYNNE

THE GUILFORD PRESS
New York • London

Library of Congress Cataloging-in-Publication Data

Campbell, James B., 1944–
    Introduction to remote sensing / James B. Campbell,
  Randolph H. Wynne. — 5th ed.
      p. cm.
  Includes bibliographical references and index.
  ISBN 978-1-60918-176-5 (hardcover)
1. Remote sensing. I. Wynne, Randolph H. II. Title.
G70.4.C23 2011
621.36′78—dc22
                                                    2011011657

To the memory of my parents,

James B. Campbell
Dorothy Ann Campbell

—J. B. C.

To my family and friends,

—R. H. W.

Contents

PART I. FOUNDATIONS

PART II. IMAGE ACQUISITION

PART III. ANALYSIS

PART IV. APPLICATIONS

Preface

Readers of earlier editions will notice immediately that among the changes for this new, fifth edition, the most significant is the addition of a coauthor—a change that will benefit readers by the contribution of additional expertise and of another perspective on the subject matter.

During the interval that has elapsed since the release of the fourth edition, the pace of changes within the field of remote sensing and within the geospatial realm in general has continued to escalate at a pace that in earlier decades would have seemed unimaginable, even for committed visionaries. Likewise, few could have anticipated the development of the analytical tools and techniques available for examination of remotely sensed data, the widespread availability of remotely sensed imagery, or the multiplicity of remote sensing's uses throughout society. These developments alone present a challenge for any text on this subject.

Further, the tremendous volume of relevant material and the thorough access provided by the World Wide Web can cause anyone to ponder the role of a university text— isn't its content already available to any reader? The reality is that, despite the value of the World Wide Web as a resource for any student of remote sensing, its complexity and dynamic character can increase, rather than diminish, the value of an introductory text. Because of the overwhelming volume of unstructured information encountered on the Web, students require a guide to provide structure and context that enable them to select and assess the many sources at hand. This text forms a guide for the use of the many other sources available—sources that may be more comprehensive and up to date than the content of any text. Thus, we encourage students to use this volume in partnership with online materials; the text can serve as a guide that provides context, and online materials as a reference that provides additional detail where needed.

Instructors should supplement the content of this volume with material of significance in their own programs. Supplementary materials will, of course, vary greatly from one institution to the next, depending on access to facilities and equipment, as well as the varying expectations and interests of instructors, students, and curricula. It is assumed that the text will be used as the basis for readings and lectures and that most courses will include at least brief laboratory exercises that permit students to examine more images than can be presented here. Because access to specific equipment and software varies so greatly, and because of the great variation in emphasis noted earlier, this book does not include laboratory exercises. Each chapter concludes with a set of Review Questions and problems that can assist in review and assessment of concepts and material.

For students who intend to specialize in remote sensing, this text forms not only an introduction but also a framework for subjects to be studied in greater detail. Students who do plan specialization in remote sensing should consult their instructors to plan a comprehensive course of study based on work in several disciplines, as discussed in Chapter 1. This approach is reflected in the text itself: It introduces the student to principal topics of significance for remote sensing but recognizes that students will require additional depth in their chosen fields of specialization.

For those students who do not intend to pursue remote sensing beyond the introductory level, this book serves as an overview and introduction, so that they can understand remote sensing, its applications in varied disciplines, and its significance in today's world. For many, the primary emphasis may focus on study of those chapters and methods of greatest significance in the student's major field of study.

Instructors may benefit from a preview of some of the changes implemented for this fifth edition. First, we have revised the material pertaining to aerial photography to recognize the transition from analog to digital systems. Although, within the realm of commercial practice, this transition has been under way for decades, it has not been fully recognized in most instructional materials. We are aware of the difficulties in making this change; some instructors have dedicated instructional materials to the analog model and may not be ready for this new content. Further, we have been challenged to define intuitive strategies for presenting this new information to students in a logical manner. Other important changes include a thorough update and revision of the accuracy assessment chapter, a new chapter on change detection and multitemporal data analysis, and revision of material devoted to digital preprocessing.

For many chapters, we have added a short list of teaching and learning resources—principally a selection of online tutorials or short videos that provide depth or breadth to the content presented in the chapter or that simply illustrate content. These have been selected for their brevity (most are less than 3–4 minutes or so in length) and for their effectiveness in explaining or illustrating content relative to the chapter in question.

For the most part, we have excluded videos that focus on promotional content; those that do serve a promotional purpose were selected for their effectiveness in presenting technical content rather than as an endorsement of a particular product or service. In most instances, the match to chapter content is largely fortuitous, so comprehensive coverage of chapter content will not be provided. We recognize that some of these links may expire in time, but we believe that instructors can easily find replacements in related material posted after publication of this volume.

To permit instructors to tailor assignments to meet specific structures for their own courses, content is organized at several levels. At the broadest level, a rough division into four units offers a progression in the kind of knowledge presented, with occasional concessions to practicality (such as placing the "Image Interpretation" chapter under "Image Acquisition" rather than in its logical position in "Analysis"). Each division consists of two or more chapters organized as follows:

Part I. Foundations
    1. History and Scope of Remote Sensing
    2. Electromagnetic Radiation

These 21 chapters each constitute more or less independent units that can be selected as necessary to meet the specific needs of each instructor. Numbered sections within chapters form even smaller units that can be selected and combined with other material as desired by the instructor.

We gratefully acknowledge the contributions of those who assisted in identifying and acquiring images used in this book. Individuals and organizations in both private industry and governmental agencies have been generous with advice and support. Daedalus Enterprises Incorporated, EROS Data Center, Digital Globe, Environmental Research Institute of Michigan, EOSAT, GeoSpectra Corporation, IDELIX Software, and SPOT Image Corporation are among the organizations that assisted our search for suitable images. For this edition, we are grateful for the continued support of these organizations and for the assistance of the U.S. Geological Survey, RADARSAT International, GeoEye, and the Jet Propulsion Laboratory. Also, we gratefully recognize the assistance of FUGRO EarthData, Leica GeoSystems, the Jet Propulsion Laboratory, and Orbital Imaging Corporation.

Much of what is good about this book is the result of the assistance of colleagues in many disciplines in universities, corporations, and research institutions who have contributed through their correspondence, criticisms, explanations, and discussions. Students in our classes have, through their questions, mistakes, and discussions, contributed greatly to our own learning, and therefore to this volume. Faculty who use this text at other universities have provided suggestions and responded to questionnaires designed by The Guilford Press.

At Guilford, Janet Crane, Peter Wissoker, and Kristal Hawkins guided preparation of the earlier editions. For these earlier editions we are grateful for the special contributions of Chris Hall, Sorin Popescu, Russ Congalton, Bill Carstensen, Don Light, David Pitts, and Jim Merchant. Buella Prestrude and George Will assisted preparation of illustrations.

Many individuals have supported the preparation of this edition, although none are responsible for the errors and shortcomings that remain. Colleagues, including Tom Dickerson, Kirsten deBeurs, Peter Sforza, Chris North, Bill Carstensen, Steve Prisley, and Maggi Kelly, have been generous in sharing graphics and related materials for this edition. We are grateful likewise for the suggestions and corrections offered by readers. Teaching assistants, including Kristin DallaPiazza, Kellyn Montgomery, Allison LeBlanc, and Baojuan Zheng, have contributed to development of materials used in this edition. Three anonymous reviewers provided especially thorough and insightful review comments that extended well beyond the scope of the usual manuscript reviews. We have implemented most of their suggestions, although the production schedule has limited our ability to adopt them all. At Guilford, Kristal Hawkins has guided the launch of this fifth edition. Seymour Weingarten, editor in chief, continued his support of this project through the course of its five editions.

Users of this text can inform the authors of errors, suggestions, and other comments at

jayhawk@vt.edu

JAMES B. CAMPBELL
RANDOLPH H. WYNNE

List of Tables

List of Figures

List of Plates

FOUNDATIONS

CHAPTER ONE

History and Scope of Remote Sensing

1.1. Introduction

A picture is worth a thousand words. Is this true, and if so, why?

Pictures concisely convey information about positions, sizes, and interrelationships between objects. By their nature, they portray information about things that we can recognize as objects. These objects in turn can convey deep levels of meaning. Because humans possess a high level of proficiency in deriving information from such images, we experience little difficulty in interpreting even those scenes that are visually complex. We are so competent in such tasks that it is only when we attempt to replicate these capabilities using computer programs, for instance, that we realize how powerful our abilities are to derive this kind of intricate information. Each picture, therefore, can truthfully be said to distill the meaning of at least a thousand words.

This book is devoted to the analysis of a special class of pictures that employ an overhead perspective (e.g., maps, aerial photographs, and similar images), including many that are based on radiation not visible to the human eye. These images have special properties that offer unique advantages for the study of the Earth's surface: We can see patterns instead of isolated points and relationships between features that otherwise seem independent. They are especially powerful because they permit us to monitor changes over time; to measure sizes, areas, depths, and heights; and, in general, to acquire information that is very difficult to acquire by other means. However, our ability to extract this kind of information is not innate; we must work hard to develop the knowledge and skills that allow us to use images (Figure 1.1).

Specialized knowledge is important because remotely sensed images have qualities that differ from those we encounter in everyday experience:

- Image presentation
- Unfamiliar scales and resolutions
- Overhead views from aircraft or satellites
- Use of several regions of the electromagnetic spectrum

This book explores these and other elements of remote sensing, including some of its many practical applications. Our purpose in Chapter 1 is to briefly outline its content, origins, and scope as a foundation for the more specific chapters that follow.

FIGURE 1.1. Two examples of visual interpretation of images. Humans have an innate ability to derive meaning from the complex patterns of light and dark that form this image—we can interpret patterns of light and dark as people and objects. At another, higher, level of understanding, we learn to derive meaning beyond mere recognition of objects, to interpret the arrangement of figures, to notice subtle differences in posture, and to assign meaning not present in the arbitrary pattern of light and dark. Thus this picture tells a story. It conveys a meaning that can be received only by observers who can understand the significance of the figures, the statue, and their relationship.

1.2. Definitions

The field of remote sensing has been defined many times (Table 1.1). Examination of common elements in these varied definitions permits identification of the topic's most important themes. From a cursory look at these definitions, it is easy to identify a central concept: the gathering of information at a distance. This excessively broad definition, however, must be refined if it is to guide us in studying a body of knowledge that can be approached in a single course of study.

FIGURE 1.1. (cont). So it is also with this second image, a satellite image of southwestern Virginia. With only modest effort and experience, we can interpret these patterns of light and dark to recognize topography, drainage, rivers, and vegetation. There is a deeper meaning here as well, as the pattern of white tones tells a story about the interrelated human and natural patterns within this landscape—a story that can be understood by those prepared with the necessary knowledge and perspective. Because this image employs an unfamiliar perspective and is derived from radiation outside the visible portion of the electromagnetic spectrum, our everyday experience and intuition are not adequate to interpret the meaning of the patterns recorded here, so it is necessary to consciously learn and apply acquired knowledge to understand the meaning of this pattern.

The kind of remote sensing to be discussed here is devoted to observation of the Earth's land and water surfaces by means of reflected or emitted electromagnetic energy. This more focused definition excludes applications that could be reasonably included in broader definitions, such as sensing the Earth's magnetic field or atmosphere or the temperature of the human body. For our purposes, the definition can be based on modification of concepts given in Table 1.1:

TABLE 1.1. Remote Sensing: Some Definitions

Remote sensing has been variously defined but basically it is the art or science of telling something about an object without touching it. (Fischer et al., 1976, p. 34)

Remote sensing is the acquisition of physical data of an object without touch or contact. (Lintz and Simonett, 1976, p. 1)

Imagery is acquired with a sensor other than (or in addition to) a conventional camera through which a scene is recorded, such as by electronic scanning, using radiations outside the normal visual range of the film and camera—microwave, radar, thermal, infrared, ultraviolet, as well as multispectral, special techniques are applied to process and interpret remote sensing imagery for the purpose of producing conventional maps, thematic maps, resources surveys, etc., in the fields of agriculture, archaeology, forestry, geography, geology, and others. (American Society of Photogrammetry)

Remote sensing is the observation of a target by a device separated from it by some distance. (Barrett and Curtis, 1976, p. 3)

The term "remote sensing" in its broadest sense merely means "reconnaissance at a distance." (Colwell, 1966, p. 71)

Remote sensing, though not precisely defined, includes all methods of obtaining pictures or other forms of electromagnetic records of the Earth's surface from a distance, and the treatment and processing of the picture data. . . . Remote sensing then in the widest sense is concerned with detecting and recording electromagnetic radiation from the target areas in the field of view of the sensor instrument. This radiation may have originated directly from separate components of the target area; it may be solar energy reflected from them; or it may be reflections of energy transmitted to the target area from the sensor itself. (White, 1977, pp. 1–2)

"Remote sensing" is the term currently used by a number of scientists for the study of remote objects (earth, lunar, and planetary surfaces and atmospheres, stellar and galactic phenomena, etc.) from great distances. Broadly defined . . . , remote sensing denotes the joint effects of employing modern sensors, data-processing equipment, information theory and processing methodology, communications theory and devices, space and airborne vehicles, and large-systems theory and practice for the purposes of carrying out aerial or space surveys of the earth's surface. (National Academy of Sciences, 1970, p. 1)

Remote sensing is the science of deriving information about an object from measurements made at a distance from the object, i.e., without actually coming in contact with it. The quantity most frequently measured in present-day remote sensing systems is the electromagnetic energy emanating from objects of interest, and although there are other possibilities (e.g., seismic waves, sonic waves, and gravitational force), our attention . . . is focused upon systems which measure electromagnetic energy. (D. A. Landgrebe, quoted in Swain and Davis, 1978, p. 1)

Remote sensing is the practice of deriving information about the Earth's land and water surfaces using images acquired from an overhead perspective, using electromagnetic radiation in one or more regions of the electromagnetic spectrum, reflected or emitted from the Earth's surface.

This definition serves as a concise expression of the scope of this volume. It is not, however, universally applicable, and is not intended to be so, because practical constraints limit the scope of this volume. So, although this text must omit many interesting topics (e.g., meteorological or extraterrestrial remote sensing), it can review knowledge and perspectives necessary for pursuit of topics that cannot be covered in full here.

1.3. Milestones in the History of Remote Sensing

The scope of the field of remote sensing can be elaborated by examining its history to trace the development of some of its central concepts. A few key events can be offered to trace the evolution of the field (Table 1.2). More complete accounts are given by Stone (1974), Fischer (1975), Simonett (1983), and others.

Early Photography and Aerial Images Prior to the Airplane

Because the practice of remote sensing focuses on the examination of images of the Earth's surface, its origins lie in the beginnings of the practice of photography. The first attempts to form images by photography date from the early 1800s, when a number of scientists, now largely forgotten, conducted experiments with photosensitive chemicals. In 1839 Louis Daguerre (1789–1851) publicly reported results of his experiments with photographic chemicals; this date forms a convenient, although arbitrary, milestone for the birth of photography. Acquisition of the first aerial photograph has been generally credited to Gaspard-Félix Tournachon (1829–1910), known also by his pseudonym, Nadar. In 1858, he acquired an aerial photo from a tethered balloon in France. Nadar's aerial photos have been lost, although other early balloon photographs survive. In succeeding years numerous improvements were made in photographic technology and in methods of acquiring photographs of the Earth from balloons and kites. These aerial images of the

TABLE 1.2. Milestones in the History of Remote Sensing

1800	Discovery of infrared by Sir William Herschel
1839	Beginning of practice of photography
1847	Infrared spectrum shown by A. H. L. Fizeau and J. B. L. Foucault to share properties with visible light
1850–1860	Photography from balloons
1873	Theory of electromagnetic energy developed by James Clerk Maxwell
1909	Photography from airplanes
1914–1918	World War I: aerial reconnaissance
1920–1930	Development and initial applications of aerial photography and photogrammetry
1929–1939	Economic depression generates environmental crises that lead to governmental applications of aerial photography
1930–1940	Development of radars in Germany, United States, and United Kingdom
1939–1945	World War II: applications of nonvisible portions of electromagnetic spectrum; training of persons in acquisition and interpretation of airphotos
1950–1960	Military research and development
1956	Colwell's research on plant disease detection with infrared photography
1960–1970	First use of term *remote sensing*
	TIROS weather satellite
	Skylab remote sensing observations from space
1972	Launch of Landsat 1
1970–1980	Rapid advances in digital image processing
1980–1990	Landsat 4: new generation of Landsat sensors
1986	SPOT French Earth observation satellite
1980s	Development of hyperspectral sensors
1990s	Global remote sensing systems, lidars

Earth are among the first to fit the definition of remote sensing given previously, but most must be regarded as curiosities rather than as the basis for a systematic field of study.

Early Uses of the Airplane for Photography

The use of powered aircraft as platforms for aerial photography forms the next milestone. In 1909 Wilbur Wright piloted the plane that acquired motion pictures of the Italian landscape near Centocelli; these are said to be the first aerial photographs taken from an airplane. The maneuverability of the airplane provided the capability of controlling speed, altitude, and direction required for systematic use of the airborne camera. Although there were many attempts to combine the camera with the airplane, the instruments of this era were clearly not tailored for use with each other (Figure 1.2).

World War I

World War I (1914–1918) marked the beginning of the acquisition of aerial photography on a routine basis. Although cameras used for aerial photography during this conflict were designed specifically for use with the airplane, the match between the two instruments was still rather rudimentary by the standards of later decades (Figure 1.3). The value of aerial photography for military reconnaissance and surveillance became increasingly clear as the war continued, and its applications became increasingly sophisticated. By the conclusion of the conflict, aerial photography's role in military operations was recognized, although training programs, organizational structures, and operational doctrine had not yet matured.

FIGURE 1.2. Early aerial photography by the U.S. Navy, 1914. This photograph illustrates difficulties encountered in early efforts to match the camera with the airplane—neither is well-suited for use with the other. From U.S. Navy, National Archives and Records Administration, ARC 295605.

FIGURE 1.3. Aerial photography, World War I. By the time of World War I, attempts to match the camera and the airplane had progressed only to a modest extent, as illustrated by this example. The aircraft (with the biplane design typical of this era) has a port for oblique photography to allow the photographer to aim the camera through the port from within the fuselage to avoid the disadvantages of leaning over the edge of the cockpit. The photographer wears a chest-mounted microphone for communication with the pilot and is shown holding a supply of extra plates for the camera. From U.S. National Archives and Records Administration, Still Pictures, E-4156.

Interwar Years: 1919–1939

Numerous improvements followed from these beginnings. Camera designs were improved and tailored specifically for use in aircraft. The science of *photogrammetry*—the practice of making accurate measurements from photographs—was applied to aerial photography, with the development of instruments specifically designed for analysis of aerial photos. Although the fundamentals of photogrammetry had been defined much earlier, the field developed toward its modern form in the 1920s, with the application of specialized photogrammetric instruments. From these origins, another landmark was established: the more or less routine application of aerial photography in government programs, initially for topographic mapping but later for soil survey, geologic mapping, forest surveys, and agricultural statistics.

Many of the innovations during this era were led by visionary pioneers who established successful niches in private industry to develop civil applications of aerial mapping. Sherman Fairchild (1896–1971) founded numerous companies, including Fairchild Surveys and Fairchild Camera and Instruments, that became leaders in aviation and in aerial camera design. Talbert Abrams (1895–1990) led many innovations in aerial survey, in aviation, in camera design, in training, and in worldwide commercial operations. During this period, the well-illustrated volume by Lee (1922), *The Face of the Earth as Seen from the Air*, surveyed the range of possible applications of aerial photography in a variety of disciplines from the perspective of those early days. Although the applications that Lee envisioned were achieved at a slow pace, the expression of governmental interest ensured continuity in the scientific development of the acquisition and analysis of aerial photography, increased the number of photographs available, and trained many people in uses of aerial photography. Nonetheless, the acceptance of the use of aerial photography in most governmental and scientific activities developed slowly because of resistance among traditionalists, imperfections in equipment and technique, and genuine uncertainties regarding the proper role of aerial photography in scientific inquiry and practical applications.

The worldwide economic depression of 1929–1939 was not only an economic and financial crisis but also for many nations an environmental crisis. National concerns about social and economic impacts of rural economic development, widespread soil erosion, reliability of water supplies, and similar issues led to some of the first governmental applications of aerial surveys to record and monitor rural economic development. In the United States, the U.S. Department of Agriculture and the Tennessee Valley Authority led efforts to apply aerial photography to guide environmental planning and economic development. Such efforts formed an important contribution to the institutionalization of the use of aerial photography in government and to the creation of a body of practical experience in applications of aerial photography (Figure 1.4).

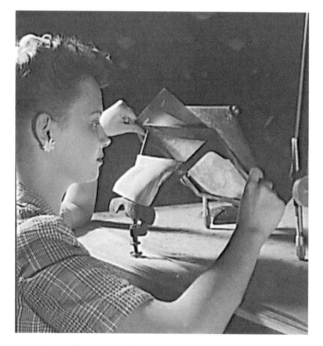

FIGURE 1.4. Progress in applications of aerial photography, 1919–1939. During the interval between World War I and World War II (1919–1939) integration of the camera and the airplane progressed, as did institutionalization of aerial photography in government and industry. By June 1943, the date of this photograph, progress on both fronts was obvious. Here an employee of the U.S. Geological Survey uses a specialized instrument, the *Oblique Sketchmaster*, to match detail on an aerial photograph to an accurate map. By the time of this photograph, aerial photography formed an integral component of the U.S. Geological Survey operations. From U.S. Geological Survey and U.S. Library of Congress, fsa 8d38549.

World War II

These developments led to the eve of World War II (1939–1945), which forms the next milestone in our history. During the war years, use of the electromagnetic spectrum was extended from almost exclusive emphasis on the visible spectrum to other regions, most notably the infrared and microwave regions (far beyond the range of human vision). Knowledge of these regions of the spectrum had been developed in both basic and applied sciences during the preceding 150 years (see Table 1.2). However, during the war years, application and further development of this knowledge accelerated, as did dissemination of the means to apply it. Although research scientists had long understood the potential of the nonvisible spectrum, the equipment, materials, and experience necessary to apply it to practical problems were not at hand. Wartime research and operational experience provided both the theoretical and the practical knowledge required for everyday use of the nonvisible spectrum in remote sensing.

Furthermore, the wartime training and experience of large numbers of pilots, camera operators, and photointerpreters created a large pool of experienced personnel who were able to transfer their skills and experience into civilian occupations after the war. Many of these people assumed leadership positions in the efforts of business, scientific, and governmental programs to apply aerial photography and remote sensing to a broad range of problems.

During World War II, the expansion of aerial reconnaissance from the tactical, local focus of World War I toward capabilities that could reach deep within enemy territory to understand industrial and transportation infrastructure and indicate long-term capabilities and plans greatly increased the significance of aerial intelligence. Perhaps the best publicized example of this capability is the role of British photointerpreters who provided key components of the intelligence that detected the German V-1 and V-2 weapons well in advance of their deployment, thereby eliminating any advantage of surprise and enabling rapid implementation of countermeasures (Babington-Smith, 1957).

World War II also saw an expansion of the thematic scope of aerial reconnaissance. Whereas photointerpreters of the World War I era focused on identification and examination of military equipment and fortifications, their counterparts in World War II also examined topography, vegetation, trafficability, and other terrain features, thereby expanding the scope and knowledge base and the practice of photointerpretation.

Cold War Era

The successes of strategic photointerpretation during World War II set the stage for continued interest in aerial surveillance during the cold war era. Initially, technological trends established during World War II were continued and improved. However, as the nature of the cold war conflict became more clearly defined, strategic photointerpretation was seen as one of the few means of acquiring reliable information from within the closed societies of the Soviet bloc. Thus the U-2 aircraft and camera systems were developed to extend aviation and optical systems far beyond their expected limits; and later (1960) the CORONA strategic reconnaissance satellite (see Day, Logsdon, and Latell, 1998, and Chapter 6, this volume) provided the ability to routinely collect imagery from space. The best known contribution of photoreconnaissance within the cold war conflict came during the Cuban Missile Crisis. In 1962 U.S. photo interpreters were able to detect with confidence the early stages of Soviet introduction of missiles into Cuba far sooner than

Soviet strategists had anticipated, thereby setting the stage for defusing one of the most serious incidents of the cold war era (Brugioni, 1991).

The cold war era saw a bifurcation of capabilities, with a wide separation between applications within civil society and those within the defense and security establishments. The beginnings of the cold war between the Western democracies and the Soviet Union created the environment for further development of advanced reconnaissance techniques (Figure 1.5), which were often closely guarded as defense secrets and therefore not immediately available for civil applications. As newer, more sophisticated instruments were developed, the superseded technologies were released for wider, nondefense applications in the civilian economy (Figure 1.6).

Robert Colwell's Research in Applying Color Infrared Film

Among the most significant developments in the civilian sphere was the work of Robert Colwell (1956), who applied color infrared film (popularly known as "camouflage detection film," developed for use in World War II) to problems of identifying small-grain cereal crops and their diseases and other problems in the plant sciences. Although many of the basic principles of his research had been established earlier, his systematic investigation of their practical dimensions forms a clear milestone in the development of the field of remote sensing. Even at this early date, Colwell delineated the outlines of modern remote sensing and anticipated many of the opportunities and difficulties of this field of inquiry.

FIGURE 1.5. A U.S. Air Force intelligence officer examines aerial photography, Korean conflict, July 1951. From U.S. Air Force, National Archives and Records Administration, ARC 542288.

FIGURE 1.6. A photograph from the 1950s shows a forester examining aerial photography to delineate landscape units. By the 1950s, aerial photography and related forms of imagery had become integrated into day-to-day operations of a multitude of businesses and industries throughout the world. From Forest History Society, Durham, North Carolina. Reproduced by permission.

Civil Applications of Aerial Imagery

By the late 1950s aerial photography had been institutionalized in applications in government and civil society as a source of cartographic information (Figures 1.7 and 1.8). The 1960s saw a series of important developments occur in rapid sequence. The first meteorological satellite (TIROS-1) was launched in April 1960. This satellite was designed for climatological and meteorological observations but provided the basis for later development of land observation satellites. During this period, some of the remote sensing instruments originally developed for military reconnaissance and classified as defense secrets were released for civilian use as more advanced designs became available for military application. These instruments extended the reach of aerial observation outside the visible spectrum into the infrared and microwave regions.

Remote Sensing

It was in this context that the term *remote sensing* was first used. Evelyn Pruitt, a scientist working for the U.S. Navy's Office of Naval Research, coined this term when she recognized that the term *aerial photography* no longer accurately described the many forms of imagery collected using radiation outside the visible region of the spectrum. Early in the 1960s the U.S. National Aeronautics and Space Administration (NASA) established a research program in remote sensing—a program that, during the next decade, was to support remote sensing research at institutions throughout the United States. During this same period, a committee of the U.S. National Academy of Sciences (NAS) studied opportunities for application of remote sensing in the field of agriculture and forestry. In

FIGURE 1.7. A stereoscopic plotting instrument used to derive accurate elevation data from aerial photography, 1957. During much of the twentieth century, photogrammetric analyses depended on optical–mechanical instruments such as the one shown here, designed to extract information by controlling the physical orientation of the photograph and optical projection of the image. By the end of the century, such processes were conducted in the digital domain using electronic instruments. From Photographic Library, U.S. Geological Survey, Denver, Colorado. Photo by E. F. Patterson, no. 223.

FIGURE 1.8. A cartographic technician uses an airbrush to depict relief, as interpreted from aerial photographs, 1961. Within a few decades, computer cartography and GIS could routinely create this effect by applying hill-shading algorithms to digital elevation models. From Photographic Library, U.S. Geological Survey, Denver, Colorado. Photo by E.F. Patterson, no. 1024.

1970, the NAS reported the results of their work in a document that outlined many of the opportunities offered by this emerging field of inquiry.

Satellite Remote Sensing

In 1972, the launch of Landsat 1, the first of many Earth-orbiting satellites designed for observation of the Earth's land areas, marked another milestone. Landsat provided, for the first time, systematic repetitive observation of the Earth's land areas. Each Landsat image depicted large areas of the Earth's surface in several regions of the electromagnetic spectrum, yet provided modest levels of detail sufficient for practical applications in many fields. Landsat's full significance may not yet be fully appreciated, but it is possible to recognize three of its most important contributions. First, the routine availability of multispectral data for large regions of the Earth's surface greatly expanded the number of people who acquired experience and interest in analysis of multispectral data. Multispectral data had been acquired previously but were largely confined to specialized research laboratories. Landsat's data greatly expanded the population of scientists with interests in multispectral analysis.

Landsat's second contribution was to create an incentive for the rapid and broad expansion of uses of digital analysis for remote sensing. Before Landsat, image analyses were usually completed visually by examining prints and transparencies of aerial images. Analyses of digital images by computer were possible mainly in specialized research institutions; personal computers, and the variety of image analysis programs that we now regard as commonplace, did not exist. Although Landsat data were initially used primarily as prints or transparencies, they were also provided in digital form. The routine availability of digital data in a standard format created the context that permitted the growth in popularity of digital analysis and set the stage for the development of image analysis software that is now commonplace. During this era, photogrammetic processes originally implemented using mechanical instruments were redefined as digital analyses, leading to improvements in precision and in streamlining the acquisition, processing, production, and distribution of remotely sensed data. A third contribution of the Landsat program was its role as a model for development of other land observation satellites designed and operated by diverse organizations throughout the world.

By the early 1980s, a second generation of instruments for collecting satellite imagery provided finer spatial detail at 30-m, 20-m, and 10-m resolutions and, by the 1990s, imagery at meter and submeter resolutions. Finally, by the late 1990s, development of commercial capabilities (e.g., Geoeye and IKONOS) for acquiring fine-resolution satellite imagery (initially at spatial resolutions of several meters but eventually submeter detail) opened new civil applications formerly available only through uses of aerial photography. It is important to note that such progress in the field of remote sensing advanced in tandem with advances in geographic information systems (GIS), which provided the ability to bring remotely sensed data and other geospatial data into a common analytical framework, thereby enhancing the range of products and opening new markets—mapping of urban infrastructure, supporting precision agriculture, and support of floodplain mapping, for example.

Hyperspectral Remote Sensing

During the 1980s, scientists at the Jet Propulsion Laboratory (Pasadena, California) began, with NASA support, to develop instruments that could create images of the Earth at unprecedented levels of spectral detail. Whereas previous multispectral sensors collected data in a few rather broadly defined spectral regions, these new instruments could collect 200 or more very precisely defined spectral regions. These instruments created the field of *hyperspectral remote sensing*, which is still developing as a field of inquiry. Hyperspectral remote sensing will advance remote sensing's analytical powers to new levels and will form the basis for a more thorough understanding of how to best develop future remote sensing capabilities.

Global Remote Sensing

By the 1990s, satellite systems had been designed specifically to collect remotely sensed data representing the entire Earth. Although Landsat had offered such a capability in principle, in practice effective global remote sensing requires sensors and processing techniques specifically designed to acquire broad-scale coverage, which requires coarse spatial detail, often at coarse details of several kilometers. Such capabilities had existed on an *ad hoc* basis since the 1980s, primarily based on the synoptic scope of meteorological satellites. By December 1999, NASA had launched Terra-1, the first satellite of a system specifically designed to acquire global coverage to monitor changes in the nature and extent of Earth's ecosystems. The deployment of these systems marks the beginning of an era of broad-scale remote sensing of the Earth, which has provided some of the scientific foundations for documenting spatial patterns of environmental changes during recent decades.

Geospatial Data

Beginning in the 1980s but not maturing fully until the mid-2000s, several technologies began to converge to create systems with a unique synergy that each enhanced and reinforced the value of the others to create *geospatial data*, a term applied collectively to several technologies—primarily remote sensing, GIS, and global positioning systems (GPS). Many of these systems are implemented in the digital domain, replacing manual and mechanical applications developed in previous decades (Figure 1.9). Although these technologies had developed as interrelated technologies, by the first decade of the 2000s, they came to form integrated systems that can acquire imagery of high positional accuracy and that enable the integration of varied forms of imagery and data. Thus, during this interval, collection and analysis of geospatial data transformed from relying on several separate, loosely connected technologies to forming fully integrated and synergistic instruments that each reinforce the value of the others. These increases in quality and flexibility of geospatial data, linked also to decentralization of the information technology (IT) services and acquisition of aerial imagery, increased the significance of a broad population of entrepreneurs motivated to develop innovative products tailored to specific markets. Driving concerns included needs for planning and construction, civil needs in agriculture and forestry, hydrology, and consumer services and products (but, it should be noted, often in very specific niches within each of these areas). Widespread availability of fine-resolution satellite data has opened markets within news media, nonprofit and nongovernmental organizations, and behind-the-scenes continuing markets within the military and national security communities. After the events associated with 9/11 and Hurricane Katrina, the value of satellite imagery for homeland security was more for-

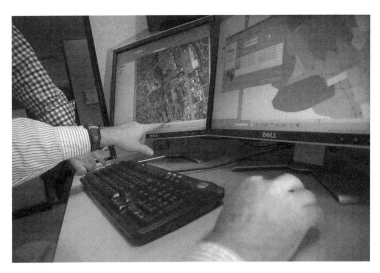

FIGURE 1.9. Analysts examine digital imagery displayed on a computer screen using specialized software, 2008. From Virginia Tech Center for Geospatial Information and Technologies.

mally recognized and integrated into emergency response and homeland security planning at national, regional, and local levels.

Public Remote Sensing

During the first decade of the 21st century, the increasing power of the Internet began to influence public access to remotely sensed imagery, in part through design of image-based products made for distribution though the Internet and in part through design of consumer products and services that relied on remotely sensed imagery presented in a map-like format. Whereas much of the previous history of remote sensing can be seen as the work of specialists to produce specialized products for the use of other specialists, these developments hinged on designing products for the use of the broader public.

Google Earth, released in 2005, forms a virtual representation of the Earth's surface as a composite of varied digital images, using basic concepts developed by Keyhole, Inc., which had been acquired by Google in 2004. Google Earth was designed to communicate with a broadly defined audience—a public without the kind of specialized knowledge that previously was an assumed prerequisite for use of remote sensing imagery. The new product assumed only that users would already be familiar with the delivery mechanism—i.e., use of the Internet and World Wide Web.

A basic innovation of Google Earth is the recognition of the value of a broad population of image products formed by accurate georeferencing of composite images acquired at varied dates, scales, and resolutions. Such composites can be viewed using an intuitive interface for browsing and roaming, changing scale, orientation, and detail. Google Earth has added specialized tools for tailoring the software to needs of specific users, enhancing display options, and integrating other data within the Google Earth framework. The product is based on the insight that the Google Earth tool could appeal to a population of nonspecialists, while at the same time providing specialized capabilities for narrowly defined communities of users in defense, security, emergency response, and commerce that require immediate, simultaneous access to a common body of geospatial data.

Google Earth and similar online services represent a new class of applications of remotely sensed imagery that contrast sharply with those of earlier eras. Campbell, Hardy, and Bernard (2010) outlined the context for the development of many of these new cartographic applications of remotely sensed data, including (1) a public policy that has maintained relaxed constraints on acquisition of fine-resolution satellite data, (2) personal privacy policies that favor widespread collection of imagery, (3) availability and popularity of reliable personal navigation devices, and (4) increasing migration of applications to mobile or handheld devices. Such developments have led to a class of cartographic products derived from remotely sensed data, including, for example, those created by MapQuest and related navigation software, which rely on road networks that are systematically updated by analysis of remotely sensed imagery. Ryerson and Aronoff (2010) have outlined practical applications of many of these technologies.

1.4. Overview of the Remote Sensing Process

Because remotely sensed images are formed by many interrelated processes, an isolated focus on any single component will produce a fragmented understanding. Therefore, our initial view of the field can benefit from a broad perspective that identifies the kinds of knowledge required for the practice of remote sensing (Figure 1.10).

Consider first the *physical objects*, consisting of buildings, vegetation, soil, water, and the like. These are the objects that applications scientists wish to examine. Knowledge of the physical objects resides within specific disciplines, such as geology, forestry, soil science, geography, and urban planning.

Sensor data are formed when an instrument (e.g., a camera or radar) views the physical objects by recording electromagnetic radiation emitted or reflected from the landscape. For many, sensor data often seem to be abstract and foreign because of their unfamiliar overhead perspective, unusual resolutions, and use of spectral regions outside the visible spectrum. As a result, effective use of sensor data requires analysis and interpretation to convert data to information that can be used to address practical problems, such as siting landfills or searching for mineral deposits. These interpretations create *extracted information*, which consists of transformations of sensor data designed to reveal specific kinds of information.

Actually, a more realistic view (Figure 1.11) illustrates that the same sensor data can be examined from alternative perspectives to yield different interpretations. Therefore, a single image can be interpreted to provide information about soils, land use, or hydrology, for example, depending on the specific image and the purpose of the analysis. Finally, we proceed to the *applications*, in which the analyzed remote sensing data can be combined with other data to address a specific practical problem, such as land-use planning, mineral exploration, or water-quality mapping. When digital remote sensing data are combined with other geospatial data, applications are implemented in the context of GIS. For example, remote sensing data may provide accurate land-use information that can be combined with soil, geologic, transportation, and other information to guide the siting of a new landfill. Although specifics of this process are largely beyond the scope of

FIGURE 1.10. Overview of the remote sensing process.

this volume, they are presented in books, such as Burrough and McDonnell (1998) and Demers (2009), devoted to applications of GIS.

1.5. Key Concepts of Remote Sensing

Although later chapters more fully develop key concepts used in the practice of remote sensing, it is useful to provide a brief overview of some of the recurring themes that are pervasive across application areas.

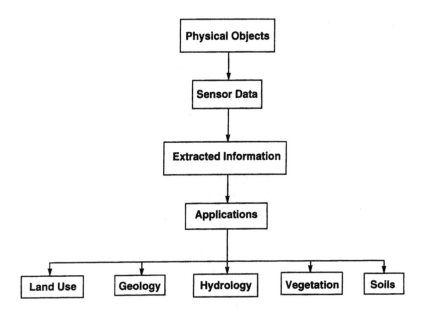

FIGURE 1.11. Expanded view of the process outlined in Figure 1.10.

Spectral Differentiation

Remote sensing depends on observed spectral differences in the energy reflected or emitted from features of interest. Expressed in everyday terms, one might say that we look for differences in the "colors" of objects, even though remote sensing is often conducted outside the visible spectrum, where "colors," in the usual meaning of the word, do not exist. This principle is the basis of *multispectral remote sensing*, the science of observing features at varied wavelengths in an effort to derive information about these features and their distributions. The term *spectral signature* has been used to refer to the spectral response of a feature, as observed over a range of wavelengths (Parker and Wolff, 1965). For the beginning student, this term can be misleading, because it implies a distinctiveness and a consistency that seldom can be observed in nature. As a result, some prefer *spectral response pattern* to convey this idea because it implies a less rigid version of the concept. Later chapters revisit these ideas in more detail.

Radiometric Differentiation

Examination of any image acquired by remote sensing ultimately depends on detection of differences in the brightness of objects and the features. The scene itself must have sufficient contrast in brightnesses, and the remote sensing instrument must be capable of recording this contrast, before information can be derived from the image. As a result, the sensitivity of the instrument and the existing contrast in the scene between objects and their backgrounds are always issues of significance in remote sensing investigations.

Spatial Differentiation

Every sensor is limited in respect to the size of the smallest area that can be separately recorded as an entity on an image. This minimum area determines the spatial detail—the fineness of the patterns—on the image. These minimal areal units, known as *pixels* ("picture elements"), are the smallest areal units identifiable on the image. Our ability to record spatial detail is influenced primarily by the choice of sensor and the altitude at which it is used to record images of the Earth. Note that landscapes vary greatly in their spatial complexity; some may be represented clearly at coarse levels of detail, whereas others are so complex that the finest level of detail is required to record their essential characteristics.

Temporal Dimension

Although a single image can easily demonstrate the value of remotely sensed imagery, its effectiveness is best demonstrated through the use of many images of the same region acquired over time. Although practitioners of remote sensing have a long history of exploiting the temporal dimension of aerial imagery, through the use of sequential aerial photography (using the archives of photography preserved over the years since aerial photography was first acquired in a systematic manner), the full value of the temporal dimension was realized later, when satellite systems could systematically observe the same regions on a repetitive basis. These later sequences, acquired by the same platforms using the same instruments under comparable conditions, have offered the ability to more fully exploit the temporal dimension of remotely sensed data.

Geometric Transformation

Every remotely sensed image represents a landscape in a specific geometric relationship determined by the design of the remote sensing instrument, specific operating conditions, terrain relief, and other factors. The ideal remote sensing instrument would be able to create an image with accurate, consistent geometric relationships between points on the ground and their corresponding representations on the image. Such an image could form the basis for accurate measurements of areas and distances. In reality, of course, each image includes positional errors caused by the perspective of the sensor optics, the motion of scanning optics, terrain relief, and Earth curvature. Each source of error can vary in significance in specific instances, but the result is that geometric errors are inherent, not accidental, characteristics of remotely sensed images. In some instances we may be able to remove or reduce locational error, but it must always be taken into account before images are used as the basis for measurements of areas and distances. Thus another recurring theme in later chapters concerns the specific geometric errors that characterize each remote sensing instrument and remedies that can be applied to provide imagery that has geometric qualities appropriate for their use.

Remote Sensing Instrumentation Acts as a System

The image analyst must always be conscious of the fact that the many components of the remote sensing process *act as a system* and therefore cannot be isolated from one another. For example, upgrading the quality of a camera lens makes little sense unless we can also use technologies that can record the improvements produced by the superior lens. Components of the system must be appropriate for the task at hand. This means that the analyst must intimately know not only the capabilities of each imaging system and how it should be deployed but also the subject matter to be examined and the specific needs of those who will use the results of the project. Successful applications of remote sensing require that such considerations be resolved to form a system with compatible components.

Role of the Atmosphere

All energy reaching a remote sensing instrument must pass through a portion of the Earth's atmosphere. For satellite remote sensing in the visible and near infrared, energy received by the sensor must pass through a considerable depth of the Earth's atmosphere. In doing so, the sun's energy is altered in intensity and wavelength by particles and gases in the Earth's atmosphere. These changes appear on the image in ways that degrade image quality or influence the accuracy of interpretations.

1.6. Career Preparation and Professional Development

For the student, a course of study in remote sensing offers opportunities to enter a field of knowledge that can contribute to several dimensions of a university education and subsequent personal and professional development. Students enrolled in introductory remote sensing courses often view the topic as an important part of their occupational and professional preparation. It is certainly true that skills in remote sensing are valuable in the initial search for employment. But it is equally important to acknowledge that this topic should form part of a comprehensive program of study that includes work in GIS and in-depth study of a specific discipline. A well-thought-out program appropriate for a student's specific interests and strengths should combine studies in several interrelated topics, such as:

- Geology, hydrology, geomorphology, soils
- Urban planning, transportation, urban geography
- Forestry, ecology; soils

Such programs are based on a foundation of supporting courses, including statistics, computer science, and the physical sciences. Students should avoid studies that provide only narrowly based, technique-oriented content. Such highly focused studies, perhaps with specific equipment or software, may provide immediate skills for entry-level positions, but they leave the student unprepared to participate in the broader assignments required for effective performance and professional advancement. Employers report that they seek employees who:

- Have a good background in at least one traditional discipline.
- Are reliable and able to follow instructions without detailed supervision.
- Can write and speak effectively.
- Work effectively in teams with others in other disciplines.
- Are familiar with common business practices.

Table 1.3 provides examples.

Because this kind of preparation is seldom encompassed in a single academic unit within a university, students often have to apply their own initiative to identify the specific courses they will need to best develop these qualities. Table 1.3 shows a selection of sample job descriptions in remote sensing and related fields, as an indication of the kinds of knowledge and skills expected of employees in the geospatial information industry.

Possibly the most important but least visible contributions are to the development of conceptual thinking concerning the role of basic theory and method, integration of knowledge from several disciplines, and proficiency in identifying practical problems in a spatial context. Although skills in and knowledge of remote sensing are very important, it is usually a mistake to focus exclusively on methodology and technique. At least two pitfalls are obvious. First, emphasis on fact and technique without consideration of basic principles and theory provides a narrow, empirical foundation in a field that is characterized by diversity and rapid change. A student equipped with a narrow background is ill prepared to compete with those trained in other disciplines or to adjust to unexpected developments in science and technology. Thus any educational experience is best perceived not as a catalog of facts to be memorized but as an experience in *how to learn* to equip oneself for independent learning later, outside the classroom. This task requires a familiarity with basic references, fundamental principles, and the content of related disciplines, as well as the core of facts that form the substance of a field of knowledge.

Second, many employers have little interest in hiring employees with shallow preparation either in their major discipline or in remote sensing. Lillesand (1982) reports that a panel of managers from diverse industries concerned with remote sensing recommended that prospective employees develop "an ability and desire to interact at a conceptual level

TABLE 1.3. Sample Job Descriptions Relating to Remote Sensing

SURVEY ENGINEER

XCELIMAGE is a spatial data, mapping, and geographic information systems (GIS) services company that provides its clients with customized products and services to support a wide range of land-use and natural resource management activities. The company collects geospatial data using a variety of airborne sensing technologies and turns that data into tools that can be used in GIS or design and engineering environments. With over 500 employees in offices nationwide, the XCELIMAGE group and affiliates represent one of the largest spatial data organizations in the world. XCELIMAGE is affiliated with six member companies and two affiliates. XCELIMAGE Aviation, located in Springfield, MA, has an immediate opening for a Survey Engineer.

XCELIMAGE Aviation supplies the aerial photography and remote sensing data from which terrain models, mapping, and GIS products are developed. XCELIMAGE Aviation operates aircraft equipped with analog, digital, and multispectral cameras; global positioning systems (GPS); a light detection and ranging system (LIDAR); a passive microwave radiometer; and thermal cameras. This position offers a good opportunity for advancement and a competitive salary and benefits package. Position requires a thorough knowledge of computer operation and applications including GIS software; a basic understanding of surveying, mapping theories, and techniques; a thorough knowledge of GPS concepts; exposure to softcopy techniques; and the ability to efficiently aid in successful implementation of new technologies and methods. This position includes involvement in all functions related to data collection, processing, analysis, and product development for aerial remote sensing clients; including support of new technology.

ECOLOGIST/REMOTE SENSING SPECIALIST

The U.S. National Survey Northern Plains Ecological Research Center is seeking an Ecologist/ Remote Sensing Specialist to be a member of the Regional Gap Analysis Project team. The incumbent's primary responsibility will be mapping vegetation and land cover for our region from analysis of multitemporal Landsat Thematic Mapper imagery and environmental data in a geographic information system. To qualify for this position, applicants must possess (1) ability to conduct digital analysis of remotely sensed satellite imagery for vegetation and land cover mapping; (2) ability to perform complex combinations and sequences of methods to import, process, and analyze data in vector and raster formats in a geographic information system; and (3) knowledge of vegetation classification, inventory, and mapping.

REMOTE SENSING/GIS AGRICULTURAL ANALYST

Position located in Washington, DC. Requires US citizenship. Unofficial abstract of the "Crop Assessment Analyst" position: An interesting semianalytic/technical position is available with the Foreign Agricultural Service working as an international and domestic agriculture commodity forecaster. The position is responsible for monitoring agricultural areas of the world, performing analysis, and presenting current season production forecasts. Tools and data used in the position include imagery data (AVHRR, Landsat TM, SPOT), vegetation indexes, crop models, GIS software (ArcView, ArcInfo), image processing s/w (Erdas, PCI), agro-meteorological data models, web browsers, web page design s/w, graphic design s/w, spreadsheet s/w, GIS software, digital image processing, weather station data, climate data, historical agricultural production data, and assessing news stories. The main crops of concern are soybeans, canola, wheat, barley, corn, cotton, peanuts, and sorghum of the major export/import countries.

A background in agronomy, geographical spatial data, good computer skills, information management, and ag economics will prove beneficial in performing the work.

REMOTE SENSING SPECIALIST POSITION, GEOSPATIAL AND INFORMATION TECHNOLOGIES INFORMATION RESOURCES UNIT

The Remote Sensing Specialist is generally responsible for implementing remote sensing technology as a tool for natural resources management. Responsibilities include planning, coordinating, and managing a regional remote sensing program, working with resource specialists at the regional level, forests and districts, to meet information needs using remotely sensed data. Tasks include working with resource and GIS analysts to implement the recently procured national image processing software, keeping users informed about the system, and assisting with installation and training. The person in this position is also the primary contact in the region for national remote sensing issues, and maintains contact with other remote sensing professionals within the Forest Service, in other agencies, and in the private and academic sectors.

The position is located in the Geospatial and Information Technologies (GIT) group of the Information Resources (IRM) unit. IRM has regional responsibility for all aspects of information and systems management. The GIT group includes the remote sensing, aerial photography, photogrammetry, cartography, database, and GIS functions.

Note. Actual notices edited to remove information identifying specific firms. Although listed skills and abilities are typical, subject area specialties are not representative of the range of applications areas usually encountered.

with other specialists" (p. 290). Campbell (1978) quotes other supervisors who are also concerned that students receive a broad preparation in remote sensing and in their primary field of study:

> It is essential that the interpreter have a good general education in an area of expertise. For example, you can make a geologist into a good photo geologist, but you cannot make an image interpreter into a geologist.
>
> Often people lack any real philosophical understanding of why they are doing remote sensing, and lack the broad overview of the interrelationships of all earth science and earth-oriented disciplines (geography, geology, biology, hydrology, meteorology, etc.). This often creates delays in our work as people continue to work in small segments of the (real) world and don't see the interconnections with another's research. (p. 35)

These same individuals have recommended that those students who are interested in remote sensing should complete courses in computer science, physics, geology, geography, biology, engineering, mathematics, hydrology, business, statistics, and a wide variety of other disciplines. No student could possibly take all recommended courses during a normal program of study, but it is clear that a haphazard selection of university courses, or one that focused exclusively on remote sensing courses, would not form a substantive background in remote sensing. In addition, many organizations have been forceful in stating that they desire employees who can write well, and several have expressed an interest in persons with expertise in remote sensing who have knowledge of a foreign language. The key point is that educational preparation in remote sensing should be closely coordinated with study in traditional academic disciplines and should be supported by a program of courses carefully selected from offerings in related disciplines.

Students should consider joining a professional society devoted to the field of remote sensing. In the United States and Canada, the American Society for Photogrammetry and Remote Sensing (ASPRS; 5410 Grosvenor Lane, Suite 210, Bethesda, MD 20814-2160; 301-493-0290; *www.asprs.org*) is the principal professional organization in this field. ASPRS offers students discounts on membership dues, publications, and meeting registration and conducts job fairs at its annual meetings. ASPRS is organized on a regional basis, so local chapters conduct their own activities, which are open to student participation. Other professional organizations often have interest groups devoted to applications of remote sensing within specific disciplines, often with similar benefits for student members.

Students should also investigate local libraries to become familiar with professional journals in the field. The field's principal journals include:

Photogrammetric Engineering and Remote Sensing

Remote Sensing of Environment

International Journal of Remote Sensing

IEEE Transactions on Geoscience and Remote Sensing

Computers and Geosciences

GIScience and Remote Sensing

ISPRS Journal of Photogrammetry and Remote Sensing

Although beginning students may not yet be prepared to read research articles in detail, those who make the effort to familiarize themselves with these journals will have prepared the way to take advantage of their content later. In particular, students may find *Photogrammetric Engineering and Remote Sensing* useful because of its listing of job opportunities, scheduled meetings, and new products.

The world of practical remote sensing has changed dramatically in recent decades. Especially since the early 1990s, commercial and industrial applications of remote sensing have expanded dramatically to penetrate well beyond the specialized applications of an earlier era—to extend, for example, into marketing, real estate, and agricultural enterprises. Aspects of remote sensing that formerly seemed to require a highly specialized knowledge became available to a much broader spectrum of users as data became less expensive and more widely available and as manufacturers designed software for use by the nonspecialist.

These developments have created a society in which remote sensing, GIS, GPS, and related technological systems have become everyday tools within the workplace and common within the daily lives of many people. In such a context, each citizen, especially the nonspecialist, must be prepared to use spatial data effectively and appropriately and to understand its strengths and limitations. Because of the need to produce a wide variety of ready-to-use products tailored for specific populations of users, the remote sensing community will continue to require specialists, especially those who can link a solid knowledge of remote sensing with subject-area knowledge (e.g., hydrology, planning, forestry, etc.). People who will work in this field will require skills and perspectives that differ greatly from those of previous graduates of only a few years earlier.

1.7. Some Teaching and Learning Resources

AmericaView

AmericaView is a nonprofit organization funded chiefly through the U.S. Geological Survey to promote uses of satellite imagery and to distribute imagery to users. It comprises a coalition of state-based organizations (such as OhioView, VirginiaView, WisconsinView, AlaskaView, and other "StateViews"), each composed of coalitions of universities, businesses, state agencies, and nonprofit organizations and each led by a university within each state. StateViews distribute image data to their members and to the public, conduct research to apply imagery to local and state problems, conduct workshops and similar educational activities, support research, and, in general, promote use of remotely sensed imagery within the widest possible range of potential users. Each StateView has its own website, with a guide to its members, activities, and data resources. AmericaView's Education Committee provides access to a variety of educational resources pertaining to teaching of remote sensing in grades K–12 through university levels, accessible at

www.americaview.org

Geospatial Revolution Project

The Geospatial Revolution Project (*geospatialrevolution.psu.edu*), developed by Penn State Public Broadcasting, has an integrated public media and outreach initiative focused on the world of digital mapping and how it is shaping the way our society functions. The project features a Web-based serial release of several video episodes, each focused on an specific narrative illuminating the role of geospatial knowledge in our world. These episodes will combine to form a full-length documentary. The project also will include an outreach initiative in collaboration with partner organizations, a chaptered program DVD, and downloadable outreach materials. The 5-minute trailer available at the website provides an introduction to the significance of geospatial data.

geospatialrevolution.psu.edu

www.youtube.com/watch?v=ZdQjc30YPOk

American Society for Photogrammetry and Remote Sensing

The American Society for Photogrammetry and Remote Sensing (*www.asprs.org*) has produced a series of short videos (each about 2 minutes or less in length) focusing on topics relevant to this chapter. The link to "ASPRS films" (under the "Education and Professional Development" tab) (*www.asprs.org/films/index.html*) connects to a series of short videos highlighting topics relating to ASPRS history and its contributions to the fields of photogrammetry and remote sensing.

Videos in this series that are especially relevant to this chapter include:

- Aerial Survey Pioneers
  www.youtube.com/watch?v=wW-JTtwNC_4&fmt=22

- Geospatial Intelligence in WWII
  www.youtube.com/watch?v=hQu0wxXN6U4&fmt=22

- Role of Women
  www.youtube.com/watch?v=kzgrwmaurKU&fmt=22

- Photogrammetry in Space Exploration
  www.youtube.com/watch?v=KVVbhqq6SRg&fmt=22

- Evolution of Analog to Digital Mapping
  www.youtube.com/watch?v=4jABMysbNbc&fmt=22

This list links to YouTube versions; the ASPRS site provides links to versions of higher visual quality.

Britain from Above

The British Broadcasting Corporation project Britain from Above has multiple short video units (usually about 3–5 minutes each) that are of interest to readers of this book. At *www.bbc.co.uk/britainfromabove*, go to "See Your World from Above" ("Browse stories"), then "Stories Overview." Then from "Stories Overview," select the "Secrets" tab and play the "The Dudley Stamp Maps." Or, from the "Stories Overview" section, select the "Behind the Scenes" tab, then play the "Archive." There are many other videos of interest at this site.

Canada Centre for Remote Sensing

The Canada Centre for Remote Sensing (*www.ccrs.nrcan.gc.ca/resource/tutor/fundam/index_e.php*) has a broad range of resources pertain to remote sensing, including introductory level tutorials and materials for instructors, including quizzes and exercises.

- Earth Resources Technology Satellite (ERTS)–1973
  www.youtube.com/watch?v=6isYzkXlTHc&feature=related
- French Kite Aerial Photography Unit, WWI
  www.youtube.com/watch?v=O5dJ9TwaIt4
- 1940 British Aerial Reconnaissance
  www.youtube.com/watch?v=PhXd6uMlPVo
- Forgotten Aircraft: The Abrams Explorer
  www.youtube.com/watch?v=gsaAeLaNr60

Review Questions

1. Aerial photography and other remotely sensed images have found rather slow acceptance into many, if not most, fields of study. Imagine that you are director of a unit engaged in geological mapping in the early days of aerial photography (e.g., in the 1930s). Can you suggest reasons why you might be reluctant to devote your efforts and resources to use of aerial photography rather than to continue use of your usual procedures?

2. Satellite observation of the Earth provides many advantages over aircraft-borne sensors. Consider fields such as agronomy, forestry, or hydrology. For one such field of study, list as many of the advantages as you can. Can you suggest some disadvantages?

3. Much (but not all) information derived from remotely sensed data is derived from spectral information. To understand how spectral data may not always be as reliable as one might first think, briefly describe the spectral properties of a maple tree and a cornfield. How might these properties change over the period of a year? Or a day?

4. All remotely sensed images observe the Earth from above. Can you list some advantages to the overhead view (as opposed to ground-level views) that make remote sensing images inherently advantageous for many purposes? List some disadvantages to the overhead view.

5. Remotely sensed images show the combined effects of many landscape elements, including vegetation, topography, illumination, soil, drainage, and others. In your view, is this diverse combination an advantage or a disadvantage? Explain.

6. List ways in which remotely sensed images differ from maps. Also list advantages and disadvantages of each. List some of the tasks for which each might be more useful.

7. Chapter 1 emphasizes how the field of remote sensing is formed by knowledge and perspectives from many different disciplines. Examine the undergraduate catalog for your college or university and prepare a comprehensive program of study in remote

sensing from courses listed. Identify gaps—courses or subjects that would be desirable but are not offered.

8. At your university library, find copies of *Photogrammetric Engineering and Remote Sensing*, *International Journal of Remote Sensing*, and *Remote Sensing of Environment*, some of the most important English-language journals reporting remote sensing research. Examine some of the articles in several issues of each journal. Although titles of some of these articles may now seem rather strange, as you progress through this course you will be able to judge the significance of most. Refer to these journals again as you complete the course.

9. Inspect library copies of some of the remote sensing texts listed in the references for Chapter 1. Examine the tables of contents, selected chapters, and lists of references. Many of these volumes may form useful references for future study or research in the field of remote sensing.

10. Examine some of the journals mentioned in Question 8, noting the affiliations and institutions of authors of articles. Be sure to look at issues that date back for several years, so you can identify some of the institutions and agencies that have been making a continuing contribution to remote sensing research.

References

Alföldi, T., P. Catt, and P. Stephens. 1993. Definitions of Remote Sensing. *Photogrammetric Engineering and Remote Sensing*, Vol. 59, pp. 611–613.

Arthus-Bertrand, Y. 1999. *Earth from Above*. New York: Abrams, 414 pp.

Avery, T. E., and G. L. Berlin. 1992. *Fundamentals of Remote Sensing and Airphoto Interpretation*. Upper Saddle River, NJ: Prentice-Hall, 472 pp.

Babington-Smith, C. 1957. *Air Spy: The Story of Photo intelligence in World War II*. New York: Ballantine, 190 pp. (Reprinted 1985. Bethesda, MD: American Society for Photogrammetry and Remote Sensing, 266 pp.)

Barrett, E. C., and C. F. Curtis. 1976. *Introduction to Environmental Remote Sensing*. New York: Macmillan, 472 pp.

Bossler, J. D. (ed.). 2010. *Manual of Geospatial Science and Technology* (2nd ed.). London: Taylor & Francis, 808 pp.

Brugioni, D. A. 1991. *Eyeball to Eyeball: The Inside Story of the Cuban Missile Crisis*. New York: Random House, 622 pp.

Burrough, P. A., and R. A. McDonnell. 1998. *Principles of Geographical Information Systems*. New York: Oxford. 333 pp.

Campbell, J. B. 1978. Employer Needs in Remote Sensing in Geography. *Remote Sensing Quarterly*, Vol. 5, No. 2, pp. 52–65.

Campbell, J. B. 2008. Origins of Aerial Photographic Interpretation, U.S. Army, 1916–1918. *Photogrammetric Engineering and Remote Sensing*, Vol. 74, pp. 77–93.

Campbell, J. B., and V. V. Salomonson. 2010. Remote Sensing—A Look to the Future. Chapter 25 in *Manual of Geospatial Science and Technology* (2nd ed.). (J. D. Bossler, ed.) London: Taylor & Francis, pp. 487–510.

Campbell, Joel, T. Hardy, and R. C. Barnard. 2010. Emerging Markets for Satellite and Aerial Imagery. Chapter 22 in *Manual of Geospatial Science and Technology* (2nd ed.). (J. D. Bossler, ed.). London: Taylor & Francis, pp. 423–437.

Collier, P. 2002. The Impact on Topographic Mapping of Developments in Land and Air Survey 1900–1939. *Cartography and Geographic Information Science*, Vol. 29, pp. 155–174.

Colwell, R. N. 1956. Determining the Prevalence of Certain Cereal Crop Diseases by Means of Aerial Photography. *Hilgardia*, Vol. 26, No. 5, pp. 223–286.

Colwell, R. N. 1966. Uses and Limitations of Multispectral Remote Sensing. In *Proceedings of the Fourth Symposium on Remote Sensing of Environment*. Ann Arbor: Institute of Science and Technology, University of Michigan, pp. 71–100.

Colwell, R. N. (ed.). 1983. *Manual of Remote Sensing* (2nd ed.). Falls Church, VA: American Society of Photogrammetry, 2 vols., 2240 pp.

Curran, P. 1985. *Principles of Remote Sensing*. New York: Longman, 282 pp.

Curran, P. 1987. Commentary: On Defining Remote Sensing. *Photogrammetric Engineering and Remote Sensing*, Vol. 53, pp. 305–306.

Day, D. A., J. M. Logsdon, and B. Latell (eds.). 1998. *Eye in the Sky: The Story of the Corona Spy Satellites*. Washington, DC: Smithsonian Institution Press, 303 pp.

Demers, M. N. 2009. *Fundamentals of Geographic Information Systems*. New York: Wiley, 442 pp.

Estes, J. E., J. R. Jensen, and D. S. Simonett. 1977. The Impact of Remote Sensing on United States' Geography: The Past in Perspective, Present Realities, Future Potentials. In *Proceedings of the Eleventh International Symposium on Remote Sensing of Environment*. Ann Arbor: Institute of Science and Technology, University of Michigan, pp. 101–121.

Executive Office of the President. 1973. *Report of the Federal Mapping Task Force on Mapping, Charting, Geodesy, and Surveying*, Washington: Government Printing Office.

Fischer, W. A. (ed.). 1975. History of Remote Sensing. Chapter 2 in *Manual of Remote Sensing* (R. G. Reeves, ed.). Falls Church, VA: American Society of Photogrammetry, pp. 27–50.

Fischer, W. A., W. R. Hemphill, and A. Kover. 1976. Progress in Remote Sensing. *Photogrammetria*, Vol. 32, pp. 33–72.

Foody, G. M., T. A. Warner, and M. D. Nellis. 2009. A Look to the Future. Chapter 34 in *The Sage Handbook of Remote Sensing* (T. A. Warner, M. D. Nellis, and G. M. Foody, eds.). Washington, DC: Sage, pp. 475–481.

Fussell, J., D. Rundquist, and J. A. Harrington. 1986. On Defining Remote Sensing. *Photogrammetric Engineering and Remote Sensing*, Vol. 52, pp. 1507–1511.

Hall, S. S. 1993. *Mapping the Next Millennium: How Computer-Driven Cartography Is Revolutionizing the Face of Science*. New York: Random House, 360 pp.

Jensen, J. R. 2002. Remote sensing—Future considerations. Chapter 23 in *Manual of Geospatial Sciences and Technology* (J. D. Bossler, Ed.). London: Taylor & Francis, pp. 389–398.

Jensen, J. R. 2007. *Remote Sensing of the Environment: An Earth Resource Perspective*. Upper Saddle River, NJ: Prentice-Hall, 608 pp.

Lee, W. T. 1922. *The Face of the Earth as Seen from the Air* (American Geographical Society Special Publication No. 4). New York: American Geographical Society.

Lillesand, T. M. 1982. Trends and Issues in Remote Sensing Education. *Photogrammetric Engineering and Remote Sensing*, Vol. 48, pp. 287–293.

Lillesand, T. M., R. W. Kiefer, and J. W. Chipman. 2008. *Remote Sensing and Image Interpretation*. New York: Wiley, 756 pp.

Lintz, J., and D. S. Simonett. 1976. *Remote Sensing of Environment*. Reading, MA: Addison-Wesley. 694 pp.

Madden, M. (ed.). 2009. *Manual of Geographic Information Systems*. Bethesda, MD: American Society for Photogrammetry and Remote Sensing, 1352 pp.

Monmonier, M. 2002. Aerial Photography at the Agricultural Adjustment Administration: Acreage Controls, Conservation Benefits, and Overhead Surveillance in the 1930s. *Photogrammetric Engineering and Remote Sensing*, Vol. 68, pp. 1257–1261.

National Academy of Sciences. 1970. *Remote Sensing with Special Reference to Agriculture and Forestry*. Washington, DC: National Academy of Sciences, 424 pp.

Parker, D. C., and M. F. Wolff. 1965. Remote Sensing. *International Science and Technology*, Vol. 43, pp. 20–31.

Pedlow, G. W. and D. E. Welzenbach. 1992. *The Central Intelligence Agency and Overhead Reconnaissance: The U-2 and OXCART Programs, 1954–1974*. Washington, DC: Central Intelligence Agency.

Ray, R. G. 1960. *Aerial Photographs in Geological Interpretation and Mapping* (U.S. Geological Survey Professional Paper 373). Washington, DC: U.S. Geological Survey, 230 pp.

Reeves, R. G. (ed.). 1975. *Manual of Remote Sensing*. Falls Church, VA: American Society of Photogrammetry, 2 vols., 2144 pp.

Ryerson, B., and S. Aronoff. 2010. *Why "Where" Matters: Understanding and Profiting from GPS, GIS, and Remote Sensing*. Manotick, ON: Kim Geomatics, 378 pp.

Simonett, D. S. (ed.). 1983. Development and Principles of Remote Sensing. Chapter 1 in *Manual of Remote Sensing* (R. N. Colwell, ed.). Falls Church, VA: American Society of Photogrammetry, pp. 1–35.

Swain, P. H., and S. M. Davis. 1978. Remote Sensing: The Quantitative Approach. New York: McGraw-Hill, 396 pp.

Warner, T. A., M. D. Nellis, and G. M. Foody. 2009. *The Sage Handbook of Remote Sensing*. Washington, DC: Sage, 504 pp.

Wetherholt, W. A., and B. C. Rudquist. 2010. A Survey of Ethics Content in College-Level Remote Sensing Courses in the United States. *Journal of Geography*. Vol. 109, pp. 75–86.

White, L. P. 1977. Aerial Photography and Remote Sensing for Soil Survey. Oxford: Clarendon Press, 104 pp.

CHAPTER TWO

Electromagnetic Radiation

2.1. Introduction

With the exception of objects at absolute zero, all objects emit electromagnetic radiation. Objects also reflect radiation that has been emitted by other objects. By recording emitted or reflected radiation and applying knowledge of its behavior as it passes through the Earth's atmosphere and interacts with objects, remote sensing analysts develop knowledge of the character of features such as vegetation, structures, soils, rock, or water bodies on the Earth's surface. Interpretation of remote sensing imagery depends on a sound understanding of electromagnetic radiation and its interaction with surfaces and the atmosphere. The discussion of electromagnetic radiation in this chapter builds a foundation that will permit development in subsequent chapters of the many other important topics within the field of remote sensing.

The most familiar form of electromagnetic radiation is visible light, which forms only a small (but very important) portion of the full electromagnetic spectrum. The large segments of this spectrum that lie outside the range of human vision require our special attention because they may behave in ways that are quite foreign to our everyday experience with visible radiation.

2.2. The Electromagnetic Spectrum

Electromagnetic energy is generated by several mechanisms, including changes in the energy levels of electrons, acceleration of electrical charges, decay of radioactive substances, and the thermal motion of atoms and molecules. Nuclear reactions within the Sun produce a full spectrum of electromagnetic radiation, which is transmitted through space without experiencing major changes. As this radiation approaches the Earth, it passes through the atmosphere before reaching the Earth's surface. Some is reflected upward from the Earth's surface; it is this radiation that forms the basis for photographs and similar images. Other solar radiation is absorbed at the surface of the Earth and is then reradiated as thermal energy. This thermal energy can also be used to form remotely sensed images, although they differ greatly from the aerial photographs formed from reflected energy. Finally, man-made radiation, such as that generated by imaging radars, is also used for remote sensing.

Electromagnetic radiation consists of an electrical field (E) that varies in magnitude in a direction perpendicular to the direction of propagation (Figure 2.1). In addition, a

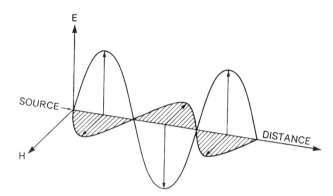

FIGURE 2.1. Electric (E) and magnetic (H) components of electromagnetic radiation. The electric and magnetic components are oriented at right angles to one another and vary along an axis perpendicular to the axis of propagation.

magnetic field (H) oriented at right angles to the electrical field is propagated in phase with the electrical field.

Electromagnetic energy can be characterized by several properties (Figure 2.2):

1. *Wavelength* is the distance from one wave crest to the next. Wavelength can be measured in everyday units of length, although very short wavelengths have

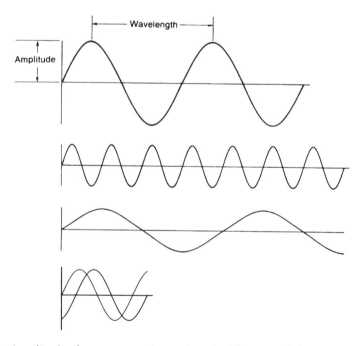

FIGURE 2.2. Amplitude, frequency, and wavelength. The second diagram represents high frequency, short wavelength; the third, low frequency, long wavelength. The bottom diagram illustrates two waveforms that are out of phase.

such small distances between wave crests that extremely short (and therefore less familiar) measurement units are required (Table 2.1).

2. *Frequency* is measured as the number of crests passing a fixed point in a given period of time. Frequency is often measured in *hertz*, units each equivalent to one cycle per second (Table 2.2), and multiples of the hertz.

3. *Amplitude* is equivalent to the height of each peak (see Figure 2.2). Amplitude is often measured as energy levels (formally known as *spectral irradiance*), expressed as watts per square meter per micrometer (i.e., as energy level per wavelength interval).

4. In addition, the *phase* of a waveform specifies the extent to which the peaks of one waveform align with those of another. Phase is measured in angular units, such as degrees or radians. If two waves are aligned, they oscillate together and are said to be "in phase" (a phase shift of 0 degrees). However, if a pair of waves are aligned such that the crests match with the troughs, they are said to be "out of phase" (a phase shift of 180 degrees) (see Figure 2.2).

The speed of electromagnetic energy (c) is constant at 299,792 kilometers (km) per second. Frequency (v) and wavelength (λ) are related:

$$c = \lambda v \qquad\qquad \text{(Eq. 2.1)}$$

Therefore, characteristics of electromagnetic energy can be specified using either frequency or wavelength. Varied disciplines and varied applications follow different conventions for describing electromagnetic radiation, using either wavelength (measured in Angström units [Å], microns, micrometers, nanometers, millimeters, etc., as appropriate) or frequency (using hertz, kilohertz, megahertz, etc., as appropriate). Although there is no authoritative standard, a common practice in the field of remote sensing is to define regions of the spectrum on the basis of wavelength, often using micrometers (each equal to one one-millionth of a meter, symbolized as µm), millimeters (mm), and meters (m) as units of length. Departures from this practice are common; for example, electrical engineers who work with microwave radiation traditionally use frequency to designate subdivisions of the spectrum. In this book we usually employ wavelength designations. The student should, however, be prepared to encounter different usages in scientific journals and in references.

TABLE 2.1. Units of Length Used in Remote Sensing

Unit	Distance	
Kilometer (km)	1,000 m	
Meter (m)	1.0 m	
Centimeter (cm)	0.01 m	$= 10^{-2}$ m
Millimeter (mm)	0.001 m	$= 10^{-3}$ m
Micrometer (µm)[a]	0.000001 m	$= 10^{-6}$ m
Nanometer (nm)		10^{-9} m
Ångstrom unit (Å)		10^{-10} m

[a]Formerly called the "micron" (µ); the term "micrometer" is now used by agreement of the General Conference on Weights and Measures.

TABLE 2.2. Frequencies Used in Remote Sensing

Unit	Frequency (cycles per second)
Hertz (Hz)	1
Kilohertz (kHz)	10^3 (= 1,000)
Megahertz (MHz)	10^6 (= 1,000,000)
Gigahertz (GHz)	10^9 (= 1,000,000,000)

2.3. Major Divisions of the Electromagnetic Spectrum

Major divisions of the electromagnetic spectrum (Table 2.3) are, in essence, arbitrarily defined. In a full spectrum of solar energy there are no sharp breaks at the divisions as indicated graphically in Figure 2.3. Subdivisions are established for convenience and by traditions within different disciplines, so do not be surprised to find different definitions in other sources or in references pertaining to other disciplines.

Two important categories are not shown in Table 2.3. The *optical spectrum*, from 0.30 to 15 µm, defines those wavelengths that can be reflected and refracted with lenses and mirrors. The *reflective spectrum* extends from about 0.38 to 3.0 µm; it defines that portion of the solar spectrum used directly for remote sensing.

The Ultraviolet Spectrum

For practical purposes, radiation of significance for remote sensing can be said to begin with the ultraviolet region, a zone of short-wavelength radiation that lies between the X-ray region and the limit of human vision. Often the ultraviolet region is subdivided into the *near ultraviolet* (sometimes known as *UV-A*; 0.32–0.40 µm), the *far ultraviolet* (*UV-B*; 0.32–0.28 µm), and the *extreme ultraviolet* (*UV-C*; below 0.28 µm). The ultraviolet region was discovered in 1801 by the German scientist Johann Wilhelm Ritter (1776–1810). Literally, *ultraviolet* means "beyond the violet," designating it as the region just outside the violet region, the shortest wavelengths visible to humans. Near ultraviolet radiation is known for its ability to induce *fluorescence*, emission of visible radiation, in some materials; it has significance for a specialized form of remote sensing (see Section

TABLE 2.3. Principal Divisions of the Electromagnetic Spectrum

Division	Limits
Gamma rays	< 0.03 nm
X-rays	0.03–300 nm
Ultraviolet radiation	0.30–0.38 µm
Visible light	0.38–0.72 µm
Infrared radiation	
Near infrared	0.72–1.30 µm
Mid infrared	1.30–3.00 µm
Far infrared	7.0–1,000 µm (1 mm)
Microwave radiation	1 mm–30 cm
Radio	≥ 30 cm

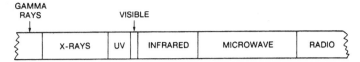

FIGURE 2.3. Major divisions of the electromagnetic spectrum. This diagram gives only a schematic representation—sizes of divisions are not shown in correct proportions. (See Table 2.3.)

2.6). However, ultraviolet radiation is easily scattered by the Earth's atmosphere, so it is not generally used for remote sensing of Earth materials.

The Visible Spectrum

Although the visible spectrum constitutes a very small portion of the spectrum, it has obvious significance in remote sensing. Limits of the visible spectrum are defined by the sensitivity of the human visual system. Optical properties of visible radiation were first investigated by Isaac Newton (1641–1727), who during 1665 and 1666 conducted experiments that revealed that visible light can be divided (using prisms, or, in our time, diffraction gratings) into three segments. Today we know these segments as the *additive primaries*, defined approximately from 0.4 to 0.5 µm (blue), 0.5 to 0.6 µm (green), and 0.6 to 0.7 µm (red) (Figure 2.4). *Primary colors* are defined such that no single primary can be formed from a mixture of the other two and that all other colors can be formed by mixing the three primaries in appropriate proportions. Equal proportions of the three additive primaries combine to form white light.

The color of an object is defined by the color of the light that it reflects (Figure 2.4). Thus a "blue" object is "blue" because it reflects blue light. Intermediate colors are formed when an object reflects two or more of the additive primaries, which combine to create the sensation of "yellow" (red and green), "purple" (red and blue), or other colors.

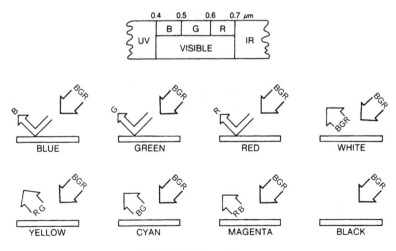

FIGURE 2.4. Colors.

The additive primaries are significant whenever we consider the colors of light, as, for example, in the exposure of photographic films.

In contrast, *representations* of colors in films, paintings, and similar images are formed by combinations of the three *subtractive primaries* that define the colors of pigments and dyes. Each of the three subtractive primaries absorbs a third of the visible spectrum (Figure 2.4). *Yellow* absorbs blue light (and reflects red and green), *cyan* (a greenish-blue) absorbs red light (and reflects blue and green), and *magenta* (a bluish red) absorbs green light (and reflects red and blue light). A mixture of equal proportions of pigments of the three subtractive primaries yields black (complete absorption of the visible spectrum). The additive primaries are of interest in matters concerning radiant energy, whereas the subtractive primaries specify colors of the pigments and dyes used in reproducing colors on films, photographic prints, and other images.

The Infrared Spectrum

Wavelengths longer than the red portion of the visible spectrum are designated as the infrared region, discovered in 1800 by the British astronomer William Herschel (1738–1822). This segment of the spectrum is very large relative to the visible region, as it extends from 0.72 to 15 µm—making it more than 40 times as wide as the visible light spectrum. Because of its broad range, it encompasses radiation with varied properties. Two important categories can be recognized here. The first consists of *near infrared* and *midinfrared* radiation—defined as those regions of the infrared spectrum closest to the visible. Radiation in the near infrared region behaves, with respect to optical systems, in a manner analogous to radiation in the visible spectrum. Therefore, remote sensing in the near infrared region can use films, filters, and cameras with designs similar to those intended for use with visible light.

The second category of infrared radiation is the *far infrared* region, consisting of wavelengths well beyond the visible, extending into regions that border the microwave region (Table 2.3). This radiation is fundamentally different from that in the visible and the near infrared regions. Whereas near infrared radiation is essentially solar radiation reflected from the Earth's surface, far infrared radiation is emitted by the Earth. In everyday language, the far infrared consists of "heat," or "thermal energy." Sometimes this portion of the spectrum is referred to as the *emitted infrared*.

Microwave Energy

The longest wavelengths commonly used in remote sensing are those from about 1 mm to 1 µm in wavelength. The shortest wavelengths in this range have much in common with the thermal energy of the far infrared. The longer wavelengths of the microwave region merge into the radio wavelengths used for commercial broadcasts. Our knowledge of the microwave region originates from the work of the Scottish physicist James Clerk Maxwell (1831–1879) and the German physicist Heinrich Hertz (1857–1894).

2.4. Radiation Laws

The propagation of electromagnetic energy follows certain physical laws. In the interests of conciseness, some of these laws are outlined in abbreviated form, because our interest here is the basic relationships they express rather than the formal derivations that are available to the student in more comprehensive sources.

Isaac Newton was among the first to recognize the dual nature of light (and, by extension, all forms of electromagnetic radiation), which simultaneously displays behaviors associated with both discrete and continuous phenomena. Newton maintained that light is a stream of minuscule particles ("corpuscles") that travel in straight lines. This notion is consistent with the modern theories of Max Planck (1858–1947) and Albert Einstein (1879–1955). Planck discovered that electromagnetic energy is absorbed and emitted in discrete units called *quanta*, or *photons*. The size of each unit is directly proportional to the frequency of the energy's radiation. Planck defined a constant (h) to relate frequency (v) to radiant energy (Q):

$$Q = hv \qquad \text{(Eq. 2.2)}$$

His model explains the photoelectric effect, the generation of electric currents by the exposure of certain substances to light, as the effect of the impact of these discrete units of energy (quanta) on surfaces of certain metals, causing the emission of electrons.

Newton knew of other phenomena, such as the refraction of light by prisms, that are best explained by assuming that electromagnetic energy travels in a wave-like manner. James Clerk Maxwell was the first to formally define the wave model of electromagnetic radiation. His mathematical definitions of the behavior of electromagnetic energy are based on the assumption from classical (mechanical) physics that light and other forms of electromagnetic energy propagate as a series of waves. The wave model best explains some aspects of the observed behavior of electromagnetic energy (e.g., refraction by lenses and prisms and diffraction), whereas quantum theory provides explanations of other phenomena (notably, the photoelectric effect).

The rate at which photons (quanta) strike a surface is the *radiant flux* (ϕ_e), measured in watts (W); this measure specifies energy delivered to a surface in a unit of time. We also need to specify a unit of area; the *irradiance* (E_e) is defined as radiant flux per unit area (usually measured as watts per square meter). Irradiance measures radiation that strikes a surface, whereas the term *radiant exitance* (M_e) defines the rate at which radiation is emitted from a unit area (also measured in watts per square meter).

All objects with temperatures above absolute zero have temperature and emit energy. The amount of energy and the wavelengths at which it is emitted depend on the temperature of the object. As the temperature of an object increases, the total amount of energy emitted also increases, and the wavelength of maximum (peak) emission becomes shorter. These relationships can be expressed formally using the concept of the *blackbody*. A blackbody is a hypothetical source of energy that behaves in an idealized manner. It absorbs all incident radiation; none is reflected. A blackbody emits energy with perfect efficiency; its effectiveness as a radiator of energy varies only as temperature varies.

The blackbody is a hypothetical entity because in nature all objects reflect at least a small proportion of the radiation that strikes them and thus do not act as perfect reradiators of absorbed energy. Although truly perfect blackbodies cannot exist, their behavior can be approximated using laboratory instruments. Such instruments have formed the basis for the scientific research that has defined relationships between the temperatures of objects and the radiation they emit. *Kirchhoff's law* states that the ratio of emitted radiation to absorbed radiation flux is the same for all blackbodies at the same temperature.

This law forms the basis for the definition of *emissivity* (ε), the ratio between the emittance of a given object (M) and that of a blackbody at the same temperature (M_b):

$$\varepsilon = M/M_b \qquad \text{(Eq. 2.3)}$$

The emissivity of a true blackbody is 1, and that of a perfect reflector (a *whitebody*) would be 0. Blackbodies and whitebodies are hypothetical concepts, approximated in the laboratory under contrived conditions. In nature, all objects have emissivities that fall between these extremes (*graybodies*). For these objects, emissivity is a useful measure of their effectiveness as radiators of electromagnetic energy. Those objects that tend to absorb high proportions of incident radiation and then to reradiate this energy will have high emissivities. Those that are less effective as absorbers and radiators of energy have low emissivities (i.e., they return much more of the energy that reaches them). (In Chapter 9, further discussion of emissivity explains that emissivity of an object can vary with its temperature.)

The *Stefan–Boltzmann law* defines the relationship between the total emitted radiation (W) (often expressed in watts · cm^{-2}) and temperature (T) (absolute temperature, K):

$$W = \sigma T^4 \qquad \text{(Eq. 2.4)}$$

Total radiation emitted from a blackbody is proportional to the fourth power of its absolute temperature. The constant (σ) is the Stefan–Boltzmann constant (5.6697×10^{-8}) (watts · m^{-2} · K^{-4}), which defines unit time and unit area. In essence, the Stefan–Boltzmann law states that hot blackbodies emit more energy per unit area than do cool blackbodies.

Wien's displacement law specifies the relationship between the wavelength of radiation emitted and the temperature of a blackbody:

$$\lambda = 2{,}897.8/T \qquad \text{(Eq. 2.5)}$$

where λ is the wavelength at which radiance is at a maximum and T is the absolute temperature (K). As blackbodies become hotter, the wavelength of maximum emittance shifts to shorter wavelengths (Figure 2.5).

All three of these radiation laws are important for understanding electromagnetic radiation. They have special significance later in discussions of detection of radiation in the far infrared spectrum (Chapter 9).

2.5. Interactions with the Atmosphere

All radiation used for remote sensing must pass through the Earth's atmosphere. If the sensor is carried by a low-flying aircraft, effects of the atmosphere on image quality may be negligible. In contrast, energy that reaches sensors carried by Earth satellites (Chapter 6) must pass through the *entire depth* of the Earth's atmosphere. Under these conditions, atmospheric effects may have substantial impact on the quality of images and data that the sensors generate. Therefore, the practice of remote sensing requires knowledge of interactions of electromagnetic energy with the atmosphere.

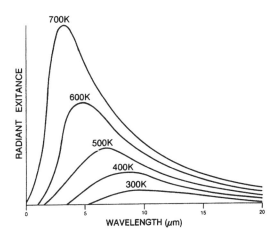

FIGURE 2.5. Wien's displacement law. For blackbodies at high temperatures, maximum radiation emission occurs at short wavelengths. Blackbodies at low temperatures emit maximum radiation at longer wavelengths.

In cities we often are acutely aware of the visual effects of dust, smoke, haze, and other atmospheric impurities due to their high concentrations. We easily appreciate their effects on brightnesses and colors we see. But even in clear air, visual effects of the atmosphere are numerous, although so commonplace that we may not recognize their significance. In both settings, as solar energy passes through the Earth's atmosphere, it is subject to modification by several physical processes, including (1) scattering, (2) absorption, and (3) refraction.

Scattering

Scattering is the redirection of electromagnetic energy by particles suspended in the atmosphere or by large molecules of atmospheric gases (Figure 2.6). The amount of scattering that occurs depends on the sizes of these particles, their abundance, the wavelength of the radiation, and the depth of the atmosphere through which the energy is traveling. The effect of scattering is to redirect radiation so that a portion of the incoming solar beam is directed back toward space, as well as toward the Earth's surface.

A common form of scattering was discovered by the British scientist Lord J. W. S. Rayleigh (1824–1919) in the late 1890s. He demonstrated that a perfectly clean atmosphere, consisting only of atmospheric gases, causes scattering of light in a manner such that the amount of scattering increases greatly as wavelength becomes shorter. *Rayleigh scattering* occurs when atmospheric particles have diameters that are very small relative to the wavelength of the radiation. Typically, such particles could be very small specks of dust or some of the larger molecules of atmospheric gases, such as nitrogen (N_2) and oxygen (O_2). These particles have diameters that are much smaller than the wavelength (λ) of visible and near infrared radiation (on the order of diameters less than λ).

Because Rayleigh scattering can occur in the absence of atmospheric impurities, it is sometimes referred to as *clear atmosphere scattering*. It is the dominant scattering process high in the atmosphere, up to altitudes of 9–10 km, the upper limit for atmospheric scattering. Rayleigh scattering is *wavelength-dependent*, meaning that the amount of scattering changes greatly as one examines different regions of the spectrum (Figure 2.7). Blue light is scattered about four times as much as is red light, and ultraviolet light is scattered almost 16 times as much as is red light. *Rayleigh's law* states

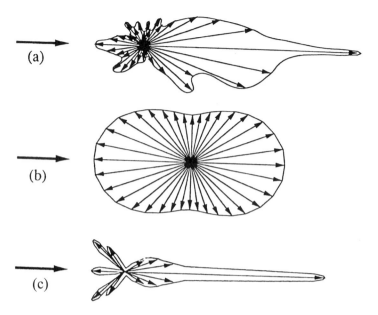

FIGURE 2.6. Scattering behaviors of three classes of atmospheric particles. (a) Atmospheric dust and smoke form rather large irregular particles that create a strong forward-scattering peak, with a smaller degree of backscattering. (b) Atmospheric molecules are more nearly symmetric in shape, creating a pattern characterized by preferential forward- and backscattering, but without the pronounced peaks observed in the first example. (c) Large water droplets create a pronounced forward-scattering peak, with smaller backscattering peaks. From Lynch and Livingston (1995). Reprinted with the permission of Cambridge University Press.

that this form of scattering is in proportion to the inverse of the fourth power of the wavelength.

Rayleigh scattering is the cause of both the blue color of the sky and the brilliant red and orange colors often seen at sunset. At midday, when the sun is high in the sky, the atmospheric path of the solar beam is relatively short and direct, so an observer at the Earth's surface sees mainly the blue light preferentially redirected by Rayleigh scatter. At sunset, observers on the Earth's surface see only those wavelengths that pass through the longer atmospheric path caused by the low solar elevation; because only the longer wavelengths penetrate this distance without attenuation by scattering, we see only the reddish component of the solar beam. Variations of concentrations of fine atmospheric dust or of tiny water droplets in the atmosphere may contribute to variations in atmospheric clarity and therefore to variations in colors of sunsets.

Although Rayleigh scattering forms an important component of our understanding of atmospheric effects on transmission of radiation in and near the visible spectrum, it applies only to a rather specific class of atmospheric interactions. In 1906 the German physicist Gustav Mie (1868–1957) published an analysis that describes atmospheric scattering involving a broader range of atmospheric particles. *Mie scattering* is caused by large atmospheric particles, including dust, pollen, smoke, and water droplets. Such particles may seem to be very small by the standards of everyday experience, but they are many times larger than those responsible for Rayleigh scattering. Those particles that cause Mie scattering have diameters that are roughly equivalent to the wavelength of the scattered radiation. Mie scattering can influence a broad range of wavelengths in and near the visible spectrum; Mie's analysis accounts for variations in the size, shape, and composition of such particles. Mie scattering is wavelength-dependent, but not in the simple manner of Rayleigh scattering; it tends to be greatest in the lower atmosphere (0 to 5 km), where larger particles are abundant.

Nonselective scattering is caused by particles that are much larger than the wavelength of the scattered radiation. For radiation in and near the visible spectrum, such particles might be larger water droplets or large particles of airborne dust. "Nonselective" means that scattering is not wavelength-dependent, so we observe it as a whitish or grayish haze—all visible wavelengths are scattered equally.

Effects of Scattering

Scattering causes the atmosphere to have a brightness of its own. In the visible portion of the spectrum, shadows are not jet black (as they would be in the absence of scattering) but are merely dark; we can see objects in shadows because of light redirected by particles in the path of the solar beam. The effects of scattering are also easily observed in vistas of landscapes—colors and brightnesses of objects are altered as they are positioned at locations more distant from the observer. Landscape artists take advantage of this effect, called *atmospheric perspective*, to create the illusion of depth by painting more distant features in subdued colors and those in the foreground in brighter, more vivid colors.

For remote sensing, scattering has several important consequences. Because of the wavelength dependency of Rayleigh scattering, radiation in the blue and ultraviolet regions of the spectrum (which is most strongly affected by scattering) is usually not considered useful for remote sensing. Images that record these portions of the spectrum tend to record the brightness of the atmosphere rather than the brightness of the scene itself. For this reason, remote sensing instruments often exclude short-wave radiation (blue and ultraviolet wavelengths) by use of filters or by decreasing sensitivities of films to these wavelengths. (However, some specialized applications of remote sensing, not discussed here, do use ultraviolet radiation.) Scattering also directs energy from outside the sensor's

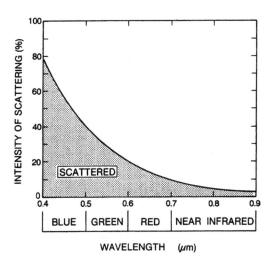

FIGURE 2.7. Rayleigh scattering. Scattering is much higher at shorter wavelengths.

field of view toward the sensor's aperture, thereby decreasing the spatial detail recorded by the sensor. Furthermore, scattering tends to make dark objects appear brighter than they would otherwise be, and bright objects appear darker, thereby decreasing the *contrast* recorded by a sensor (Chapter 3). Because "good" images preserve the range of brightnesses present in a scene, scattering degrades the quality of an image.

Some of these effects are illustrated in Figure 2.8. Observed radiance at the sensor, I, is the sum of I_S, radiance reflected from the Earth's surface, conveying information about surface reflectance; I_O, radiation scattered from the solar beam directly to the sensor without reaching the Earth's surface, and I_D, diffuse radiation, directed first to the ground, then to the atmosphere, before reaching the sensor. Effects of these components are additive within a given spectral band (Kaufman, 1984):

$$I = I_S + I_O + I_D$$ (Eq. 2.6)

I_S varies with differing surface materials, topographic slopes and orientation, and angles of illumination and observation. I_O is often assumed to be more or less constant over large areas, although most satellite images represent areas large enough to encompass atmospheric differences sufficient to create variations in I_O. Diffuse radiation, I_D, is expected to be small relative to other factors, but it varies from one land surface type to another, so in practice it would be difficult to estimate. We should note the special case presented by shadows in which $I_S = 0$, because the surface receives no direct solar radiation. However, shadows have their own brightness, derived from I_D, and their own spectral patterns, derived from the influence of local land cover on diffuse radiation. Remote sensing is

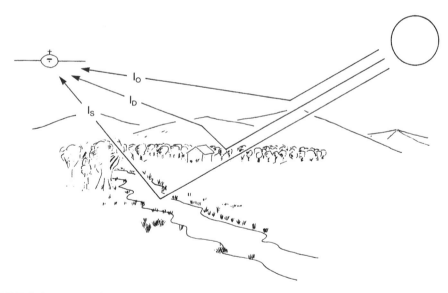

FIGURE 2.8. Principal components of observed brightness. I_S represent radiation reflected from the ground surface, I_O is energy scattered by the atmosphere directly to the sensor, and I_D represents diffuse light directed to the ground, then to the atmosphere, before reaching the sensor. This diagram describes behavior of radiation in and near the visible region of the spectrum. From Campbell and Ran (1993). Copyright 1993 by Elsevier Science Ltd. Reproduced by permission.

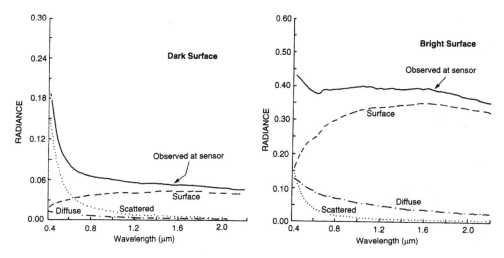

FIGURE 2.9. Changes in reflected, diffuse, scattered, and observed radiation over wavelength for dark (left) and bright (right) surfaces. The diagram shows the magnitude of the components illustrated in Figure 2.8. Atmospheric effects constitute a larger proportion of observed brightness for dark objects than for bright objects, especially at short wavelengths. Radiance has been normalized; note also the differences in scaling of the vertical axes for the two diagrams. Redrawn from Kaufman (1984). Reproduced by permission of the author and the Society of Photo-Optical Instrumentation Engineers.

devoted to the examination of I_S at different wavelengths to derive information about the Earth's surface. Figure 2.9 illustrates how I_D, I_S, and I_O vary with wavelength for surfaces of differing brightness.

Refraction

Refraction is the bending of light rays at the contact area between two media that transmit light. Familiar examples of refraction are the lenses of cameras or magnifying glasses (Chapter 3), which bend light rays to project or enlarge images, and the apparent displacement of objects submerged in clear water. Refraction also occurs in the atmosphere as light passes through atmospheric layers of varied clarity, humidity, and temperature. These variations influence the density of atmospheric layers, which in turn causes a bending of light rays as they pass from one layer to another. An everyday example is the shimmering appearances on hot summer days of objects viewed in the distance as light passes through hot air near the surface of heated highways, runways, and parking lots. The index of refraction (n) is defined as the ratio between the velocity of light in a vacuum (c) to its velocity in the medium (c_n):

$$n = c/c_n \qquad\qquad (Eq.\ 2.7)$$

Assuming uniform media, as the light passes into a denser medium it is deflected toward the *surface normal*, a line perpendicular to the surface at the point at which the light ray enters the denser medium, as represented by the solid line in Figure 2.10. The angle that defines the path of the refracted ray is given by *Snell's law*:

$$n \sin \theta = n' \sin \theta' \qquad \text{(Eq. 2.8)}$$

where n and n' are the indices of refraction of the first and second media, respectively, and θ and θ' are angles measured with respect to the surface normal, as defined in Figure 2.10.

Absorption

Absorption of radiation occurs when the atmosphere prevents, or strongly attenuates, transmission of radiation or its energy through the atmosphere. (Energy acquired by the atmosphere is subsequently reradiated at longer wavelengths.) Three gases are responsible for most absorption of solar radiation. Ozone (O_3) is formed by the interaction of high-energy ultraviolet radiation with oxygen molecules (O_2) high in the atmosphere (maximum concentrations of ozone are found at altitudes of about 20–30 km in the stratosphere). Although naturally occurring concentrations of ozone are quite low (perhaps 0.07 parts per million at ground level, 0.1–0.2 parts per million in the stratosphere), ozone plays an important role in the Earth's energy balance. Absorption of the high-energy, short-wavelength portions of the ultraviolet spectrum (mainly less than 0.24 µm) prevents transmission of this radiation to the lower atmosphere.

Carbon dioxide (CO_2) also occurs in low concentrations (about 0.03% by volume of a dry atmosphere), mainly in the lower atmosphere. Aside from local variations caused by volcanic eruptions and mankind's activities, the distribution of CO_2 in the lower atmosphere is probably relatively uniform (although human activities that burn fossil fuels have apparently contributed to increases during the past 100 years or so). Carbon dioxide is important in remote sensing because it is effective in absorbing radiation in the mid and far infrared regions of the spectrum. Its strongest absorption occurs in the region from about 13 to 17.5 µm in the mid infrared.

Finally, water vapor (H_2O) is commonly present in the lower atmosphere (below about 100 km) in amounts that vary from 0 to about 3% by volume. (Note the distinction between *water vapor*, discussed here, and droplets of *liquid* water, mentioned pre-

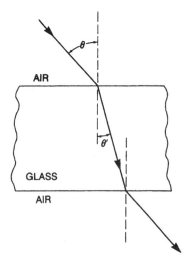

FIGURE 2.10. Refraction. This diagram represents the path of a ray of light as is passes from one medium (air) to another (glass), and again as it passes back to the first.

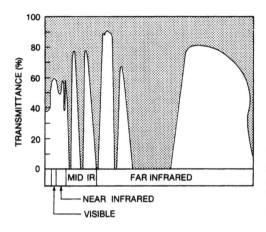

FIGURE 2.11. Atmospheric windows. This is a schematic representation that can depict only a few of the most important windows. The shaded region represents absorption of electromagnetic radiation.

viously.) From everyday experience we know that the abundance of water vapor varies greatly from time to time and from place to place. Consequently, the role of atmospheric water vapor, unlike those of ozone and carbon dioxide, varies greatly with time and location. It may be almost insignificant in a desert setting or in a dry air mass but may be highly significant in humid climates and in moist air masses. Furthermore, water vapor is several times more effective in absorbing radiation than are all other atmospheric gases combined. Two of the most important regions of absorption are in several bands between 5.5 and 7.0 μm and above 27.0 μm; absorption in these regions can exceed 80% if the atmosphere contains appreciable amounts of water vapor.

Atmospheric Windows

Thus the Earth's atmosphere is by no means completely transparent to electromagnetic radiation because these gases together form important barriers to transmission of electromagnetic radiation through the atmosphere. It selectively transmits energy of certain wavelengths; those wavelengths that are relatively easily transmitted through the atmosphere are referred to as *atmospheric windows* (Figure 2.11). Positions, extents, and effec-

TABLE 2.4. Major Atmospheric Windows

Ultraviolet and visible	0.30–0.75 μm
	0.77–0.91 μm
Near infrared	1.55–1.75 μm
	2.05–2.4 μm
Thermal infrared	8.0–9.2 μm
	10.2–12.4 μm
Microwave	7.5–11.5 mm
	20.0+ mm

Note. Data selected from Fraser and Curran (1976, p. 35). Reproduced by permission of Addison-Wesley Publishing Co., Inc.

tiveness of atmospheric windows are determined by the absorption spectra of atmospheric gases. Atmospheric windows are of obvious significance for remote sensing—they define those wavelengths that can be used for forming images. Energy at other wavelengths, not within the windows, is severely attenuated by the atmosphere and therefore cannot be effective for remote sensing. In the far infrared region, the two most important windows extend from 3.5 to 4.1 µm and from 10.5 to 12.5 µm. The latter is especially important because it corresponds approximately to wavelengths of peak emission from the Earth's surface. A few of the most important atmospheric windows are given in Table 2.4; other, smaller windows are not given here but are listed in reference books.

Overview of Energy Interactions in the Atmosphere

Remote sensing is conducted in the context of all the atmospheric processes discussed thus far, so it is useful to summarize some of the most important points by outlining a perspective that integrates much of the preceding material. Figure 2.12 is an idealized diagram of interactions of shortwave solar radiation with the atmosphere; values are based on typical, or average, values derived from many places and many seasons, so they are by no means representative of values that might be observed at a particular time and place. This diagram represents only the behavior of "shortwave" radiation (defined loosely here to include radiation with wavelengths less than 4.0 µm). It is true that the Sun emits a broad spectrum of radiation, but the maximum intensity is emitted at approximately 0.5 µm within this region, and little solar radiation at longer wavelengths reaches the ground surface.

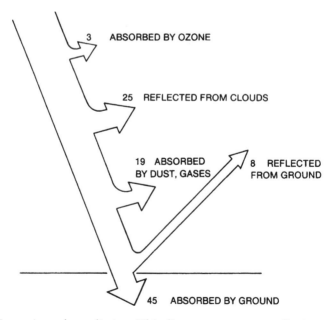

FIGURE 2.12. Incoming solar radiation. This diagram represents radiation at relatively short wavelengths, in and near the visible region. Values represent approximate magnitudes for the Earth as a whole—conditions at any specific place and time would differ from those given here.

FIGURE 2.13. Outgoing terrestrial radiation. This diagram represents radiation at relatively long wavelengths—what we think of as sensible heat, or thermal radiation. Because the Earth's atmosphere absorbs much of the radiation emitted by the Earth, only those wavelengths that can pass through the atmospheric windows can be used for remote sensing.

Of 100 units of shortwave radiation that reach the outer edge of the Earth's atmosphere, about 3 units are absorbed in the stratosphere as ultraviolet radiation interacts with oxygen (O_2) to form ozone (O_3). Of the remaining 97 units, about 25 are reflected from clouds, and about 19 are absorbed by dust and gases in the lower atmosphere. About 8 units are reflected from the ground surface (this value varies greatly with different surface materials), and about 45 units (about 50%) are ultimately absorbed at the Earth's surface. For remote sensing in the visible spectrum, it is the portion reflected from the Earth's surface that is of primary interest (see Figure 2.4), although knowledge of the quantity scattered is also important.

The 45 units that are absorbed are then reradiated by the Earth's surface. From Wien's displacement law (Eq. 2.5), we know that the Earth, being much cooler than the Sun, must emit radiation at much longer wavelengths than does the Sun. The Sun, at 6,000 K, has its maximum intensity at 0.5 µm (in the green portion of the visible spectrum); the Earth, at 300 K, emits with maximum intensity near 10 µm, in the far infrared spectrum.

Terrestrial radiation, with wavelengths longer than 10 µm, is represented in Figure 2.13. There is little, if any, overlap between the wavelengths of solar radiation, depicted in Figure 2.12, and the terrestrial radiation, shown in Figure 2.13. This diagram depicts the transfer of 143 units of long-wave radiation (as defined above) from the ground surface to the atmosphere, in three separate categories.

About 8 units are transferred from the ground surface to the atmosphere by "turbulent transfer" (heating of the lower atmosphere by the ground surface, which causes upward movement of air, then movement of cooler air to replace the original air). About 22 units are lost to the atmosphere by evaporation of moisture in the soil, water bodies, and vegetation (this energy is transferred as the latent heat of evaporation). Finally, about 113 units are radiated directly to the atmosphere.

Because atmospheric gases are very effective in absorbing this long-wave (far infrared) radiation, much of the energy that the Earth radiates is retained (temporarily) by the atmosphere. About 15 units pass directly through the atmosphere to space; this is energy emitted at wavelengths that correspond to atmospheric windows (chiefly 8–13 μm). Energy absorbed by the atmosphere is ultimately reradiated to space (49 units) and back to the Earth (98 units). For meteorology, it is these reradiated units that are of interest because they are the source of energy for heating of the Earth's atmosphere. For remote sensing, it is the 15 units that pass through the atmospheric windows that are of significance, as it is this radiation that conveys information concerning the radiometric properties of features on the Earth's surface.

2.6. Interactions with Surfaces

As electromagnetic energy reaches the Earth's surface, it must be reflected, absorbed, or transmitted. The proportions accounted for by each process depend on the nature of the surface, the wavelength of the energy, and the angle of illumination.

Reflection

Reflection occurs when a ray of light is redirected as it strikes a nontransparent surface. The nature of the reflection depends on sizes of surface irregularities (roughness or smoothness) in relation to the wavelength of the radiation considered. If the surface is smooth relative to wavelength, *specular* reflection occurs (Figure 2.14a). Specular reflection redirects all, or almost all, of the incident radiation in a single direction. For such surfaces, the angle of incidence is equal to the angle of reflection (i.e., in Eq. 2.8, the two media are identical, so $n = n$, and therefore $\theta = \theta'$). For visible radiation, specular reflection can occur with surfaces such as a mirror, smooth metal, or a calm water body.

If a surface is rough relative to wavelength, it acts as a *diffuse*, or *isotropic*, reflector. Energy is scattered more or less equally in all directions. For visible radiation, many natural surfaces might behave as diffuse reflectors, including, for example, uniform grassy surfaces. A perfectly diffuse reflector (known as a *Lambertian surface*) would have equal brightnesses when observed from any angle (Figure 2.14b).

The idealized concept of a perfectly diffuse reflecting surface is derived from the work of Johann H. Lambert (1728–1777), who conducted many experiments designed to describe the behavior of light. One of Lambert's laws of illumination states that the per-

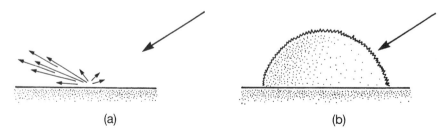

(a)   (b)

FIGURE 2.14. Specular (a) and diffuse (b) reflection. Specular reflection occurs when a smooth surface tends to direct incident radiation in a single direction. Diffuse reflection occurs when a rough surface tends to scatter energy more or less equally in all directions.

ceived brightness (radiance) of a perfectly diffuse surface does not change with the angle of view. This is Lambert's cosine law, which states that the observed brightness (I) of such a surface is proportional to the cosine of the incidence angle (θ), where I is the brightness of the incident radiation as observed at zero incidence:

$$I' = I/\cos \theta \qquad \text{(Eq. 2.9)}$$

This relationship is often combined with the equally important inverse square law, which states that observed brightness decreases according to the square of the distance from the observer to the source:

$$I' = (I/D^2)\,(\cos \theta) \qquad \text{(Eq. 2.10)}$$

(Both the cosine law and the inverse square law are depicted in Figure 2.15.)

Bidirectional Reflectance Distribution Function

Because of its simplicity and directness, the concept of a Lambertian surface is frequently used as an approximation of the optical behavior of objects observed in remote sensing. However, the Lambertian model does not hold precisely for many, if not most, natural surfaces. Actual surfaces exhibit complex patterns of reflection determined by details of surface geometry (e.g., the sizes, shapes, and orientations of plant leaves). Some surfaces may approximate Lambertian behavior at some incidence angles but exhibit clearly non-Lambertian properties at other angles.

Reflection characteristics of a surface are described by the *bidirectional reflectance distribution function* (BRDF). The BRDF is a mathematical description of the optical behavior of a surface with respect to angles of illumination and observation, given that it has been illuminated with a parallel beam of light at a specified azimuth and elevation. (The function is "bidirectional" in the sense that it accounts both for the angle of illumination and the angle of observation.) The BRDF for a Lambertian surface has the shape depicted in Figure 2.14b, with even brightnesses as the surface is observed from any angle. Actual surfaces have more complex behavior. Description of BRDFs for actual, rather than idealized, surfaces permits assessment of the degrees to which they approach the ideals of specular and diffuse surfaces (Figure 2.16).

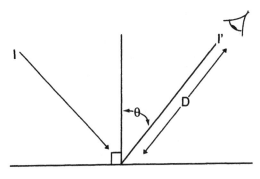

FIGURE 2.15. Inverse square law and Lambert's cosine law.

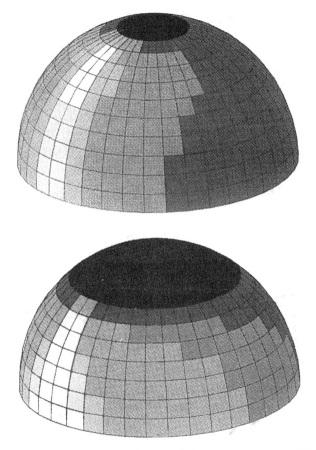

FIGURE 2.16. BRDFs for two surfaces. The varied shading represents differing intensities of observed radiation. (Calculated by Pierre Villeneuve.)

Transmission

Transmission of radiation occurs when radiation passes through a substance without significant attenuation (Figure 2.17). From a given thickness, or depth, of a substance, the ability of a medium to transmit energy is measured as the transmittance (t):

$$t = \frac{\text{Transmitted radiation}}{\text{Incident radiation}} \qquad \text{(Eq. 2.11)}$$

In the field of remote sensing, the transmittance of films and filters is often important. With respect to naturally occurring materials, we often think only of water bodies as capable of transmitting significant amounts of radiation. However, the transmittance of many materials varies greatly with wavelengths, so our direct observations in the visible spectrum do not transfer to other parts of the spectrum. For example, plant leaves are generally opaque to visible radiation but transmit significant amounts of radiation in the infrared.

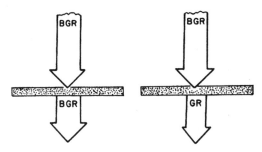

FIGURE 2.17. Transmission. Incident radiation passes through an object without significant attenuation (left), or may be selectively transmitted (right). The object on the right would act as a yellow ("minus blue") filter, as it would transmit all visible radiation except for blue light.

Fluorescence

Fluorescence occurs when an object illuminated with radiation of one wavelength emits radiation at a different wavelength. The most familiar examples are some sulfide minerals, which emit visible radiation when illuminated with ultraviolet radiation. Other objects also fluoresce, although observation of fluorescence requires very accurate and detailed measurements that are not now routinely available for most applications. Figure 2.18 illustrates the fluorescence of healthy and senescent leaves, using one axis to describe the spectral distribution of the illumination and the other to show the spectra of the emitted energy. These contrasting surfaces illustrate the effectiveness of fluorescence in revealing differences between healthy and stressed leaves.

Polarization

The *polarization* of electromagnetic radiation denotes the orientation of the oscillations within the electric field of electromagnetic energy (Figure 2.19). A light wave's electric field (traveling in a vacuum) is typically oriented perpendicular to the wave's direction of travel (i.e., the energy propagates as a *transverse* wave); the field may have a preferred orientation, or it may rotate as the wave travels. Although polarization of electromagnetic

FIGURE 2.18. Fluorescence. Exitation and emission are shown along the two horizontal axes (with wavelengths given in nanometers). The vertical axes show strength of fluorescence, with the two examples illustrating the contrast in fluorescence between healthy (*a*) and senescent (*b*) leaves. From Rinker (1994).

HORIZONTALLY   VERTICALLY
POLARIZED   POLARIZED

FIGURE 2.19. Schematic representation of horizontally and vertically polarized radiation. The smaller arrows signify orientations of the electric fields.

radiation is too complex for a full discussion here, it is possible to introduce some of the basics and highlight its significance.

An everyday example of the effect of polarized light is offered by polarizing sunglasses, which are specifically designed to reduce glare. Typically, sunlight within the atmosphere has a mixture of polarizations; when it illuminates surfaces at steep angles (i.e., when the sun is high in the sky), the reflected radiation tends to also have a mixture of polarizations. However, when the sun illuminates a surface at low angles (i.e., the sun is near the horizon), many surfaces tend to preferentially reflect the horizontally polarized component of the solar radiation. The polarizing sunglasses are manufactured with lenses that include molecules that can preferentially absorb the horizontally polarized bright radiation, thereby reducing glare. Polarization has broader significance in the practice of remote sensing. Within the atmosphere, polarization of light is related to the nature and abundance of atmospheric aerosols and atmospheric clarity. Chapter 7 introduces the use of polarized radiation in the design of active microwave sensors.

Reflectance

For many applications of remote sensing, the brightness of a surface is best represented not as irradiance but rather as *reflectance*. Reflectance (R_{rs}) is expressed as the relative brightness of a surface as measured for a specific wavelength interval:

$$\text{Reflectance} = \frac{\text{Observed brightness}}{\text{Irradiance}} \qquad \text{(Eq. 2.12)}$$

As a ratio, it is a dimensionless number (between 0 and 1), but it is commonly expressed as a percentage. In the usual practice of remote sensing, R_{rs} is not directly measurable, because normally we can observe only the observed brightness and must estimate irradiance. Strategies devised for estimation of reflectance are discussed in Chapter 11.

Spectral Properties of Objects

Remote sensing consists of the study of radiation emitted and reflected from features at the Earth's surface. In the instance of emitted (far infrared) radiation, the object itself is the immediate source of radiation. For reflected radiation, the source may be the Sun, the atmosphere (by means of scattering of solar radiation), or man-made radiation (chiefly imaging radars).

A fundamental premise in remote sensing is that we can learn about objects and features on the Earth's surface by studying the radiation reflected and/or emitted by these features. Using cameras and other remote sensing instruments, we can observe the brightnesses of objects over a range of wavelengths, so that there are numerous points of comparison between brightnesses of separate objects. A set of such observations or measurements constitutes a spectral response pattern, sometimes called the *spectral signature* of an object (Figure 2.20). In the ideal, detailed knowledge of a spectral response pattern might permit identification of features of interest, such as separate kinds of crops, forests, or minerals. This idea has been expressed as follows:

> Everything in nature has its own unique distribution of reflected, emitted, and absorbed radiation. These spectral characteristics can—if ingeniously exploited—be used to distinguish one thing from another or to obtain information about shape, size, and other physical and chemical properties. (Parker and Wolff, 1965, p. 21)

This statement expresses the fundamental concept of the spectral signature, the notion that features display unique spectral responses that would permit clear identification, from spectral information alone, of individual crops, soils, and so on from remotely

FIGURE 2.20. Spectral response curves for vegetation and water. These curves show the contrasting relationships between brightness (vertical axis) and wavelength (horizontal axis) for two common surfaces—living vegetation and open water. The sketches represent schematic views of a cross-section of a living leaf (left) and a pond with clear, calm water (right). The large arrows represent incident radiation from the Sun, the small lateral arrows represent absorbed radiation, the downward arrows represent transmitted energy, and the upward arrows represent energy directed upward to the sensor (known as "reflectance") that form the spectral response patterns illustrated at the top of the diagram. Fuller discussions of both topics are presented in later chapters.

sensed images. In practice, it is now recognized that spectra of features change both over time (e.g., as a cornfield grows during a season) and over distance (e.g., as proportions of specific tree species in a forest change from place to place).

Nonetheless, the study of the spectral properties of objects forms an important part of remote sensing. Some research has been focused on examination of spectral properties of different classes of features. Thus, although it may be difficult to define unique signatures for specific kinds of vegetation, we can recognize distinctive spectral patterns for vegetated and nonvegetated areas and for certain classes of vegetation, and we can sometimes detect the existence of diseased or stressed vegetation. In other instances, we may be able to define spectral patterns that are useful within restricted geographic and temporal limits as a means of studying the distributions of certain plant and soil characteristics. Chapter 15 describes how very detailed spectral measurements permit application of some aspects of the concept of the spectral signature.

2.7. Summary: Three Models for Remote Sensing

Remote sensing typically takes one of three basic forms, depending on the wavelengths of energy detected and on the purposes of the study. In the simplest form one records the reflection of solar radiation from the Earth's surface (Figure 2.21). This is the kind of remote sensing most nearly similar to everyday experience. For example, film in a camera records radiation from the Sun after it is reflected from the objects of interest, regardless of whether one uses a simple handheld camera to photograph a family scene or a complex aerial camera to photograph a large area of the Earth's surface. This form of remote sensing mainly uses energy in the visible and near infrared portions of the spectrum. Key variables include atmospheric clarity, spectral properties of objects, angle and intensity of the solar beam, choices of films and filters, and others explained in Chapter 3.

A second strategy for remote sensing is to record radiation *emitted* (rather than *reflected*) from the Earth's surface. Because emitted energy is strongest in the far infrared spectrum, this kind of remote sensing requires special instruments designed to record these wavelengths. (There is no direct analogue to everyday experience for this kind of remote sensing.) Emitted energy from the Earth's surface is mainly derived from short-

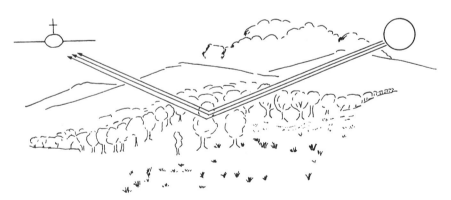

FIGURE 2.21. Remote sensing using reflected solar radiation. The sensor detects solar radiation that has been reflected from features at the Earth's surface. (See Figure 2.12.)

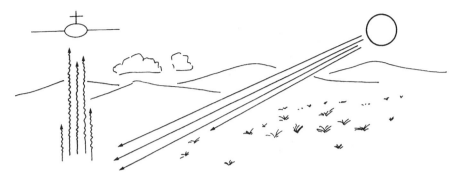

FIGURE 2.22. Remote sensing using emitted terrestrial radiation. The sensor records solar radiation that has been absorbed by the Earth, and then reemitted as thermal infrared radiation. (See Figures 2.12 and 2.13.)

wave energy from the Sun that has been absorbed, then reradiated at longer wavelengths (Figure 2.22).

Emitted radiation from the Earth's surface reveals information concerning thermal properties of materials, which can be interpreted to suggest patterns of moisture, vegetation, surface materials, and man-made structures. Other sources of emitted radiation (of secondary significance here, but often of primary significance elsewhere) include geothermal energy and heat from steam pipes, power plants, buildings, and forest fires. This example also represents "passive" remote sensing, because it employs instruments designed to sense energy emitted by the Earth, not energy generated by a sensor.

Finally, sensors belonging to a third class of remote sensing instruments generate their own energy, then record the reflection of that energy from the Earth's surface (Figure 2.23). These are "active" sensors—"active" in the sense that they provide their own energy, so they are independent of solar and terrestrial radiation. As an everyday analogy, a camera with a flash attachment can be considered to be an active sensor. In practice, active sensors are best represented by imaging radars and lidars (Chapters 7 and 8), which transmit energy toward the Earth's surface from an aircraft or satellite, then receive the reflected energy to form an image. Because they sense energy provided directly

FIGURE 2.23. Active remote sensing. The sensor illuminates the terrain with its own energy, then records the reflected energy as it has been altered by the Earth's surface.

by the sensor itself, such instruments have the capability to operate at night and during cloudy weather.

2.8. Some Teaching and Learning Resources

- Quantum Mechanics (Chapter 1b of 6)
  www.youtube.com/watch?v=l_t8dn4c6_g

- XNA Atmospheric Scattering
  www.youtube.com/watch?v=W0ocgQd_huU

- Atmospheric Rayleigh Scattering in 3D with Unity 3D
  www.youtube.com/watch?v=PNBnfqUycto

- How a Sunset Works
  www.youtube.com/watch?v=BdNQ1xB34QI&feature=related

- Why the Sky Is Blue
  www.youtube.com/watch?v=reBvvKBliMA&feature=related

- Tour of the EMS 04—Infrared Waves
  www.youtube.com/watch?v=i8caGm9Fmh0

- IR Reflection
  www.youtube.com/watch?v=h2n9WQCH1ds&feature=related

- What the Heck Is Emissivity? (Part 1)
  www.youtube.com/watch?v=QHszoA5Cy1I

- Emissivity Makes a Temperature Difference: Blackbody Calibrator
  www.youtube.com/watch?v=JElKE-ADXr8&feature=related

Review Questions

1. Using books provided by your instructor or available through your library, examine reproductions of landscape paintings to identify artistic use of atmospheric perspective. Perhaps some of your own photographs of landscapes illustrate the optical effects of atmospheric haze.

2. Some streetlights are deliberately manufactured to provide illumination with a reddish color. From material presented in this chapter, can you suggest why?

3. Although this chapter has largely dismissed ultraviolet radiation as an important aspect of remote sensing, there may well be instances in which it might be effective, despite the problems associated with its use. Under what conditions might it prove practical to use ultraviolet radiation for remote sensing?

4. The human visual system is most nearly similar to which model of remote sensing as described in the last sections of this chapter?

5. Can you identify analogues from the animal kingdom for each of the models for remote sensing discussed in Section 2.7?

6. Examine Figures 2.12 and 2.13, which show the radiation balance of the Earth's atmosphere. Explain how it can be that 100 units of solar radiation enter at the outer edge of the Earth's atmosphere, yet 113 units are emitted from the ground.

7. Examine Figures 2.12 and 2.13 again. Discuss how the values in this figure might change in different environments, including (a) desert, (b) the Arctic, (c) an equatorial climate. How might these differences influence our ability to conduct remote sensing in each region?

8. Spectral signatures can be illustrated using values indicating the brightness in several spectral regions.

	UV	Blue	Green	Red	NIR
Forest	28	29	36	27	56
Water	22	23	19	13	8
Corn	53	58	59	60	71
Pasture	40	39	42	32	62

Assume for now that these signatures are influenced by effects of the atmosphere. Can all categories be reliably separated, based on these spectral values? Which bands are most useful for distinguishing between these classes?

9. Describe ideal atmospheric conditions for remote sensing.

References

Bohren, C. F. 1987. *Clouds in a Glass of Beer: Simple Experiments in Atmospheric Physics.* New York: Wiley, 195 pp.

Campbell, J. B., and L. Ran. 1993. CHROM: A C Program to Evaluate the Application of the Dark Object Subtraction Technique to Digital Remote Sensing Data. *Computers and Geosciences*, Vol. 19, pp. 1475–1499.

Chahine, M. T. 1983. Interaction Mechanisms within the Atmosphere. Chapter 5 in *Manual of Remote Sensing* (R. N. Colwell, ed.). Falls Church, VA: American Society of Photogrammetry, pp. 165–230.

Chameides, W. L., and D. D. Davis. 1982. Chemistry in the Troposphere. *Chemical and Engineering News*, Vol. 60, pp. 39–52.

Clark, R. N., and T. L. Roush. 1984. Reflectance Spectroscopy: Quantitative Analysis Techniques for Remote Sensing Applications. *Journal of Geophysical Research*, Vol. 89, pp. 6329-

Estes, J. E. 1978. The Electromagnetic Spectrum and Its Use in Remote Sensing. Chapter 2 in *Introduction to Remote Sensing of Environment* (B. F. Richason, ed.). Dubuque, IA: Kendall-Hunt, pp. 15–39.

Fraser, R. S., and R. J. Curran. 1976. Effects of the Atmosphere on Remote Sensing. Chapter 2 in *Remote Sensing of Environment* (C. C. Lintz and D. S. Simonett, eds.). Reading, MA: Addison-Wesley, pp. 34–84.

Goetz, A. F. H., J. B. Wellman, and W. L. Barnes. 1985. Optical Remote Sensing of the Earth. *Proceedings of the IEEE*, Vol. 73, pp. 950–969.

Hariharan, T. A. 1969. Polarization of reflected solar radiation over land, sea and cloud surfaces. *Pure and Applied Physics*, Vol. 77, pp. 151–159.

Kaufman, Y. J. 1984. Atmospheric Effects on Remote Sensing of Surface Reflectance. Special issue: *Remote Sensing* (P. N. Slater, ed.). *Proceedings, SPIE*, Vol. 475, pp. 20–33.

Kaufman, Y. J. 1989. The Atmospheric Effect on Remote Sensing and Its Correction. Chapter 9 in *Theory and Applications of Optical Remote Sensing* (G. Asrar, ed.). New York: Wiley, pp. 336–428.

Lynch, D. K., and W. Livingston. 1995. *Color and Light in Nature*. New York: Cambridge University Press, 254 pp.

Minnaert, M. 1954. *The Nature of Light and Color* (revision by H. M. Kremer-Priest; translation by K. E. Brian Jay). New York: Dover, 362 pp.

Mobley, C. D. 1999. Estimation of the remote-sensing reflectance from above-surface measurements. *Applied Optics*, Vol. 38, pp. 7442–7455.

Parker, D. C., and M. F. Wolff. 1965. Remote Sensing. *International Science and Technology*, Vol. 43, pp. 20–31.

Rinker, J. N. 1994. ISSSR Tutorial I: Introduction to Remote Sensing. In *Proceedings of the International Symposium on Spectral Sensing Research '94*. Alexandria, VA: U.S. Army Topographic Engineering Center, pp. 5–43.

Schaepman-Strub, G., M. E. Schaepaman, J. Martonchik, T. Painter, and S. Dangel. 2009. Radiometry and Reflectance: From Terminology Concepts to Measured Quantities. Chapter 15 in *Sage Handbook of Remote Sensing* (T. M. Warner, M. D. Nellis, and G. M. Foody, Eds.). London: Sage, pp. 215–228.

Slater, P. N. 1980. *Remote Sensing: Optics and Optical Systems*. Reading, MA: Addison-Wesley, 575 pp.

Stimson, Allen. 1974. *Photometry and Radiometry for Engineers*. New York: Wiley, 446 pp.

Turner, R. E., W. A. Malila, and R. F. Nalepka. 1971. Importance of Atmospheric Scattering in Remote Sensing. In *Proceedings of the 7th International Symposium on Remote Sensing of Environment*. Ann Arbor, MI: Willow Run Laboratories, University of Michigan, pp. 1651–1697.

PART TWO

IMAGE ACQUISITION

CHAPTER THREE

Mapping Cameras

3.1. Introduction

This chapter introduces sensors used for acquiring aerial photographs. Although cameras are the oldest form of remote sensing instrument, they have changed dramatically in recent decades, yet nonetheless they exhibit continuity with respect to their fundamental purposes. Cameras designed for use in aircraft capture imagery that provides high positional accuracy and fine spatial detail. Despite the many other forms of remotely sensed imagery in use today, aerial photography remains the most widely used form of aerial imagery, employed for a wide variety of tasks by local and state governments, private businesses, and federal agencies to gather information to support planning, environmental studies, construction, transportation studies, routing of utilities, and many other tasks. The versatility of these images accounts for a large part of their enduring utility over the decades, even as fundamental technological shifts have transformed the means by which these images are acquired and analyzed. It is noteworthy that especially in the United States there is a large archive of aerial photographs acquired over the decades that form an increasingly valuable record of landscape changes since the 1930s.

During recent decades, the cameras, films, and related components that long formed the basis for traditional photographic systems (known as *analog* technologies) are rapidly being replaced by *digital* instruments that provide imagery with comparable characteristics that is acquired using electronic technologies. Here we introduce basic concepts that apply to these sensors, which are characterized by their use of aircraft as a platform and of the visible and near infrared spectrum and by their ability to produce imagery with fine detail and robust geometry. Although the majority of this chapter presents broad, generally applicable concepts without reference to specific instruments, it does introduce a selection of specific systems now used for acquiring aerial imagery. The transition from analog to digital aerial cameras has been under way for several decades and is nearing completion with respect to collection of imagery and for analysis, storage, and distribution. Yet digital systems are still evolving, with a variety of systems in use and under development, with uncertain standards, and with debate concerning relative merits of alternative systems. The following sections, therefore, present a snapshot of the transition from analog to digital technologies and the development of digital systems, with an outline of important principles established in the analog era to establish context.

3.2. Fundamentals of the Aerial Photograph

Systems for acquiring aerial images rely on the basic components common to the familiar handheld cameras we all have used for everyday photography: (1) a lens to gather light to form an image; (2) a light-sensitive surface to record the image; (3) a shutter that controls entry of light; and (4) a camera body—a light-tight enclosure that holds the other components together in their correct positions (Figure 3.1). Aerial cameras include these components in a structure that differs from that encountered in our everyday experience with cameras: (1) a film magazine, (2) a drive mechanism, and (3) a lens cone (Figure 3.1). This structure characterizes the typical design for the analog aerial camera that has been used (in its many variations) for aerial photography starting in the early 1900s. Although alternative versions of analog cameras were tailored to optimize specific capabilities, for our discussion, it is the *metric*, or *cartographic*, camera that has the greatest significance. Whereas other cameras may have been designed to acquire images (for example) of very large areas or under unfavorable operational conditions, the design of the metric camera is optimized to acquire high-quality imagery of high positional fidelity; it is the metric camera that forms the current standard for aerial photography.

For most of the history of remote sensing, aerial images were recorded as photographs or photograph-like images. A photograph forms a physical record—paper or film with chemical coatings that portray the patterns of the images. Such images are referred to as *analog* images because the brightnesses of a photograph are proportional (i.e., analogous) to the brightness in a scene. Although photographic media have value for recording images, in the context of remote sensing, their disadvantages, including difficulties of storage, transmission, searching, and analysis, set the stage for replacement by digital media. Digital technologies, in contrast, record image data as arrays of individual values that convey the pattern of brightnesses within an image.

Although a digital aerial camera shares many of the components and characteristics outlined above, in detail its design differs significantly from that of the analog camera.

FIGURE 3.1. Schematic diagram of an aerial camera, cross-sectional view.

(a)

(b)

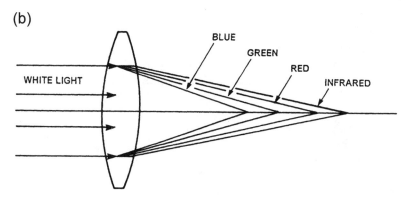

FIGURE 3.2. (a) Cross-sectional view of an image formed by a simple lens. (b) Chromatic aberration. Energy of differing wavelengths is brought to a focus at varying distances from the lens. More complex lenses are corrected to bring all wavelengths to a common focal point.

Because the image is captured by digital technology, digital cameras do not require the film and the complex mechanisms for manipulating the film. Further, digital cameras often include many capabilities not fully developed during the analog era, including links to positional and navigational systems and elaborate systems for annotating images.

The Lens

The *lens* gathers reflected light and focuses it on the focal plane to form an image. In its simplest form, a lens is a glass disk carefully ground into a shape with nonparallel curved surfaces (Figure 3.2). The change in optical densities as light rays pass from the atmosphere to the lens and back to the atmosphere causes refraction of light rays; the sizes, shapes, arrangements, and compositions of lenses are carefully designed to control refraction of light to maintain color balance and to minimize optical distortions. Optical characteristics of lenses are determined largely by the refractive index of the glass (see Chapter 2) and the degree of curvature. The quality of a lens is determined by the quality of its glass, the precision with which that glass is shaped, and the accuracy with which it is positioned within a camera. Imperfections in lens shape contribute to *spherical aberration*, a source of error that distorts images and causes loss of image clarity. For modern aerial photography, spherical aberration is usually not a severe problem because most modern aerial cameras use lenses of very high quality. Figure 3.2a shows the simplest of all lenses: a simple positive lens. Such a lens is formed from a glass disk with equal curvature on both sides; light rays are refracted at both edges to form an image.

Most aerial cameras use *compound lenses*, formed from many separate lenses of varied sizes, shapes, and optical properties. These components are designed to correct for errors that may be present in any single component, so the whole unit is much more accurate than any single element. For present purposes, consideration of a simple lens will be sufficient to define the most important features of lenses, even though a simple lens differs greatly from those actually used in modern aerial cameras. The *optical axis* joins the centers of curvature of the two sides of the lens. Although refraction occurs throughout a lens, a plane passing through the center of the lens, known as the *image principal plane*, is considered to be the center of refraction within the lens (Figure 3.2a). The image principal plane intersects the optical axis at the *nodal point*.

Parallel light rays reflected from an object at a great distance (at an "infinite" distance) pass through the lens and are brought to focus at the principal *focal point*—the point at which the lens forms an image of the distant object. The *chief ray* passes through the nodal point without changing direction; the paths of all other rays are deflected by the lens. A plane passing through the focal point parallel to the image principal plane is known as the *focal plane*. For handheld cameras, the distance from the lens to the object is important because the image is brought into focus at distances that increase as the object is positioned closer to the lens. For such cameras, it is important to use lenses that can be adjusted to bring each object to a correct focus as the distance from the camera to the object changes. For aerial cameras, the scene to be photographed is always at such large distances from the camera that the focus can be fixed at infinity, with no need to change the focus of the lens.

In a simple positive lens, the focal length is defined as the distance from the center of the lens to the focal point, usually measured in inches or millimeters. (For a compound lens, the definition is more complex.) For a given lens, the focal length is not identical for all wavelengths. Blue light is brought to a focal point at a shorter distance than are red or infrared wavelengths (Figure 3.2b). This effect is the source of *chromatic aberration*. Unless corrected by lens design, chromatic aberration would cause the individual colors of an image to be out of focus in the photograph. Chromatic aberration is corrected in high-quality aerial cameras to assure that the radiation used to form the image is brought to a common focal point.

The field of view of a lens can be controlled by a *field stop*, a mask positioned just in front of the focal plane. An *aperture stop* is usually positioned near the center of a compound lens; it consists of a mask with a circular opening of adjustable diameter (Figure 3.3). An aperture stop can control the intensity of light at the focal plane but does not influence the field of view or the size of the image. Manipulation of the aperture stop controls only the brightness of the image without changing its size. Usually aperture size is measured as the diameter of the adjustable opening that admits light to the camera.

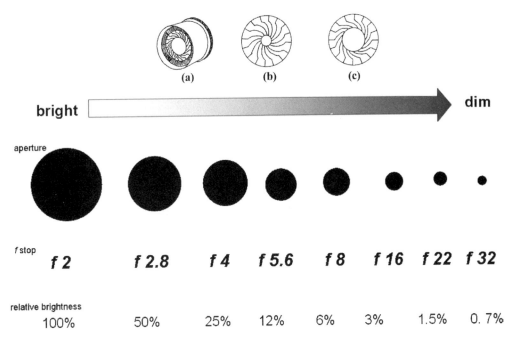

FIGURE 3.3. Diaphragm aperture stop. (a) Perspective view. (b) Narrow aperture. (c) Wide aperture. f stops represented below.

Relative aperture is defined as

$$f = \frac{\text{Focal length}}{\text{Aperture size}} \qquad \text{(Eq. 3.1)}$$

where focal length and aperture are measured in the same units of length and f is the f *number*, the relative aperture. A large f number means that the aperture opening is small relative to focal length; a small f number means that the opening is large relative to focal length.

Why use f numbers rather than direct measurements of aperture? One reason is that standardization of aperture with respect to focal length permits specification of aperture sizes using a value that is independent of camera size. Specification of an aperture as "23 mm" has no practical meaning unless we also know the size (focal length) of the camera. Specification of an aperture as "f 4" has meaning for cameras of all sizes; we know that it is one-fourth of the focal length for any size camera.

The standard sequence of apertures is: f 1, f 1.4, f 2, f 2.8, f 4, f 5.6, f 8, f 11, f 16, f 22, f 32, f 64, and so forth. This sequence is designed to change the amount of light by a factor of 2 as the f-stop is changed by one position. For example, a change from f 2 to f 2.8 halves the amount of light entering the camera; a change from f 11 to f 8 doubles the amount of light. A given lens, of course, is capable of using only a portion of the range of apertures mentioned above.

Lenses for aerial cameras typically have rather wide fields of view. As a result, light reaching the focal plane from the edges of the field of view is typically dimmer than light reflected from objects positioned near the center of the field of view. This effect creates a dark rim around the center of the aerial photograph—an effect known as *vignetting*. It is possible to employ an *antivignetting filter*, darker at the center and clearer at the periphery, that can be partially effective in evening brightnesses across the photograph. Digital systems can also employ image processing algorithms, rather than physical filters, to compensate for vignetting.

The Shutter

The *shutter* controls the length of time that the film is exposed to light. The simplest shutters are often metal blades positioned between elements of the lens, forming "intralens," or "between-the-lens," shutters. An alternative form of shutter is the focal plane shutter, consisting of a metal or fabric curtain positioned just in front of the detector array, near the focal plane. The curtain is constructed with a number of slits; the choice of shutter speed by the operator selects the opening that produces the desired exposure. Although some analog aerial cameras once used focal plane shutters, the between-the-lens shutter is preferred for most aerial cameras. The between-the-lens shutter subjects the entire focal plane to illumination simultaneously and presents a clearly defined perspective that permits use of the image as the basis for precise measurements.

Image Motion Compensation

High-quality aerial cameras usually include a capability known as *image motion compensation* (or *forward motion compensation*) to acquire high-quality images. Depending on the sensitivity of the recording media (either analog or digital), the forward motion of the aircraft can subject the image to blur when the aircraft is operated at low altitudes and/or high speeds. In the context of analog cameras, image motion compensation is achieved by mechanically moving the film focal plane at a speed that compensates for the apparent motion of the image in the focal plane. In the context of digital systems, image motion compensation is achieved electronically. Use of image motion compensation widens the range of conditions (e.g., lower altitudes and faster flight speeds) that can be used, while preserving the detail and clarity of the image.

3.3. Geometry of the Vertical Aerial Photograph

This section presents the basic geometry of a vertical aerial photograph as acquired by a classic framing camera. Not all portions of this discussion apply directly to all digital cameras, but the concepts and terminology presented here do apply to a broad range of optical systems used for remote sensing instruments described both in this chapter and in later sections.

Aerial photographs can be classified according to the orientation of the camera in relation to the ground at the time of exposure (Figure 3.4). *Oblique* aerial photographs have been acquired by cameras oriented toward the side of the aircraft. *High oblique* photographs (Figure 3.4a and Figure 3.5) show the horizon; *low oblique* photographs (Figure 3.4b) are acquired with the camera aimed more directly toward the ground surface and

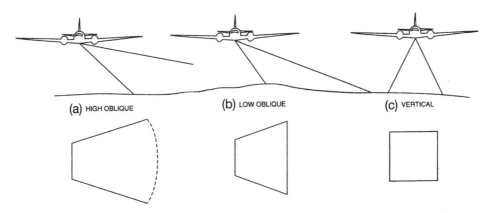

FIGURE 3.4. Oblique and vertical aerial photographs. Oblique perspectives provide a more intuitive perspective for visual interpretation but present large variation in image scale. Vertical photography presents a much more coherent image geometry, although objects are presented from unfamiliar perspectives and thus it can be more challenging to interpret.

do not show the horizon. Oblique photographs have the advantage of showing very large areas in a single image. Often those features in the foreground are easily recognized, as the view in an oblique photograph may resemble that from a tall building or mountain peak. However, oblique photographs are not widely used for analytic purposes, primarily because the drastic changes in scale that occur from foreground to background prevent convenient measurement of distances, areas, and elevations.

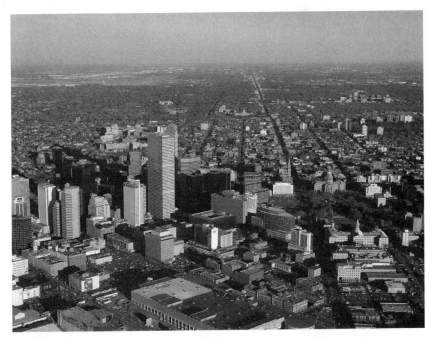

FIGURE 3.5. High oblique aerial photograph. From authors' photographs.

Vertical photographs are acquired by a camera aimed directly at the ground surface from above (Figures 3.4c and 3.6). Although objects and features are often difficult to recognize from their representations on vertical photographs, the map-like view of the Earth and the predictable geometric properties of vertical photographs provide practical advantages. It should be noted that few, if any, aerial photographs are truly vertical; most have some small degree of tilt due to aircraft motion and other factors. The term *vertical photograph* is commonly used to designate aerial photographs that are within a few degrees of a corresponding (hypothetical) truly vertical aerial photograph.

Because the geometric properties of vertical and nearly vertical aerial photographs are well understood and can be applied to many practical problems, they form the basis for making accurate measurements using aerial photographs. The science of making accurate measurements from aerial photographs (or from any photograph) is known as *photogrammetry*. The following paragraphs outline some of the most basic elements of introductory photogrammetry; the reader should consult a photogrammetry text (e.g., Wolf, 1983) for complete discussion of this subject.

Analog aerial cameras are manufactured to include adjustable index marks attached rigidly to the camera so that the positions of the index marks are recorded on the photograph during exposure. These *fiducial marks* (usually four or eight in number) appear as silhouettes at the edges and/or corners of the photograph (Figure 3.7). Lines that connect opposite pairs of fiducial marks intersect to identify the *principal point*, defined as the intersection of the optical axis with the focal plane, which forms the optical center of the image. The *ground nadir* is defined as the point on the ground vertically

FIGURE 3.6. Vertical aerial photograph. From U.S. Geological Survey.

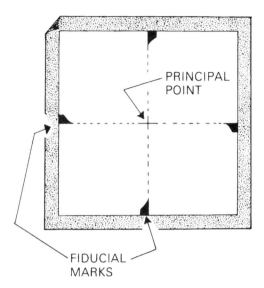

PRINCIPAL
POINT

FIDUCIAL
MARKS

FIGURE 3.7. Fiducial marks and principal point.

beneath the center of the camera lens at the time the photograph was taken (Figure 3.8). The *photographic nadir* is defined by the intersection with the photograph of the vertical line that intersects the ground nadir and the center of the lens (i.e., the image of the ground nadir). Accurate evaluation of these features depends on systematic and regular *calibration* of aerial cameras—the camera's internal optics and positioning of fiducial marks are assessed and adjusted to ensure the optical and positional accuracy of imagery for photogrammetric applications. Calibration can be achieved by using the cameras to photograph a standardized target designed to evaluate the quality of the imagery, as well as by internal measurements of the camera's internal geometry (Clarke and Fryer, 1998).

The *isocenter* can be defined informally as the focus of tilt. Imagine a truly vertical photograph that was taken at the same instant as the real, almost vertical, image. The almost vertical image would intersect with the (hypothetical) perfect image along a line that would form a "hinge"; the isocenter is a point on this hinge. On a truly vertical photograph, the isocenter, the principal point, and the photographic nadir coincide. The most important positional, or geometric, errors in the vertical aerial photograph can be summarized as follows.

1. *Optical distortions* are errors caused by an inferior camera lens, camera malfunction, or similar problems. These distortions are probably of minor significance in most modern photography flown by professional aerial survey firms.

2. *Tilt* is caused by displacement of the focal plane from a truly horizontal position by aircraft motion (Figure 3.8). The focus of tilt, the isocenter, is located at or near the principal point. Image areas on the upper side of the tilt are displaced farther away from the ground than is the isocenter; these areas are therefore depicted at scales smaller than the nominal scale. Image areas on the lower side of the tilt are displaced down; these areas are depicted at scales larger than the nominal scale. Therefore, because all photographs have some degree of tilt, measurements confined to one portion of the image run the risk of including systematic error caused by tilt

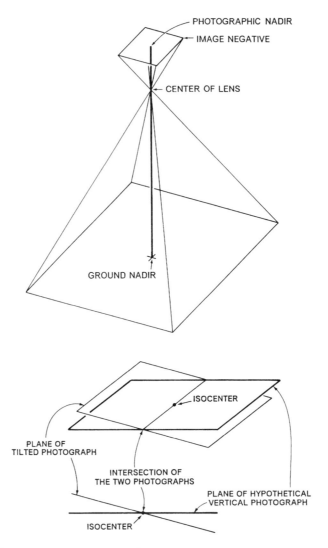

FIGURE 3.8. Schematic representation of terms to describe geometry of vertical aerial photographs.

(i.e., measurements may be consistently too large or too small). To avoid this effect, it is a good practice to select distances used for scale measurements (Chapter 5) as lines that pass close to the principal point; then errors caused by the upward tilt compensate for errors caused by the downward tilt. The resulting value for image scale is not, of course, precisely accurate for either portion of the image, but it will not include the large errors that can arise in areas located further from the principal point.

3. Because of routine use of high-quality cameras and careful inspection of photography to monitor image quality, today the most important source of positional error in vertical aerial photography is probably *relief displacement* (Figure

3.9). Objects positioned directly beneath the center of the camera lens will be photographed so that only the top of the object is visible (e.g., object A in Figure 3.9). All other objects are positioned such that both their tops and their sides are visible from the position of the lens. That is, these objects appear to lean outward from the central perspective of the camera lens. Correct planimetric positioning of these features would represent only the top view, yet the photograph shows both the top and sides of the object. For tall features, it is intuitively clear that the base and the top cannot both be in their correct planimetric positions.

This difference in apparent location is due to the height (*relief*) of the object and forms an important source of positional error in vertical aerial photographs. The direction of relief displacement is radial from the nadir; the amount of displacement depends on (1) the height of the object and (2) the distance of the object from the nadir. Relief displacement increases with increasing heights of features and with increasing distances from the nadir. (It also depends on focal length and flight altitude, but these may be regarded as constant for a selection of sequential photographs.) Relief displacement can form the basis of measurements of heights of objects, but its greatest significance is its role as a source of positional error. Uneven terrain can create significant relief displacement, so all measurements made directly from uncorrected aerial photographs are suspect. We should note that this is a source of positional error, but it is not the kind of error that can be corrected by selection of better equipment or more careful operation—it is an error that is caused by the central perspective of the lens and so is inherent to the choice of basic technology.

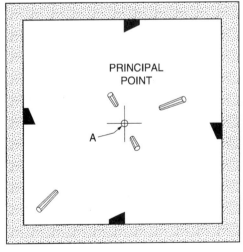

PLANIMETRICALLY CORRECT POSITIONS ⟶

FIGURE 3.9. Relief displacement. The diagram depicts a vertical aerial photograph of an idealized flat terrain with five towers of equal height located at different positions with respect to the principal point. Images of the tops of towers are displaced away from the principal point along lines that radiate from the nadir, as discussed in the text.

3.4. Digital Aerial Cameras

Digital imagery is acquired using a family of instruments that can systematically view portions of the Earth's surface, recording photons reflected or emitted from individual patches of ground, known as *pixels* ("picture elements"), that together compose the array of discrete brightnesses that form an image. Thus a digital image is composed of a matrix of many thousands of pixels, each too small to be individually resolved by the human eye. Each pixel represents the brightness of a small region on the Earth's surface, recorded digitally as a numeric value, often with separate values for each of several regions of the electromagnetic spectrum (Figure 3.10).

Although the lens of any camera projects an image onto the focal plane, the mere formation of the image does not create a durable image that can be put to practical use. To record the image, it is necessary to position a light-sensitive material at the focal plane. Analog cameras record images using the photosensitive chemicals that coated the surfaces of photographic films, as previously described. In contrast, digital cameras use an array of detectors positioned at the focal plane to capture an electronic record of the image. Detectors are light-sensitive substances that generate minute electrical currents when they intercept photons from the lens, thereby creating an image from the matrix of brightnesses that is proportional to the strengths of the electrical charges that reach the focal plane. Detectors in digital aerial cameras apply either of two alternative designs— charged-coupled devices (CCDs) or complementary metal oxide semiconductor (CMOS) chips. Each strategy offers its own advantages and disadvantages.

A CCD (Figure 3.11) is formed from light-sensitive material embedded in a silicon chip. The *potential well* receives photons from the scene through an optical system designed to collect, filter, and focus radiation. The sensitive components of CCDs can be manufactured to be very small, perhaps as small as 1µm in diameter, and sensitive to selected regions within the visible and near infrared spectra. These elements can be connected to each other using microcircuitry to form *arrays*. Detectors arranged in a single line form a *linear array;* detectors arranged in multiple rows and columns form *two-dimensional arrays*. Individual detectors are so small that a linear array shorter than 2 cm in length might include several thousand separate detectors. Each detector within a CCD collects photons that strike its surface and accumulates a charge proportional to the intensity of the radiation it receives. At a specified interval, charges accumulated at each detector pass through a *transfer gate*, which controls the flow of data from the

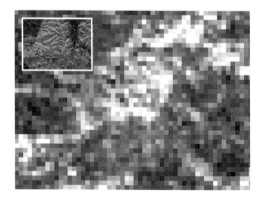

FIGURE 3.10. Pixels. A complete view of an image is represented in the inset; the larger image shows an enlargement of a small section to illustrate the pixels that convey variations in brightnesses.

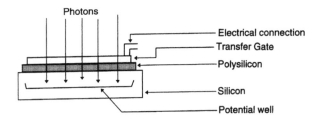

FIGURE 3.11. Schematic diagram of a charged-coupled device.

detectors. Microcircuits connect detectors within an array to form *shift registers*. Shift registers permit charges received at each detector to be passed to adjacent elements (in a manner analogous to a bucket brigade), temporarily recording the information until it is convenient to transfer it to another portion of the instrument. Through this process, information read from the shift register is read sequentially.

A CCD, therefore, scans electronically without the need for mechanical motion. Moreover, relative to other sensors, CCDs are compact, efficient in detecting photons (so CCDs are especially effective when intensities are dim), and respond linearly to brightness. As a result, CCD-based linear arrays have been used for remote sensing instruments that acquire imagery line by line as the motions of the aircraft or satellite carry the field of view forward along the flight track (Figure 3.12). As a result, over the past several decades, CCD technology has established a robust, reliable track record for scientific imaging.

An alternative imaging technology, CMOS, is often used for camcorders and related consumer products to record digital imagery and less often for acquiring aerial imagery. CMOS-based instruments often provide fine detail at low cost and at low power require-

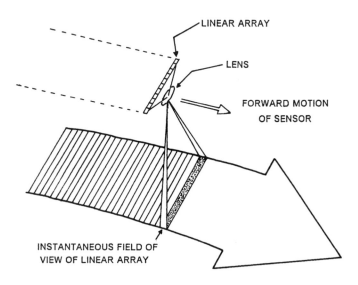

FIGURE 3.12. Schematic diagram of a linear array.

ments. Whereas CCDs expose all pixels at the same instant, then read these values as the next image is acquired, CMOS instruments expose a single line at a time, then expose the next line in the image while data for the previous line is transferred. Therefore, pixels within a CMOS image are not exposed at the same instant. This property of CMOS technology, plus the advantage of the low noise that characterizes CCD imagery, often favors use of CCDs for digital remote sensing instruments. Currently, the large linear or area arrays necessary for metric mapping cameras are CCD-based. However, CMOS technology is rapidly evolving, and the current CCD ascendancy is likely temporary.

Digital Camera Designs

In the digital realm, there are several alternative strategies for acquiring images, each representing a different strategy for forming digital images that are roughly equivalent to the 9 in. × 9 in. size of analog aerial photographs that became a commonly accepted standard in the United States after the 1930s. Although this physical size offers certain advantages with respect to convenience and standardization during the analog era, there is no technical reason to continue use of this format in the digital era, and indeed some digital cameras use slightly different sizes. In due course, a new standard or set of standards may well develop as digital systems mature to establish their own conventions.

Practical constraints of forming the large arrays of detectors necessary to approximate this standard size have led to camera designs that differ significantly from those of analog cameras described earlier. Analog cameras captured images frame by frame, meaning that each image was acquired as a single image corresponding to the single image projected into the focal plane at the time the shutter closed. This area, known as the camera *format*, varied in size and shape depending on the design of the camera, although, as mentioned above, a common standard for mapping photography used the 9 in. × 9 in., now defined by its metric equivalent, 230 mm × 230 mm. This photographic frame, acquired at a single instant, forms the fundamental unit for the image—every such image is a frame, a portion of a frame, or a composite of several frames. Such cameras are therefore designated as *framing cameras*, or as *frame array cameras*, which have formed the standard for analog aerial camera designs.

However, the framing camera design does not transfer cleanly into the digital domain. A principal reason for alternative designs for digital cameras is that the use of the traditional 230 mm × 230 mm film format for mapping cameras would require a nearly 660 megapixel array—a size that, currently, is much too large (i. e, too expensive) for most civilian applications. This situation requires some creative solutions for large-format digital cameras. One solution is to use multiple-area CCDs (and thus multiple lens systems) to acquire images of separate quadrants within the frame, then to stitch the four quadrants together to form a single image. Such composites provide an image that is visually equivalent to that of an analog mapping camera but that will have its own distinctive geometric properties; for example, such an image will have a nadir for each lens that might be used, and its brightnesses will be altered when the images are processed to form the composite. Another design solution for a digital aerial camera is to employ linear rather than area arrays. One such design employs a camera with separate lens systems to view (1) the nadir, (2) the forward-looking, and (3) the aft-looking position. At any given instant, the camera views only a few lines at the nadir, at the forward-looking position, and at the aft-viewing positions. However, as the aircraft moves forward along its flight track, each lens accumulates a separate set of imagery; these separate images can be digitally assembled to provide complete coverage from several perspectives in a single pass of the aircraft.

The following paragraphs highlight several state-of the-art examples of digital camera designs, including the line sensor-based Hexagon/Leica Geosystems ADS40, the Vexcel UltraCamX, and the large-format frame-based digital modular camera (DMC) from Intergraph, to illustrate general concepts. By necessity, these descriptions can outline only basic design strategies; readers who require detailed specifications should refer to the manufacturers' complete design specifications to see both the full details of the many design variations and the up-to-date information describing the latest models, which offer many specific applications of the basic strategies outlined here.

Area Arrays: The Intergraph Digital Modular Camera

The Intergraph Digital Modular Camera (DMC; Figure 3.13) is a large-format-frame digital camera. It uses four high-resolution panchromatic camera heads (focal length 120 mm) in the center and four multispectral camera heads (focal length 25 mm) on the periphery. The panchromatic CCD arrays are 7,000 × 4,000 pixels, resulting in a resolution of 3,824 pixels across track and 7,680 pixels along track (Boland et al., 2004). The multispectral arrays are 3,000 × 2,000 pixels with wavelength ranges as follows: blue (0.40–0.58 μm), green (0.50–0.65 μm), red (0.59–0.675 μm), and near infrared (0.675–0.85 or 0.74–0.85 μm). Note that, although this type of band overlap does not produce the ideal brightness value vectors, it is the best way to produce visually pleasing images and is not dissimilar from film color sensitivity. Standard photogrammetric processing (and packages) can be used with the postprocessed (virtual) stereo images from the DMC, but the base-to-height ratio is a bit lower than that of standard large-format film cameras.

Area Arrays: The Vexcel UltraCamX

The Vexcel UltraCamX employs multiple lens systems and CCDs positioned in the same plane, but with timing of the exposures to offset exposures slightly such that they view the scene from the same perspective center. Together, they form a system of eight CCDs— four panchromatic CCDs at fine resolution and four multispectral CCDs at coarser reso-

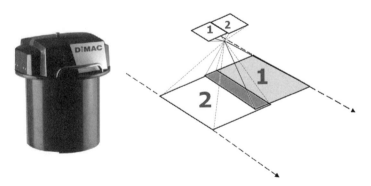

FIGURE 3.13. DMC area array. A single composite image is composed of two separate images acquired by independent lens systems with overlapping fields of view. From Intergraph.

lution to image each frame. The panchromatic images together form a master image that represents the entire frame, forming a high-resolution spatial framework for stitching images from the four multispectral cameras together to form a single multispectral image of the frame. The composite image forms a rectangular shape with the long axis oriented in the across-track dimension. This design does not replicate the optical and radiometric properties of analog framing cameras, as it employs multiple lens systems and various processing and interpolation procedures to produce the full-frame image and to ensure the proper registration of the separate bands (Figure 3.14).

Linear Arrays

The Leica ADS 40 (Figure 3.15) views the Earth with several linear arrays, each oriented to collect imagery line by line, separately from forward-viewing, nadir-viewing, and aft-viewing orientations. As an example, one of the most common versions of this design (known as the SH52), employs one forward-viewing, two nadir-viewing, and one aft-viewing panchromatic linear arrays, with the two nadir-viewing arrays offset slightly to provide a high-resolution image. In addition, multispectral arrays acquire nadir-viewing and aft-viewing images in the blue, green, red, and near infrared regions. The red, green, and blue bands are optically coregistered, but the near infrared band requires additional postprocessing for coregistration. Individual linear segments can be assembled to create four images in the blue, green, red, and near infrared regions, as well as the panchromatic (Boland et al., 2004). Thus, in one pass of the aircraft, using one instrument, it is possible to acquire, for a given region, multispectral imagery from several perspectives (Figure 3.16). One distinctive feature of this configuration is that the nadir for imagery collected by this system is, in effect, line connecting the nadirs of each linear array, rather than the center of the image, as is the case for an image collected by a framing camera. Therefore, each image displays relief displacement along track as a function only of object height, whereas in the across-track dimension relief displacement resembles that of a framing camera (i.e., relief displacement is lateral from the nadir). This instrument, like many

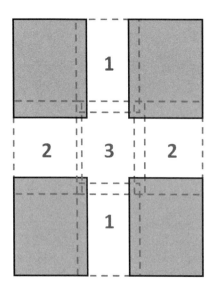

FIGURE 3.14. Schematic diagram of composite image formed from separate areal arrays.

NIR B R P G

GREEN PAN RED BLUE NIR

FIGURE 3.15. Schematic diagram of a linear array applied for digital aerial photography. The nadir-viewing red linear array acquires imagery in the red region; two aft-viewing arrays acquire green and panchromatic data in the blue and panchromatic channels, and forward-viewing arrays acquire imagery in the green and NIR. Because of the camera's continuous forward motion, each field of view acquires a strip of imagery of the flight line. See Figure 3.16. From Leica Geosystems.

others, requires high-quality positioning data, and, to date, data from this instrument require processing by software provided by the manufacturer. We can summarize by saying that this linear array solution is elegant and robust and is now used by photogrammetric mapping organizations throughout the world.

3.5. Digital Scanning of Analog Images

The value of the digital format for digital analysis has led to scanning of images originally acquired in analog form to create digital versions, which offer advantages for storage,

FIGURE 3.16. Schematic diagram of imagery collection by a linear array digital camera. See Figure 3.15. From Leica Geosystems.

transmission, and analysis. Although the usual scanners designed for office use provide, for casual use, reasonable positional accuracy and preserve much of the detail visible in the original, they are not satisfactory for scientific or photogrammetric applications. Such applications require scanning of original positives or transparencies to preserve positional accuracy of images and to accurately capture colors and original spatial detail using specialized high-quality flat-bed scanners, which provide large scanning surfaces, large CCD arrays, and sophisticated software to preserve information recorded in the original. Although there are obvious merits to scanning archived imagery, scanning of imagery for current applications must be regarded as an improvisation relative to original collection of data in a digital format.

3.6. Comparative Characteristics of Digital and Analog Imagery

Advantages of digital systems include:

1. Access to the advantages of digital formats without the need for film scanning, and therefore a more direct path to analytical processing.
2. Economies of storage, processing, and transmission of data.
3. Economies of operational costs.
4. Versatility in applications and in the range of products that can be derived from digital imagery.
5. The greater range of brightnesses of digital imagery, which facilitates interpretation and analysis.
6. True multispectral coverage.

Disadvantages include:

1. Varied camera designs do not replicate the optical and radiometric properties of an analog framing camera and employ various processing and interpolation procedures to produce the full-frame image, to adjust brightnesses across an image, and to ensure the proper registration of the separate bands. Thus many experts consider the geometry and radiometry of digital imagery to be inferior to that of high-quality metric cameras.
2. The typically smaller footprints of digital images require more stereomodels (i.e., more images) relative to analog systems.
3. Linear systems are especially dependent on high-quality airborne GPS/inertial measurement unit (AGPS/IMU) data.
4. Linear scanners also have sensor models that are less widely supported in softcopy photogrammetric software.
5. Digital systems require high initial investments.
6. There is less inherent stability than in metric film cameras (which can require reflying).
7. Component-level calibration and quality control can be difficult. Whereas analog

cameras can be calibrated by organizations such as the U.S. Geological Survey (USGS Optical Sciences Laboratory; *calval.cr.usgs.gov/osl*), which offers a centralized and objective source of expertise, at present, most digital cameras have highly specialized designs, experience changes in design and specification, and vary so greatly from manufacturer to manufacturer that calibration of most metric digital cameras must be completed by the specific manufacturer rather than by a centralized service.

Although these lists may suggest that the digital format is problematic, the transition from analog to digital imagery has not been driven solely by characteristics of digital cameras but rather by characteristics of the overall imaging system, including flight planning, processing capabilities, data transmission and storage, and image display, to name a few of many considerations. In this broader context, digital systems offer clear advantages when the full range of trade-offs are considered. Most organizations have concluded that the uncertain aspects of digital systems will be overcome as the technologies mature, and in the meantime they can exploit the many immediate business advantages relative to the older analog technology.

3.7. Spectral Sensitivity

Just as analog cameras used color films to capture the spectral character of a scene, so detectors can be configured to record separate regions of the spectrum as separate bands, or channels. CCDs and CMOS arrays have sensitivities determined by the physical properties of the materials used to construct sensor chips and the details of their manufacture. The usual digital sensors have spectral sensitivities that encompass the visible spectrum (with a maximum in the green region) and extend into the near infrared. Although arrays used for consumer electronics specifically filter to exclude the near infrared (NIR) radiation, aerial cameras can use this sensitivity to good advantage.

Color films use emulsions that are sensitive over a range of wavelengths, so even if their maximum sensitivity lies in the red, green, or blue regions, they are sensitive to radiation beyond the desired limits. In contrast, digital sensors can be designed to have spectral sensitivities cleanly focused in a narrow range of wavelengths and to provide high precision in measurement of color brightness. Therefore, digital sensors provide better records of the spectral characteristics of a scene—a quality that is highly valued by some users of aerial imagery. If the sensor chip is designed as separate arrays for each region of the spectrum, it acquires color images as separate planar arrays for each region of the spectrum. Although such designs would be desirable, they are not practical for current aerial cameras. Such large-array sizes are extremely expensive, difficult to manufacture to the required specifications, and may require long readout times to retrieve the image data. In due course, these costs will decline, and the manufacturing and technical issues will improve. In the meantime, aerial cameras use alternative strategies for simulating the effect of planar color data.

One alternative strategy uses a single array to acquire data in the three primaries using a specialized filter, known as a *Bayer filter*, to select the wavelengths that reach each pixel. A Bayer filter is specifically designed to allocate 50% of the pixels in an array to receive the green primary and 25% each to the red and blue primaries (Figure 3.17). (The

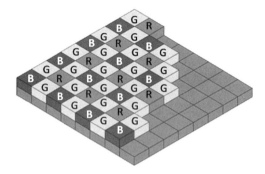

FIGURE 3.17. Bayer filter. The "B," "G," and "R" designations each signify cells with blue, green, and red filters, as explained in the text. Cells for each color are separately interpolated to produce individual layers for each primary.

rationale is that the human visual system has higher sensitivity in the green region, and, as mentioned in Chapter 2, peak radiation in the visible region lies in the green region.) In effect, this pattern samples the distribution of colors within the image, then the CCD chip processes pixel values to extrapolate, or interpolate, to estimate the missing values for the omitted pixels for each color. For example, the complete blue layer is formed by using the blue pixels to estimate the blue brightnesses omitted in the array, and similarly for the red and green primaries. This basic strategy has been implemented in several variations of the Bayer filter that have been optimized for various applications, and variations on the approach have been used to design cameras sensitive to the near infrared region.

This strategy, widely used in consumer electronics, produces an image that is satisfactory for visual examination because of the high density of detectors relative to the patterns recorded in most images and the short distances of the interpolation. However, this approach is less satisfactory for scientific applications and for aerial imagery—contexts in which the sharpness and integrity of each pixel may be paramount and artifacts of the interpolation process may be significant. Further, the Bayer filter has the disadvantages that the color filters reduce the amount of energy reaching the sensor and that interpolation required to construct the several bands reduces image sharpness.

An alternative strategy—Foveon technology (*www.foveon.com*)—avoids these difficulties by exploiting the differential ability of the sensor's silicon construction to absorb light. Foveon detectors (patented as the X3 CMOS design) are designed as three separate detector layers encased in silicon—blue-sensitive detectors at the surface, green-sensitive below, and red-sensitive detectors below the green. As light strikes the surface of the detector, blue light is absorbed near the chip's surface, green penetrates below the surface, and red radiation below the green. Thus each pixel can be represented by a single point that portrays all three primaries without the use of filters. This design has been employed for consumer cameras and may well find a role in aerial systems. At present, however, there are concerns that colors captured deeper in the chip may receive weaker intensities of the radiation and may have higher noise levels.

3.8. Band Combinations: Optical Imagery

Effective display of an image is critical for effective practice of remote sensing. *Band combinations* is the term that remote sensing practitioners use to refer to the assignment of colors to represent brightnesses in different regions of the spectrum. Although there are many ways to assign colors to represent different regions of the spectrum, experience shows that some are proven to be more useful than others. A key constraint for the display of any multispectral image is that human vision is sensitive only to the three additive primaries—blue, green, and red. Because our eyes can distinguish between brightnesses in these spectral regions, we can distinguish not only between blue, green, and red surfaces but also between intermediate mixtures of the primaries, such as yellow, orange, and purple. Color films and digital displays portray the effect of color by displaying pixels that vary the mixtures of the blue, green, and red primaries. Although photographic films must employ a single, fixed strategy for portraying colors, image processing systems and digital displays offer the flexibility to use any of many alternative strategies for assigning colors to represent different regions of the spectrum. These alternative choices define the band selection task; that is, how to decide which primary colors to select to best portray on the display screen specific features represented on imagery.

If the imagery at hand is limited to three spectral regions (as is the case for normal everyday color imagery), then the band selection task is simple, that is, display radiation from blue objects in nature as blue on the screen, green as green, and red as red. However, once we have bands available from outside the visible spectrum, as is common for remotely sensed imagery, then the choice of color assignment must have an arbitrary dimension. For example, there can be no logical choice for the primary we might use to display energy from the near infrared region. The common choices for the band selection problem, then, are established in part by conventions that have been defined by accepted use over the decades and in part by practice that has demonstrated that certain combinations are effective for certain purposes.

Here we introduce the band combinations most common for optical aerial imagery. Others are presented in Chapter 4, as other instruments are introduced.

Black-and-White Infrared Imagery

Imagery acquired in the near infrared region, because it is largely free of effects of atmospheric scattering, shows vegetated regions and land–water distinctions; it is one of the most valuable regions of the spectrum (Figure 3.18; see also Figure 3.28). An image representing the near infrared is formed using an optical sensor that has filtered the visible portion of the spectrum, so the image is prepared using only the brightness of the near infrared region (Figure 3.18). Examples are presented in Figures 3.20b and 3.28.

Panchromatic Imagery

Panchromatic means "across the colors," indicating that the visible spectrum is represented as a single channel (without distinguishing between the three primary colors). A panchromatic view provides a black-and-white image that records the brightnesses using radiation from the visible region but without separating the different colors (Figures 3.19, 3.20a). (This model is sometimes designated by the abbreviation "PAN.") Digital remote sensing systems often employ a panchromatic band that substitutes spatial detail for a color representation; that is, the instrument is designed to capture a detailed version of the scene using the data capacity that might have been devoted to recording the three primaries. That is, a decision has been made that the added detail provides more valuable information than would a color representation.

Because the easily scattered blue radiation will degrade the quality of an aerial image,

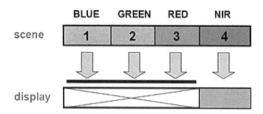

FIGURE 3.18. Diagram representing black-and-white infrared imagery. Visible radiation is filtered to isolate the near infrared radiation used to form the image.

some instruments are designed to capture radiation across the green, red, and near infrared regions of the spectrum, thereby providing a sharper, clearer image than would otherwise be the case. Therefore, even though the traditional definition of a panchromatic image is restricted to those based only on visible radiation, the term has a long history of use within the field of remote sensing to designate a broader region extending into the near infrared. If the single band encompasses the entire visible and the NIR, it can be designated sometimes as VNIR—signifying the use of visible radiation and the NIR region together. Other versions of this approach use only the green, red, and NIR, as illustrated in Figure 3.20b. For many applications, panchromatic aerial imagery is completely satisfactory, especially for imagery of urban regions in which color information may not be essential and added spatial detail is especially valuable.

Natural-Color Model

In everyday experience, our visual system applies band combinations in what seems to be a totally obvious manner—we see blue as blue, green as green, and red as red. The usual color films, color displays, and television screens apply this same strategy for assigning colors, often known as the *natural-color* assignment model (Figure 3.21 and Plate 1), or sometimes as the RGB (i.e., red–green–blue) model. Although natural-color imagery has value for its familiar representation of a scene, it suffers from a disadvantage outlined in Chapter 2—that the blue region of the spectrum is subject to atmospheric scattering, thereby limiting the utility of natural color images acquired at high altitudes.

Although remote sensing instruments collect radiation across many regions of the spectrum, outside the visible region humans are limited by our visual system to perceive

FIGURE 3.19. Two forms of panchromatic imagery. Left: visible spectrum only. Right: alternative form using green, red, and NIR radiation.

FIGURE 3.20. Panchromatic (left) and black-and-white infrared (right) imagery. From U.S. Geological Survey.

only the blue, green, and red primaries. Because our visual system is sensitive only in the visible region and can use only the three primaries, in remote sensing, we must make color assignments that depart from the natural-color model. These create *false-color* images—false in the sense that the colors on the image do not match their true colors in nature. Analysts select specific combinations of three channels to represent those patterns on the imagery needed to attain specific objectives. When some students first encounter this concept, it often seems nonsensical to represent an object using any color other than its natural color. Because the field of remote sensing uses radiation outside the visible spectrum, the use of the false-color model is a necessary task in displaying remote sensed imagery. The assignment of colors in this context is arbitrary, as there can be no correct way to represent the appearance of radiation outside the visible spectrum but simply a collection of practices that have proven to be effective for certain purposes.

Color Infrared Model

One of the most valuable regions of the spectrum is the NIR region, characterized by wavelengths just longer than the longest region of the visible spectrum. This region carries important information about vegetation and is not subject to atmospheric scattering,

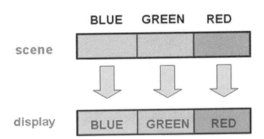

FIGURE 3.21. Natural-color model for color assignment.

so it is a valuable adjunct to the visible region. Use of the NIR region adds a fourth spectral channel to the natural-color model. Because we can recognize only three primaries, adding an NIR channel requires omission of one of the visible bands.

The color infrared model (CIR) (Figures 3.22 and Plate 1) creates a three-band color image by discarding the blue band from the visible spectrum and adding a channel in the NIR. This was a widely used model implemented in color infrared films, initially developed in World War II as *camouflage detection film* (i. e., designed to use NIR radiation to detect differences between actual vegetation and surfaces painted to resemble vegetation to the eye), and later as *CIR film*, now commonly used for displays of digital imagery. It shows living vegetation and water bodies very clearly and greatly reduces atmospheric effects compared with the natural-color model, so it is very useful for high-altitude aerial photography, which otherwise is subject to atmospheric effects that degrade the image. This band combination is important for studies in agriculture, forestry, and water resources, to list only a few of many.

Later chapters extend this discussion of band selection beyond those bands that apply primarily to aerial photography to include spectral channels that are acquired by other instruments.

3.9. Coverage by Multiple Photographs

A flight plan usually calls for acquisition of vertical aerial photographs by flying a series of parallel flight lines that together build up complete coverage of a specific region. For framing cameras, each flight line consists of individual frames, usually numbered in sequence (Figure 3.23). The camera operator can view the area to be photographed through a viewfinder and can manually trigger the shutter as aircraft motion brings predesignated landmarks into the field of view or can set controls to automatically acquire photographs at intervals tailored to provide the desired coverage. Individual frames form ordered strips, as shown in Figure 3.23a. If the plane's course is deflected by a crosswind, the positions of ground areas shown by successive photographs form the pattern shown in Figure 3.23b, known as *drift*. *Crab* (Figure 3.23c) is caused by correction of the flight path to compensate for drift without a change in the orientation of the camera.

Usually flight plans call for a certain amount of *forward overlap* (Figure 3.24) to duplicate coverage by successive frames in a flight line, usually by about 50–60% of each frame. If forward overlap is 50% or more, then the image of the principal point of one photograph is visible on the next photograph in the flight line. These are known as *conjugate principal points* (Figure 3.24). When it is necessary to photograph large areas,

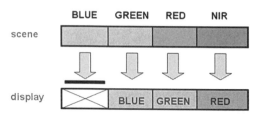

FIGURE 3.22. Color infrared model for color assignment.

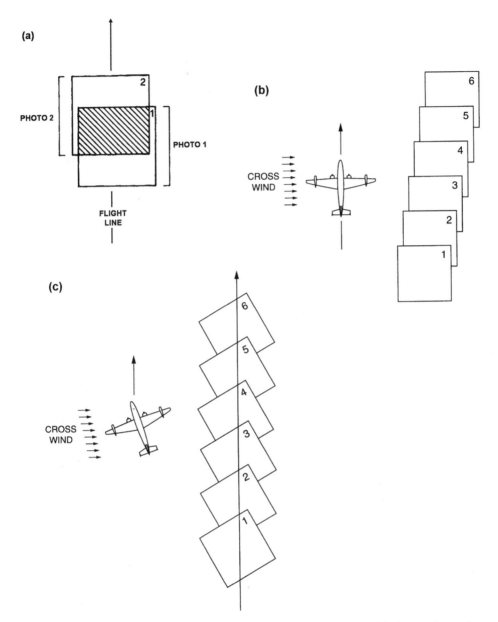

FIGURE 3.23. Aerial photographic coverage for framing cameras. (a) forward overlap, (b) drift, and (c) crab.

coverage is built up by means of several parallel strips of photography; each strip is called a *flight line*. Sidelap between adjacent flight lines may vary from about 5 to 15%, in an effort to prevent gaps in coverage of adjacent flight lines.

However, as pilots collect complete photographic coverage of a region, there may still be gaps (known as *holidays*) in coverage due to equipment malfunction, navigation errors, or cloud cover. Sometimes photography acquired later to cover holidays differs

FIGURE 3.24. Forward overlap and conjugate principal points.

noticeably from adjacent images with respect to sun angle, vegetative cover, and other qualities. For planning flight lines, the number of photographs required for each line can be estimated using the relationship:

$$\text{Number of photos} = \frac{\text{Length of flight line}}{(gd \text{ of photo}) \times (1 - \text{overlap})} \qquad \text{(Eq. 3.4)}$$

where *gd* is the ground distance represented on a single frame, measured in the same units as the length of the planned flight line. For example, if a flight line is planned to be 33 mi. in length, if each photograph is planned to represent 3.4 mi. on a side, and if forward overlap is to be 0.60, then 33/ [3.4 × (1 − .60)] = 33/(1.36) = 24.26; about 25 photographs are required. (Chapter 5 shows how to calculate the coverage of a photograph for a given negative size, focal length, and flying altitude.)

Stereoscopic Parallax

If we have two photographs of the same area taken from different perspectives (i.e., from different camera positions), we observe a displacement of images of objects from one image to the other (to be discussed further in Chapter 5). The reader can observe this effect now by simple observation of nearby objects. Look up from this book at nearby objects. Close one eye, then open it and close the other. As you do this, you observe a change in the appearance of objects from one eye to the next. Nearby objects are slightly different in appearance because one eye tends to see, for example, only the front of an object, whereas the other, because of its position (about 2.5 in.) from the other, sees the front and some of the side of the same object. This difference in appearances of objects due to change in perspective is known as *stereoscopic parallax*. The amount of parallax decreases as objects increase in distance from the observer (Figure 3.25). If you repeat the experiment looking out the window at a landscape you can confirm this effect by noting that distant objects display little or no observable parallax.

Stereoscopic parallax can therefore be used as a basis for measuring distance or height. Overlapping aerial photographs record parallax due to the shift in position of the camera as aircraft motion carries the camera forward between successive exposures. If forward overlap is 50% or more, then the entire ground area shown on a given frame can be viewed in stereo using three adjacent frames (a *stereo triplet*). Forward overlap

FIGURE 3.25. Stereoscopic parallax. These two photographs of the same scene were taken from slightly different positions. Note the differences in the appearances of objects due to the difference in perspective; note also that the differences are greatest for objects nearest the camera and least for objects in the distance. From author's photographs.

of 50–60% is common. This amount of overlap doubles the number of photographs required but ensures that the entire area can be viewed in stereo, because each point on the ground will appear on two successive photographs in a flight line.

Displacement due to stereo parallax is always parallel to the flight line. Tops of tall objects nearer to the camera show more displacement than do shorter objects, which are more distant from the camera. Measurement of parallax therefore provides a means of estimating heights of objects. Manual measurement of parallax can be accomplished as follows. Tape photographs of a stereo pair to a work table so the axis of the flight line is oriented from right to left (Figure 3.26). For demonstration purposes, distances can be measured with an engineer's scale.

1. Measure the distance between two principal points (X).
2. Measure the distance between separate images of the base of the object as represented on the two images (Y). Subtract this distance from that found in (1) to get P.
3. Measure top-to-top distances (B) and base-to-base distances (A), then subtract to find dp.

In practice, parallax measurements can be made more conveniently using devices that permit accurate measurement of small amounts of parallax.

Orthophotos and Orthophotomaps

Aerial photographs are not planimetric maps because they have geometric errors, most notably the effects of tilts and relief displacement, in the representations of the features they show. That is, objects are not represented in their correct planimetric positions, and as a result the images cannot be used as the basis for accurate measurements.

$$h = \frac{H \times dp}{P + dp}$$

H = FLYING HEIGHT OF AIRCRAFT

dp = DIFFERENTIAL PARALLAX = A - B

P = X - Y

h = HEIGHT OF THE OBJECT

FIGURE 3.26. Measurement of stereoscopic parallax.

Stereoscopic photographs and terrain data can be used to generate a corrected form of an aerial photograph known as an *orthophoto* that shows photographic detail without the errors caused by tilt and relief displacement. During the 1970s, an optical–mechanical instrument known as an *orthophotoscope* was developed to optically project a corrected version of a very small portion of an aerial photograph. An orthophotoscope, instead of exposing an entire image from a central perspective (i.e., through a single lens), systematically exposes a small section of an image individually in a manner that corrects for the elevation of that small section. The result is an image that has orthographic properties rather than those of the central perspective of the original aerial photograph. Digital versions of the orthophotoscope, developed in the mid-1980s, are capable of scanning an entire image piece by piece to generate a corrected version of that image. The result is an image that shows the same detail as the original aerial photograph but without the geometric errors introduced by tilt and relief displacement. Orthophotos form the basis for *orthophotomaps*, which show the image in its correct planimetric form, together with place names, symbols, and geographic coordinates. Thus they form digital map products that can be used in GIS as well as traditional maps because they show correct planimetric position and preserve consistent scale throughout the image.

Orthophotomaps are valuable because they show the fine detail of an aerial photograph without the geometric errors that are normally present and because they can be compiled much more quickly and cheaply than the usual topographic maps. Therefore, they can be very useful as map substitutes in instances in which topographic maps are not available or as map supplements when maps are available but the analyst requires the finer detail, and more recent information, provided by an image. Because of their digital format, fine detail, and adherence to national map accuracy standards, orthophotomaps are routinely used in GIS.

Digital Orthophoto Quadrangles

Digital orthophoto quadrangles (DOQs) are orthophotos prepared in a digital format designed to correspond to the 7.5-minute quadrangles of the U.S. Geological Survey (USGS). DOQs are presented either as black-and-white or color images that have been processed to attain the geometric properties of a planimetric map (Figure 3.27).

DOQs are prepared from National Aerial Photography Program (NAPP) photography (high-altitude photography described in Section 3.11) at scales of 1:40,000, supplemented by other aerial photography as needed. The rectification process is based on the use of digital elevation models (DEMs) to represent variations in terrain elevation. The final product is presented (as either panchromatic or CIR imagery) to correspond to the matching USGS 7.5-minute quadrangle, with a supplementary border of imagery representing 50–300 m beyond the limits of the quadrangle, to facilitate matching and mosaicking with adjacent sheets. A related product, the *digital orthophoto quarter-quadrangle* (DOQQ), formatted to provide a more convenient unit, represents one-fourth of the area of a DOQ at a finer level of detail, is available for some areas. DOQs provide image detail equivalent to 2 m or so for DOQs presented in the quadrangle format and finer detail (about 1 m) for DOQQs. The USGS has responsibility for leading the U.S. federal government's effort to prepare and disseminate digital cartographic data. The USGS

FIGURE 3.27. Digital orthophoto quarter quad, Platte River, Nebraska. From USGS.

has a program to prepare DOQs for many regions of the United States, especially urbanized regions, and the U.S. Department of Agriculture supports preparation of DOQs for agricultural regions (see Section 3.11). For more information on DOQs, visit the USGS website at

   http://edc.usgs.gov/glis/hyper/guide/usgs_doq.

3.10. Photogrammetry

Photogrammetry is the science of making accurate measurements from photographs. Photogrammetry applies the principles of optics and knowledge of the interior geometry of the camera and its orientation to reconstruct dimensions and positions of objects represented within photographs. Therefore, its practice requires detailed knowledge of specific cameras and the circumstances under which they were used and accurate measurements of features within photographs. Photographs used for analog photogrammetry have traditionally been prepared on glass plates or other dimensionally stable materials (i.e., materials that do not change in size as temperature and humidity change).

Photogrammetry can be applied to any photograph, provided the necessary supporting information is at hand to reconstruct the optical geometry of the image. However, by far the most frequent application of photogrammetry is the analysis of stereo aerial photography to derive estimates of topographic elevation for topographic mapping. With the aid of accurate locational information describing key features within a scene (*ground control*), photogrammetrists estimate topographic relief using stereo parallax for an array of points within a region. Although stereo parallax can be measured manually, it is far more practical to employ specialized instruments designed for stereoscopic analysis.

Initially, such instruments, known as *analytical stereoplotters*, first designed in the 1920s, reconstruct the orientations of photographs at the time they were taken using optical and mechanical instruments to reconstruct the geometry of the images at the time they were acquired (see Figure 1.7 for an example of an optical–mechanical photogrammetric instrument). Operators could then view the image in stereo; by maintaining constant parallax visually, they could trace lines of uniform elevation. The quality of information derived from such instruments depends on the quality of the photography, the accuracy of the data, and the operator's skill in setting up the stereo model and tracing lines of uniform parallax. As the design of instruments improved, it eventually became possible to automatically match corresponding points on stereo pairs and thereby identify lines of uniform parallax with limited assistance from the operator.

With further advances in instrumentation, it became possible to extend automation of the photogrammetric process to conduct the stereo analysis completely within the digital domain. With the use of GPS (*airborne global positioning systems [AGPS]*) to acquire accurate, real-time positional information and the use of data recorded from the aircraft's navigational system (*inertial navigational systems [INS]*) to record orientations of photographs, it then became feasible to reconstruct the geometry of the image using precise positional and orientation data gathered as the image was acquired. This process forms the basis for *softcopy photogrammetry*, so named because it does not require the physical (hardcopy) form of the photograph necessary for traditional photogrammetry. Instead, the digital (softcopy) version of the image is used as input for a series of mathematical models that reconstruct the orientation of each image to create planimetrically correct representations. This process requires specialized computer software installed in workstations (see Figure 5.18) that analyzes digital data specifically acquired for the purpose of photogrammetric analysis. Softcopy photogrammetry, now the standard for photogrammetric production, offers advantages of speed and accuracy and generates output data that are easily integrated into other production and analytical systems, including GIS.

The application of photogrammetric principles to imagery collected by digital cameras described above differs from that tailored for the traditional analog framing camera. Because each manufacturer has specific designs, each applying a different strategy for collecting and processing imagery, the current photogrammetric analyses are matched to the differing cameras. One characteristic common to many of these imaging systems is the considerable redundancy within imagery they collect—that is, each pixel on the ground can be viewed many times, each from a separate perspective. Because these systems each collect so many independent views of the same features (due to the use of several lenses, or several linear arrays, as outlined previously), it is possible to apply *multiray photogrammetry*, which can exploit these redundancies to extract highly detailed positional and elevation data beyond that which was possible using analog photography. Because, in the digital domain, these additional views do not incur significant additional costs, photogrammetric firms can provide high detail and a wide range of image products without the increased costs of acquiring additional data.

3.11. Sources of Aerial Photography

Aerial photography can (1) be acquired by the user or (2) purchased from organizations that serve as repositories for imagery flown by others (*archival imagery*). In the first instance, aerial photography can be acquired by contract with firms that specialize in high-quality aerial photography. Such firms are listed in the business sections of most metropolitan phone directories. Customers may be individuals, governmental agencies, or other businesses that use aerial photography. Such photography is, of course, customized to meet the specific needs of customers with respect to date, scale, film, and coverage. As a result, costs may be prohibitive for many noncommercial uses.

Thus, for pragmatic reasons, many users of aerial photography turn to archival photography to acquire the images they need. Although such photographs may not exactly match users' specifications with respect to scale or date, the inexpensive costs and ease of access may compensate for any shortcomings. For some tasks that require reconstruction of conditions at earlier dates (such as the Environmental Protection Agency's search for abandoned toxic waste dumps), the archival images may form the only source of information (e.g., Erb et al., 1981; Lyon, 1987).

It is feasible to take "do-it-yourself" aerial photographs. Many handheld cameras are suitable for aerial photography. Often the costs of local air charter services for an hour or so of flight time are relatively low. Small-format cameras, such as the usual 35-mm cameras, can be used for aerial photography if the photographer avoids the effects of aircraft vibration. (Do not rest the camera against the aircraft!) A high-wing aircraft offers the photographer a clear view of the landscape, although some low-wing aircraft are satisfactory. The most favorable lighting occurs when the camera is aimed away from the sun. Photographs acquired in this manner (e.g., Figure 3.5) may be useful for illustrative purposes, although for scientific or professional work the large-format, high-quality work of a specialist or an aerial survey firm may be required.

EROS Data Center

The EROS Data Center (EDC) in Sioux Falls, South Dakota, is operated by the USGS as a repository for aerial photographs and satellite images acquired by NASA, the USGS, and many other federal agencies. A computerized database at EDC provides an indexing system for information pertaining to aerial photographs and satellite images. For more information contact:

> Customer Services
> U.S. Geological Survey
> Earth Resources Observation and Science (EROS)
> 47914 252nd Street
> Sioux Falls, SD 57198-0001
> Tel: 800-252-4547 or 605-594-6151
> Fax: 605-594-6589
> E-mail: *custserv@usgs.gov*
> Website: *eros.usgs.gov*

Earth Science Information Centers

The Earth Science Information Centers (ESIC) are operated by the USGS as a central source for information pertaining to maps and aerial photographs.

> *http://ask.usgs.gov/sils_index.html*

ESIC has a special interest in information pertaining to federal programs and agencies but also collects data pertaining to maps and photographs held by state and local governments. The ESIC headquarters is located at Reston, Virginia, but ESIC also maintains seven other offices throughout the United States, and other federal agencies have affiliated offices. ESIC can provide information to the public concerning the availability of maps and remotely sensed images. The following sections describe two programs administered by ESIC that can provide access to archival aerial photography.

National Aerial Photography Program

NAPP acquires aerial photography for the coterminous United States, according to a systematic plan that ensures uniform standards. NAPP was initiated in 1987 by the USGS as a replacement for the National High-Altitude Aerial Photography Program (NHAP), begun in 1980 to consolidate the many federal programs that use aerial photography. The USGS manages the NAPP, but it is funded by the federal agencies that are the primary users of its photography. Program oversight is provided by a committee of representatives from the USGS, the Bureau of Land Management, the National Agricultural Statistics Service, the National Resources Conservation Service (NRCS; previously known as the Soil Conservation Service), the Farm Services Agency (previously known as the Agricultural Stabilization and Conservation Service), the U.S. Forest Service, and the Tennessee Valley Authority. Light (1993) and Plasker and TeSelle (1988) provide further details.

Under NHAP, photography was acquired under a plan to obtain complete coverage of the coterminous 48 states, then to update coverage as necessary to keep pace with requirements for current photography. Current plans call for updates at intervals of 5 years, although the actual schedules are determined in coordination with budgetary constraints. NHAP flight lines were oriented north–south, centered on each of four quadrants systematically positioned within USGS 7.5-minute quadrangles, with full stereoscopic coverage at 60% forward overlap and sidelap of at least 27%. Two camera systems were used to acquire simultaneous coverage: black-and-white coverage was acquired at scales of about 1:80,000, using cameras with focal lengths of 6 in. Color infrared coverage was acquired at 1:58,000, using a focal length of 8.25 in. Plate 2 shows a high-altitude CIR image illustrating the broad-scale coverage provided by this format.

Dates of NHAP photography varied according to geographic region. Flights were timed to provide optimum atmospheric conditions for photography and to meet specifications for sun angle, snow cover, and shadowing, with preference for autumn and winter seasons to provide images that show the landscape without the cover of deciduous vegetation.

Specifications for NAPP photographs differ from those of NHAP. NAPP photographs are acquired at 20,000-ft. altitude using a 6-in. focal length lens. Flight lines are centered on quarter quads (1:24,000-scale USGS quadrangles). NAPP photographs are planned for 1:40,000, black-and-white or color infrared film, depending on specific requirements for each area.

Photographs are available to all who may have an interest in their use. Their detail and quality permit use for land-cover surveys and assessment of agricultural, mineral, and forest resources, as well as examination of patterns of soil erosion and water quality. Further information is available at

http://edc.usgs.gov/guides/napp.html

http://eros.usgs.gov/products/aerial/napp.php

National Agricultural Imagery Program

The National Agriculture Imagery Program (NAIP) acquires aerial imagery during the agricultural growing seasons in the continental United States. The NAIP program focuses on providing digital orthophotography freely to governmental agencies and the public, usually as color or CIR imagery at about 1 m resolution. The DOQQ format means that the images are provided in a ready-to-use format (i.e., digital and georeferenced). An important difference between NAIP imagery and other programs (such as NHAP) is that NAIP imagery is acquired during the growing season (i.e., "leaf-on"), so it forms a valuable resource not only for agricultural applications but also for broader planning and resources assessment efforts. Further information is available at

www.fsa.usda.gov/FSA/apfoapp?area=home&subject=prog&topic=nai

Two other important sources of archival aerial photography include the U.S. Department of Agriculture (USDA) Aerial Photography Field Office:

www.apfo.usda.gov

and the U.S. National Archives and Records Administration:

www.archives.gov

3.12. Summary

Aerial photography offers a simple, reliable, flexible, and inexpensive means of acquiring remotely sensed images. The transition from the analog systems that formed the foundation for aerial survey in the 20th century to digital systems is now basically complete, although the nature of the digital systems that will form the basis for the field in the 21st century is not yet clear. The migration to digital formats has reconstituted, even rejuvenated aerial imagery's role in providing imagery for state and local applications. Although aerial photography is useful mainly in the visible and near infrared portions of the spectrum, it applies optical and photogrammetric principles that are important throughout the field of remote sensing.

Aerial photographs form the primary source of information for compilation of large-scale maps, especially large-scale topographic maps. Vertical aerial photographs are valuable as map substitutes or as map supplements. Geometric errors in the representation of location prevent direct use of aerial photographs as the basis for measurement of distance or area. But, as these errors are known and are well understood, it is possible for photogrammetrists to use photographs as the basis for reconstruction of correct positional relationships and the derivation of accurate measurements. Aerial photographs record complex detail of the varied patterns that constitute any landscape. Each image interpreter must develop the skills and knowledge necessary to resolve these patterns by disciplined examination of aerial images.

3.13. Some Teaching and Learning Resources

- Additive Color vs Subtractive Color
  www.youtube.com/watch?v=ygUchcpRNyk&feature=related

- What Are CMYK And RGB Color Modes?
  www.youtube.com/watch?v=0K8fqf2XBaY&feature=related

- Evolution of Analog to Digital Mapping
  www.youtube.com/watch?v=4jABMysbNbc

- Aerial Survey Photography Loch Ness Scotland G-BKVT
  www.youtube.com/watch?v=-YsDflbXMHk

- Video of the day; Aerial photography
  www.youtube.com/watch?v=VwtSTvF_Q2Q&NR=1

- How a Pixel Gets its Color; Bayer Sensor; Digital Image
  www.youtube.com/watch?v=2-stCNB8jT8

- Photography Equipment and Info: Explanation of Camera Lens Magnification
  www.youtube.com/watch?v=YEG93Hp3y4w&feature=fvst

- How a Digital Camera Works—CMOS Chip
  www.youtube.com/watch?v=VP__-EKrkbk&feature=related

- Digital Camera Tips: How a Compact Digital Camera Works
  www.youtube.com/watch?v=eyyMu8UEAVc&NR=1

- Aero Triangulation
  www.youtube.com/watch?v=88KFAU6I_jg

Review Questions

1. List several reasons why time of day might be very important in flight planning for aerial imagery.

2. Outline advantages and disadvantages of high-altitude photography. Explain why routine high-altitude aerial photography was not practical before infrared imagery was available.

3. List several problems that you would encounter in acquiring and interpreting large-scale aerial imagery of a mountainous region.

4. Speculate on the likely progress of aerial photography since 1890 if George Eastman (Chapter 1) had not been successful in popularizing the practice of photography to the general public.

5. Should an aerial photograph be considered as a "map"? Explain.

6. Assume you have recently accepted a position as an employee of an aerial survey company; your responsibilities include preparation of flight plans for the company's customers. What are the factors that you must consider as you plan each mission?

7. List some of the factors you would consider in selection of band combinations described in this chapter.

8. Suggest circumstances in which oblique aerial photography might be more useful than vertical photographs.

9. It might seem that large-scale aerial images might always be more useful than small-scale aerial photographs; yet larger scale images are not always the most useful. What are the disadvantages to the use of large-scale images?

10. A particular object will not always appear the same when images by an aerial camera. List some of the factors that can cause the appearance of an object to change from one photograph to the next.

References

Aber, J. S., S. W. Aber, and F. Pavri. 2002. Unmanned Small-Format Aerial Photography from Kites for Acquiring Large-Scale, High-Resolution Multiview-Angle Imagery. *Pecora 15/ Land Satellite Information IV/ISPRS Commission I/FIEOS 2002 Conference Proceedings*. Bethesda, MD: American Society of Photogrammetry and Remote Sensing.

Aber, J. S., I. Marzolff, and J. Ries. 2010. *Small-Format Aerial Photography: Principals, Techniques, and Geosciences Applications.* Amsterdam: Elsevier, 268 pp.

Boland, J., T. Ager, E. Edwards, E. Frey, P. Jones, R. K. Jungquiet, A. G. Lareau, J. Lebarron, C. S. King, K. Komazaki, C. Toth, S. Walker, E. Whittaker, P. Zavattero, and H. Zuegge. 2004. Cameras and Sensing Systems. Chapter 8 in the *Manual of Photogrammetry* (J. C. McGlone, E. M Mikhail, J. Bethel, and R. Mullen, eds.). Bethesda, MD: American Society for Photogrammetry and Remote Sensing, pp. 581–676.

Boreman, G. D. 1998. *Basic Electro-Optics for Electrical Engineers.* SPIE Tutorial Texts in Optical Engineering. Vol. TT32. Bellingham, WA: Society of Photo-optical Instrumentation Engineers, 97 pp.

British Columbia Ministry of Sustainable Resource Management. 2003. *Specifications for Scanning Aerial Photographic Imagery.* Victoria: Base Mapping and Geomatic Services Branch, 26 pp.

Clarke, T. A., and J. G. Fryer. 1998. The Development of Camera Calibration Methods and Models. *Photogrammetric Record.* Vol. 91, pp. 51–66.

Eller, R. 2000. *Secrets of Successful Aerial Photography.* Buffalo, NY: Amherst Media, 104 pp.

Erb, T. L., W. R., Philipson, W. T. Tang, and T. Liang. 1981. Analysis of Landfills with Historic Airphotos. *Photogrammetric Engineering and Remote Sensing*, Vol. 47, p. 1363–1369.

Graham, R., and A. Koh. 2002. *Digital Aerial Survey: Theory and Practice.* Boca Raton, FL: CRC Press, 247 pp.

Li, R., and C. Liu. 2010. Photogrammetry for Remote Sensing. Chapter 16 in *Manual of Geospatial Science and Technology* (J. D. Bossler, ed.). New York: Taylor & Francis, pp. 285–302.

Light, D. L. 1993. The National Aerial Photography as a Geographic Information System Resource. *Photogrammetric Engineering and Remote Sensing.* Vol. 59, pp. 61–65.

Linder, W. 2006. *Digital Photogrammetry: A Practical Course* (2nd ed.). Berlin: Springer.

Lyon, J. G. 1987. Use of Maps, Aerial Photographs, and Other Remote Sensor Data for Practical Evaluations of Hazardous Waste Sites. *Photogrammetric Engineering and Remote Sensing*, Vol. 53, pp. 515–519.

Petrie, G. 2007. Airborne Digital Imaging Technology: A New Overview. *Photogrammetric Record*, Vol. 22(119), pp. 203–225.

Petrie, G. 2009. Systematic Oblique Aerial Photography Using Multiple Frame Cameras. *Photogrammetric Engineering and Remote Sensing*, Vol. 75, pp. 102–107.

Plasker, J. R., and G. w. TeSelle. 1998. Present Status and Future Applications of the National Aerial Photography Program. In Proceedings of the ACSM/ASPRS Convention. Bethesda, MD: American Society for Photogrammetry and Remote Sensing, pp. 86–92.

Sandau, R., B. Braunecker, H. Driescher, A. Eckart, S. Hilbert, J. Hutton, W. Kirchhofer, E. Lithopoulos, R. Reulke, and S. Wicki. 2000. Design principles of the LH Systems ADS40 Airborne Digital Sensor. *International Archives of Photogrammetry and Remote Sensing*, Vol. 33, Part B1, pp. 258–265.

Stimson, A. 1974. *Photometry and Radiometry for Engineers.* New York: Wiley, 446 pp.

Stow, D. A., L. L. Coulter, and C. A. Benkleman. 2009. Airborne Digital Multispectral Imaging. Chapter 11. In *The Sage Handbook of Remote Sensing* (T. A. Warner, M. Duane Nellis, and G. M. Foody, eds.). London: Sage, pp. 131–165.

Wolf, P. R. 1983. *Elements of Photogrammetry, with Air Photo Interpretation and Remote Sensing.* New York: McGraw-Hill, 628 pp.

YOUR OWN INFRARED PHOTOGRAPHS

Anyone with even modest experience with amateur photography can practice infrared photography, given the necessary materials (see Figure 3.28). Although 35-mm film cameras, the necessary filters, and infrared-sensitive films are still available for the dedicated amateur, many will prefer to consider use of digital cameras that have been specially modified to

FIGURE 3.28. Black-and-white infrared photograph (top), with a normal black-and-white photograph of the same scene shown for comparison. From author's photographs.

acquire only radiation in the near infrared region. Infrared films are essentially similar to the usual films, but they should be refrigerated prior to use and exposed promptly, as the emulsions deteriorate much more rapidly than do those of normal films. Black-and-white infrared films should be used with a deep red filter to exclude most of the visible spectrum. Black-and-white infrared film can be developed using normal processing for black-and-white emulsions, as specified by the manufacturer. Digital cameras that have been modified for infrared photography do not require use of an external filter.

CIR films are also available in 35-mm format. They should be used with a yellow filter, as specified by the manufacturer. Processing of CIR film will require the services of a photographic laboratory that specializes in customized work, rather than the laboratories that handle only the more usual films. Before purchasing the film, it is best to inquire concerning the availability and costs of processing. There are few digital cameras currently available that have been modified for color infrared photography. Models formerly available may be available within the used camera market, although expense may be high even for secondhand cameras.

Results are usually best with bright illumination. For most scenes, the photographer should take special care to face away from the sun while taking photographs. Because of differences in the reflectances of objects in the visible and the NIR spectra, the photographer should anticipate the nature of the scene as it will appear in the infrared region of the spectrum. (Artistic photographers have sometimes used these differences to create special effects.) The camera lens will bring infrared radiation to a focal point that differs from that for visible radiation, so infrared images may be slightly out of focus if the normal focus is used. Some lenses have special markings to show the correct focus for infrared films; most digital cameras modified for infrared photography have been also modified to provide the correct focus.

What's Hiding in Infrared; Make Your Own Infrared Camera
www.youtube.com/watch?v=PmpgeFoYbBI

YOUR OWN 3D PHOTOGRAPHS

You can take your own stereo photographs using a handheld camera simply by taking a pair of overlapping photographs. Two photographs of the same scene, taken from slightly different positions, create a stereo effect in the same manner in which overlapping aerial photographs provide a three-dimensional view of the terrain.

This effect can be accomplished by aiming the camera to frame the desired scene, taking the first photograph, moving the camera laterally a short distance, then taking a second photograph that overlaps the field of view of the first. The lateral displacement need only be a few inches (equivalent to the distance between the pupils of a person's eyes), but a displacement of a few feet will often provide a modest exaggeration of depth that can be useful in distinguishing depth (Figure 3.29). However, if the displacement is too great, the eye cannot fuse the two images to simulate the effect of depth.

Prints of the two photographs can then be mounted side by side to form a stereo pair that can be viewed with a stereoscope, just as a pair of aerial photos can be viewed in stereo. Stereo images can provide three-dimensional ground views that illustrate conditions encountered within different regions delineated on aerial photographs. Section 5.10 provides more information about viewing stereo photographs.

FIGURE 3.29. Ground-level stereo photographs acquired with a personal camera. From author's photographs.

How to Take 3D Photos Using a Basic Camera
www.youtube.com/watch?v=ih37yJjzqAI&feature=related

No 3D Glasses Required—Amazing 3D Stereoscopic Images
www.youtube.com/watch?v=Eq3MyjDS1co&feature=related

YOUR OWN KITE PHOTOGRAPHY

Although success requires persistence and attention, do-it-yourself kite photography is within the reach of most who have the interest. The main perquisites are access to a small digital camera, a reasonably robust kite, and the skill to fabricate a homemade mount for the camera (Figure 3.30). Aside from experience, the main obstacle for most beginners will be devising the mount to permit the camera's field of view to face the ground at the desired orientation. There is an abundance of books and websites that can provide design and instructions. The motion of the kite will cause the camera to swing from side to side, thereby producing a number of unsatisfactory photographs that must be screened to find those that are most suitable. These effects can be minimized by use of more elaborate mounts for cameras and possibly by attention to the choice of kite.

www.arch.ced.berkeley.edu/kap/kaptoc.html

http://scott haefner.com/360panos/kap

Make Podcast: Weekend Projects—Make a Kite Aerial Photograph
www.youtube.com/watch?v=kEprozoxnLY&feature=fvw

Maker Workshop—Kite Aerial Photography on MAKE:television
www.youtube.com/watch?v=swqFA9Mvq5M

FIGURE 3.30. Kite photography. Left: Example of a handmade mount for rigging a digital camera for kite photography. The mount allows the camera's view to maintain a field of view that is oriented toward the ground. Right: Sample photograph of an agricultural field marked to delineate sample sites. Precise orientation of the camera is problematic with simple camera mounts, so it can be difficult to acquire systematic photography that has a consistent orientation. From J. B. Campbell and T. Dickerson.

CHAPTER FOUR

Digital Imagery

4.1. Introduction

For much of the history of aerial survey and remote sensing, images were recorded as photographs or photograph-like images. A photographic image forms a physical record—pieces of paper or film with chemical coatings that record the patterns of the images. Such images are referred to as *analog* images because the brightnesses within a photograph are proportional (i.e., analogous) to the brightnesses within a scene. Although photographic media have enduring value for recording images, in the context of remote sensing their disadvantages, including difficulties of storage, transmission, searching, and analysis, form liabilities.

In contrast, digital image formats represent images as arrays of many individual values, known as *pixels* ("picture elements"), that together form an image. When an image is represented as discrete numbers, it acquires qualities that offer many advantages over earlier analog formats. Digital values can be added, subtracted, multiplied, and, in general, subjected to statistical manipulation that is not feasible if images are presented only in analog format. Digital images are also easy to store in compact formats and easily transmitted, and storage and retrieval is inexpensive and effective. Thus the digital format greatly increases our ability to display, examine, and analyze remotely sensed data. However, we should note that digital formats have their own limitations, not always recognized. Imagery can be only as secure as the media on which they are stored, so just as analog images are subject to deterioration by aging, mishandling, and wear of physical media, so digital data are subject to corruption, damage to disk drives, magnetic fields, and deterioration of the physical media. Equally significant are changes in the formats of digital storage media, which can render digital copies inaccessible because of obsolescence of the hardware necessary to read the digital media.

This chapter introduces some of the fundamental concepts underlying applications of digital data for remote sensing and expands on some of the concepts first introduced in Chapter 3. It addressees collection of digital data, representation of digital values, alternative formats for storing digital data, display of digital data, and image processing software systems.

4.2. Electronic Imagery

Digital data can be created by a family of instruments that can systematically scan portions of the Earth's surface, recording photons reflected or emitted from individual patches of ground, known as pixels. A digital image is composed of many thousands of pixels, usually each too small to be individually resolved by the human eye, each representing the brightness of a small region on the Earth's surface, recorded digitally as a numeric value, usually with separate values for each of several regions of the electromagnetic spectrum (Figure 4.1). Color images are composed of several such arrays of the same ground area, each representing brightnesses in a separate region of the spectrum.

Digital images can be generated by several kinds of instruments. Chapter 3 has already introduced some of the most important technologies for digital imaging: CCDs and CMOS. Another technology—*optical–mechanical scanning*—is older, but has proven to be reliable, and is still important in several realms of remote sensing practice. Here, we focus on a specific form of the optical–mechanical scanner that is designed to acquire imagery in several spectral regions—the *multispectral scanner*, which has formed an enduring technology for the practice of remote sensing. Other forms of optical–mechanical scanners have been used for collecting thermal imagery (thermal scanners; Chapter 9) and hyperspectral imagery (hyperspectral scanners; Chapter 15).

As noted in Chapter 3, CCDs can be positioned in the focal plane of a sensor such that they view a thin rectangular strip oriented at right angles to the flight path (Figure 4.2a). The forward motion of the aircraft or satellite moves the field of view forward along the flight path, building up coverage, a process known as *pushbroom scanning*— the linear array of pixels slides forward along the flight path in a manner analogous to the motion of a janitor's pushbroom along a floor. In contrast, mechanical scanning can be visualized by analogy to a whiskbroom, in which the side-to-side motion of the scanner constructs the lateral dimension of the image (Figure 4.2b), as the forward motion of the aircraft or satellite creates its longitudinal dimension.

Optical–mechanical scanners physically move mirrors, or prisms, to systematically

FIGURE 4.1. Multispectral pixels.

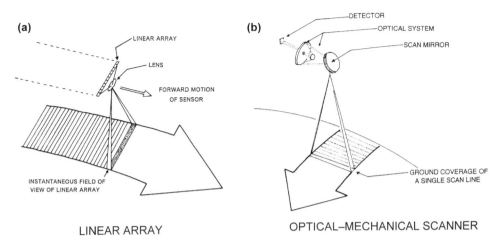

FIGURE 4.2. Optical–mechanical scanner. Whereas a linear array (a) acquires imagery line by line as its field of view slides forward along the ground track, an optical–mechanical scanner (b) oscillates from side to side to build coverage pixel by pixel as the field of view progresses forward.

aim the field of view over the Earth's surface. The scanning mirror scans across the field of view at a right angle to the flight path of the sensor, directing radiation from the Earth's surface to a secondary optical system and eventually to detectors that generate an electrical current that varies in intensity as the land surface varies in brightness (Figure 4.2b). Filters or diffraction gratings (to be discussed later) split the radiation into several segments to define separate spectral channels, so the instrument generates several signals, each carrying information about the brightness in a separate region of the spectrum. The electrical current provides an electronic version of the brightness of the terrain but is still in analog form—it provides a continuous record of brightnesses observed by the sensor's optics. To create a digital version, the electrical signal must be subdivided into distinct units to create the discrete values necessary for digital analysis. This conversion from the continuously varying analog signal to the discrete values is accomplished by sampling the current at a uniform interval, a process known as analog-to-digital, or A-to-D, conversion (Figure 4.3). Because the values within this interval are represented as a single average, all variation within this interval is lost. The process of subdivision and averaging the continuous signal corresponds to sampling the terrain at a set spatial interval, so the choice of sampling interval establishes the spatial detail recorded by the image.

The *instantaneous field of view* (IFOV) of an optical–mechanical scanner refers to the area viewed by the instrument if it were possible to suspend the motion of the aircraft and the scanning of the sensor for an instant (Figure 4.4). The IFOV therefore defines the smallest area viewed by the scanner and establishes a limit for the level of spatial detail that can be represented in a digital image. Although data in the final image can be aggregated so that an image pixel represents a ground area larger than the IFOV, it is not possible for pixels to carry information about ground areas smaller than the IFOV. More generally, the concept of the *ground resolved distance* (GRD) specifies the estimated dimension of the size of the smallest feature that can be reliably resolved by an imaging system.

FIGURE 4.3. Analog-to-digital conversion.

Electronic sensors must be operated within the limits of their design capabilities. Altitudes and speeds of aircraft and satellites must be selected to match the sensitivities of the sensors, so that detectors view a given ground area (pixel) long enough to accumulate enough photons to generate reliable signals (this interval is known as *dwell time*). If designed and operated effectively, the imaging system should provide a linear response to scene brightness, such that the values within an image will display consistent, predicable relationships with brightness on the ground. Although most sensors have good performance under normal operating conditions, they will be subject to failures under extreme conditions.

The lower end of an instrument's sensitivity is subject to the *dark current* signal (or *dark current noise*) (Figure 4.5). At low levels of brightness, a CCD can record low levels of brightness even when there is none in the scene, due to energy within the CCD's structure that is captured by the potential well and presented as brightness even though none is present in the scene. Thus very dark features will not be represented at their correct brightness. Likewise, at an instrument's upper threshold of sensitivity, bright targets *saturate* the sensor's response—the instrument fails to record the full magnitude of the target's brightness. As an example in the context of remote sensing, saturation might be encountered in images representing glaciers or snowfields, which may exceed a sensor's ability to record the full range of brightnesses in the optical region of the spectrum. For

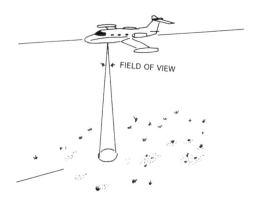

FIGURE 4.4. Instantaneous field of view.

FIGURE 4.5. Dark current, saturation, and dynamic range.

instruments using CCDs, saturation can sometimes manifest itself as *streaking* or *blooming*, as the excess charges that accumulate at a specific pixel site spill over to influence the charges at adjacent locations, creating bright streaks or patches unrelated to the actual features with the scene.

Between these limits, sensors are designed to generate signals that have predictable relationships with scene brightness; these relationships are established by careful design, manufacture, and calibration of each instrument. These characteristics define the upper and lower limits of the system's sensitivity to brightness and the range of brightnesses over which a system can generate measurements with consistent relationships to scene brightness.

The range of brightnesses that can be accurately recorded is known as the sensor's *dynamic range*. The lower limit of an instrument's dynamic range is set during calibration at a level above the minimum illustrated in Figure 4.5, known as the *offset*. In general, electronic sensors have large dynamic ranges compared with those of photographic films, computer displays, or the human visual system. Therefore, photographic representations of electronic imagery tend to lose information at the upper and/or lower ranges of brightness. Because visual interpretation forms such an important dimension of our understanding of images, the way that image displays and image-enhancement methods (discussed subsequently) handle this problem forms an important dimension of the field of image analysis.

The slope of the line depicted in Figure 4.5 defines the *gain* of a sensor, expressing the relationship between the brightness in the original scene and its representation in the image. Figure 4.5 represents the behavior of an instrument that portrays a range of brightnesses that are approximately proportional to the range of brightnesses in the original scene (i.e., the slope of the line representing the relationship between scene brightness and image brightness is oriented at approximately 45°). A slope of 1 means that a given range of brightness in the scene is assigned the same range in the image, whereas a steeper slope (*high gain*) indicates that a given range of brightness in the scene is expanded to assume a larger range in the scene. In contrast, Figure 4.6 shows two hypothetical instruments—one with *high gain* (i.e., it portrays a given range of brightness in the scene as a larger range in the image) and another with *low gain* (i.e., the instrument creates an image with a narrower range of brightness than is observed in the scene). The gain for a sensor is usually fixed by the design of an instrument, although some may have alternative settings (*high gain* or *low gain*), to accommodate varied scenes or operational conditions.

Each sensor creates responses unrelated to target brightness—that is, *noise*, created in part by accumulated electronic errors from various components of the sensor. (In this context "noise" refers specifically to noise generated by the sensor, although the noise that the analyst receives originates not only in the sensor but also in the atmosphere, the interpretation process, etc.). For effective use, instruments must be designed so that their noise levels are small relative to the signal (brightness of the target). This is measured as the signal-to-noise ratio (S/N or SNR) (Figure 4.7). Analysts desire signals to be large relative to noise, so the SNR should be large not only for bright targets when the signal is large but also over the entire dynamic range of the instrument, especially at the lower levels of sensitivity when the signal is small relative to noise. Engineers who design sensors must balance the radiometric sensitivity of the instrument with pixel size, dynamic range, operational altitude, and other factors to maintain acceptable SNRs.

4.3. Spectral Sensitivity

Optical sensors often use prisms and filters to separate light into spectral regions. Filters are pieces of specialized glass that selectively pass certain wavelengths and block or

FIGURE 4.6. Examples of sensors characterized by high and low gain.

FIGURE 4.7. Signal-to-noise (S/N) ratio. At the bottom, a hypothetical scene is composed of two cover types. The signal records this region, with only a small difference in brightness between the two classes. Atmospheric effects, sensor error, and other factors contribute to noise, which is added to the signal. The sensor then records a combination of signal and noise. When noise is small relative to the signal (left: high S/N ratio), the sensor conveys the difference between the two regions. When the signal is small relative to noise (right, low S/N ratio) the sensor cannot portray the difference in brightness between the two regions.

absorb those that the designer desired to exclude. The most precise (and therefore most expensive) filters are manufactured by adding dyes to glass during manufacture. Less precise, and less durable, filters are manufactured by coating the surface of glass with a film that absorbs the desired wavelengths. Usually filters are manufactured by firms that produce a suite of filters, each with their own systems for defining and designating filters, specifically tailored for the needs of certain communities of customers with specialized needs.

Because of the scattering of shorter (ultraviolet and blue) wavelengths, filters are often used when recording visible radiation to screen out blue light (Figure 4.8a; see also Figure 3.18). Such a filter creates an image within the visible region that excludes the shorter wavelengths that can degrade the visual quality of the image. Often it is desirable to exclude all visible radiation, to create an image that is based entirely on near infrared radiation (see Figure 3.19). A deep red filter (Figure 4.8b) blocks visible radiation but allows infrared radiation to pass. An image recorded in the near infrared region (see Figure 3.28) is quite different from its representation in the visible spectrum. For example, living vegetation is many times brighter in the near infrared portion of the spectrum than it is in the visible portion, so vegetated areas appear bright white on the black-and-white infrared image.

Although filters can be used in the collection of digital imagery, electronic sensors often use *diffraction gratings*, considered more efficient because of their effectiveness, small size, and light weight. Diffraction gratings are closely spaced transmitting slits cut into a flat surface (a *transmission grating*) or grooves cut into a polished surface (a *reflection grating*). Effective transmission gratings must be very accurately and consistently spaced and must have very sharp edges. Light from a scene is passed though a *collimating lens*, designed to produce a beam of parallel rays of light that is oriented to strike the diffraction grating at an angle (Figure 4.9).

Light striking a diffraction grating experiences both destructive and constructive interference as wavefronts interact with the grating. *Destructive interference* causes some

FIGURE 4.8. Transmission curves for two filters. (a) Pale yellow filter (Kodak filter 2B) to prevent ultraviolet light from reaching the focal plane; it is frequently used to acquire panchromatic images. (b) Kodak 89B filter used to exclude visible light, used for infrared images. (Shaded portions of the diagrams signify that the filter is blocking transmission of radiation at specified wavelengths.) Copyright Eastman Kodak Company. Permission has been granted to reproduce this material from KODAK Photographic Filters Handbook (Code B-3), courtesy of Eastman Kodak Company.

wavelengths to be suppressed, whereas *constructive interference* causes others to be reinforced. Because the grating is oriented at an angle with respect to the beam of light, different wavelengths are diffracted at different angles, and the radiation can be separated spectrally. This light then illuminates detectors to achieve the desired spectral sensitivity.

Because the various filters and diffraction gratings that instruments use to define the spectral limits (i.e., the "colors" that they record) do not define discrete limits, spectral sensitivity varies across a specific defined interval. For example, an instrument designed to record radiation in the green region of the spectrum will not exhibit equal sensitivity across the green region but will exhibit greater sensitivity near the center of the region than at the transitions to the red and blue regions on either side (Figure 4.10). Defining the spectral sensitivity to be the extreme limits of the energy received would not be satisfactory because it is clear that the energy received at the extremes is so low that the effective sensitivity of the instrument is defined by a much narrower wavelength interval.

As a result, the spectral sensitivity of an instrument is often specified using the defi-

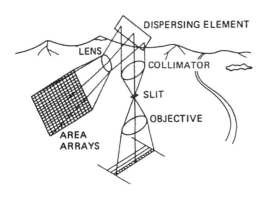

FIGURE 4.9. Diffraction grating and collimating lens (NASA diagram).

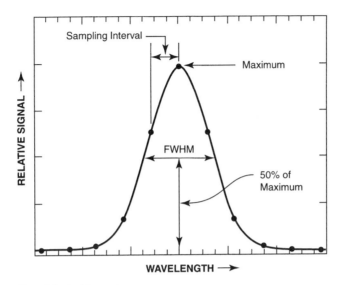

FIGURE 4.10. Full width, half maximum.

nition of *full width, half maximum* (FWHM)—the spectral interval measured at the level at which the instrument's response reaches one-half of its maximum value (Figure 4.10). Thus FWHM forms a definition of *spectral resolution*, the narrowest spectral interval that can be resolved by an instrument. (Even though the instrument is sensitive to radiation at the extreme limits, beyond the limits of FWHM, the response is so weak and unreliable at these limits that FWHM forms a measure of functional sensitivity.) Figure 4.10 also illustrates the definition of the *spectral sampling interval* (known also as *spectral bandwidth*), which specifies the spectral interval used to record brightness in relation to wavelength.

4.4. Digital Data

Output from electronic sensors reaches the analyst as a set of numeric values. Each digital value is recorded as a series of binary values known as *bits*. Each bit records an exponent of a power of 2, with the value of the exponent determined by the position of the bit in the sequence. As an example, consider a system designed to record 7 bits for each digital value. This means (for unsigned integers) that seven binary places are available to record the brightness sensed for each band of the sensor. The seven values record, in sequence, successive powers of 2. A "1" signifies that a specific power of 2 (determined by its position within the sequence) is to be evoked; a "0" indicates a value of zero for that position. Thus the 7-bit binary number "1111111" signifies $2^6 + 2^5 + 2^4 + 2^3 + 2^2 + 2^1 + 2^0 = 64 + 32 + 16 + 8 + 4 + 2 + 1 = 127$. And "1001011" records $2^6 + 0^5 + 0^4 + 2^3 + 0^2 + 2^1 + 2^0 = 64 + 0 + 0 + 8 + 0 + 2 + 1 = 75$. Figure 4.11 shows different examples. Eight bits constitute a *byte*, intended to store a single character. Larger amounts of memory can be indicated in terms of *kilobytes* (KB), 1,024 (2^{10}) bytes; *megabytes* (MB), 1,048,576 (2^{20}) bytes; and *gigabytes* (GB), 1,073,741,824 (2^{30}) bytes (Table 4.1).

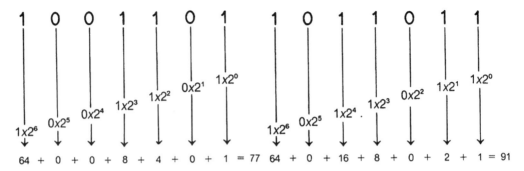

FIGURE 4.11. Digital representation of values in 7 bits.

In this manner, discrete digital values for each pixel are recorded in a form suitable for storage on disks and for analysis. These values are popularly known as "digital numbers" (DNs), "brightness values" (BVs), or "digital counts," in part as a means of signifying that these values do not record true brightnesses (known as *radiances*) from the scene but rather are scaled values that represent relative brightness within each scene. The number of brightness values within a digital image is determined by the number of bits available. The 7-bit example given above permits a maximum range of 128 possible values (0–127) for each pixel. A decrease to 6 bits would decrease the range of brightness values to 64 (0–63); an increase to 8 bits would extend the range to 256 (0–255). Thus, given a constant noise level, the number of bits minus a reserved sign bit, if used, determines the radiometric resolution (Chapter 10) of a digital image. The number of bits available is determined by the design of the system, especially the sensitivity of the sensor and its capabilities for recording and transmitting data (each added bit increases transmission requirements). If we assume that transmission and storage resources are fixed, then increasing the number of bits for each pixel means that we will have fewer pixels per image and that pixels would each represent a larger ground area. Thus technical specifications for remote sensing systems requires trade-offs between image coverage and radiometric, spectral, and spatial resolutions.

Radiances

The brightness of radiation reflected from the Earth's surface is measured as brightness (watts) per wavelength interval (micrometer) per angular unit (steradian) per square meter from which it was reflected; thus the measured brightness is defined with respect to wavelength (i.e., "color"), spatial area (angle), intensity (brightness), and area. Radiances record actual brightnesses, measured in physical units, represented as real values

TABLE 4.1. Terminology for Computer Storage

Bit	A binary digit (0 or 1)	
Byte	8 bits, 1 character	
Kilobyte (K or KB)	1,024 bytes	(2^{10} bytes)
Megabyte (MB)	1,048,576 bytes	(2^{20} bytes)
Gigabyte (GB)	1,073,741,824 bytes	(2^{30} bytes)
Terabyte (TB)	1,099,511,627,776 bytes	(2^{40} bytes)

(i.e., to include decimal fractions). Use of DNs facilitates the design of instruments, data communications, and the visual display of image data. For visual comparison of different scenes, or analyses that examine relative brightnesses, use of DNs is satisfactory. However, because a DN from one scene does not represent the same brightness as the same DN from another scene, DNs are not comparable from scene to scene if an analysis must examine actual scene brightnesses for purposes that require use of original physical units. Such applications include comparisons of scenes of the same area acquired at different times, or matching adjacent scenes to make a mosaic.

For such purposes, it is necessary to convert the DNs to the original radiances or to use reflectance (Chapters 2 and 11), which are comparable from scene to scene and from one instrument to another. Calculation of radiances and reflectances from DNs requires knowledge of calibration data specific to each instrument. To ensure that a given sensor provides an accurate measure of brightness, it must be calibrated against targets of known brightness. The sensitivities of electronic sensors tend to drift over time, so to maintain accuracy, they must be recalibrated on a systematic schedule. Although those sensors used in aircraft can be recalibrated periodically, those used in satellites are not available after launch for the same kind of recalibration. Typically, such sensors are designed so that they can observe calibration targets onboard the satellite, or they are calibrated by viewing landscapes of uniform brightness (e.g., the moon or desert regions). Nonetheless, calibration errors, such as those described in Chapter 11, sometimes remain.

4.5. Data Formats

Digital image analysis is usually conducted using *raster* data structures in which each image is treated as an array of values. Additional spectral channels form additional arrays that register to one another. Each pixel is treated as a separate unit, which can always be located within the image by its row and column coordinates. In most remote sensing analysis, coordinates originate in the upper left-hand corner of an image and are referred to as *rows* and *columns*, or as *lines* and *pixels*, to measure position down and to the right, respectively.

Raster data structures offer advantages for manipulation of pixel values by image processing systems, as it is easy to find and locate pixels and their values. The disadvantages are usually apparent only when we need to represent not the individual pixels, but areas of pixels, as discrete patches or regions. Then the alternative structure, *vector format*, becomes more attractive. The vector format uses polygonal patches and their boundaries as the fundamental units for analysis and manipulation. The vector format is not appropriate for digital analysis of remotely sensed data, although sometimes we may wish to display the results of our analysis using a vector format. Almost always, equipment and software for digital processing of remotely sensed data must be tailored for a raster format.

Digital remote sensing data are typically organized according to one of three alternative strategies for storing images. Consider an image consisting of four spectral channels, which together can be visualized as four superimposed images, with corresponding pixels in one band registering exactly to those in the other bands.

One of the earliest formats for digital data was *band interleaved by pixel* (BIP). Data are organized in sequence values for line 1, pixel 1, band 1; then for line 1, pixel 1, band 2; then for line 1, pixel 1, band 3; and finally for line 1, pixel 1, band 4. Next are the four bands for line 1, pixel 2, and so on (Figure 4.12). Thus values for all four bands are written before values for the next pixel are represented. Any given pixel, once located within the data, is found with values for all four bands written in sequence one directly after the other. This arrangement is advantageous for many analyses in which the brightness value (or digital number) vector is queried or used to calculate another quantity. However, it is an unwieldy format for image display.

The *band interleaved by line* (BIL) format treats each line of data as a separate unit (Figure 4.13). In sequence, the analyst encounters line 1 for band 1, line 1 for band 2, line 1 for band 3, line 1 for band 4, line 2 for band 1, line 2 for band 2, and so on. Each line is represented in all four bands before the next line is encountered. A common variation on the BIL format is to group lines in sets of 3 or 7, for example, rather than to consider each single line as the unit.

A third convention for recording remotely sensed data is the *band sequential* (BSQ) format (Figure 4.14). All data for band 1 are written in sequence, followed by all data for band 2, and so on. Each band is treated as a separate unit. For many applications, this format is the most practical, as it presents data in the format that most closely resembles the data structure used for display and analysis. However, if areas smaller than the entire scene are to be examined, the analyst must read all four images before the subarea can be identified and extracted.

Actual data formats used to distribute digital remote sensing data are usually variations on these basic alternatives. Exact details of data formats are specific to particular organizations and to particular forms of data, so whenever an analyst acquires data, he or she must make sure to acquire detailed information regarding the data format. Although organizations attempt to standardize formats for specific kinds of data, it is also true that data formats change as new mass storage media come into widespread use and as user communities employ new kinds of hardware or software.

The "best" data format depends on immediate context and often on the specific software and equipment available. If all bands for an entire image must be used, then the BSQ and BIL formats are useful because they are convenient for reconstructing the entire scene in all four bands. If the analyst knows beforehand the exact position on the image of the subarea that is to be studied, then the BIP format is useful because values for all bands are found together and it is not necessary to read through the entire data set to find a specific region. In general, however, the analyst must be prepared to read the data in the

FIGURE 4.12. Band interleaved by pixel format. In effect, each band is subdivided such that pixels from the several bands are collected together and written to digital storage in neighboring positions. Pixels from each band are intermingled as illustrated.

FIGURE 4.13. Band interleaved by line format. Lines of pixels from each band are selected, then written to digital storage such that the lines for separate bands are positioned in sequence. Lines from each band are intermingled as illustrated.

format in which they are received and to convert them into the format most convenient for use at a specific laboratory.

Other formats are less common in everyday applications but are important for applications requiring use of long sequences of multispectral images. *Hierarchical data format* (HDF) is a specialized data structure developed and promoted by the National Center for Supercomputing Applications (*http://hdf.ncsa.uiuc.edu/index.html*) and designed specifically to promote effective management of scientific data. Whereas the formats discussed thus far organize data conveyed by a specific image, HDF and related structures provide frameworks for organizing collections of images. For example, conventional data formats become awkward when it is necessary to portray three-dimen-

FIGURE 4.14. Band sequential (BSQ) format. The structure of each band is retained in digital storage—all pixels for each band are written in their entirety before the next band is written. There is no intermingling of pixels from separate bands.

sional data structures as they might vary over time. Although such structures might typically portray complex atmospheric data as it varies hourly, daily, seasonally, or yearly, they also lend themselves to recording large sequences of multispectral images. HDF therefore enables effective analysis and visualization of such large, multifaceted data structures.

A related but distinctly different format, *network common data form* (NetCDF), which also provides structures tailored for handling dynamic, array-oriented data, is specifically designed to be compatible with a wide variety of computer platforms, so that it can facilitate sharing of data over the World Wide Web. NetCDF was designed specifically for the Unidata system (*www.unidata.ucar.edu/software/netcdf*), designed for rapid transmission of meteorological data to a wide range of users.

Although HDF and NetCDF structures are unlikely to be encountered in usual remote sensing applications, they are becoming more common in advanced applications requiring the handling of very large sequences of images, such as those encountered in geophysics, meteorology, and environmental modeling—applications that often include remotely sensed data.

Data compression reduces the amount of digital data required to store or transmit information by exploiting the redundancies within a data set. If data arrays contain values that are repeated in sequence, then compression algorithms can exploit that repetition to reduce the size of the array, while retaining the ability to restore the array to its original form. When the complete array is needed for analysis, then the original version can be restored by *decompression*. Because remotely sensed images require large amounts of storage and usually are characterized by modest levels of redundancies, data compression is an important tool for effective storage and transmission of digital remote sensing data. Compression and decompression are accomplished, for example, by executing computer programs that receive compressed data as input and produce a decompressed version as output.

The *compression ratio* compares the size of the original image with the size of the compressed image. A ratio of 2:1 indicates that the compressed image is one-half the size of the original. *Lossless compression* techniques restore compressed data to their exact original form; *lossy* techniques degrade the reconstructed image, although in some applications the visual impact of a lossy technique may be imperceptible. For digital satellite data, lossless compression techniques can achieve ratios from 1.04:1 to 1.9 to 1. For digitized cartographic data, ratios of 24:1 using lossy techniques have been reported to exhibit good quality.

It is beyond the scope of this discussion to describe the numerous techniques and algorithms available for image compression. Probably the most well-known compression standard is the JPEG (Joint Photographic Experts Group) format, a lossy technique that applies the discrete cosine transform (DCT) as a compression–decompression algorithm. Although the JPEG algorithm has been widely accepted as a useful technique for compression of continuous-tone photographs, it is not likely to be ideal for remotely sensed images, nor for geospatial data in general. A modification to the basic JPEG format, JPEG2000 (*www.jpeg.org/*) has been recognized as providing a high compression rate with high fidelity (Liu et al., 2005). Depending on its application, JPEG2000 can be either lossy or nonlossy. Generally stated, lossy compression techniques should not be applied to data intended for analysis or as archival copies. Lossy compression may be appropriate for images used to present visual records of results of analytical process, provided they do not form input to other analyses.

4.6. Band Combinations: Multispectral Imagery

Effective display of an image is critical for effective practice of remote sensing. *Band combinations* is the term that remote sensing practitioners use to refer to the assignment of colors to represent brightnesses in different regions of the spectrum. Although there are many ways to assign colors to represent different regions of the spectrum, experience shows that some have proven to be more useful than others. A key constraint for the display of any multispectral image is that human vision portrays differences in the colors of surfaces through our eyes' ability to detect differences in brightnesses in three additive primaries—blue, green, and red. Because our eyes can distinguish between brightnesses in these spectral regions, we can distinguish not only between blue, green, and red surfaces but also between intermediate mixtures of the primaries, such as yellow, orange, and purple.

Color films, digital displays, and the like portray the effect of color by varying the mixtures of the blue, green, and red primaries. Although films must employ a single strategy for portraying colors, image processing systems and digital displays offer the flexibility to use any of many alternative strategies for assigning colors to represent different regions of the spectrum. These alternative choices then define the band selection task; that is, how to decide which primary colors to select to portray on the display screen specific radiation collected by remote sensing systems.

If imagery at hand is limited to three spectral regions (as is the case with normal everyday color imagery), then the band selection task is simple: display radiation from blue in nature as blue on the screen, green as green, red as red. However, once we have more than three channels at hand, as is common for remotely sensed imagery, then the choice of assignment can have only arbitrary answers, as (for example) there can be no logical choice for the primary we might use to display energy from the near infrared region. The common choices for the band selection problem, then, are established in part on conventions that have been accepted by use over the decades and in part by practice that has demonstrated that certain combinations are effective for certain purposes. An important theme for band combinations is that bands that are close to one another tend to replicate information in their adjacent regions of the spectrum. Therefore, the most effective band combinations are often (but not always!) formed from spectral regions that have different locations on the spectrum, because they tend to provide independent representations of the same landscape.

Other color assignment models are used often but do not have widely accepted names. Here we discuss a few that are designated by the band numbers used for the Landsat Thematic Mapper (discussed in more detail in Chapter 6).

742

The 742 combination uses one region from the visible spectrum, one from the near infrared, and one from the mid infrared region (Figure 4.15). It portrays landscapes using "false" colors, but in a manner that resembles their natural appearance. Living, healthy vegetation appears in bright greens, barren soil as pink, dry vegetation and sparsely vegetated areas as oranges and browns, and open water as blue. This combination is often employed for geologic analyses, especially in desert landscapes, as differing mineralogies of surface soils appear as distinctive colors. Applications also include agriculture, wet-

FIGURE 4.15. 742 band combination.

lands, and, in forestry, for fire management for postfire analysis of burned and unburned areas.

451

A 451 combination uses blue and mid infrared radiation, together with a band from the near infrared region (Figure 4.16). Deep, clear water bodies will appear very dark using this choice of bands—shallow or turbid water appears as shades of lighter blues. Healthy vegetation is represented in reds, browns, and oranges. Greens and browns often represent bare soils; white, cyan, and gray colors often represent urban features.

754

The 754 combination uses three bands from outside the visible region (Figure 4.17), used often for geological analysis. Because it employs longer wavelengths, it is free of effects of atmospheric scattering. Coastlines are clearly and sharply defined. Textural and moisture characteristics of soils can often be discerned.

543

A 543 combination uses the near infrared, mid infrared, and red regions (Figure 4.18). Edges of water bodies are sharply defined. It is effective in displaying variations in vegetation type and status as browns, greens, and oranges. It is sensitive to variations in soil moisture and is useful for analysis of soil and vegetation conditions. Wetter surface soils appear in darker tones.

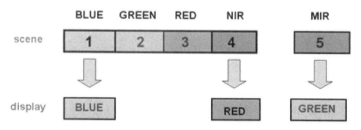

FIGURE 4.16. 451 band combination.

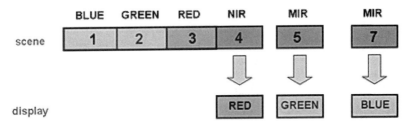

FIGURE 4.17. 754 band combination.

4.7. Image Enhancement

Image enhancement is the process of improving the visual appearance of digital images. Image enhancement has increasing significance in remote sensing because of the growing importance of digital analyses. Although some aspects of digital analysis may seem to reduce or replace traditional image interpretation, many of these procedures require analysts to examine images on computer displays, doing tasks that require many of the skills outlined in earlier sections of this chapter.

Most image-enhancement techniques are designed to improve the visual appearance of an image, often as evaluated by narrowly defined criteria. Therefore, it is important to remember that enhancement is often an arbitrary exercise—what is successful for one purpose may be unsuitable for another image or for another purpose. In addition, image enhancement is conducted without regard for the integrity of the original data; the original brightness values will be altered in the process of improving their visual qualities, and they will lose their relationships to the original brightnesses on the ground. Therefore, enhanced images should not be used as input for additional analytical techniques; rather, any further analysis should use the original values as input.

Contrast Enhancement

Contrast refers to the range of brightness values present on an image. Contrast enhancement is required because sensors often generate brightness ranges that do not match the capabilities of the human visual system. Therefore, for analysts to view the full range of information conveyed by digital images, it is usually necessary to rescale image brightnesses to ranges that can be accommodated by human vision, photographic films, and computer displays. For example, if the maximum possible range of values is 0–255 (i.e.,

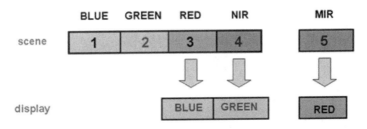

FIGURE 4.18. 543 band combination.

8 bits) but the display can show only the range from 0 to 63 (6 bits), then the image will have poor contrast, and important detail may be lost in the values that cannot be shown on the display (Figure 4.19a). Contrast enhancement alters each pixel value in the old image to produce a new set of values that exploits the full range of 256 brightness values (Figure 4.19b).

Figure 4.20 illustrates the practical effect of image enhancement. Before enhancement (left), detail is lost in the darker regions of the image. After enhancement has stretched the histogram of brightness values to take advantage of the capabilities of the display system, the detail is more clearly visible to the eye.

Many alternative approaches have been proposed to improve the quality of the displayed image. The appropriate choice of technique depends on the image, the previous experience of the user, and the specific problem at hand. The following paragraphs illustrate a few of the simpler and more widely used techniques.

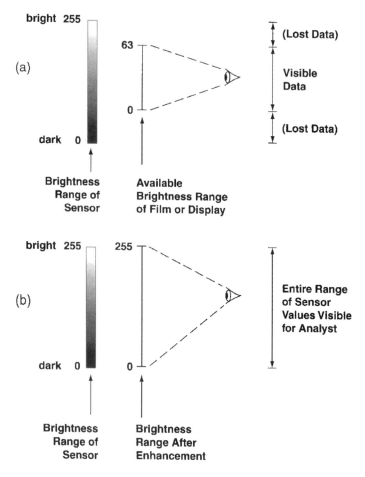

FIGURE 4.19. Schematic representation of the loss of visual information in display of digital imagery. (a) Often the brightness range of digital imagery exceeds the ability of the image display to represent it to the human visual system. (b) Image enhancement rescales the digital values to more nearly match the capabilities of the display system.

FIGURE 4.20. Pair of images illustrating effect of image enhancement. By altering the distribution of brightness values, as discussed in the text, the analyst is able to view detail formerly hidden by the ineffective distribution of image brightnesses.

Linear Stretch

Linear stretch converts the original digital values into a new distribution, using new minimum and maximum values specified, often plus or minus two standard deviations from the mean. The algorithm then matches the old minimum to the new minimum and the old maximum to the new maximum. All of the old intermediate values are scaled proportionately between the new minimum and maximum values (Figure 4.21). *Piecewise linear stretch* means that the original brightness range was divided into segments before each segment was stretched individually. This variation permits the analyst to emphasize

FIGURE 4.21. Linear stretch spreads the brightness values over a broader range, allowing the eye to see detail formerly concealed in the extremely dark or bright tones.

certain segments of the brightness range that might have more significance for a specific application.

Histogram Equalization

Histogram equalization reassigns digital values in the original image such that brightnesses in the output image are equally distributed among the range of output values (Figure 4.22). Unlike contrast stretching, histogram equalization is achieved by applying a nonlinear function to reassign the brightnesses in the input image such that the output image approximates a uniform distribution of intensities. The histogram peaks are broadened, and the valleys are made shallower. Histogram equalization has been widely used for image comparison processes (because it is effective in enhancing image detail) and for adjustment of artifacts introduced by digitizers or other instruments.

Density Slicing

Density slicing is accomplished by arbitrarily dividing the range of brightnesses in a single band into intervals, then assigning each interval a color (Figure 4.23 and Plates 3 and 15). Density slicing may have the effect of emphasizing certain features that may be represented in vivid colors, but, of course, it does not convey any more information than does the single image used as the source.

Edge Enhancement

Edge enhancement is an effort to reinforce the visual transitions between regions of contrasting brightness. Typically, the human interpreter prefers sharp edges between adjacent parcels, whereas the presence of noise, coarse resolution, and other factors often tend to blur or weaken the distinctiveness of these transitions. Edge enhancement in effect magnifies local contrast—enhancement of contrast within a local region. A typical

FIGURE 4.22. Histogram equalization spreads the range of brightness values but preserves the peaks and valleys in the histogram.

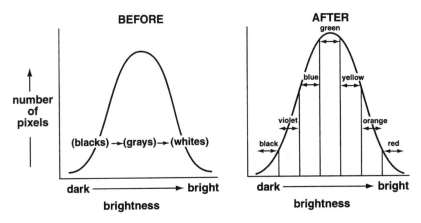

FIGURE 4.23. Density slicing assigns colors to specific intervals of brightness values. See also Plates 3 and 15.

edge-enhancement algorithm consists of a usually square window that is systematically moved through the image, centered successively on each pixel. There are many edge enhancement filters, but one of the most common is a variant of the Laplacian that works as follows for a 3 × 3 window: (1) the brightness value of each input pixel under the moving window, except for the center pixel, is multiplied by –1; (2) the center pixel is multiplied by 8; and (3) the center pixel in the output image is then given the value of the sum of all nine products resulting from (1) and (2).

Rohde et al. (1978) describe an edge-enhancement procedure that can illustrate some of the specifics of this approach. A new (*output*) digital value is calculated using the original (*input*) value and the local average of five adjacent pixels. A constant can be applied to alter the effect of the enhancement as necessary in specific situations. The output value is the difference between twice the input value and the local average, thereby increasing the brightness of those pixels that are already brighter than the local average and decreasing the brightness of pixels that are already darker than the local average. Thus the effect is to accentuate differences in brightnesses, especially at places ("edges") at which a given value differs greatly from the local average (Figure 4.24).

4.8. Image Display

For remote sensing analysis, the image display is especially important, because the analyst must be able to examine images and to inspect results of analyses, which often are themselves images. At the simplest level, an image display can be thought of as a high-quality television screen, although those tailored specifically for image processing have image-display processors, which are special computers designed to very rapidly receive digital data from the main computer, then display them as brightnesses on the screen. The capabilities of an image display are determined by several factors. First is the *size* of the image it can display, usually specified by the number of rows and columns it can show at any one time. Second, a display has a given *radiometric resolution* (Chapter 10)—that is, for each pixel, it has a capability to show a range of brightnesses. One-bit resolution

WITHOUT SHARPENING   **WITH SHARPENING**

FIGURE 4.24. Edge enhancement and image sharpening. A sample image is shown with and without enhancement. In this example, the effect of sharpening is especially noticeable at the edges of some of the larger shadows.

would give the capability to represent either black or white—certainly not enough detail to be useful for most purposes. In practice, most modern displays use 256 brightness levels for each of the primary colors of light (red, green, and blue).

A third factor controls the *rendition of color* in the displayed image. The method of depicting color is closely related to the design of the image display and the display processor. Image-display data are held in the *frame buffer*, a large segment of computer memory dedicated to handling data for display. The frame buffer provides one or more bits to record the brightness of each pixel to be shown on the screen (the "bit plane"); thus the displayed image is generated, bit by bit, in the frame buffer. The more bits that have been designed in the frame buffer for each pixel, the greater the range of brightnesses that can be shown for that pixel, as explained earlier. For actual display on the screen, the digital value for each pixel is converted into an electrical signal that controls the brightness of the pixel on the screen. This requires a digital-to-analog (D-to-A) converter that translates discrete digital values into continuous electrical signals (the opposite function of the A-to-D converter mentioned previously).

Three strategies have been used for designing image displays, each outlined here in abbreviated form.

The *cathode ray tube* (CRT) dates from early 1940s, when it formed the basis for the first television displays. A CRT is formed from a large glass tube, wide at one end (the "screen") and narrow at the other. The inside of the wide end is coated with phosphor atoms. An electron gun positioned at the narrow end directs a stream of electrons against the inside of the wide end of the tube. As the electrons strike the phosphor coating, it glows, creating an image as the intensity of the electron beam varies according to the strength of the video signal. Electromagnets positioned on four sides of the narrow portion of the tube control the scan of the electron stream across the face of the tube, left to right, top to bottom.

As each small region of the screen is illuminated from the inside by the stream of electrons, it glows. Because the gun directs the stream of electrons systematically and very rapidly, it creates the images we can see on a computer display or television screen. Because the electron gun scans very rapidly (30–70 times each second), it can return to refresh, or update, the brightness at each pixel before the phosphor coating fades. In this manner, the image appears to the eye as a continuous image.

CRTs produce very clear images, but because an increase in the size of the screen requires a commensurate increase in the depth of the tube (so the gun can illuminate the entire width of the screen), the size and weight of a CRT display forms a major inconvenience, and today CRTs have been supplanted by alternative technologies.

An alternative display technology was developed in the early 1970s. *Liquid crystal displays* (LCDs) depend on *liquid crystals*, substances that are intermediate between solid and liquid phases. The state assumed at a specific time depends on the temperature. An electrical current can change the orientation of the molecules within a liquid crystal and thereby block or transmit light at each pixel.

LCD displays use two sheets of polarizing materials that enclose a liquid crystal solution between them. When the video signal sends an electrical current to the display, the crystals align to block the passage of light between them. In effect, the liquid crystal at each pixel acts like a shutter, either blocking or passing the light. Color LCD displays use either of two alternative strategies; the best quality is provided by *active matrix displays*, also known as thin-film transistors (TFTs), which permit rapid refresh of the image. LCDs are used in watches, alarm clocks, and similar consumer products, but also for the flat-panel displays in portable computers and the compact displays now used for desktop computers.

A third display technology, *plasma display* represents each pixel using three tiny fluorescent lights. These fluorescent lights are small sealed glass tubes containing an internal phosphor coating, an inert gas (mercury), and two electrodes. As an electrical current flows across the electrodes, it vaporizes some of the mercury. The electrical current also raises the energy levels of some of the mercury atoms; when they return to their original state, they emit photons in the ultraviolet portion of the spectrum The ultraviolet light strikes the phosphor coating on the tube, creating visible light used to make an image. Variations in the coatings can create different colors. The positions of the tiny fluorescent light can be referenced as intersections in a raster grid, so the tube required for the CRT is not necessary for a plasma display, and the screen can be much more compact. Plasma displays are suitable for large, relatively compact image displays, but they are expensive, so they are not now preferred for analytical use.

Advanced Image Display

Remote sensing typically generates very large images portraying fine levels of detail. Conventional systems permit the users to examine regions in fine detail only by zooming in to display the region of interest at the cost of losing the broader context. Or users can discard the finer detail and examine broad regions at coarse detail. Although this trade-off sometimes causes little or no inconvenience, in other situations the sacrifice of one quality for the other means that some of the most valuable qualities of the data are discarded. As a result, there are incentives to design display systems that can simultaneously represent fine detail and large image size. Two alternative strategies have each found roles for viewing remotely sensed imagery.

"Fisheye," or *"focus + context,"* displays enable the analyst to simultaneously view selected detail without discarding the surrounding context. Fisheye displays use existing display hardware, but with a simulated magnifier that can roam over the image to selectively enlarge selected regions within the context of the coarser-resolution display (Figure 4.25). Software "lenses" locally magnify a subset while maintaining a continuous visual connection to the remainder of the unmagnified image. Such capabilities can be linked to existing analytical software to enable the analyst to annotate, measure, and delineate regions to improve the functionality of existing software. The same effects can be achieved in a different manner by linking two windows within the display—one for detail, one for the broader context—and providing the analyst with the capability to alter the size of the window. This approach is known as "multiple linked views" or the "overview + detail" display.

Multiple-monitor systems (sometimes referred to as *tiled displays*) are formed as arrays of flat-panel monitors (or rear-projection displays) that can display very large images at high levels of spatial detail (Figure 4.26). The highest quality tiled displays can project between 250 and 300 million pixels; within the near future, gigapixel displays (1 billion pixels) will likely be attempted. Multiple-monitor systems enable the computer's operating system to use the display areas from two or more display devices to create a single display. The rear-projection systems (sometimes referred to as "power walls") do not have the seams between tiles that characterize the LCD tiled systems, so they have greater visual continuity, but the tiled LCDs have better visual quality and are cheaper.

FIGURE 4.25. Example of "fisheye"-type image display. Left: Image of aircraft parked in a storage area for out-of-service aircraft. Right: Same image as viewed with a fisheye-type display that magnifies the central region of the image, while preserving the context of the surrounding image. From Pliable Display Technology: IDELIX Software Inc, Vancouver, BC, Canada (*www.idelix.com/imageintel.shtml*); satellite image from Digital Globe, Longmont, Colorado (*www.digitalglobe.com/*).

FIGURE 4.26. Example of tiled image display. From Chris North, Virginia Tech Center for Human–Computer Interaction.

Multiple-monitor systems became practical as the costs of random access memory (RAM) and LCD displays decreased to enable economical development of larger displays. Tiled displays are formed as a mosaic of screens, supported by operating systems configured to support multiple displays. Innovations in titled displays became possible when LCD technology permitted assembly of multiple flat-panel displays into tiled displays. Tiled displays assembled from several LCDs use special mounts to hold multiple displays provided by commercial vendors and specialized software systems to permit integrated use of the several monitors. Analysts can display images across the seams formed by the edges of the displays and can zoom, create multiple windows, run multiple applications, and arrange windows as most effective for specific tasks.

Obvious applications include remotely sensed and GIS images that require the analyst to simultaneously exploit fine detail and broad areal coverage. Analysts sometimes desire to run two or more analytical programs simultaneously, with independent displays of the images generated by each analysis. Both fisheye and multiple-monitor systems are emerging technologies in the sense that they are both proven to be successful but are still under investigation to explore how they can be most effectively used in specific applications. Multiple-monitor systems have value for emergency management systems, which benefit from having the capability for many people to simultaneously view large images representing states, counties, or similar regions as single images, with the ability to manipulate the display as needed to assist in evaluation of complex situations.

4.9. Image Processing Software

Digital remote sensing data can be interpreted by computer programs that manipulate the data recorded in pixels to yield information about specific subjects, as described in subsequent chapters. This kind of analysis is known as *image processing*, a term that encompasses a very wide range of techniques. Image processing requires a system of specialized computer programs tailored to the manipulation of digital image data. Although such programs vary greatly in purpose and in detail, it is possible to identify the major components likely to be found in most image processing systems.

A separate specific portion of the system is designed to read image data, usually from CD-ROM or other storage media, and to reorganize the data into the form to be used by the program. For example, many image processing programs manipulate the data in BSQ format. Thus the first step may be to read BIL or BIP data and then reformat the data into the BSQ format required for the analytical components of the system. Another portion of the system may permit the analyst to subdivide the image into subimages; to merge, superimpose, or mosaic separate images; and in general to prepare the data for analysis, as described later in Chapter 11. The heart of the system consists of a suite of programs that analyze, classify (Chapter 12), and manipulate data to produce output images and the statistics and data that may accompany them. Finally, a section of the image processing system must prepare data for display and output, either to the display processor or to the line printer. In addition, the program requires "housekeeping" subprograms that monitor movement and labeling of files from one portion of the program to another, generate error messages, and provide online documentation and assistance to the analysts.

Widely used image processing systems run on personal computers (PCs), Macintoshes, or workstations. More elaborate systems can be supported by peripheral equipment, including extra mass storage, digitizers, scanners, color printers, disk drives, and related equipment. Almost all such systems are directed by menus and graphic user interfaces that permit the analyst to select options from a list on the screen.

Although many good image processing systems are available, some of the most commonly used are:

- ERDAS ER Mapper (ERDAS)
  Tel: 877-463-7327
  Website: *www.erdas.com/tabid/84/currentid/1052/default.aspx*

- EASI/PACE
  PCI Geomatics Headquarters
  50 West Wilmot Street
  Richmond Hill, Ontario
  L4B 1M5, Canada
  Tel: 905-764-0614
  Website: *www.pcigeomatics.com*

- ENVI
  ITT Visual Systems Solutions
  Website: *www.ittvis.com*

- ERDAS Imagine (part of the Hexagon Group, Sweden)
  ERDAS, Inc., Worldwide Headquarters
  5051 Peachtree Corners Circle
  Norcross, GA 30092-2500
  Tel: 770-776-3400
  Website: *www.erdas.com*

- GRASS GIS
  Open Source Geospatial Foundation
  Website: *grass.itc.it/index.php*

- IDRISI
  IDRISI Project
  Clark Labs
  Clark University
  950 Main Street
  Worcester, MA 01610-1477
  Tel: 508-793-7526
  Website: *www.idrisi.clarku.edu*

Lemmens (2004) provides a point-by-point comparison of image processing systems designed for remote sensing applications; Voss et al. (2010) list image processing systems, with point-by-point comparisons of their capabilities. The specific systems listed here can be considered general-purpose image processing systems; others have been designed specifically to address requirements for specific kinds of analysis (e.g., geology, hydrology); and some of the general purpose systems have added optional modules that focus on more specific topics. Further details of image analysis systems are given by user manuals or help files for specific systems. Authors of image processing systems typically upgrade their systems to add new or improved capabilities, accommodate new equipment, or address additional application areas.

Several image processing systems are available to the public either without cost or at minimal cost. For students, some of these systems offer respectable capabilities to illustrate basics of image processing. A partial list includes:

ISIS: *http://isis.astrogeology.usgs.gov/index.html*

IVICS: *www.nsstc.uah.edu/ivics*

MultiSpec: *http://dynamo.ecn.purdue.edu/~biehl/MultiSpec*

TNTlite: *www.microimages.com*

Image Viewers and Online Digital Image Archives

Image viewers (or, sometimes, *map viewers*) are programs designed to provide basic capabilities to view and navigate through digital maps and images. Some image viewers are available commercially; others are available online at minimal cost or as freeware. They provide a convenient means of examining digital maps, GIS data, and aerial imagery. Although most image viewers do not offer analytical capabilities, they do permit users to examine a wide range of spatial data by searching, roaming, magnifying, and applying a variety of projection and coordinate systems. For example, GIS Viewer 4.0 (University of California, Berkeley; *http://gcmd.nasa.gov/records/GIS_Viewer.html*) provides an illustration of the basic functions of an image viewer. Image viewers are closely connected to the idea of *digital imagery archives* or *libraries*, which provide collections of digital imagery in standardized formats, such that viewers can easily retrieve and navigate through the collection. Microsoft's Research Maps (MSR) (*http://msrmaps.com/Default.aspx*) is one of the most comprehensive online archives of digital imagery. It includes digital aerial photographs and maps of large portions of the United States, with an ability to search by place names and to roam across a landscape. (See also *www.terraserver.com* for another such resource.) Such systems may well form prototypes for design of more sophisticated image archive systems. *Google Earth* (*http://earth.google.com*) provides a comprehensive coverage of the Earth's land areas, with the ability to roam and change scale, orientation, and detail.

4.10. Summary

Although digital data provide multiple advantages for practitioners of remote sensing, these advantages can be exploited only if the analyst has mastered the underlying concepts and how they influence applications of remote sensing to specific problems. The information presented here resurfaces in later chapters, as the fundamental nature of digital data underlie the basic design and application of remote sensing systems. Each instrument must be operated within its design constraints and applied to an appropriate task. Although this chapter cannot provide the details necessary to make such assessments, it can equip the reader with the perspective to seek the specifics that will permit a sound assessment of the question at hand.

4.11. Some Teaching and Learning Resources

- Diffraction Gratings
  www.youtube.com/watch?v=5D8EVNZdyy0

- Laser Diffraction Gratings Tutorial
  www.youtube.com/watch?v=gxgfthefKbg&feature=related

- Navigating a 13.3 Gigapixel Image on a 22 Megapixel Display Wall
  www.youtube.com/watch?v=8bHWuvzBtJo

- DOG (Difference of Gaussians)
  www.youtube.com/watch?v=Fe-pubQw5Xc

- Contrast Enhancement by IMG
  www.youtube.com/watch?v=5XS80BcUqhA

- PLC Tutorial—Analog Controller A to D Converter I/O
  www.youtube.com/watch?v=rWMGY3ChaZU

Review Questions

1. It may be useful to practice conversion of some values from digital to binary form as confirmation that you understand the concepts. Convert the following digital numbers to 8-bit binary values:

a. 100	c. 24	e. 2	g. 256
b. 15	d. 31	f. 111	h. 123

2. Convert the following values from binary to digital form:

a. 10110	c. 10111	e. 0011011
b. 11100	d. 1110111	f. 1101101

3. Consider the implications of selecting the appropriate number of bits for recording remotely sensed data. One might be tempted to say, "Use a large number of bits to

be sure that all values are recorded precisely." What would be the disadvantage of using, for example, seven bits to record data that are accurate only to five bits?

4. Describe in a flow chart or diagram steps required to read data in a BIP format, then organize them in a BSQ structure.

5. What is the minimum number of bits required to represent the following values precisely?

   a. 1,786   c. 689   e. 17   g. 29
   b. 32   d. 32,000 f. 3

6. Why are enhanced images usually not used as input for other analyses?

7. Density slicing produces an image that uses a range of contrasting colors. Examine the examples in Plates 3 and 15, and prepare a list of advantages and disadvantages of this form of image enhancement.

8. Do you expect that it possible to estimate a sensor's SNR by visual examination of an image? How?

9. The digital format is now becoming the de facto standard for recording and storing aerial imagery. Discuss some of the advantages and disadvantages that accompany this change.

10. Some students find it highly illogical to use band combinations in which, for example, radiation in the red region of the spectrum is displayed using another color. Explain in a few concise sentences why such band combinations are useful.

References

Ball, R., and C. North. 2005 Effects of Tiled High-Resolution Display on Basic Visualization and Navigation Tasks. In *Proceedings, Extended Abstracts of ACM Conference on Human Factors in Computing Systems* (CHI 2005), Portland, OR: Association for Computing Machinery, pp. 1196–1199.

Jensen, J. R. 2004. *Introductory Digital Image Processing, A Remote Sensing Perspective* (3rd ed.). Upper Saddle River, NJ: Prentice-Hall, 544 pp.

Jensen, J. R., and R. R. Jensen. 2002. Remote Sensing Digital Image Processing System Hardware and Software Considerations. *Manual of Geospatial Science and Technology* (J. Bossler, ed.). London: Taylor & Francis, pp. 325–348.

Lemmens, M. 2004. Remote Sensing Processing Software. *GIM International*, Vol. 18, pp. 53–57. (See also: 2010. Remote Sensing Image Processing Software. *GIM International*. (June 2010, pp. 38–39.)

Liu, J.-K., H. Wu, and T. Shih. 2005. Effects of JPEG2000 on the Information and Geometry Content of Aerial Photo Compression. *Photogrammetric Engineering and Remote Sensing*, Vol. 71, pp. 157–167.

Rohde, W. G., J. K. Lo, and R. A. Pohl. 1978. EROS Data Center Landsat Digital Enhancement Techniques and Imagery Availability, 1977. *Canadian Journal of Remote Sensing*, Vol. 4, pp. 63–76.

Voss, M., R. Sugumaran, and D. Ershov. 2010. Software for Processing Remotely Sensed Data. Chapter 20 in *Manual of Geospatial Science and Technology* (J. D. Bossler, ed.). London: Taylor & Francis, pp. 391–402.

CHAPTER FIVE

Image Interpretation

5.1. Introduction

Earlier chapters have defined our interest in remote sensing as focused primarily on images of the Earth's surface—map-like representations of the Earth's surface based on the reflection of electromagnetic energy from vegetation, soil, water, rocks, and man-made structures. From such images we learn much that cannot be derived from other sources. Yet such information is not presented to us directly: The information we seek is encoded in the varied tones and textures we see on each image. To translate images into information, we must apply a specialized knowledge—knowledge that forms the field of *image interpretation*, which we can apply to derive useful information from the raw uninterpreted images we receive from remote sensing systems. Proficiency in image interpretation is formed from three separate kinds of knowledge, of which only one—the final one listed here—falls within the scope of this text.

Subject

Knowledge of the *subject* of our interpretation—the kind of information that motivates us to examine the image—is the heart of the interpretation. Accurate interpretation requires familiarity with the subject of the interpretation. For example, interpretation of geologic information requires education and experience in the field of geology. Yet narrow specializations are a handicap, because each image records a complex mixture of many kinds of information, requiring application of broad knowledge that crosses traditional boundaries between disciplines. For example, accurate interpretation of geologic information may require knowledge of botany and the plant sciences as a means of understanding how vegetation patterns on an image reflect geologic patterns that may not be directly visible. As a result, image interpreters should be equipped with a broad range of knowledge pertaining to the subjects at hand and their interrelationships.

Geographic Region

Knowledge of the specific geographic *region* depicted on an image can be equally significant. Every locality has unique characteristics that influence the patterns recorded on an image. Often the interpreter may have direct experience within the area depicted on the image that can be applied to the interpretation. In unfamiliar regions the interpreter may find it necessary to make a field reconnaissance or to use maps and books that describe analogous regions with similar climate, topography, or land use.

Remote Sensing System

Finally, knowledge of the *remote sensing system* is obviously essential. The interpreter must understand how each image is formed and how each sensor portrays landscape features. Different instruments use separate portions of the electromagnetic spectrum, operate at different resolutions, and use different methods of recording images. The image interpreter must know how each of these variables influences the image to be interpreted and how to evaluate their effects on his or her ability to derive useful information from the imagery. This chapter outlines how the image interpreter derives useful information from the complex patterns of tone and texture on each image.

5.2. The Context for Image Interpretation

Human beings are well prepared to examine images, as our visual system and experience equip us to discern subtle distinctions in brightness and darkness, to distinguish between various image textures, to perceive depth, and to recognize complex shapes and features. Even in early childhood we apply such skills routinely in everyday experience so that few of us encounter difficulties as we examine, for example, family snapshots or photographs in newspapers. Yet image analysis requires a conscious, explicit effort not only to learn about the subject matter, geographic setting, and imaging systems (as mentioned above) in unfamiliar contexts but also to develop our innate abilities for image analysis.

Three issues distinguish interpretation of remotely sensed imagery from interpretation conducted in everyday experience. First, remotely sensed images usually portray an *overhead view*—an unfamiliar perspective. Training, study, and experience are required to develop the ability to recognize objects and features from this perspective. Second, many remote sensing images use *radiation outside the visible portion* of the spectrum—in fact, use of such radiation is an important advantage that we exploit as often as possible. Even the most familiar features may appear quite different in nonvisible portions of the spectrum than they do in the familiar world of visible radiation. Third, remote sensing images often portray the Earth's surface at *unfamiliar scales and resolutions.* Commonplace objects and features may assume strange shapes and appearances as scale and resolution change from those to which we are accustomed.

This chapter outlines the art of image interpretation as applied to aerial photography. Students cannot expect to become proficient in image analysis simply by reading about image interpretation. Experience forms the only sure preparation for skillful interpretation. Nonetheless, this chapter can highlight some of the issues that form the foundations for proficiency in image analysis.

In order to discuss this subject at an early point in the text, we must confine the discussion to interpretation of aerial photography, the only form of remote sensing imagery discussed thus far. But the principles, procedures, and equipment described here are equally applicable to other kinds of imagery acquired by the sensors described in later chapters. Manual image interpretation is discussed in detail by Paine and Kiser (2003), Avery and Berlin (2003), Philipson (1996), and Campbell (2005); older references that may also be useful are the text by Lueder (1959) and the *Manual of Photographic Interpretation* (Colwell, 1960).

5.3. Image Interpretation Tasks

The image interpreter must routinely conduct several kinds of tasks, many of which may be completed together in an integrated process. Nonetheless, for purposes of clarification, it is important to distinguish between these separate functions (Figure 5.1).

Classification

Classification is the assignment of objects, features, or areas to classes based on their appearance on the imagery. Often a distinction is made between three levels of confidence and precision. *Detection* is the determination of the presence or absence of a feature. *Recognition* implies a higher level of knowledge about a feature or object, such that the object can be assigned an identity in a general class or category. Finally, *identification* means that the identity of an object or feature can be specified with enough confidence and detail to place it in a very specific class. Often an interpreter may qualify his or her confidence in an interpretation by specifying the identification as "possible" or "probable."

Enumeration

Enumeration is the task of listing or counting discrete items visible on an image. For example, housing units can be classified as "detached single-family home," "multifamily

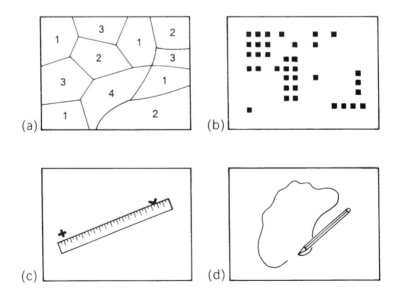

FIGURE 5.1. Image interpretation tasks. (a) Classification. (b) Enumeration. (c) Mensuration. (d) Delineation.

complex," "mobile home," and "multistory residential," and then reported as numbers present within a defined area. Clearly, the ability to conduct such an enumeration depends on an ability to accurately identify and classify items as discussed above.

Measurement

Measurement, or *mensuration*, is an important function in many image interpretation problems. Two kinds of measurement are important. First is the measurement of distance and height and, by extension, of volumes and areas as well. The practice of making such measurements forms the subject of *photogrammetry* (Chapter 3), which applies knowledge of image geometry to the derivation of accurate distances. Although, strictly speaking, photogrammetry applies only to measurements from photographs, by extension it has analogs for the derivation of measurements from other kinds of remotely sensed images.

A second form of measurement is quantitative assessment of image brightness. The science of *photometry* is devoted to measurement of the intensity of light and includes estimation of scene brightness by examination of image tone, using special instruments known as *densitometers*. If the measured radiation extends outside the visible spectrum, the term *radiometry* applies. Both photometry and radiometry apply similar instruments and principles, so they are closely related.

Delineation

Finally, the interpreter must often delineate, or outline, regions as they are observed on remotely sensed images. The interpreter must be able to separate distinct areal units that are characterized by specific tones and textures and to identify edges or boundaries between separate areas. Typical examples include delineation of separate classes of forest or of land use—both of which occur only as areal entities (rather than as discrete objects). Typical problems include: (1) selection of appropriate levels of generalization (e.g., when boundaries are intricate, or when many tiny but distinct parcels are present); and (2) placement of boundaries when there is a gradation (rather than a sharp edge) between two units.

The image analyst may simultaneously apply several of these skills in examining an image. Recognition, delineation, and mensuration may all be required as the interpreter examines an image. Yet specific interpretation problems may emphasize specialized skills. Military photo interpretation often depends on accurate recognition and enumeration of specific items of equipment, whereas land-use inventory emphasizes delineation, although other skills are obviously important. Image analysts therefore need to develop proficiency in all of these skills.

5.4. Elements of Image Interpretation

By tradition, image interpreters are said to employ some combination of the eight *elements of image interpretation*, which describe characteristics of objects and features as they appear on remotely sensed images. Image interpreters quite clearly use these characteristics together in very complex, but poorly understood, processes as they examine images. Nonetheless, it is convenient to list them separately as a way of emphasizing their significance.

Image Tone

Image tone denotes the lightness or darkness of a region within an image (Figure 5.2). For black-and-white images, tone may be characterized as "light," "medium gray," "dark gray," "dark," and so on, as the image assumes varied shades of white, gray, or black. For color or CIR imagery, image tone refers simply to "color," described informally perhaps in such terms as "dark green," "light blue," or "pale pink."

Image tone can also be influenced by the intensity and angle of illumination and by the processing of the film. Within a single aerial photograph, vignetting (Section 3.2) may create noticeable differences in image tone due solely to the position of an area within a frame of photography: The image becomes darker near the edges. Thus the interpreter must employ caution in relying solely on image tone for an interpretation, as it can be influenced by factors other than the absolute brightness of the Earth's surface. Analysts should also remember that very dark or very bright regions on an image may be exposed in the nonlinear portion of the characteristic curve (Chapter 3), so they may not be represented in their correct relative brightnesses. Also, nonphotographic sensors may record such a wide range of brightness values that they cannot all be accurately represented on photographic film—in such instances digital analyses (Chapter 4) may be more accurate.

Experiments have shown that interpreters tend to be consistent in interpretation of tones on black-and-white imagery but less so in interpretation of color imagery (Cihlar and Protz, 1972). Interpreters' assessment of image tone is much less sensitive to subtle differences in tone than are measurements by instruments (as might be expected). For the range of tones used in the experiments, human interpreters' assessment of tone expressed a linear relationship with corresponding measurements made by instruments. Cihlar and Protz's results imply that a human interpreter can provide reliable estimates of relative differences in tone but not be capable of accurate description of absolute image brightness.

Image Texture

Image texture refers to the apparent roughness or smoothness of an image region. Usually texture is caused by the pattern of highlighted and shadowed areas created when an irregular surface is illuminated from an oblique angle. Contrasting examples (Figure 5.3) include the rough textures of a mature forest and the smooth textures of a mature wheat

DARK   **LIGHT**

FIGURE 5.2. Varied image tones, dark to light (left to right). From USDA.

SMOOTH; EVEN   **MOTTLED**   **COARSE**

FIGURE 5.3. Varied image textures, with descriptive terms. From USDA.

field. The human interpreter is very good at distinguishing subtle differences in image texture, so it is a valuable aid to interpretation—certainly equal in importance to image tone in many circumstances.

Image texture depends not only on the surface itself but also on the angle of illumination, so it can vary as lighting varies. Also, good rendition of texture depends on favorable image contrast, so images of poor or marginal quality may lack the distinct textural differences so valuable to the interpreter.

Shadow

Shadow is an especially important clue in the interpretation of objects. A building or vehicle, illuminated at an angle, casts a shadow that may reveal characteristics of its size or shape that would not be obvious from the overhead view alone (Figure 5.4). Because military photointerpreters often are primarily interested in identification of individual items of equipment, they have developed methods to use shadows to distinguish subtle differences that might not otherwise be visible. By extension, we can emphasize this role of shadow in interpretation of any man-made landscape in which identification of separate kinds of structures or objects is significant.

Shadow is of great significance also in interpretation of natural phenomena, even though its role may not be as obvious. For example, Figure 5.5 depicts an open field in which scattered shrubs and bushes are separated by areas of open land. Without shadows, the individual plants might be too small (as seen from above) and too nearly similar

(a)   (b)   (c)

FIGURE 5.4. Examples of significance of shadow in image interpretation, as illustrated by (a) fuel storage tanks, (b) military aircraft on a runway, and (c) a water tower. From USDA.

FIGURE 5.5. Significance of shadow for image interpretation, as illustrated by the characteristic pattern caused by shadows of shrubs cast on open field. Shadows at the edge of a forest enhance the boundary between two different land covers. From USDA.

in tone to their background to be visible. Yet their shadows are large enough and dark enough to create the streaked pattern on the imagery typical of this kind of land. A second example is also visible in Figure 5.5—at the edges between the trees in the hedgerows and the adjacent open land, trees cast shadows that form a dark strip that enhances the boundary between the two zones, as seen on the imagery.

Pattern

Pattern refers to the arrangement of individual objects into distinctive recurring forms that facilitate their recognition on aerial imagery (Figure 5.6). Pattern on an image usually follows from a functional relationship between the individual features that compose the pattern. Thus the buildings in an industrial plant may have a distinctive pattern due to their organization to permit economical flow of materials through the plant, from receiving raw material to shipping of the finished product. The distinctive spacing of trees in an orchard arises from careful planting of trees at intervals that prevent competition between individual trees and permit convenient movement of equipment through the orchard.

(a)   (b)   (c)   (d)

FIGURE 5.6. Significance of distinctive image pattern, as illustrated by (a) structures in a suburban residential neighborhood, (b) an orchard, (c) a highway interchange, and (d) a rural trailer park. From VDOT (a, d); USDA (b); USGS(c).

Association

Association specifies the occurrence of certain objects or features, usually without the strict spatial arrangement implied by pattern. In the context of military photointerpretation, association of specific items has great significance, as, for example, when the identification of a specific class of equipment implies that other, more important, items are likely to be found nearby.

Shape

Shapes of features are obvious clues to their identities (Figure 5.7). For example, individual structures and vehicles have characteristic shapes that, if visible in sufficient detail, provide the basis for identification. Features in nature often have such distinctive shapes that shape alone might be sufficient to provide clear identification. For example, ponds, lakes, and rivers occur in specific shapes unlike others found in nature. Often specific agricultural crops tend to be planted in fields that have characteristic shapes (perhaps related to the constraints of equipment used or the kind of irrigation that the farmer employs).

Size

Size is important in two ways. First, the relative size of an object or feature in relation to other objects on the image provides the interpreter with an intuitive notion of its scale and resolution, even though no measurements or calculations may have been made. This intuition is achieved via recognition of familiar objects (dwellings, highways, rivers, etc.) followed by extrapolation to use the sizes of these known features in order to estimate the sizes and identities of those objects that might not be easily identified. This is probably the most direct and important function of size.

Second, absolute measurements can be equally valuable as interpretation aids. Measurements of the size of an object can confirm its identification based on other factors, especially if its dimensions are so distinctive that they form definitive criteria for specific items or classes of items. Furthermore, absolute measurements permit derivation of quantitative information, including lengths, volumes, or (sometimes) even rates of movement (e.g., of vehicles or ocean waves as they are shown in successive photographs).

(a)   (b)   (c)   (d)

FIGURE 5.7. Significance of shape for image interpretation, as illustrated by (a) athletic fields; (b) aircraft parked on a runway, (c) automobiles in a salvage yard, and (d) a water treatment plant. From USDA.

Site

Site refers to topographic position. For example, sewage treatment facilities are positioned at low topographic sites near streams or rivers to collect waste flowing through the system from higher locations. Orchards may be positioned at characteristic topographic sites—often on hillsides (to avoid cold air drainage to low-lying areas) or near large water bodies (to exploit cooler spring temperatures near large lakes to prevent early blossoming).

5.5. Collateral Information

Collateral, or ancillary, *information* refers to nonimage information used to assist in the interpretation of an image. Actually, all image interpretations use collateral information in the form of the implicit, often intuitive, knowledge that every interpreter brings to an interpretation in the form of everyday experience and formal training. In its narrower meaning, it refers to the explicit, conscious effort to employ maps, statistics, and similar material to aid in analysis of an image. In the context of image interpretation, use of collateral information is permissible, and certainly desirable, provided two conditions are satisfied. First, the use of such information is to be explicitly acknowledged in the written report; and second, the information must not be focused on a single portion of the image or map to the extent that it produces uneven detail or accuracy in the final map. For example, it would be inappropriate for an interpreter to focus on acquiring detailed knowledge of tobacco farming in an area of mixed agriculture if he or she then produced highly detailed, accurate delineations of tobacco fields but mapped other fields at lesser detail or accuracy.

Collateral information can consist of information from books, maps, statistical tables, field observations, or other sources. Written material may pertain to the specific geographic area under examination, or, if such material is unavailable, it may be appropriate to search for information pertaining to analogous areas—similar geographic regions (possibly quite distant from the area of interest) characterized by comparable ecology, soils, landforms, climate, or vegetation.

5.6. Imagery Interpretability Rating Scales

Remote sensing imagery can vary greatly in quality due to both environmental and to technical conditions influencing acquisition of the data. In the United States, some governmental agencies use rating scales to evaluate the suitability of imagery for specific purposes. The National Imagery Interpretability Rating Scale (NIIRS) has been developed for single-channel and panchromatic imagery, and the Multispectral Imagery Interpretability Rating Scale (MSIIRS; Erdman et al., 1994) has been developed for multispectral imagery. Such scales are based on evaluations using a large number of experienced interpreters to independently evaluate images of varied natural and manmade features, as recorded by images of varying characteristics. They provide a guide for evaluation of whether a specific form of imagery is likely to be satisfactory for specific purposes.

5.7. Image Interpretation Keys

Image interpretation keys are valuable aids for summarizing complex information portrayed as images. They have been widely used for image interpretation (e.g., Coiner and Morain, 1971). Such keys serve either or both of two purposes: (1) they are a means of training inexperienced personnel in the interpretation of complex or unfamiliar topics, and (2) they are a reference aid for experienced interpreters to organize information and examples pertaining to specific topics.

An *image interpretation key* is simply reference material designed to permit rapid and accurate identification of objects or features represented on aerial images. A key usually consists of two parts: (1) a collection of annotated or captioned images or stereograms and (2) a graphic or word description, possibly including sketches or diagrams. These materials are organized in a systematic manner that permits retrieval of desired images by, for example, date, season, region, or subject.

Keys of various forms have been used for many years in the biological sciences, especially botany and zoology. These disciplines rely on complex taxonomic systems that are so extensive that even experts cannot master the entire body of knowledge. The key, therefore, is a means of organizing the essential characteristics of a topic in an orderly manner. It must be noted that scientific keys of all forms require a basic familiarity with the subject matter. A key is not a substitute for experience and knowledge but a means of systematically ordering information so that an informed user can learn it quickly.

Keys were first routinely applied to aerial images during World War II, when it was necessary to train large numbers of inexperienced photointerpreters in the identification of equipment of foreign manufacture and in the analysis of regions far removed from the experience of most interpreters. The interpretation key formed an effective way of organizing and presenting the expert knowledge of a few individuals. After the war ended, interpretation keys were applied to many other subjects, including agriculture, forestry, soils, and landforms. Their use has been extended from aerial photography to other forms of remotely sensed imagery. Today interpretation keys are still used for instruction and training, but they may have somewhat wider use as reference aids. Also, it is true that construction of a key tends to sharpen one's interpretation skills and encourages the interpreter to think more clearly about the interpretation process.

Keys designed solely for use by experts are referred to as *technical keys*. *Nontechnical keys* are those designed for use by those with a lower level of expertise. Often it is more useful to classify keys by their formats and organizations. *Essay keys* consist of extensive written descriptions, usually with annotated images as illustrations. A *file key* is essentially a personal image file with notes; its completeness reflects the interests and knowledge of the compiler. Its content and organization suit the needs of the compiler, so it may not be organized in a manner suitable for use by others.

5.8. Interpretive Overlays

Often in resource-oriented interpretations it is necessary to search for complex associations of several related factors that together define the distribution or pattern of interest. For example, often soil patterns may be revealed by distinctive relationships between separate patterns of vegetation, slope, and drainage. The *interpretive overlays* approach to image interpretation is a way of deriving information from complex interrelationships between separate distributions recorded on remotely sensed images. The correspondence between several separate patterns may reveal other patterns not directly visible on the image (Figure 5.8).

The method is applied by means of a series of individual overlays for each image to be examined. The first overlay might show the major classes of vegetation, perhaps consisting of dense forest, open forest, grassland, and wetlands. A second overlay maps slope classes, including perhaps level, gently sloping, and steep slopes. Another shows the drainage pattern, and still others might show land use and geology. Thus, for each image, the interpreter may have as many as five or six overlays, each depicting a separate pattern. By superimposing these overlays, the interpreter can derive information presented by the coincidence of several patterns. From his or her knowledge of the local terrain, the interpreter may know that certain soil conditions can be expected where the steep slopes and the dense forest are found together and that others are expected where the dense forest matches to the gentle slopes. From the information presented by several patterns, the interpreter can resolve information not conveyed by any single pattern.

5.9. The Significance of Context

In Chapter 1 the discussion of Figure 1.1 introduced the significance of *context* in deriving meaning from an image. That is, a purely visual understanding of an image does not necessarily lead to an understanding of its underlying meaning. This topic deserves further exploration in the context of image interpretation.

Most of us are familiar with the kind of visual illusion illustrated in Figure 5.9, the *Rubin illusion*, in which the viewer sees either a white vase against a black background or

FIGURE 5.8. Interpretive overlays. Image interpretation produces several separate overlays that can combine to permit interpretation of another feature that is not directly visible on the image.

FIGURE 5.9. Rubin face/vase illusion.

two faces in silhouette facing each other against a white background. The success of the illusion depends on its ability to confuse the viewer's capacity to assess the *figure–ground relationship*. To make visual sense of an image, our visual system must decide which part of a scene is the *figure* (the feature of interest) and which is the *ground* (the background that simply outlines the figure).

Normally, our visual system expects the background to constitute the larger proportion of a scene. The Rubin illusion, like most visual illusions, is effective because it is contrived to isolate the viewer's perception of the scene—in this instance, by designing the illustration so that figure and ground constitute equal proportions of the scene. The viewer's visual system cannot resolve the ambiguity, so the viewer experiences difficulty in interpreting the meaning of the scene.

Although such contrived images are not encountered in day-to-day practice, the principles that they illustrate apply to situations that are frequently encountered. For example, *relief inversion* occurs when aerial images of shadowed terrain are oriented in a manner that confuses our intuitive expectations. Normally, we expect to see terrain illuminated from the upper right (Figure 5.10, left); most observers see such images in their

FIGURE 5.10. Photographs of landscapes with pronounced shadowing are usually perceived in correct relief when shadows fall toward the observer. Left: When shadows fall toward the observer, relief is correctly perceived. Right: When the image is rotated so that shadows fall in the opposite direction, away from the observer, topographic relief appears to be reversed. From USGS.

correct relief. If the image is oriented so that the illumination appears to originate from the lower right, most observers tend to perceive the relief as inverted (Figure 5.10, right). Experimentation with conditions that favor this effect confirms the belief that, like most illusions, relief inversion is perceived only when the context has confined the viewer's perspective to present an ambiguous visual situation.

Image analysts can encounter many situations in which visual ambiguities can invite misleading or erroneous interpretations. When NASA analysts examined the 1976 Viking Orbiter images of the Cydonia region of Mars, they noticed, with some amusement, the superficial resemblance of an imaged feature to a humanoid face (Figure 5.11, left). Once the images were released to the public, however, the issue became a minor sensation, as many believed the images conveyed clear evidence of intelligent design and openly speculated about the origin and meaning of a feature that appeared to have such visually obvious significance. In 1998 the same region of Mars was imaged again, this time by the Mars Global Surveyor, a spacecraft with instruments that provided images with much higher spatial resolution (Figure 5.11, right). These images reveal that the region in question does not in fact offer a striking resemblance to a face.

Photointerpreters should remember that the human visual system has a powerful drive to impose its own interpretation on the neurological signals it receives from the eye and can easily create plausible interpretations of images when the evidence is uncertain, confused, or absent. Image analysts must strive always to establish *several independent lines of evidence and reasoning* to set the context that establishes the meaning of an image. When several lines of evidence and reasoning converge, then an interpretation can carry authority and credibility. When multiple lines of evidence and reasoning do not converge or are absent, then the interpretation must be regarded with caution and suspicion.

Image interpretation's successes illustrate the significance of establishing the proper context to understand the meaning of an image. The use of photointerpretation to identify the development and monitor the deployment of the German V-1 and V-2 missiles in

FIGURE 5.11. Mars face illusion. Left: This 1976 Viking Orbiter image of the Cydonia region of Mars was considered by some to present a strong resemblance to a human face, causing speculation that the feature was created by intelligent beings. From NASA. Right: In 1998 The Mars Global Surveyor reimaged the Cydonia region at much higher spatial resolution. With greater spatial detail available to the eye, the previous features are seen to be interesting but without any convincing resemblance to an artificial structure. From NASA.

World War II (Babbington-Smith, 1957; Irving, 1965) and to identify at an early stage the deployment of Soviet missiles in Cuba during the 1962 Cuban Missile Crisis (Brugioni, 1991) was successful because it provided information that could be examined and evaluated in a broader context. Image interpretation proved to be less successful in February 2003 when U.S. Secretary of State Colin Powell presented images to the United Nations to document the case for an active threat from weapons of mass destruction in Iraq; later it became quite clear that there was insufficient information at hand to establish the proper meaning of the images.

5.10. Stereovision

Stereoscopy is the ability to derive distance information (or in the case of aerial photography, height information) from two images of the same scene. (Section 3.9 introduced the manner in which aerial cameras can collect duplicate coverage of a single region using overlapping images.) Stereovision contributes a valuable dimension to information derived from aerial photography. Full development of its concepts and techniques are encompassed by the field of photogrammetry (Wolf, 1974), here we can introduce some of its applications by describing some simple instruments.

Stereoscopes are devices that facilitate stereoscopic viewing of aerial photographs. This simplest and most common is the *pocket stereoscope* (Figure 5.12). This simple, inexpensive, instrument forms an important image interpretation aid that can be employed in a wide variety of situations, and introduce concepts that underlie more advanced instruments. Its compact size and inexpensive cost make it one of the most widely used remote sensing instruments, even in era of digital imagery. Other kinds of stereoscopes include the *mirror stereoscope* (Figure 5.13), which permits stereoscopic viewing of large areas, usually at low magnification, and the *binocular stereoscope* (Figure 5.14), designed primarily for viewing film transparencies on light tables. Often the binocular stereoscope has adjustable magnification that enables enlargement of portions of the image up to 20 or 40 times.

FIGURE 5.12. A U.S. Geological Survey geologist uses a pocket stereoscope to examine vertical aerial photography, 1957. From USGS Photographic Library. Photograph by E. F. Patterson, No. 22.

FIGURE 5.13. Image interpretation equipment, Korean conflict, March 1952. A U.S. Air Force image interpreter uses a tube magnifier to examine an aerial photograph in detail. A mirror stereoscope is visible in the foreground. From U.S. Air Force, U.S. National Archives and Records Administration, ARC 542277.

The pocket stereoscope consists of a body holding two low-power lenses attached to a set of collapsible legs that can be folded so that the entire instrument can be stored in a space a bit larger than a deck of playing cards. The body is usually formed from two separate pieces, each holding one of the two lenses, which can be adjusted to control the spacing between the two lenses to accommodate the individual user.

Although at first glance, the stereoscope appears designed to magnify images, magnification is really an incidental feature of the instrument. In fact, the purpose of the stereoscope is to assist the analyst in maintaining parallel lines of sight. Stereoscopic vision is based on the ability of our visual system to detect *stereoscopic parallax*, the difference in the appearance of objects due to differing in perspectives. So, when we view a scene using only the right eye, we see a slightly different view than we do using only the left eye—this difference is stereoscopic parallax. Because stereoscopic parallax is greater for nearby objects than is for more distant objects, our visual system can use this information to make accurate judgments about distance.

Stereoscopic aerial photographs are acquired in sequences designed to provide overlapping views of the same terrain—that is, they provide two separate perspectives of the same landscape, just as our eyes provide two separate images of a scene. So we can use a stereo pair of aerial photographs to simulate a stereoscopic view of the terrain, provided we can maintain parallel lines of sight, just as we would normally do in viewing a distant object (Figure 5.15a). Parallel lines of sight assure that the right and left eyes each see

FIGURE 5.14. Binocular stereoscopes permit stereoscopic viewing of images at high magnification. From U.S. Air Force.

FIGURE 5.15. The role of the stereoscope in stereoscopic vision. (a) To acquire the two independent views of the same scene required for stereoscopic vision, we must maintain parallel lines of sight. (b) Normally, when we view nearby objects, our lines of sight converge, preventing us from acquiring the stereo effect. (c) The stereoscope is an aid to assist in maintaining parallel lines of sight even when the photographs are only a few inches away from the viewer.

independent views of the same scene, to provide the parallax needed for the stereoscopic illusion. However, when we view objects that are nearby, our visual system instinctively recognizes that the objects are close, so our lines of sight converge (Figure 5.15b), depriving our visual system of the two independent views needed for stereoscopic vision. Therefore, the purpose of the stereoscope is to assist us in maintaining the parallel lines of sight that enable the stereoscopic effect (Figure 5.15c).

Although many students will require the assistance of the instructor as they learn to use the stereoscope, the following paragraphs may provide some assistance for beginners. First, stereo photographs must be aligned so that the flight line passes left to right (as shown in Figure 5.16). Check the photo numbers to be sure that the photographs have been selected from adjacent positions on the flight line. Usually (but not always) the numbers and annotations on photos are placed on the *leading edge* of the image: the edge of the image nearest the front of the aircraft at the time the image was taken. Therefore, these numbers should usually be oriented in sequence from left to right, as shown in Figure 5.16. If the overlap between adjacent photos does not correspond to the natural positions of objects on the ground, then the photographs are incorrectly oriented.

Next, the interpreter should identify a distinctive feature on the image within the zone of stereoscopic overlap. The photos should then be positioned so that the duplicate images of this feature (one on each image) are approximately 64 mm (2.5 in.) apart. This distance represents the distance between the two pupils of a person of average size (referred to as the *interpupillary distance*), but for many it may be a bit too large or too small, so the spacing of photographs may require adjustment as the interpreter follows the procedure outlined here. The pocket stereoscope should be opened so that its legs are locked in place to position the lens at their correct height above the photographs. The two segments of the body of the stereoscope should be adjusted so that the centers of the eyepieces are about 64 mm (2.5 in) apart (or a slightly larger or smaller distance, as mentioned above).

Then the stereoscope should be positioned so that the centers of the lenses are positioned above the duplicate images of the distinctive feature selected previously. Looking through the two lenses, the analyst sees two images of this feature; if the images are

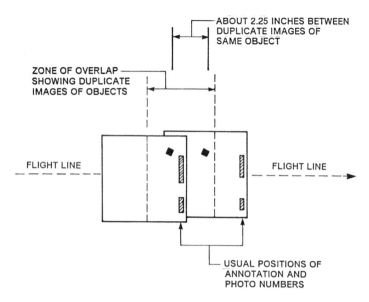

FIGURE 5.16. Positioning aerial photographs for stereoscopic viewing. The flight line must be oriented laterally in front of the viewer.

properly positioned, the two images will appear to "float" or "drift." The analyst can, with some effort, control the apparent positions of the two images so that they fuse into a single image; as this occurs, the two images should merge into a single image that is then visible in three dimensions. Usually aerial photos show exaggerated heights, due to the large separation (relative to distance to the ground) between successive photographs as they were taken along the flight line. Although exaggerated heights can prevent convenient stereo viewing in regions of high relief, it can be useful in interpretations of subtle terrain features that might not otherwise be noticeable. The student who has successfully used the stereoscope to examine a section of the photo should then practice moving the stereoscope over the image to view the entire region within the zone of overlap. As long as the axis of the stereoscope is oriented parallel to the flight line, it is possible to retain stereo vision while moving the stereoscope. If the stereoscope is twisted with respect to the flight line, the interpreter loses stereo vision. By lifting the edge of one of the photographs, it is possible to view the image regions near the edges of the photos. Although the stereoscope is a valuable instrument for examining terrain, drainage, and vegetation patterns, it does not provide the detailed measurements within the realm of photogrammetry and more sophisticated instruments.

The stereoscope is only one of several devices designed to present separate images intended for each eye to create the stereo effect. Its way of doing this is known as the *optical separation* technique. Left and right images are presented side by side and an optical device is used to separate the analyst's view of the left and right images. The *red/blue anaglyph* presents images intended for each eye in separate colors, blues for the left eye, reds for the right eye, and shades of magenta for those portions of the image common to both eyes. The analyst views the image using special glasses with a red lens for the left eye and blue for the right eye. The colored lenses cause the image intended for the other eye to blend into the background; the image intended for its own eye will appear as black. The anaglyph has been widely used for novelties, less often as an analytical device.

The use of *polarized lenses* for stereovision is based on the projection of images for each eye through separate polarizing filters (e.g., horizontal for the left eye, vertical for the right eye). The combined image must be viewed through special glasses that use orthogonal polarizations for the left and right lenses. This technique is one of the most effective means of stereoviewing for instructional and analytical applications. A proprietary variation of this technique (CrystalEyes®) displays left- and right-eye views of a digital image in sequential refresh scans on a monitor, then uses synchronized polarized shutter glasses to channel the correct image to the correct eye. This technique forms the basis for stereographic images in many virtual reality display environments. There are many other techniques that are effective to varying degrees for stereovision (including random dot stereograms, Magic Eye® images, and others)—most are less effective for scientific and analytical applications than stereoscopes and polarized lenses.

5.11. Data Transfer

Analysts often need to transfer information from one map or image to another to ensure accurate placement of features, to update superseded information, or to bring several kinds of information into a common format. Traditionally, these operations have been accomplished by optical projection of maps or images onto a working surface, from which they could be traced onto an overlay that registered to another image. Manipulation of the optical system permitted the operator to change scale and to selectively enlarge or reduce portions of the image to correct for tilt and other geometric errors. As digital analysis has become more important, such devices are designed to digitize imagery and match them to other data, with computational adjustments for positional errors.

5.12. Digital Photointerpretation

Increasing use of digital photography and softcopy photogrammetry (Section 3.10) has blurred a previously distinct separation between manual and digital photointerpretation. Analyses that previously were conducted by visual examination of photographic prints or transparencies can now be completed by examination of digital images viewed on computer screens. Analysts record the results of their interpretations as onscreen annotations, using the mouse and cursor to outline and label images. Figure 5.17 illustrates a digital record of interpreted boundaries recorded by onscreen digitization (left) and the outlines shown without the image backdrop (center). The right-hand image shows an enlargement of a portion of the labeled region, illustrating the raster structure of the image and the boundaries.

Some systems employ photogrammetric software to project image detail in its correct planimetric location, without the positional or scale errors that might be present in the original imagery. Further, the digital format enables the analyst to easily manipulate image contrast to improve interpretability of image detail. Digital photogrammetric workstations (Figure 5.18), often based on the usual PC or UNIX operating systems, can accept scanned film imagery, airborne digital imagery, or digital satellite data. The full

FIGURE 5.17. A digital record of image interpretation, showing the outlines as traced by the analyst using onscreen digitization (left), the outlines without the image backdrop (center), and a detail of the raster structure of the digitized outlines (right).

range of photogrammetric processes can be implemented digitally, including triangulation, compilation of digital terrain models (DTMs), feature digitization, construction of orthophotos, mosaics, and flythroughs. Analysts can digitize features onscreen ("heads-up" digitization), using the computer mouse, to record and label features in digital format.

5.13. Image Scale Calculations

Scale is a property of all images. Knowledge of image scale is essential for making measurements from images and for understanding the geometric errors present in all remotely sensed images. *Scale* is an expression of the relationship of the *image distance* between two points and the *actual distance* between the two corresponding points on the ground. This relationship can be expressed in several ways.

The *word statement* sets a unit distance on the map or photograph equal to the correct corresponding distance on the ground—for example, "One inch equals one mile" or, just as correctly, "One centimeter equals five kilometers." The first unit in the statement in the expression specifies the map distance; the second, the corresponding ground distance. A second method of specifying scale is the *bar scale*, which simply labels a line with subdivisions that show ground distances. The third method, the *representative fraction* (RF), is more widely used and often forms the preferred method of reporting image

FIGURE 5.18. Digital photogrammetric workstation. The operator is depicted using polarizing stereovision glasses.

scale. The RF is the ratio between image distance and ground distance. It usually takes the form "1:50,000" or "1/50,000," with the numerator set equal to 1 and the denominator equal to the corresponding ground distance.

The RF has meaning in any unit of length as long as both the numerator and the denominator are expressed in the same units. Thus, "1:50,000" can mean "1 in. on the image equals 50,000 in. on the ground" or "1 cm on the image equals 50,000 cm on the ground." A frequent source of confusion is converting the denominator into the larger units that we find more convenient to use for measuring large ground distance. With metric units, the conversion is usually simple; in the example given above, it is easy to see that 50,000 cm is equal to 0.50 km and that 1 cm on the map represents 0.5 km on the ground. With English units, the same process is not quite so easy. It is necessary to convert inches to miles to derive "1 in. equals 0.79 mi." from 1:50,000. For this reason, it is useful to know that 1 mi. equals 63,360 in. Thus, 50,000 in. is equal to 50,000/63,360 = 0.79 mi.

A typical scale problem requires estimation of the scale of an individual photograph. One method is to use the focal length and altitude method (Figure 5.19):

$$RF = \frac{\text{Focal length}}{\text{Altitude}} \qquad \text{(Eq. 5.1)}$$

Both values must be expressed in the same units. Thus, if a camera with a 6-in. focal length is flown at 10,000 ft., the scale is 0.5/10,000 = 1:20,000. (Altitude always specifies the flying height above the terrain, *not* above sea level.) Because a given flying altitude is seldom the exact altitude at the time the photography was done, and because of the several sources that contribute to scale variations within a given photograph (Chapter 3), we must always regard the results of such calculations as an approximation of the scale of any specific portion of the image. Often such values are referred to as the "nominal"

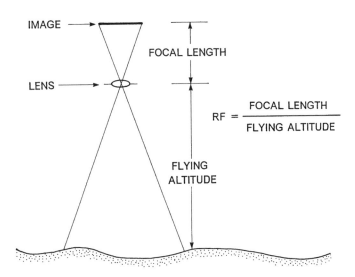

FIGURE 5.19. Estimating image scale by focal length and altitude.

scale of an image, meaning that it is recognized that the stated scale is an approximation and that image scale will vary within any given photograph.

A second method is the use of a *known ground distance*. We identify two points at the same elevation on the aerial photograph that are also represented on a map. For example, in Figure 5.20, the image distance between points *A* and *B* is measured to be approximately 2.2 in. (5.6 cm). From the map, the same distance is determined to correspond to a ground distance of 115,000 in. (about 1.82 mi.). Thus the scale is found to be:

$$RF = \frac{\text{Focal length}}{\text{Ground distance}} = \frac{2.2 \text{ in.}}{1.82 \text{ mi.}} = \frac{2.2 \text{ in.}}{115,000 \text{ in.}} = \frac{1}{52,273} \qquad \text{(Eq. 5.2)}$$

In instances in which accurate maps of the area represented on the photograph may not be available, the interpreter may not know focal length and altitude. Then an approximation of image scale can be made if it is possible to identify an object or feature of known dimensions. Such features might include a football field or baseball diamond; measurement of a distance from these features as they are shown on the image provides the "image distance" value needed to use the relationship given above. The "ground distance" is derived from our knowledge of the length of a football field or the distance between bases on a baseball diamond. Some photointerpretation manuals provide tables of standard dimensions of features commonly observed on aerial images, including sizes of athletic fields (soccer, field hockey, etc.), lengths of railroad boxcars, distances between telephone poles, and so on, as a means of using the known ground distance method.

FIGURE 5.20. Measurement of image scale using a map to derive ground distance.

A second kind of scale problem is the use of a known scale to measure a distance on the photograph. Such a distance might separate two objects on the photograph but not be represented on the map, or the size of a feature may have changed since the map was compiled. For example, we know that image scale is 1:15,000. A pond not shown on the map is measured on the image as 0.12 in. in width. Therefore, we can estimate the actual width of the pond to be:

$$\frac{1}{15,000} = \frac{\text{Image distance}}{\text{Ground distance}} \qquad \text{(Eq. 5.3)}$$

$$\frac{0.12 \text{ in.}}{\text{Unknown GD}} = \frac{\text{Image distance}}{\text{Ground distance}}$$

$$\text{GD} = 0.12 \times 15,000 \text{ in.}$$

$$\text{GD} = 1,800 \text{ in., or } 150 \text{ ft.}$$

This example can illustrate two other points. First, because image scale varies throughout the image, we cannot be absolutely confident that our distance for the width of the pond is accurate; it is simply an estimate, unless we have high confidence in our measurements and in the image scale at this portion of the photo. Second, measurements of short image distances are likely to have errors due simply to our inability to make accurate measurements of very short distances (e.g., the 0.12-in. distance measured above). As distances become shorter, our errors constitute a greater proportion of the estimated length. Thus an error of 0.005 in. is 0.08% of a distance of 6 in. but 4% of the distance of 0.12 in. mentioned above. Thus the interpreter should exercise a healthy skepticism regarding measurements made from images unless he or she has taken great care to ensure maximum accuracy and consistency.

5.14. Summary

Image interpretation was once practiced entirely within the realm of photographic prints and transparencies, using equipment and techniques outlined in the preceding sections. As digital analyses have increased in significance, so has the importance of interpretation of imagery presented on computer displays. Although such interpretations are based on the same principles outlined here for traditional imagery, digital data have their own characteristics that require special treatment in the context of visual interpretation.

Despite the increasing significance of digital analysis in all aspects of remote sensing, image interpretation still forms a key component in the way that humans understand images. Analysts must evaluate imagery, either as paper prints or as displays on a computer monitor, using the skills outlined in this chapter. The fundamentals of manual image interpretation were developed for application to aerial photographs at an early date in the history of aerial survey, although it was not until the 1940s and 1950s that they were formalized in their present form. Since then, these techniques have been applied, without substantial modification, to other kinds of remote sensing imagery. As a result, we have a long record of experience in their application and comprehensive knowledge of their advantages and limitations.

Interesting questions remain. In what ways might image interpretation skills be modified in the context of interpretation using computer monitors? What new skills might be necessary? How have analysts already adjusted to new conditions? How might equipment and software be improved to facilitate interpretation in this new context?

5.15. Some Teaching and Learning Resources

- Introduction to Photo Interpretation
  www.youtube.com/watch?v=LlBDGBopt_g
  This 1955 film has separate sections addressing photointerpretation for hydrology, soils, geology, and forestry. Although the production methods and presentation are clearly dated, many of the basic principles are effectively presented. Many viewers can safely skip the introduction, which concludes at about 4:30.

- Map and Compass Basics: Understanding Map Scale
  www.youtube.com/watch?v=jC1w2jb13GQ

- Map Reading: Understanding Scale
  www.youtube.com/watch?v=93xYDoEA7CQ&feature=related

- 3D-Test #2 - yt3d:enable=true
  www.youtube.com/watch?v=oLfXZs0dyWcd
  [intended for viewing with blue and red anaglyph glasses]

Review Questions

1. A vertical aerial photograph was acquired using a camera with a 9-in. focal length at an altitude of 15,000 ft. Calculate the nominal scale of the photograph.

2. A vertical aerial photograph shows two objects to be separated by 63/4 in. The corresponding ground distance is 91/2 mi. Calculate the nominal scale of the photograph.

3. A vertical aerial photograph shows two features to be separated by 4.5 in. A map at 1:24,000 shows the same two features to be separated by 9.3 in. Calculate the scale of the photograph.

4. Calculate the area represented by a 9 in. × 9 in. vertical aerial photograph taken at an altitude of 10,000 ft. using a camera with a 6-in. focal length.

5. You plan to acquire coverage of a county using a camera with 6-in. focal length and a 9 in. × 9 in. format. You require an image scale of 4 in. equal to 1 mi., 60% forward overlap, and sidelap of 10%. Your county is square in shape, measuring 15.5 mi. on a side. How many photographs are required? At what altitude must the aircraft fly to acquire these photos?

6. You have a flight line of 9 in. × 9 in. vertical aerial photographs taken by camera

with a 9-in. focal length at an altitude of 12,000 ft. above the terrain. Forward overlap is 60%. Calculate the distance (in miles) between ground nadirs of successive photographs.

7. You require complete stereographic coverage of your study area, which is a rectangle measuring 1.5 mi. × 8 mi. How many 9 in. × 9 in. vertical aerial photographs at 1:10,000 are required?

8. You need to calculate the scale of a vertical aerial photograph. Your estimate of the ground distance is 2.8 km. Your measurement of the corresponding image distance is 10.4 cm. What is your estimate of the image scale?

9. You have very little information available to estimate the scale of a vertical aerial photograph, but you are able to recognize a baseball diamond among features in an athletic complex. You use a tube magnifier to measure the distance between first and second base, which is 0.006 ft. What is your estimate of the scale of the photo?

10. Assume you can easily make an error of 0.001 in your measurement for Question 9. Recalculate the image scale to estimate the range of results produced by this level of error. Now return to Question 3 and assume that the same measurement error applies (do not forget to consider the different measurement units in the two questions). Calculate the effect on your estimates of the image scale. The results should illustrate why it is always better whenever possible to use long distances to estimate image scale.

11. Visual Search Exercise. [see pp. 154–155]

12. Identification Exercise. [see p. 156]

References

Avery, T. E., and G. L. Berlin. 2003. *Fundamentals of Remote Sensing and Airphoto Interpretation* (6th ed.). New York: Macmillan, 540 pp.

Brugioni, D. 1991. Eyeball to Eyeball: The Inside History of the Cuban Missile Crisis. New York: Random House, 622 pp.

Brugioni, D. 1996. The Art and Science of Photoreconnaissance. *Scientific American*, Vol. 274, pp. 78–85.

Campbell, J. B. 2005. Visual Interpretation of Aerial Imagery. Chapter 10 in *Remote Sensing for GIS Managers* (S. Aronoff, ed.). Redlands, CA: ESRI Press, pp. 259–283.

Campbell, J. B. 2010. Information Extraction from Remotely Sensed Data. Chapter 19 in *Manual of Geospatial Science and Technology* (J. D. Bossler, ed.). London: Taylor & Francis, pp. 363–389.

Cihlar, J., and R. Protz. 1972. Perception of Tone Differences from Film Transparencies. *Photogrammetria*, Vol. 8, pp. 131–140.

Coburn, C., A. Roberts, and K. Bach. 2001. Spectral and Spatial Artifacts from the Use of Desktop Scanners for Remote Sensing. *International Journal of Remote Sensing*, Vol. 22, pp. 3863–3870.

Coiner, J. C., and S. A. Morain. 1972. SLAR Image Interpretation Keys for Geographic Analysis (Technical Report 177-19). Lawrence, KS: Center for Research, Inc., 110 pp.

11. **Visual Search Exercise.** This exercise develops skills in matching pasterns and developing a sense of spatial context in for image interpretation. Examine the each image on the facing page, and match it to the correct location on the source image shown below, using the coordinates marked at the edge of the source image. As an example, image 1 has been identified.

Answers:

1. <u>B-3</u>   3. ____   5. ____   7. ____   9. ____   11. ____
2. ____   4. ____   6. ____   8. ____   10. ____   12. ____

12. **Identification exercise.** Identify the principal features depicted in each image, using the elements of image interpretation listed in the text. Be prepared to identify the key elements important for each image.

Colwell, R. N. (ed.). 1960. *Manual of Photographic Interpretation.* Falls Church, VA: American Society of Photogrammetry, 868 pp.

Erdman, C., K. Riehl, L. Mayer, J. Leachtenauer, E. Mohr, J. Odenweller, R. Simmons, and D. Hothem. 1994. Quantifying Multispectral Imagery Interpretability. In *International Symposium on Spectra Sensing Research,* Vol. 1, pp. 468–476. Alexandria, VA: U.S. Corps of Engineers.

Haack, B., and S. Jampoler. 1995. Colour Composite Comparisons for Agricultural Assessments. *International Journal of Remote Sensing,* Vol. 16, pp. 1589–1598.

Lillesand, T., R. Keifer, and J. W. Chipman. 2008. *Remote Sensing and Image Interpretation* (6th ed.). New York: Wiley, 756 pp.

Lueder, D. R. 1959. *Aerial Photographic Interpretation: Principles and Applications.* New York: McGraw-Hill, 462 pp.

Paine, D. P., and J. D. Kiser. 2003. *Aerial Photography and Image Interpretation.* New York: Wiley, 648 pp.

Philipson, W. R. (ed.). 1996. *Manual of Photographic Interpretation* (2nd ed.). Bethesda, MD: American Society for Photogrammetry and Remote Sensing, 689 pp.

Wolf, P. R. 1974. *Elements of Photogrammetry.* New York: McGraw-Hill, 562 pp.

CHAPTER SIX

Land Observation Satellites

6.1. Satellite Remote Sensing

Today many corporations and national governments operate satellite remote sensing systems specifically designed for observation of the Earth's surface to collect information concerning topics such as crops, forests, water bodies, land use, cities, and minerals. Satellite sensors offer several advantages over aerial platforms: They can provide a synoptic view (observation of large areas in a single image), fine detail, and systematic, repetitive coverage. Such capabilities are well suited to creating and maintaining a worldwide cartographic infrastructure and to monitoring changes in the many broad-scale environmental issues that the world faces today, to list two of many pressing concerns.

Because of the large number of satellite observation systems in use and the rapid changes in their design, this chapter cannot list or describe all systems currently in use or planned. It can, however, provide readers with the basic framework they need to understand key aspects of Earth observation satellites in general as a means of preparing readers to acquire knowledge of specific satellite systems as they become available. Therefore, this chapter outlines essential characteristics of the most important systems—past, present, and future—as a guide for understanding other systems not specifically discussed here.

Today's land observation satellites have evolved from earlier systems. The first Earth observation satellite, the Television and Infrared Observation Satellite (TIROS), was launched in April 1960 as the first of a series of experimental weather satellites designed to monitor cloud patterns. TIROS was the prototype for the operational programs that now provide meteorological data for daily weather forecasts throughout the world. Successors to the original TIROS vehicle have seen long service in several programs designed to acquire meteorological data. Although data from meteorological satellites have been used to study land resources (to be discussed in Chapters 17, 20, and 21), this chapter focuses on satellite systems specifically tailored for observation of land resources, mainly via passive sensing of radiation in the visible and near infrared regions of the spectrum. Because of the complexity of this topic, the multiplicity of different systems, and the frequency of organizational changes, any attempt to survey the field is likely to be incomplete or out of date. The reader is advised to use this account as a framework and to consult online sources for up-to-date details.

6.2. Landsat Origins

Early meteorological sensors had limited capabilities for land resources observation. Although their sensors were valuable for observing cloud patterns, most had coarse spatial resolution, so they could provide only rudimentary detail concerning land resources. Landsat ("land satellite") was designed in the 1960s and launched in 1972 as the first satellite tailored specifically for broad-scale observation of the Earth's land areas—to accomplish for land resource studies what meteorological satellites had accomplished for meteorology and climatology. Today the Landsat system is important both in its own right—as a remote sensing system that has contributed greatly to Earth resources studies—and as an introduction to similar land observation satellites operated by other organizations.

Landsat was proposed by scientists and administrators in the U.S. government who envisioned application of the principles of remote sensing to broad-scale, repetitive surveys of the Earth's land areas. (Initially, Landsat was known as the "Earth Resources Technology Satellite," "ERTS" for short.) The first Landsat sensors recorded energy in the visible and near infrared spectra. Although these regions of the spectrum had long been used for aircraft photography, it was by no means certain that they would also prove practical for observation of Earth resources from satellite altitudes. Scientists and engineers were not completely confident that the sensors would work as planned, that they would prove to be reliable, that detail would be satisfactory, or that a sufficient proportion of scenes would be free of cloud cover. Although many of these problems were encountered, the feasibility of the basic concept was demonstrated, and Landsat became the model for similar systems now operated by other organizations.

The Landsat system consists of spacecraft-borne sensors that observe the Earth and then transmit information by microwave signals to ground stations that receive and process data for dissemination to a community of data users. Early Landsat vehicles carried two sensor systems: the return beam vidicon (RBV) and the multispectral scanner subsystem (MSS) (Table 6.1). The RBV was a camera-like instrument designed to provide, relative to the MSS, high spatial resolution and geometric accuracy but lower spectral and radiometric detail. That is, positions of features would be accurately represented

TABLE 6.1. Landsat Missions

Satellite	Launched	End of service[a]	Principal sensors[b]
Landsat 1	23 July 1972	6 January 1978	MSS, RBV
Landsat 2	22 January 1975	25 January 1982	MSS, RBV
Landsat 3	5 March 1978	3 March 1983	MSS, RBV
Landsat 4	16 July 1982	[c]	TM, MSS
Landsat 5	1 March 1984		TM, MSS
Landsat 6	5 October 1993	Lost at launch	ETM
Landsat 7	15 April 1999	[d]	ETM+

Note. See *http://geo.arc.nasa.gov/sge/landsat/lpchron.html* for a complete chronology.

[a]Satellite systems typically operate on an intermittent or standby basis for considerable periods prior to formal retirement from service.

[b]Sensors are discussed in the text. MSS, multispectral scanner subsystem; RBV, return beam vidicon; TM, thematic mapper; ETM, enhanced TM; ETM+, enhanced TM plus.

[c]Transmission of TM data failed in August 1993.

[d]Malfunction of TM scan line corrector has limited the quality of imagery since May 2003.

but without fine detail concerning their colors and brightnesses. In contrast, the MSS was designed to provide finer detail concerning spectral characteristics of the Earth but less positional accuracy. Because technical difficulties restricted RBV operation, the MSS soon became the primary Landsat sensor. A second generation of Landsat vehicles (Landsat 4 and 5) added the thematic mapper (TM)—a more sophisticated version of the MSS. The current family of Landsats carry ETM+, an advanced version of the thematic mapper (discussed later in the chapter).

6.3. Satellite Orbits

Satellites are placed into orbits tailored to match the objectives of each satellite mission and the capabilities of the sensors they carry. For simplicity, this section describes *normal* orbits, based on the assumption that the Earth's gravitational field is spherical, although in fact satellites actually follow *perturbed* orbits, due in part to distortion of the Earth's gravitational field by the Earth's *oblate* shape (flattened at the poles, and bulging at the equator) and in part to lunar and solar gravity, tides, solar wind, and other influences.

A normal orbit forms an ellipse with the center of the Earth at one focus, characterized by an *apogee* (A; point farthest from the Earth), *perigee* (P; point closest to the Earth), *ascending node* (AN; point where the satellite crosses the equator moving south to north), and *descending node* (DN; point where the satellite crosses the equator passing north to south). For graphical simplicity, the *inclination* (i) is shown in Figure 6.1 as the angle that a satellite track forms with respect to the equator at the descending node. (More precisely, the inclination should be defined as the angle between the Earth's axis at the North Pole and a line drawn perpendicular to the plane of the satellite orbit, viewed such that the satellite follows a counterclockwise trajectory.)

The time required for a satellite to complete one orbit (its period) increases with altitude. At an altitude of about 36,000 km, a satellite has the same period as the Earth's surface, so (if positioned in the equatorial plane) it remains stationary with respect to the Earth's surface—it is in a *geostationary* orbit. Geostationary orbits are ideal for mete-

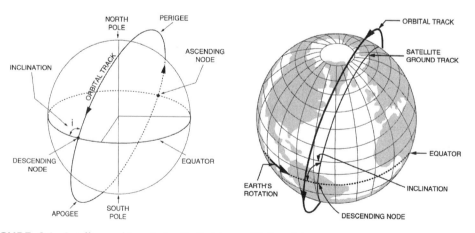

FIGURE 6.1. Satellite orbits. Left: Definitions. Right: Schematic representation of a sun-synchronous orbit.

orological or communications satellites designed to maintain a constant position with respect to a specific region on the Earth's surface. Earth observation satellites, however, are usually designed to satisfy other objectives.

Ideally, all remotely sensed images acquired by satellite would be acquired under conditions of uniform illumination, so that brightnesses of features within each scene would reliably indicate conditions on the ground rather than changes in the conditions of observation. In reality, brightnesses recorded by satellite images are not directly indicative of ground conditions because differences in latitude, time of day, and season lead to variations in the nature and intensity of light that illuminates each scene.

Sun-synchronous orbits are designed to reduce variations in illumination by systematically moving (*precessing*) the orbital track such that it moves westward 360° each year. Illumination observed under such conditions varies throughout the year, but repeats on a yearly basis. The *hour angle* (*h*) describes the difference in longitude between a point of interest and that of the direct solar beam (Figure 6.2). The value of *h* can be found using the formula

$$h = [(GMT - 12.0) \times 15] - \text{longitude} \qquad \text{(Eq. 6.1)}$$

where GMT is Greenwich mean time and *longitude* is the longitude of the point in question. Because *h* varies with longitude, to maintain uniform local sun angle, it is necessary to design satellite orbits that acquire each scene at the same local sun time. Careful selection of orbital height, eccentricity, and inclination can take advantage of the gravitational effect of the Earth's equatorial bulge to cause the plane of the satellite's orbit to rotate with respect to the Earth to match the seasonal motion of the solar beam. That is, the nodes of the satellite's orbit will move eastward about 1° each day, so that over a year's time the orbit will move through the complete 360° cycle. A satellite placed in a sun-synchronous orbit will observe each part of the Earth within its view at the same local

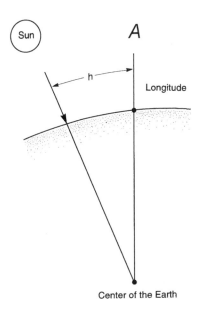

FIGURE 6.2. Hour angle. The hour angle (*h*) measures the difference in longitude between a point of interest (*A*) of known longitude and the longitude of the direct solar beam.

sun time each day (i.e., constantly), thereby removing time of day as a source of variation in illumination.

Although the optimum local sun time varies with the objectives of each project, most Earth observation satellites are placed in orbits designed to acquire imagery between 9:30 and 10:30 a.m. local sun time—a time that provides an optimum trade-off between ideal illumination for some applications and time of minimum cloud cover in tropical regions. Although it may initially seem that the characteristics of satellite orbits should be very stable and precisely known, in fact they are subject to numerous effects that disturb actual orbits and cause them to deviate from their idealized forms. Uncertainties in orbital path, timing, and orientation can lead to significant errors in estimation of satellite position, in aiming the sensor, and other variables that determine the geometric accuracy of an image. For example, a hypothetical satellite in an equatorial orbit that has a pointing (orientation) error of 1° will not be aimed at the intended point on the Earth's surface. The pointing error can be estimated as (altitude) × sin (angle). For an altitude of 800 km, the 1° pointing error could create a 14-km positional error in the location of a point on an image. Although 1° might seem to be a small error, the 14 km result is obviously far too large for practical purposes, so it is clear that pointing errors for remote sensing satellites must be much smaller than 1°. (This example does not include contributions from other effects, such as uncertainties in knowledge of the orbital path, or in time, which determines the satellite's position along the orbital path.)

6.4. The Landsat System

Although the first generation of Landsat sensors is no longer in service, those satellites acquired a large library of images that are available as a baseline reference of environmental conditions for land areas throughout the world. Therefore, knowledge of these early Landsat images is important both as an introduction to later satellite systems and as a basis for work with the historical archives of images from Landsats 1, 2, and 3.

From 1972 to 1983, various combinations of Landsats 1, 2, and 3 orbited the Earth in sun-synchronous orbits every 103 minutes—14 times each day. After 252 orbits—completed every 18 days—Landsat passed over the same place on the Earth to produce repetitive coverage (Figure 6.3). When two satellites were both in service, their orbits were tailored to provide repetitive coverage every 9 days. Sensors were activated to acquire images only at scheduled times, so this capability was not always used. In addition, equipment malfunctions and cloud cover sometimes prevented acquisition of planned coverage.

As a function of the Earth's rotation on its axis from west to east, each successive north-to south pass of the Landsat platform was offset to the west by 2,875 km (1,786 mi.) at the equator (Figure 6.4). Because the westward longitudinal shift of adjacent orbital tracks at the equator was approximately 159 km (99 mi.), gaps between tracks were incrementally filled during the 18-day cycle. Thus, on Day 2, orbit number 1 was displaced 159 km to the west of the path of orbit number 1 on Day 1. On the 18th day, orbit number 1 was identical to that of orbit number 1, Day 1. The first orbit on the 19th day coincided with that of orbit number 1 on Day 2, and so on. Therefore, the entire surface of the Earth between 81° N and 81° S latitude was subject to coverage by Landsat sensors once every 18 days (every 9 days, if two satellites were in service).

FIGURE 6.3. Coverage cycle, Landsats 1, 2, and 3. Each numbered line designates a north-east-to-southwest pass of the satellite. In a single 24-hour interval the satellite completes 14 orbits; the first pass on the next day (orbit 15) is immediately adjacent to pass 1 on the preceding day.

Support Subsystems

Although our interest here is primarily focused on the sensors that these satellites carried, it is important to briefly mention the *support subsystems*, units that are necessary to maintain proper operation of the sensors. Although this section refers specifically to the Landsat system, all Earth observation satellites require similar support systems. The *attitude control subsystem* (ACS) maintained orientation of the satellite with respect to the Earth's surface and with respect to the orbital path. The *orbit adjust subsystem* (OAS) maintained the orbital path within specified parameters after the initial orbit was attained. The OAS also made adjustments throughout the life of the satellite to maintain the planned repeatable coverage of imagery. The *power subsystem* supplied electrical

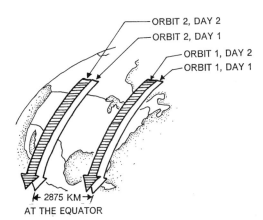

FIGURE 6.4. Incremental increases in Landsat 1 coverage. On successive days, orbital tracks begin to fill in the gaps left by the displacement of orbits during the preceding day. After 18 days, progressive accumulation of coverage fills in all gaps left by coverage acquired on Day 1.

power required to operate all satellite systems by means of two solar array panels and eight batteries. The batteries were charged by energy provided by the solar panels while the satellite was on the sunlit side of the Earth, then provided power when the satellite was in the Earth's shadow. The *thermal control subsystem* controlled the temperatures of satellite components by means of heaters, passive radiators (to dissipate excess heat), and insulation. The *communications and data-handling subsystem* provided microwave communications with ground stations for transmitting data from the sensors, commands to satellite subsystems, and information regarding satellite status and location.

Data from sensors were transmitted in digital form by microwave signal to ground stations equipped to receive and process data. Direct transmission from the satellite to the ground station as the sensor acquired data was possible only when the satellite had direct line-of-sight view of the ground antenna (a radius of about 1,800 km from the ground stations). In North America, stations at Greenbelt, Maryland; Fairbanks, Alaska; Goldstone, California; and Prince Albert, Saskatchewan, Canada, provided this capability for most of the United States and Canada. Elsewhere a network of ground stations has been established over a period of years through agreements with other nations. Areas outside the receiving range of a ground station could be imaged only by use of the two tape recorders on board each of the early Landsats. Each tape recorder could record about 30 minutes of data; then, as the satellite moved within range of a ground station, the recorders could transmit the stored data to a receiving station. Thus these satellites had, within the limits of their orbits, a capability for worldwide coverage. Unfortunately, the tape recorders proved to be one of the most unreliable elements of the Landsat system, and when they failed the system was unable to image areas beyond the range of the ground stations. These unobservable areas became smaller as more ground stations were established, but there were always some areas that Landsat could not observe. Later satellites were able to avoid this problem by use of communications relay satellites.

Return Beam Vidicon

The RBV camera system generated high-resolution television-like images of the Earth's surface. Its significance for our discussion is that it was intended to apply the remote sensing technology of the day for a spacecraft to use from orbital rather than aircraft altitudes. Thus it provided three spectral channels, in the green, red, and near infrared, to replicate the information conveyed by color infrared film. The RBV was designed to provide a camera-like perspective, using a shutter and a electronic record of the image projected into the focal plane, to provide an image that could be used for application of photogrammetry, in much the same way that photogrammetry was used to analyze aerial photographs acquired at aircraft altitudes.

On Landsats 1 and 2, the RBV system consisted of three independent cameras that operated simultaneously, with each sensing a different segment of the spectrum (Table 6.2). All three instruments were aimed at the same region beneath the satellite, so the images they acquired registered to one another to form a three-band multispectral representation of a 185 km × 170 km ground area, known as a *Landsat scene* (Figure 6.5). This area matched the area represented by the corresponding MSS scene. The RBV shutter was designed to open briefly, in the manner of a camera shutter, to simultaneously view the entire scene. Technical difficulties prevented routine use of the RBV, so attention was then directed to imagery from the other sensor on the Landsat, the MSS, then considered to be an experimental system of yet untested capabilities, but a system that proved

TABLE 6.2. Landsats 1–5 Sensors

Sensor	Band	Spectral sensitivity
	Landsats 1 and 2	
RBV	1	0.475–0.575 μm (green)
RBV	2	0.58–0.68 μm (red)
RBV	3	0.69–0.83 μm (near infrared)
MSS	4	0.5–0.6 μm (green)
MSS	5	0.6–0.7 μm (red)
MSS	6	0.7–0.8 μm (near infrared)
MSS	7	0.8–1.1 μm (near infrared)
	Landsat 3	
RBV		0.5–0.75 μm (panchromatic response)
MSS	4	0.5–0.6 μm (green)
MSS	5	0.6–0.7 μm (red)
MSS	6	0.7–0.8 μm (near infrared)
MSS	7	0.8–1.1 μm (near infrared)
MSS	8	10.4–12.6 μm (far infrared)
	Landsats 4 and 5[a]	
TM	1	0.45–0.52 μm (blue-green)
TM	2	0.52–0.60 μm (green)
TM	3	0.63–0.69 μm (red)
TM	4	0.76–0.90 μm (near infrared)
TM	5	1.55–1.75 μm (mid infrared)
TM	6	10.4–12.5 μm (far infrared)
TM	7	2.08–2.35 μm (mid infrared)
MSS	1	0.5–0.6 μm (green)
MSS	2	0.6–0.7 μm (red)
MSS	3	0.7–0.8 μm (near infrared)
MSS	4	0.8–1.1 μm (near infrared)

[a]On Landsats 4 and 5 MSS bands were renumbered, although the spectral definitions remained the same.

to be highly successful and that over the next decades formed the model for collection of imagery from space.

Multispectral Scanner Subsystem

As a result of malfunctions in the RBV sensors early in the missions of Landsats 1 and 2, the MSS became the primary Landsat sensor. Whereas the RBV was designed to capture images with known geometric properties, the MSS was tailored to provide multispectral data without as much concern for positional accuracy. In general, MSS imagery and data were found to be of good quality—indeed, much better than many expected—and clearly demonstrated the merits of satellite observation for acquiring Earth resources data. The economical routine availability of MSS digital data has formed the foundation for a sizeable increase in the number and sophistication of digital image processing capabilities available to the remote sensing community. A version of the MSS was placed on Landsats 4 and 5; later systems have been designed with an eye toward maintaining the continuity of MSS data.

The MSS (Figure 6.6) is a scanning instrument utilizing a flat oscillating mirror

FIGURE 6.5. Schematic diagram of a Landsat scene.

to scan from west to east to produce a ground swath of 185 km (100 nautical mi.) perpendicular to the orbital track. The satellite motion along the orbital path provides the along-track dimension to the image. Solar radiation reflected from the Earth's surface is directed by the mirror to a telescope-like instrument that focuses the energy onto fiber optic bundles located in the focal plane of the telescope. The fiber optic bundles then transmit energy to detectors sensitive to four spectral regions (Table 6.2).

Each west-to-east scan of the mirror covers a strip of ground approximately 185 km long in the east–west dimension and 474 m wide in the north–south dimension. The 474-m distance corresponds to the forward motion of the satellite during the interval required for the west-to-east movement of the mirror and its inactive east-to-west retrace

FIGURE 6.6. Schematic diagram of the Landsat multispectral scanner.

to return to its starting position. The mirror returns to start another active scan just as the satellite is in position to record another line of data at a ground position immediately adjacent to the preceding scan line. Each motion of the mirror corresponds to six lines of data on the image because the fiber optics split the energy from the mirror into six contiguous segments.

The *instantaneous field of view* (IFOV) of a scanning instrument can be informally defined as the ground area viewed by the sensor at a given instant in time. The nominal IFOV for the MSS is 79 m × 79 m. Slater (1980) provides a more detailed examination of MSS geometry that shows the IFOV to be approximately 76 m square (about 0.58 ha, or 1.4 acres), although differences exist between Landsats 1, 2, and 3 and within orbital paths as satellite altitude varies. The brightness from each IFOV is displayed on the image as a pixel formatted to correspond to a ground area said to be approximately 79 m × 57 m in size (about 0.45 ha, or 1.1 acres). In everyday terms, the ground area corresponding to an MSS pixel can be said to be somewhat less than that of a U.S. football field (Figure 6.7).

For the MSS instruments on board Landsat 1 and 2, the four spectral channels were located in the green, red, and infrared portions of the spectrum:

- *Band 1*: 0.5–0.6 µm (green)
- *Band 2*: 0.6–0.7 µm (red)
- *Band 3*: 0.7–0.8 µm (near infrared)
- *Band 4*: 0.8–1.1 µm (near infrared)

(Originally, the four MSS bands were designated as Bands 4, 5, 6, and 7. To maintain continuity of band designations, later band designations were changed, even though the spectral designations remained consistent.)

The Landsat 3 MSS included an additional band in the far infrared from 10.4 to 12.6 µm. Because this band included only two detectors, energy from each scan of the mirror was subdivided into only two segments, each 234 m wide. Therefore, the IFOV for the thermal band was 234 m × 234 m, much coarser than MSS images for the other bands.

Although the MSS has now been replaced by subsequent systems, it is significant because it introduces, in rather basic form, concepts used for later systems that have become too complex to be discussed in detail in an introductory text. Further, it has an important historic significance, as the techniques developed by scientists to interpret MSS imagery form the origins of practice of digital image processing, which is now the principal means of examining satellite imagery and a significant portion of imagery from other sources.

6.5. Multispectral Scanner Subsystem

Multispectral Scanner Subsystem Scene

The *MSS scene* is defined as an image representing a ground area approximately 185 km in the east–west (across-track) direction and 170 km in the north–south (along-track) direction (Figure 6.8). The across-track dimension is defined by the side-to-side motion of the MSS; the along-track dimension is defined by the forward motion of the satellite along its orbital path. If the MSS were operated continuously for an entire descending

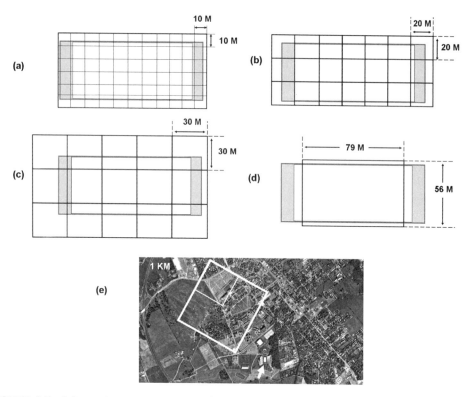

FIGURE 6.7. Schematic representations of spatial resolutions of selected satellite imaging systems. Spatial detail of satellite imagery varies greatly. This diagram attempts to depict the relative sizes of pixels with reference to the dimensions of a U.S. football field as an everyday reference. (a) SPOT HRV, panchromatic mode; (b) SPOT HRV, multispectral mode; (c) Landsat thematic mapper imagery; (d) Landsat MSS imagery (for the MSS, the orientation of the pixel is rotated 90° to match to the usual rendition of a football field; on an MSS image, the narrow ends of MSS pixels are oriented approximately north and south); (e) broad-scale remote sensing satellites depict much coarser spatial resolution, often on the order of a kilometer or so, indicated by the large square, or a quarter of a kilometer, depicted by the smaller square, against the background of an aerial photograph of a small town. For comparison, note in the lower center of image (e) that the white rectangle, (indicated by the white arrow) represents a football field of the size used as a reference in images (a) through (d). Examples (a) through (d) represent resolutions of Landsat-class systems, as discussed in the text. Example (e) represents detail portrayed by broad-scale observation systems. The fine-resolution systems discussed in the text are not represented in this illustration—such systems might represent the football field with hundreds to thousands of pixels.

pass, it would provide a continuous strip of imagery representing an area 185 km wide. The 178-km north–south dimension simply divides this strip into segments of convenient size.

The MSS scene, then, is an array of pixel values (in each of four bands) consisting of about 2,400 scan lines, each composed of 3,240 pixels (Figure 6.8). Although center points of scenes acquired at the same location at different times are intended to register with each other, there is often, in fact, a noticeable shift from one date to another (known

FIGURE 6.8. Diagram of an MSS scene.

as the temporal registration problem) in ground locations of center points due to uncorrected drift in the orbit.

There is a small overlap (about 5%, or 9 km) between scenes to the north and south of a given scene. This overlap is generated by repeating the last few lines from the preceding image, not by stereoscopic viewing of the Earth. Overlap with scenes to the east and west depends on latitude; sidelap will be a minimum of 14% (26 km) at the equator and increases with latitude to 57% at 60°, then to 85% at 80° N and S latitude. Because this overlap is created by viewing the same area of the Earth from different perspectives, the area within the overlap can be viewed in stereo. At high latitudes, this area can constitute an appreciable portion of a scene.

Image Format

MSS data are available in several image formats that have been subjected to different forms of processing to adjust for geometric and radiometric errors. The following section describes some of the basic forms for MSS data, which provide a general model for other kinds of satellite imagery, although specifics vary with each kind of data.

In its initial form a digital satellite image consists of a rectangular array of pixels in each of four bands (Figure 6.8). In this format, however, no compensation has been made for the combined effects of spacecraft movement and rotation of the Earth as the sensor acquires the image (this kind of error is known as *skew*). When these effects are removed, the image assumes the shape of a parallelogram (Figure 6.8). For convenience in recording data, *fill* pixels are added to preserve the correct shape of the image. Fill pixels, of course, convey no information, as they are simply assigned values of zero as necessary to attain the desired shape.

Each of the spectral channels of a multispectral image forms a separate image, each emphasizing landscape features that reflect specific portions of the spectrum (Figures 6.9 and 6.10). These separate images, then, record in black-and-white form the spectral

FIGURE 6.9. Landsat MSS image, band 2. New Orleans, Louisiana, 16 September 1982. Scene ID 40062-15591-2. Image reproduced by permission of EOSAT. (MSS band 2 was formerly designated as band 5; see Table 6.2.)

reflectance in the green, red, and infrared portions of the spectrum, for example. The green, red, and infrared bands can be combined into a single color image (Plate 4), known as a *false-color composite.* The near infrared band is projected onto color film through a red filter, the red band 2 through a green filter, and the green band through a blue filter. The result is a false-color rendition that uses the same assignment of colors used in conventional color infrared aerial photography (Chapter 3). Strong reflectance in the green portion of the spectrum is represented as blue on the color composite, red as green, and infrared as red. Thus living vegetation appears as bright red, turbid water as a blue color, and urban areas as a gray or sometimes pinkish gray.

On photographic prints of MSS images, the annotation block at the lower edge gives essential information concerning the identification, date, location, and characteristics of the image. During the interval since the launch of Landsat 1, the content and form of the annotation block have been changed several times, but some of the basic information can be shown in a simplified form to illustrate key items (Figure 6.11). The *date* has obvious meaning. The *format center* and *ground nadir* give, in degrees and minutes of latitude

FIGURE 6.10. Landsat MSS image, band 4. New Orleans, Louisiana, 16 September 1982. Scene ID 40062-15591-7. Image reproduced by permission of EOSAT. (MSS band 4 was formerly designated as band 7; see Table 6.2.)

and longitude, the ground location of the center point of the image. The spectral band is given in the form "MSS 1" (meaning multispectral scanner, band 1). *Sun angle* and *sun elevation* designate, in degrees, the solar elevation (above the horizon) and the azimuth of the solar beam from true north at the center of the image. Of the remaining items on the annotation block, the most important for most users is the *scene ID*, a unique number that specifies the scene and band. The *scene ID* uniquely specifies any MSS scene, so it is especially useful as a means of cataloging and indexing MSS images, as explained below. MSS imagery can be downloaded from the U.S. Geological Survey's EROS Data Center.

Worldwide Reference System

The *worldwide reference system* (WRS), a concise designation of nominal center points of Landsat scenes, is used to index Landsat scenes by location. The reference system is based on a coordinate system in which there are 233 north–south paths corresponding to orbital tracks of the satellite and 119 rows representing latitudinal center lines of Land-

FIGURE 6.11. Landsat annotation block. The annotation block has been changed several times, but all show essentially the same information.

sat scenes. The combination of a path number and a row number uniquely identifies a nominal scene center (Figure 6.12). Because of the drift of satellite orbits over time, actual scene centers may not match exactly to the path–row locations, but the method does provide a convenient and effective means of indexing locations of Landsat scenes. To outline the concept in its most basic form, this discussion has presented the Landsat 1–3 WRS; note that Landsats 4 through 7 each have WRS systems analogous to those described here, though they differ with respect to detail. Therefore, investigate the specifics of the WRS system you expect to use.

6.6. Landsat Thematic Mapper

Even before Landsat 1 was launched, it was recognized that existing technology could improve the design of the MSS. Efforts were made to incorporate improvements into a new instrument modeled on the basic design of the MSS. Landsats 4 and 5 carried a replacement for MSS known as the *thematic mapper* (TM), which can be considered as an upgraded MSS. In these satellites, both the TM and an MSS were carried on an

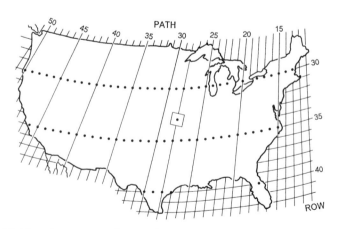

FIGURE 6.12. WRS path–row coordinates for the United States.

improved platform that maintained a high degree of stability in orientation as a means of improving geometric qualities of the imagery.

TM was based on the same principles as the MSS but had a more complex design. It provides finer spatial resolution, improved geometric fidelity, greater radiometric detail, and more detailed spectral information in more precisely defined spectral regions. Improved satellite stability and an OAS were designed to improve positional and geometric accuracy. The objectives of the second generation of Landsat instruments were to assess the performance of the TM, to provide ongoing availability of MSS data, and to continue foreign data reception.

Despite the historical relationship between the MSS and the TM, the two sensors are distinct. Whereas the MSS has four broadly defined spectral regions, the TM records seven spectral bands (Table 6.3). TM band designations do not follow the sequence of the spectral definitions (the band with the longest wavelengths is band 6, rather than band 7) because band 7 was added so late in the design process that it was not feasible to relabel bands to follow a logical sequence.

TM spectral bands were tailored to record radiation of interest to specific scientific investigations rather than the more arbitrary definitions used for the MSS. Spatial resolution is about 30 m (about 0.09 ha, or 0.22 acre) compared with the 76-m IFOV of the MSS. (TM band 6 had a coarser spatial resolution of about 120 m.) The finer spatial resolution provided a noticeable increase (relative to the MSS) in spatial detail recorded

TABLE 6.3. Summary of TM Sensor Characteristics

Band	Resolution	Spectral definition	Some applications[a]
1	30 m	Blue-green, 0.45–0.52 µm	Penetration of clear water; bathymetry; mapping of coastal waters; chlorophyll absorption; distinction between coniferous and deciduous vegetation
2	30 m	Green, 0.52–0.60 µm	Records green radiation reflected from healthy vegetation; assesses plant vigor; reflectance from turbid water
3	30 m	Red, 0.63–0.69 µm	Chlorophyll absorption important for plant-type discrimination
4	30 m	Near infrared, 0.76–0.90 µm	Indicator of plant cell structure; biomass; plant vigor; complete absorption by water facilitates delineation of shorelines
5	30 m	Mid infrared, 1.55–1.75 µm	Indicative of vegetation moisture content; soil moisture mapping; differentiating snow from clouds; penetration of thin clouds
6	120 m	Far infrared, 10.4–12.5 µm	Vegetation stress analysis; soil moisture discrimination; thermal mapping; relative brightness temperature; soil moisture; plant heat stress
7	30 m	Mid infrared, 2.08–2.35 µm	Discrimination of rock types; alteration zones for hydrothermal mapping; hydroxyl ion absorption

[a]Sample applications listed here; these are not the only applications.

by each TM image (Figure 6.7). Digital values are quantized at 8 bits (256 brightness levels), which provide (relative to the MSS) a much larger range of brightness values. These kinds of changes produced images with much finer detail than those of the MSS (Figures 6.13 and 6.14). Each scan of the TM mirror acquired 16 lines of data. Unlike the MSS, the TM scan acquires data as it moves in both the east–west and west–east directions. This feature permitted engineers to design a slower speed of mirror movement, thereby improving the length of time the detectors can respond to brightness in the scene. However, this design required additional processing to reconfigure image positions of pixels to form a geometrically accurate image. TM detectors are positioned in an array in the focal plane; as a result, there may be a slight misregistration of TM bands.

TM imagery is analogous to MSS imagery with respect to areal coverage and organization of data into several sets of multispectral digital values that overlay to form an image. In comparison with MSS images, TM imagery has a much finer spatial and radiometric resolution, so TM images show relatively fine detail of patterns on the Earth's surface.

Use of seven rather than four spectral bands and of a smaller pixel size within the same image area means that TM images consist of many more data values than do MSS images. As a result, each analyst must determine those TM bands that are most likely to

FIGURE 6.13. TM image, band 3. This image shows New Orleans, Louisiana, 16 September 1982. Scene ID 40062-15591-4.

provide the required information. Because the "best" combinations of TM bands vary according to the purpose of each study, season, geographic region, and other factors, a single selection of bands is unlikely to be equally effective in all circumstances.

A few combinations of TM bands appear to be effective for general purpose use. Use of TM bands 2, 3, and 4 creates an image that is analogous to the usual false-color aerial photograph. TM bands 1 (blue–green), 2 (green), and 3 (red) form a natural color composite, approximately equivalent to a color aerial photograph in its rendition of colors. Experiments with other combinations have shown that Bands 2, 4, and 5; 2, 4, and 7; and 3, 4, and 5 are also effective for visual interpretation. Of course, there are many other combinations of the seven TM bands that may be useful in specific circumstances. TM imagery can be downloaded from the USGS's EROS Data Center (see *http://edc.usgs.gov*).

Orbit and Ground Coverage: Landsats 4 and 5

Landsats 4 and 5 were placed into orbits resembling those of earlier Landsats. Sun-synchronous orbits brought the satellites over the equator at about 9:45 a.m., thereby maintaining approximate continuity of solar illumination with imagery from Landsats 1, 2, and 3. Data were collected as the satellite passed northeast to southwest on the sunlit

FIGURE 6.14. TM image, band 4. This image shows New Orleans, Louisiana, 16 September 1982. Scene ID 40062-15591-5.

side of the Earth. The image swath remained at 185 km. In these respects, coverage was compatible with that of the first generation of Landsat systems.

However, there were important differences. The finer spatial resolution of the TM was achieved in part by a lower orbital altitude, which required several changes in the coverage cycle. Earlier Landsats produced adjacent image swaths on successive days. However, Landsats 4 and 5 acquired coverage of adjacent swaths at intervals of 7 days. Landsat 4 completed a coverage cycle in 16 days. Successive passes of the satellite were separated at the equator by 2,752 km; gaps between successive passes were filled in over an interval of 16 days. Adjacent passes were spaced at 172 km. At the equator, adjacent passes overlapped by about 7.6 %; overlap increased as latitude increased.

A complete coverage cycle was achieved in 16 days—233 orbits. Because this pattern differs from earlier Landsat systems, Landsats 4 and later required a new WRS indexing system for labeling paths and rows (Figure 6.15). Row designations remained the same as before, but a new system of numbering paths was required. In all, there are 233 paths and 248 rows, with row 60 positioned at the equator.

6.7. Administration of the Landsat Program

Landsat was originally operated by NASA as a part of its mission to develop and demonstrate applications of new technology related to aerospace engineering. Although NASA had primary responsibility for Landsat, other federal agencies, including the USGS and the National Oceanographic and Atmospheric Administration (NOAA) contributed to the program. These federal agencies operated the Landsat system for many years, but Landsat was officially considered to be "experimental" because of NASA's mission to develop new technology (rather than assume responsibility for routine operations) and because other federal agencies were unwilling to assume responsibility for a program of such cost and complexity.

Over a period of many years, successive presidential administrations and numerous

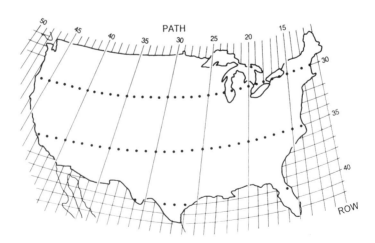

FIGURE 6.15. WRS for Landsats 4 and 5.

sessions of the U.S. Congress debated to define a model for operation of the U. S. land remote sensing policy. Although Landsat 4 was built and launched by NASA, NOAA initially oversaw satellite operations. In 1984, responsibility for Landsat 4 operations was contracted to the Earth Observation Satellite Company (EOSAT). This model was found to be unsatisfactory, and by 1998, the management of Landsat 4 (and Landsat 5) was transferred from NOAA to the USGS; operations were continued by the private sector until mid-2001, when Space Imaging (formerly EOSAT) returned the operations contract to the U.S. Government.

In 1993, when Landsat 6 (the satellite intended to continue Landsat operations into the 1990s and beyond) was lost at launch, the Landsat program was left without a replacement for Landsats 4 and 5. Additional backup instruments had not been constructed, and there was no new program under way to fund, design, construct, and launch a successor system. The instruments on board Landsats 4 and 5 have proved to be remarkably robust, continuing to acquire imagery well beyond their expected service lives. However, as time passed, they experienced a variety of malfunctions that have reduced the quantity and, in some instances, the quality of imagery. The reduced capabilities of the system raised fears of a "data gap"—that is, an interval after the inevitable failures of Landsats 4 and 5 and before operation of a successor system. The USGS has conducted studies to evaluate strategies for filling a data gap using imagery collected by other satellite systems, now in operation, with capabilities analogous to those of TM. Although such a strategy can, of course, provide the required imagery, it would not necessarily be able to maintain continuity with respect to definitions of spectral regions, spatial resolution, and instrument calibration that is required to maintain the integrity of the Landsat archive.

In 2005 NASA was directed to begin development of the Landsat Data Continuity Mission (LDCM) (previous designations include the "Landsat Continuity Mission", and the "Landsat-7 Follow-On"), which will include a successor instrument for the TM, known as the Operational Land Imager (OLI), now planned for launch in December 2012. The OLI can be thought of as a sensor sharing some characteristics of the thematic mapper but designed using current technologies (Table 6.4). Although the OLI does not itself have a thermal channel, the LDCM will include a companion instrument, the thermal infrared sensor (TIRS), with sensitivity in two channels centered at 10.8 μm and 12.0 μm and a spatial resolution of about 100 m (Jhabvala et al 2009).

The principal mission objectives of LDCM are as follows:

TABLE 6.4. LDCM OLI Characteristics

Band	Name	Spectral definition	Spatial resolution
1.	Coastal aerosol	0.443–0.433 μm	30
2.	Blue	0.450–0.515 μm	30
3.	Green	0.450–0.525 μm	30
4.	Red	0.525–0.600 μm	30
5.	NIR	0.845–0.885 μm	30
6.	SWIR1	1.560–1.660 μm	60
7.	SWIR2	2.100–2.300 μm	30
8.	Panchromatic	0.500–0.680 μm	10
9.	Cirrus	1.360–1.390 μm	30

Source: *USGS Fact Sheet 2007-3093.*

- Acquire and archive moderate-resolution (circa 30-m ground sample distance) multispectral image data affording seasonal coverage of the global land mass for a period of no less than 5 years with no credible single-point failures.
- Acquire and archive medium-low-resolution (circa 120-m ground sample distance) thermal image data affording seasonal coverage of the global land mass for a continuous period of not less than 3 years with no credible single-point failures.
- Ensure LDCM data is consistent with data from the earlier Landsat missions, in terms of acquisition geometry, calibration, coverage, spectral definitions, data quality, and data availability, to permit studies of land cover and land use change over multidecadal periods.

6.8. Current Satellite Systems

This outline of early Landsat systems sets the stage for examination of the broader range of systems now in use or scheduled for deployment in the near future. In recent years the number of such systems has increased so rapidly that it is impractical to list them, let alone describe their characteristics. Understanding satellite systems can be easier if we consider them as members of three families of satellites.

The first group consists of *Landsat-like systems*, designed for acquisition of rather broad geographic coverage at moderate levels of detail. Data from these systems have been used for an amazingly broad range of applications, which can be generally described as focused on survey and monitoring of land and water resources. A second group is formed by those satellite observation systems designed to acquire *very-broad-scale images at coarse resolutions*, intended in part to acquire images that can be aggregated to provide continental or global coverage. Such images enable scientists to monitor broad-scale environmental dynamics. Finally, a third family of satellite systems provides *fine detail for small regions* to acquire imagery that might assist in urban planning or design of highway or pipeline routes, for example. Although this categorization is imperfect, it does provide a framework that helps us understand the capabilities of a very large number of satellite systems now available. For systems not discussed here, Stoney (2008) provides a comprehensive catalog of land observation satellite systems that is reasonably up-to-date at the time of this writing.

Landsat-Class Systems

The Landsat model has formed a template for the land remote sensing systems proposed and in use by many other nations. Lauer et al. (1997) listed 33 separate Landsat or Landsat-like systems launched or planned for launch between 1972 and 2007. This number attests to the value of observational systems designed to acquire regional overviews at moderate-to-coarse levels of spatial detail. Landsat has formed the model for these systems with respect to their essential characteristics: technological design, data management, and overall purpose. For example, Landsat established the value of digital data for the general community of remote sensing practitioners, set the expectation that imagery would be available for all customers, and established a model for organizing, archiving, and cataloging imagery. Related systems have followed Landsat's lead in almost every respect.

Landsat 7 and the Enhanced Thematic Mapper Plus

Landsat's flagship sensor, the enhanced thematic mapper plus (ETM+), was placed in orbit in April 1999, with the successful launch of Landsat 7. This system was designed by NASA and is operated by NOAA and USGS. ETM+ is designed to extend the capabilities of previous TMs by adding modest improvements to the TM's original design. In the visible, near infrared, and mid infrared, its spectral channels duplicate those of earlier TMs (Table 6.5). The thermal channel has 60-m resolution, improved from the 120-m resolution of earlier TMs. The ETM+ also has a 15-m panchromatic channel. Swath width remains at 185 km, and the system is characterized by improvements in accuracy of calibration, data transmission, and other characteristics. ETM+ extends the continuity of earlier Landsat data (from both MSS and TM) by maintaining consistent spectral definitions, resolutions, and scene characteristics, while taking advantage of improved technology, calibration, and efficiency of data transmission. Details concerning Landsat 7 and ETM+ are available at the Landsat Data Access Web page (*http://geo.arc.nasa.gov/sge/landsat/daccess.html*) and at the Landsat 7 web page (*http://landsat.gsfc.nasa.gov/*).

In May 2003 the Landsat 7 TM experienced an anomaly in the functioning of the scan-line corrector (SLC), a portion of the sensor that ensures that scan lines are oriented properly in the processed imagery (see Section 6.6). As a result, the images display wedge-shaped gaps near the edges, such that only the central two-thirds of the image presents a continuous view of the scene. Although images can be processed to address the cosmetic dimension of the problem, the instrument is restricted in its ability to fulfill its mission as long as this problem persists. As a result, the aging Landsat 5 TM remains as the only fully functional Landsat system at the time of this writing. The system will be operated in a manner that will maximize its capabilities and minimize the impact of the SLC anomaly. Because there is no program under way to replace the Landsat system in a timely manner, it seems likely that there will be a gap in Landsat converge as the Landsat 5 TM approaches the end of its service. Although systems operated by other nations and other US systems can provide useful imagery in the interim, it seems clear that there will be a loss in the continuity and compatibility of the image archive.

Landsat 7 ETM+ data are distributed to the public from the USGS's EROS Data Center (EDC), the primary Landsat 7 receiving station in the United States:

Customer Services
U.S. Geological Survey
EROS Data Center

TABLE 6.5. Spectral Characteristics for ETM+

Band	Spectral range	Ground resolution
1	0.450–0.515 µm	30 m
2	0.525–0.605 µm	30 m
3	0.630–0.690 µm	30 m
4	0.75–0.90 µm	30 m
5	1.55–1.75 µm	30 m
6	10.4–12.5 µm	60 m
7	2.09–2.35 µm	30 m
Pan	0.52–0.90 µm	15 m

Note. See Table 6.3.

Sioux Falls, SD 57198-0001
Tel: 605-594-6151
Fax: 605-594-6589
E-mail: *custserv@usgs.gov*
http://glovis.usgs.gov

SPOT

SPOTs 1, 2, and 3

SPOT—Le Système pour l'Observation de la Terre (Earth Observation System)—began operation in 1986 with the launch of SPOT 1. SPOT was conceived and designed by the French Centre National d'Etudes Spatiales (CNES) in Paris, in collaboration with other European organizations. SPOT 1, launched in February 1986, was followed by SPOT 2 (January 1990), SPOT 3 (September 1993), SPOT 4 (March 1998), and SPOT 5 (May 2002), which carry sensors comparable to those of SPOT 1.

The SPOT system is designed to provide data for land-use studies, assessment of renewable resources, exploration of geologic resources, and cartographic work at scales of 1:50,000 to 1:100,000. Design requirements included provision for complete world coverage, rapid dissemination of data, stereo capability, high spatial resolution, and sensitivity in spectral regions responsive to reflectance from vegetation.

The SPOT *bus* is the basic satellite vehicle, designed to be compatible with a variety of sensors (Figure 6.16). The bus provides basic functions related to orbit control and stabilization, reception of commands, telemetry, monitoring of sensor status, and the like. The bus, with its sensors, is placed in a sun-synchronous orbit at about 832 km, with a 10:30 a.m. equatorial crossing time. For vertical observation, successive passes occur at 26-day intervals, but because of the ability of SPOT sensors to be pointed off-nadir, successive imagery can be acquired, on the average, at 2½-day intervals. (The exact interval for repeat coverage varies with latitude.) The SPOT payload consists of two identical sensing instruments, a telemetry transmitter, and magnetic tape recorders. The two sensors are known as HRV (high resolution visible) instruments. HRV sensors use pushbroom scanning, based on CCDs (discussed in Chapter 4), which simultaneously images an entire line of data in the cross-track axis (Figure 4.2a). SPOT linear arrays consist of 6,000 detectors for each

FIGURE 6.16. SPOT bus.

scan line in the focal plane; the array is scanned electronically to record brightness values (8 bits; 256 brightness values) in each line. Radiation from the ground is reflected to the two arrays by means of a moveable plane mirror. An innovative feature of the SPOT satellite is the ability to control the orientation of the mirror by commands from the ground—a capability that enables the satellite to acquire oblique images, as described below.

The HRV can be operated in either of two modes. In *panchromatic* (PN) mode, the sensor is sensitive across a broad spectral band from 0.51 to 0.73 µm. It images a 60-km swath with 6,000 pixels per line for a spatial resolution of 10 m (see Figure 6.17a). In this mode the HRV instrument provides fine spatial detail but records a rather broad spectral region.

In the panchromatic mode, the HRV instrument provides coarse spectral resolution but fine (10 m) spatial resolution.

In the other mode, the *multispectral* (XS) configuration, the HRV instrument senses three spectral regions:

- *Band 1*: 0.50–0.59 µm (green)
- *Band 2*: 0.61–0.68 µm (red; chlorophyll absorption)
- *Band 3*: 0.79–0.89 µm (near infrared; atmospheric penetration)

In this mode the sensor images a strip 60 km in width using 3,000 samples for each line at a spatial resolution of about 20 m (see Figure 6.17b). Thus, in the XS mode, the sensor records fine spectral resolution but coarse spatial resolution. The three images from the multispectral mode can be used to form false-color composites, in the manner of CIR, MSS, and TM images (Plate 5). In some instances, it is possible to "sharpen" the lower spatial detail of multispectral images by superimposing them on the fine spatial detail of high-resolution panchromatic imagery of the same area.

With respect to sensor geometry, each of the HRV instruments can be positioned in either of two configurations (Figure 6.17). For *nadir viewing* (Figure 6.17a), both sensors are oriented in a manner that provides coverage of adjacent ground segments. Because

FIGURE 6.17. Geometry of SPOT imagery: (a) Nadir viewing. (b) Off-nadir viewing.

the two 60-km swaths overlap by 3 km, the total image swath is 117 km. At the equator, centers of adjacent satellite tracks are separated by a maximum of only 108 km, so in this mode the satellite can acquire complete coverage of the Earth's surface.

An *off-nadir viewing* capability is possible by pointing the HRV field of view as much as 27° relative to the vertical in 45 steps of 0.6° each (Figure 6.17b) in a plane perpendicular to the orbital path. Off-nadir viewing is possible because the sensor observes the Earth through a pointable mirror that can be controlled by command from the ground. (Note that although mirror orientation can be changed on command, it is not a scanning mirror, as used by the MSS and the TM.) With this capability, the sensors can observe any area within a 950-km swath centered on the satellite track. The pointable mirror can position any given off-nadir scene center at 10-km increments within this swath.

When SPOT uses off-nadir viewing, the swath width of individual images varies from 60 to 80 km, depending on the viewing angle. Alternatively, the same region can be viewed from separate positions (from different satellite passes) to acquire stereo coverage. (Such stereo coverage depends, of course, on cloud-free weather during both passes.) The twin sensors are not required to operate in the identical configuration; that is, one HRV can operate in the vertical mode while the other images obliquely. Using its off-nadir viewing capability, SPOT can acquire repeated coverage at intervals of 1–5 days, depending on latitude. Reception of SPOT data is possible at a network of ground stations positioned throughout the world.

CNES processes SPOT data in collaboration with the Institut Geographique National (IGN). Image archives are maintained by the Centre de Rectification des Images Spatiales (CRIS), also operated by CNES and IGN. Four levels of processing are listed below in ascending order of precision:

- *Level 1*: Basic geometric and radiometric adjustments
  - *Level 1a*: Sensor normalization
  - *Level 1b*: Level 1a processing with the addition of simple geometric corrections
- *Level 2*: Use of ground control points (GCPs) to correct image geometry; no correction for relief displacement
- *Level 3*: Further corrections using digital elevation models.

SPOTs 4 and 5

A principal feature of the SPOT 4 (launched in March 1998) mission is the high-resolution visible and infrared (HRVIR) instrument, a modification of the HRV used for SPOTs 1, 2, and 3. HRVIR resembles the HRV, with the addition of a mid infrared band (1.58–1.75 μm) designed to provide capabilities for geological reconnaissance, for vegetation surveys, and for survey of snow cover. Whereas the HRV's 10-m resolution band covers a panchromatic range of 0.51–0.73 μm, the HRVIR's 10-m band is positioned to provide spectral coverage identical to band 2 (0.61–0.68 μm). In addition, the 10-m band is registered to match data in band 2, facilitating use of the two levels of resolution in the same analysis. SPOT 4 carries two identical HRVIR instruments, each with the ability to point 27° to either side of the ground track, providing a capability to acquire data within a 460-km swath for repeat coverage or stereo. HRVIR's monospectral (M) mode, at 10m resolution, matches HRVIR band 2's spectral range. In multispectral (X) mode, the HRVIR acquires four bands of data (1, 2, 3, and mid infrared) at 20-m resolution. Both

M and X data are compressed onboard the satellite, then decompressed on the ground to provide 8-bit resolution.

SPOT 5 (launched in May 2002) carries an upgraded version of the HRVIR, which acquires panchromatic imagery at either 2.5-m or 5-m resolution and provides a capability for along-track stereo imagery. Stereo imagery at 5-m resolution is intended to provide data for compilation of large-scale topographic maps and data. The new instrument, the *high-resolution geometrical* (HRG), has the flexibility to acquire data using the same bands and resolutions as the SPOT 4 HRVIR, thereby providing continuity with earlier systems. SPOT 5 also includes the high-resolution stereoscopic (HRS) imaging instrument, dedicated to acquiring simultaneous stereopairs of a swath 120 km in the across-track dimension (the width of the observed scene centered on the satellite ground track) and 600 km in the along-track dimension long (the maximum length of a scene). These stereopairs, acquired as panchromatic images, have a spatial resolution of 10 m.

In the United States, SPOT imagery can be purchased from:

SPOT Image Company
14595 Avion Parkway, Suite 500
Chantilly, VA 20151
Tel: 703-715-3100
Fax: 703-715-3120
E-mail: *sales@spot.com*
Website: *www.spotimage.fr/home*

India Remote Sensing

After operating two coarse-resolution remote sensing satellites in the 1970s and 1980s, India began to develop multispectral remote sensing programs in the style of the Landsat system. During the early 1990s, two India remote sensing (IRS) satellites were in service. IRS-1A, launched in 1988, and IRS-1B, launched in 1991, carried the LISS-I and LISS-II pushbroom sensors (Tables 6.6 and 6.7). These instruments collect data in four bands: blue (0.45–0.52 μm), green (0.52–0.59 μm), red (0.62–0.68 μm), and near infrared (0.77–0.86 μm), creating images of 2,400 lines in each band. LISS-I provides resolution of 72.5 m in a 148-km swath, and LISSII has 36.25-m resolution. Two LISS-II cameras acquire data from 74-km-wide swaths positioned within the field of view of LISS-I (Figure 6.18), so that four LISS-II images cover the area imaged by LISS-I, with an overlap of 1.5 km in the cross-track direction and of about 12.76 km in the along-track direction. Repeat coverage is 22 days at the equator, with more frequent revisit capabili-

TABLE 6.6. Spectral Characteristics for LISS-I and LISS-II Sensors (IRS-1A and IRS-1B)

Band	Spectral limits	Resolution LISS-I	Resolution LISS-II
1	Blue-green 0.45–0.52 μm	72.5 m	36.25 m
2	Green 0.52–0.59 μm	72.5 m	36.25 m
3	Red 0.62–0.68 μm	72.5 m	36.25 m
4	Near infrared 0.77–0.86 μm	72.5 m	36.25 m

TABLE 6.7. Spectral Characteristics of a LISS-III Sensor (IRS-1C and IRS-1D)

Band	Spectral limits	Resolution
1[a]	Blue —	
2	Green 0.52–0.59 μm	23 m
3	Red 0.62–0.68 μm	23 m
4	Near infrared 0.77–0.86 μm	23 m
5	Mid infrared 1.55–1.70 μm	70 m

[a]Band 1 is not included in this instrument, although the numbering system from earlier satellites is maintained to provide continuity.

ties at higher latitudes. In the United States, IRS imagery can be purchased from a variety of image services that can be located using the Internet.

The LISS-III instrument is designed for the IRS-1C and IRS-1D missions, launched in 1995 and 1997, respectively. These systems use tape recorders, permitting acquisition of data outside the range of the receiving stations mentioned above. They acquire data in four bands: green (0.52–0.59 μm), red (0.62–0.68 μm), near infrared (0.77–0.86 μm), and shortwave infrared (1.55–1.70 μm). LISS-III provides 23-m resolution for all bands except for the shortwave infrared, which has a 70-m resolution (Table 6.7). Swath width is 142 km for bands 2, 3, and 4 and 148 km in band 5. The satellite provides a capability for 24-day repeat coverage at the equator.

In 2003 India Remote Sensing launched Resourcesat 1, which includes three instruments of interest. The first is LISS-3, as described above. The second is LISS-4, which acquires a 23.9-km image swath in multispectral mode and a 70.3-km swath in panchromatic mode. LISS-4 has a spatial resolution of 5.5 m; and three spectral channels match-

FIGURE 6.18. Coverage diagram for the IRS LISS-II sensors. Subscenes gathered by the LISS-II sensors (right) combine to cover the area imaged by the coarser resolution of the LISS-I instrument (left).

ing to channels 1, 2, and 3 of the LISS-3 instrument. The field of view can be aimed along the across-track dimension to acquire stereoscopic coverage and provide a 5-day revisit capability. The third instrument is the Advanced Wide field Sensor (AWiFS) instrument, which provides imagery of a 760 swath at 56-m resolution, using the same four spectral channels as LISS-3.

IRS's Cartosat-1 (launched in May 2005) uses two panchromatic cameras, aimed fore and aft, to acquire stereoscopic imagery within a 30-km swath in the along-track dimension in a single panchromatic band (0.50-0.85 μm). Its objective is to acquire imagery to support detailed mapping and other cartographic applications at the cadastral level, urban and rural infrastructure development and management, as well as applications in land information systems (LIS) and geographical information systems (GIS).

Broad-Scale Coverage

This class of systems includes satellite systems that provide coarse levels of detail for very large regions. Images collected over a period of several weeks can be used to generate composites that represent large areas of the Earth without the cloud cover that would be present in any single scene. These images have opened a new perspective for remote sensing by allowing scientists to examine topics that require examination at continental or global scale that previously were outside the scope of direct observation.

AVHRR

AVHRR (advanced very-high-resolution radiometer) is a scanning system carried on NOAA's polar orbiting environmental satellites. The first version was placed in service in 1978 primarily as a meteorological satellite to provide synoptic views of weather systems and to assess the thermal balance of land surfaces in varied climate zones. The satellite makes 14 passes each day, viewing a 2,399-km swath. Unlike other many other systems described here, the system provides complete coverage of the Earth from pole to pole.

At nadir, the system provides an IFOV of about 1.1 km (see Figure 6.7). Data presented at that level of detail are referred to as local area coverage (LAC). In addition, a global areal coverage (GAC) dataset is generated by onboard averaging of the full-resolution data. GAC data are formed by the selection of every third line of data in the full-resolution dataset; for each of these lines, four out of every five pixels are used to compute an average value that forms a single pixel in the GAC dataset. This generalized GAC coverage provides pixels of about 4 km × 4 km resolution at nadir. A third AVHRR image product, HRPT (high-resolution picture transmission) data, is created by direct transmission of full-resolution data to a ground receiving station as the scanner collects the data. Like LAC data, the resolution is 1.1 km. (LAC data are stored onboard for later transmission when the satellite is within line of sight of a ground receiving station; HRPT data are directly transmitted to ground receiving stations.)

For recent versions of the AVHRR scanner, the spectral channels are:

- *Band 1*: 0.58–0.68 μm (red; matches TM Band 3)
- *Band 2*: 0.725–1.10 μm (near infrared; matches TM Band 4)
- *Band 3*: 3.55–3.93 μm (mid infrared)
- *Band 4*: 10.3–11.3 μm (thermal infrared)
- *Band 5*: 11.5–12.5 μm (thermal infrared)

Data from several passes are collected to create georeferenced composites that show large regions of the Earth without cloud cover. As is discussed in Chapter 17, scientists interested in broad-scale environmental issues can use such data to examine patterns of vegetation, climate, and temperature.

Further information describing AVHRR is available at

www.ngdc.noaa.gov/seg/globsys/avhrr.shtml.

SeaWiFS

SeaWiFS (sea-viewing wide field of view sensor) (Orbview 2), launched in August 1997, is designed to observe the Earth's oceans. The instrument was designed by NASA-Goddard Space Flight Center and was built and is operated by ORBIMAGE. ORBIMAGE is an affiliate of Orbital Sciences Corporation (Dulles, Virginia), which retains rights for commercial applications but sells data to NASA in support of scientific and research applications. Under this data-sharing agreement, most data are held privately for a 2-week interval to preserve their commercial value, then released for more general distribution to the scientific community.

SeaWiFS's primary mission is to observe ocean color. The color of the ocean surface is sensitive to changes in abundance and types of marine phytoplankton, which indicate the rate at which marine organisms convert solar energy into biomass (i.e., basic production of food for other marine organisms). Estimation of this rate leads to estimates of primary production of marine environments, an important consideration in the study of global marine systems. SeaWiFS data also contribute to studies of meteorology, climatology, and oceanography (e.g., currents, upwelling, sediment transport, shoals) and, over land bodies, to studies of broad-scale vegetation patterns.

To accomplish these objectives, SeaWiFS provides broad-scale imagery at coarse spatial resolution but at fine spectral and radiometric resolution (Plate 6). That is, it is designed to survey very broad regions but to make fine distinctions of brightness and color within those regions. It provides imagery at a spatial resolution of 1.1 km, with a swath width of 2,800 km. The satellite is positioned in a sun-synchronous orbit (with a noon equatorial crossing) that can observe a large proportion of the Earth's oceans every 48 hours. SeaWiFS uses eight spectral channels:

- *Band 1*: 0.402–0.422 µm (blue; yellow pigment/phytoplankton)
- *Band 2*: 0.433–0.453 µm (blue; chlorophyll)
- *Band 3*: 0.480–0.500 µm (blue-green; chlorophyll)
- *Band 4*: 0.500–0.520 µm (green; chlorophyll)
- *Band 5*: 0.545–0.565 µm (red; yellow pigment/phytoplankton)
- *Band 6*: 0.660–0.680 µm (red; chlorophyll)
- *Band 7*: 0.745–0.785 µm (near infrared; land–water contact, atmospheric correction, vegetation)
- *Band 8*: 0.845–0.885 µm (near infrared; land–water contact, atmospheric correction, vegetation)

For more information, see

http://seawifs.gsfc.nasa.gov/SEAWIFS.html

www.geoeye.com

VEGETATION

SPOT 4 and SPOT 5 each carry an auxiliary sensor, the VEGETATION (VGT) instrument, a joint project of several European nations. VGT is a wide-angle radiometer designed for high radiometric sensitivity and broad areal coverage to detect changes in spectral responses of vegetated surfaces. The swath width is 2,200 km, with repeat coverage on successive days at latitudes above 35° and, at the equator, coverage for 3 out of every 4 days. The VGT instrument has a CCD linear array sensitive in four spectral bands designed to be compatible with the SPOT 4 HRVIR mentioned above:

- *Band 1*: 0.45–0.50 µm (blue)
- *Band 2*: 0.61–0.68 µm (red)
- *Band 3*: 0.79–0.89 µm (near infrared)
- *Band 4*: 1.58–1.75 µm (mid-infrared)

VEGETATION concurrently provides data in two modes. In direct (regional observation) mode, VGT provides a resolution of 1 km at nadir. In recording (worldwide observation) mode, each pixel corresponds to four of the 1-km pixels, aggregated by onboard processing. In worldwide observation mode, VGT can acquire data within the region between 60° N and 40° S latitude. Data from HRVIR and VGT can be added or mixed, using onboard processing, as needed to meet requirements for specific projects.

The system provides two products. VGT-P offers full radiometric fidelity, with minimal processing, for researchers who require data as close as possible to the original values. VGT-S data have been processed to provide geometric corrections composites generated to filter out cloud cover. These images are designed for scientists who require data prepared to fit directly with applications projects.

Fine-Resolution Satellite Systems

The third class of land observation satellites consists of systems designed to provide detailed coverage for small-image footprints. In the 1970s and 1980s, Landsat, SPOT, and other systems established the technical and commercial value of the land observation satellite concept, but they also revealed the limitations of systems intended to provide general purpose data for a broad community of users who may have diverse requirements. The success of these early satellite systems revealed the demand for finer spatial resolution imagery required for tasks not satisfied by coarser resolution, especially urban infrastructure analysis, transportation planning, reconnaissance, and construction support. By the late 1980s and early 1990s, technological, analytical, and commercial infrastructure was in place to support the development of smaller, special purpose, Earth observation satellites focused on the requirements of these narrowly defined markets. By 1994, the U.S. government relaxed its restrictions on commercial applications of high-resolution satellite imagery. These changes in policy, linked with advances in technology, opened opportunities for deployment of satellites designed to provide high-resolution image data with characteristics that are distinct from the design of the other, coarser-resolution systems described here.

Relative to the moderate-resolution class of satellite systems, fine-resolution systems offer highly specialized image products tailored for a specific set of applications. They provide imagery characterized by high spatial detail, high radiometric resolution with narrow fields of view, and small footprints targeted on specified areas. They acquire data in the optical and near infrared regions of the spectrum, often with the capability to provide PN imagery at even finer spatial detail. They employ sun-synchronous orbits, so they can provide worldwide coverage, subject, of course, to the coverage gaps at high latitudes common to all such systems. Some systems offer stereo capability and may offer orthocorrected imagery (to fully account for positional effects of terrain relief within the image area). Fine-resolution systems are organized as commercial enterprises, although it is quite common for many to have strategic relationships with government organizations. Image providers often collect image data at 10- or 11-bit resolution, although some products are distributed at 8-bit resolution for customers who prefer smaller file sizes.

Sometimes fine-resolution satellite data is prepared as *pan-sharpened* (PS) imagery—that is, fused, or merged, XS channels that combine XS data with the finer detail of the corresponding PN channel (discussed further in Chapter 11). For visual interpretation, such products can provide the benefit of XS imagery at the finer spatial detail of the PN channel. However, unlike a true multiband image, the components of the pan-sharpened image are, so to speak, baked together such that they cannot be independently manipulated to change the image presentation. Therefore, they have limited utility in the analytical context, although they have value for their effectiveness in visual interpretations.

Stated broadly, image archives for fine-resolution imaging systems began in the late 1990s, so they will not have the comprehensive temporal or geographic scope of those moderate-resolution systems due to smaller footprints and acquisition strategies designed to respond to customer requests rather than to systematically build geographic coverage. The archives of the numerous commercial enterprises are not centralized, and users may find that some organizations have changed as they were reorganized or acquired by other enterprises.

Fine-resolution satellite imagery has created viable alternatives to aerial photography, particularly in remote regions in which the logistics of aerial photography may generate high costs. Fine-resolution satellite imagery has found important application niches for mapping of urban utility infrastructure, floodplain mapping, engineering and construction analysis, topographic site mapping, change detection, transportation planning, and precision agriculture.

In the following capsule descriptions of fine-resolution systems, readers should refer to individual URLs for each system for specifics concerning stereo capabilities, positional accuracy revisit times, and other characteristics that are too intricate for the concise descriptions presented here.

GeoEye-1

GeoEye-1 satellite, launched in September 2008, acquires PN imagery at 0.41-m resolution and XS imagery at 1.65 m. Its footprint is 15.2 km at nadir.

- Panchromatic: 0.450–0.800 μm
- Blue: 0.450–0.510 μm
- Green 0.510–0.580 μm
- Red: 0.655–0.690 μm
- NIR: 0.780–0.920 μm

Details are available at

www.geoeye.com

IKONOS

The IKONOS satellite system (named from the Greek word for "image") was launched in September 1999 as the first commercial satellite able to collect submeter imagery (0.80-m resolution in PN mode [0.45–0.90 µm] and 3.2-m resolution in XS mode [Figures 6.19 and 6.20 and Plate 7]:

- *Band 1*: 0.45–0.52 µm (blue)
- *Band 2*: 0.52–0.60 µm (green)
- *Band 3*: 0.63–0.69 µm (red)
- *Band 4*: 0.76–0.90 µm (near infrared)

IKONOS imagery has a image swath of 11.3 km at nadir; imagery is acquired from a sun-synchronous orbit, with a 10:30 a.m. equatorial crossing. The revisit interval varies with latitude; at 40°, repeat coverage can be acquired at about 3 days in the XS mode and at about 11–12 days in the PN mode. For details see

www.geoeye.com

QuickBird

QuickBird, launched in October 2001, collects PN imagery at 0.60-m resolution and XS imagery at 2.4-m resolution. QuickBird can revisit a region within 3.5 days, depending on latitude. Its footprint at nadir is 16.5 km. It collects imagery in PN and XS channels as follows:

- *Band 1*: 0.45–0.52. µm (blue)
- *Band 2*: 0.52–0.60 µm (green)
- *Band 3*: 0.63–0.69 µm (red)
- *Band 4*: 0.76–0.89 µm (near infrared)
- *Band 5*: 0.45–0.90 µm (panchromatic)

For more information, see

www.digitalglobe.com

WorldView-1

WorldView-1, launched in September 2007, images using a single PN band (0.04 µm–0.09 µm) at 0.50-m resolution. Its swath width is 17.6 km at nadir. The revisit interval is about 4.6 days, depending on latitude. Further details are available at

www.digitalglobe.com

WorldView-2

WorldView-2, launched in October 2009, acquires eight XS channels at 1.84 m and a PN band at 0.46 m. Swath width is 16.4 km at nadir. The eight XS channels include the blue, green, red, and near-infrared regions, as well as specific bands tailored to observe the red edge, coastal bathymetry, a yellow channel, and second near infrared channel. The PN channel is collected at 0.46 m but distributed at 0.50 m for nongovernmental custom-

FIGURE 6.19. IKONOS panchromatic scene of Washington, D.C., 30 September 1999. From Space Imaging, Inc.

FIGURE 6.20. Detail of Figure 6.19 showing the Jefferson Memorial. From Space Imaging, Inc.

ers to comply with U.S. governmental policy. Revisit time is about 3.7 days, with detail dependent on latitude. Details are available at

www.digitalglobe.com

RapidEye

RapidEye AG, a German geospatial data provider, launched a constellation of five satellites in August 2008 into a formation within the same orbital plane. The satellites carry identical XS sensors calibrated to a common standard. Therefore, an image from one satellite will be equivalent to those from any of the other four, permitting coordinated acquisition of a large amount of imagery and daily revisit to a given area.

RapidEye collects imagery at spatial data at 6.5 m resolution, formatted for delivery at pixels of 5 m GSD. It collects imagery in the blue (0.440–0.510 μm), green (520–590 μm), red (0.630–0.685 μm), and near infrared (760–850 μm) regions. It also collects a fifth channel at 760–850 μm to monitor radiation at the red edge, to capture the increase in brightness of living vegetation at the boundary between the visible and near infrared regions, significant for assessment of maturity of agricultural crops (see Chapter 17).

These satellites collect image data along a corridor 77 km wide (at a specific inclination), processed to generate uniform tiles based on a fixed grid. The RapidEye tiling system divides the world between +/– 84° into zones based the Universal Transverse Mercator (UTM) grid; within these limits, the RapidEye system can image any point on Earth each day. Each tile is defined by a respective UTM zone and then by a RapidEye-defined row and column number (analogous to the WRS system described earlier.). The tile grid defines 24-km by 24-km tiles, with a 1-km overlap, resulting in 25-km by 25-km tiles. Users can select tiles they require. For details, see

www.rapideye.de

ImageSat International

Earth resource observation satellites, often known as EROS satellites, deployed in sun-synchronous, near-polar orbits, are focused primarily on intelligence, homeland security, and national development missions but are also employed in a wide range of civilian applications, including mapping, border control, infrastructure planning, agricultural monitoring, environmental monitoring, and disaster response. EROS A (launched in December 2000) collects PN imagery at 1.9-m to 1.2-m resolution. EROS B (launched in April 2006) acquires PN imagery at 70-cm resolution.

6.9. Data Archives and Image Research

Because of the unprecedented amount of data generated by Earth observation satellite systems, computerized databases have formed the principal means of indexing satellite imagery. Each satellite system mentioned above is supported by an electronic catalog that characterizes each scene by area, date, quality, cloud cover, and other elements. Usually users can examine such archives through the World Wide Web. One example is Earth Explorer, offered by the USGS as an index for USGS and Landsat imagery and cartographic data:

http://earthexplorer.usgs.gov

Users can specify a geographic region by using geographic coordinates, by outlining a region on a map, or by entering a place name. Subsequent screens allow users to select specific kinds of data, specify desirable dates of coverage, and indicate the minimum quality of coverage. The result is a computer listing that provides a tabulation of coverage meeting the constraints specified by the user.

Another example is the USGS Global Visualization Viewer (GLOVIS):

http://glovis.usgs.gov

The GLOVIS screen allows the user to roam from one Landsat path or row to another, then to scroll through alternative dates for the path or row selected (Figure 6.21). The user can set constraints on date, cloud cover, and scene quality to display only those scenes likely to be of interest to the analyst. Effective October 2008, the USGS implemented a policy to provide Landsat data to the public. For electronic delivery through the Internet, there is no charge, although for data delivered by other media, there may be a cost for the medium and for shipping. GLOVIS users can identify and download full-resolution imagery through the GLOVIS interface.

Other organizations offer variations on the same basic strategy. Some commercial

FIGURE 6.21. GLOVIS example. GLOVIS allows the user to roam from one Landsat path/row to another (the central scene in the viewer), then to scroll through alternative dates for the path/row selected. The user can set constraints on date, cloud cover, scene quality, and so on, to display only those scenes likely to be of interest to the analyst.

organizations require users to register before they can examine their full archives. Once a user has selected scenes that meet requirements for specific projects, the scene IDs can be used to order data from the organization holding the data. If the user requires imagery to be acquired specifically on a given date, an additional change is required to plan for the dedicated acquisition, which, of course, can be acquired subject to prevailing weather for the area in question.

Descriptive records describing coverages of satellite data are examples of a class of data known as *metadata*. Metadata consist of descriptive summaries of other datasets (in this instance, the satellite images themselves). As increasingly large volumes of remotely sensed data are accumulated, the ability to search, compare, and examine metadata has become increasingly significant (Mather and Newman, 1995). In the context of remote sensing, metadata usually consist of text describing images: dates, spectral regions, quality ratings, cloud cover, geographic coverage, and so on. An increasing number of efforts are under way to develop computer programs and communication systems that will link databases together to permit searches of multiple archives.

6.10. Summary

Satellite observation of the Earth has greatly altered the field of remote sensing. Since the launch of Landsat 1, a larger and more diverse collection of scientists than ever before have conducted remote sensing research and used its applications. Public knowledge of and interest in remote sensing has increased. Digital data for satellite images has contributed greatly to the growth of image processing, pattern recognition, and image analysis (Chapters 10–14). Satellite observation systems have increased international cooperation through joint construction and operation of ground receiving stations and though collaboration in the training of scientists.

The history of the U.S. Landsat system, as well as the histories of comparable systems of other nations, illustrate the continuing difficulties experienced in defining structures for financing the development, operation, and distribution costs required for satellite imaging systems. The large initial investments and high continuing operational costs (quite unlike those of aircraft systems) resemble those of a public utility. Yet there is no comparable shared understanding that such systems generate widespread public benefits and that users therefore must contribute substantially to supporting their costs. But if such systems were to pass their full costs on to customers, the prices of imagery would be too high for development of new applications and growth of the market. In the United States, as well as in other nations, there has been increasing interest in development of policies that encourage governmental agencies, military services, and private corporations to share costs of development and operation of satellite systems. This approach (*dual use*) may represent an improvement over previous efforts to establish long-term support of satellite remote sensing, but it is now too early to determine whether it constitutes a real solution to the long-term problem.

Another concern that will attract more and more public attention is the issue of personal privacy. As systems provide more or less routine availability of imagery with submeter resolution, governmental agencies and private corporations will have direct access to data that could provide very detailed information about specific individuals and their property. Although this imagery would not necessarily provide information not already available through publicly available aerial photography, the ease of access and the standardized format open new avenues for use of such information. The real concern should focus not so much on the imagery itself as on the effects of combining such imagery with other data from marketing information, census information, and the like. The combination of these several forms of information, each in itself rather benign, could develop capabilities that many people would consider to be objectionable or even dangerous.

A third issue concerns reliable public access to fine-resolution imagery. For example, news organizations might wish to maintain their own surveillance systems to provide independent sources of information regarding military, political, and economic developments throughout the world. Will governments, as investors and/or operators of satellite imaging systems, feel that they are in a position to restrict access when questions of national security are at stake?

This chapter forms an important part of the foundation necessary to develop topics presented in subsequent chapters. The specific systems described here are significant in their own right, but they also provide the foundation for understanding other satellite systems that operate in the microwave (Chapter 7) and far infrared (Chapter 9) regions of the spectrum. Finally, it can be noted that the discussion thus far has emphasized acquisition of satellite data. Little has been said about analysis of these data and their applications to specific fields of study. Both topics will be covered in subsequent chapters (Chapters 11–16 and 17–21, respectively).

6.11. Some Teaching and Learning Resources

NASA and USGS have a wide variety of educational resources pertaining to the Landsat program and Earth observation satellites, so careful attention may be find those that are appropriate for a particular purpose. Some of the many videos have a historical perspective and so will not necessarily relate to current capabilities.

- *http://landsat.gsfc.nasa.gov/education/resources.html*

Others:

- NASA Earth Observations Landsat 5 Turns 25
  www.youtube.com/watch?v=ArLvDtsewn0

- Remote Sensing: Observing the Earth
  www.youtube.com/watch?v=oei4yOwjIyQ&feature=related

- A Landsat Flyby
  www.youtube.com/watch?v=BPbHDKgBBxA

- Earth Observation Satellite Image Gallery—Worldview-1 and QuickBird
  www.youtube.com/watch?v=6gz8tbDhAV0

- Satellite Image Gallery—GeoEye-1 and IKONOSI
  www.youtube.com/watch?v=Xs_eFKM12sU

- MichiganView Satellite Tracking Application
  http://apps.michiganview.org/satellite-tracking

Review Questions

1. Outline the procedure for identifying and ordering SPOT, IRS, or Landsat images for a study area near your home. Identify for each step the *information* and *materials* (maps, etc.) necessary to complete that step and proceed to the next. Can you anticipate some of the difficulties you might encounter?

2. In some instances it may be necessary to form a mosaic of several satellite scenes by matching several images together at the edges. List some of the problems you expect to encounter as you prepare such a mosaic.

3. What are some of the advantages (relative to use of aerial photography) of using satellite imagery? Can you identify disadvantages?

4. Manufacture, launch, and operation of Earth observation satellites is a very expensive undertaking—so large that it requires the resources of a national government to

support the many activities necessary to continue operation. Many people question whether it is necessary to spend government funds for Earth resource observation satellites and have other ideas about use of these funds. What arguments can you give to justify the costs of such programs?

5. Why are orbits of land observation satellites so low relative to those of communications satellites?

6. Prepare a short essay that outlines pros and cons of private ownership of high-resolution satellite remote sensing systems.

7. Discuss problems that would arise as engineers attempt to design multispectral satellite sensors with smaller and smaller pixels. How might some of these problems be avoided?

8. Can you suggest some of the factors that might be considered as scientists select the observation time (local sun time) for a sun-synchronous Earth observation satellite?

9. Earth observation satellites do not continuously acquire imagery; they only record individual scenes as instructed by mission control. List some factors that might be considered in planning scenes to be acquired during a given week. Design a strategy for acquiring satellite images worldwide, specifying rules for deciding which scenes are to be given priority.

10. Explain why a satellite image and an aerial mosaic of the same ground area are not equally useful, even though image scale might be the same.

11. On a small-scale map (such as a road map or similar map provided by your instructor) plot, using relevant information from the text, at the correct scale the outline (i.e., the "footprint") of a TM scene centered on a nearby city. How many different counties are covered by this area? Repeat the exercise for an IKONOS image.

References

Arnaud, M. 1995. The SPOT Programme. Chapter 2 in *TERRA 2: Understanding the Terrestrial Environment* (P. M. Mather, ed.). New York: Wiley, pp. 29–39.

Baker, J. C., K. M. O'Connell, and R. A. Williamson (eds.). 2001. *Commercial Observation Satellites at the Leading Edge of Global Transparency*. Bethesda, MD: American Society for Photogrammetry and Remote Sensing, 668 pp.

Brugioni, D. A. 1991. *Eyeball to Eyeball: The Inside Story of the Cuban Missile Crisis*. New York: Random House, 622 pp.

Brugioni, D. A. 1996. The Art and Science of Photoreconnaissance. *Scientific American*, Vol. 274, pp. 78–85.

Chevrel, M., M. Courtois, and G. Weill. 1981. The SPOT Satellite Remote Sensing Mission. *Photogrammetric Engineering & Remote Sensing*, Vol. 47, pp. 1163–1171.

Cihlar, J. 2000. Land Cover Mapping of Large Areas from Satellites: Status and Research Priorities. *International Journal of Remote Sensing*, Vol. 21, pp. 1093–1114.

Curran, P. J. 1985. *Principals of Remote Sensing*. New York: Longman, 282 pp.

Day, D. A., J. M. Logsdon, and B. Latell (eds.). 1998. *Eye in the Sky: The Story of the Corona Spy Satellites*. Washington, DC: Smithsonian Institution Press, 303 pp.

Fritz, L. W. 1996. Commercial Earth Observation Satellites. In *International Archives of Photogrammetry and Remote Sensing, ISPRS Com. IV*, pp. 273–282.

Goward, S. N., and D. L. Williams. 1997. Landsat and Earth Systems Science: Development of Terrestrial Monitoring. *Photogrammetric Engineering and Remote Sensing*, Vol. 63, pp. 887–900.

Jhabvala, M., D. Reuter, K. Choi, C. Jhabvala, and M. Sundaram. 2009. QWIP-Based Thermal Infrared Sensor for the Landsat Data Continuity Mission. *Infrared Physics and Technology*, Vol. 52, pp. 424–429.

Ju, J., and D. P. Roy. 2008. The Availability of Cloud-Free Landsat ETM+ Data over the Conterminous United States and Globally. *Remote Sensing of Environment*, Vol. 112, pp. 1196–1211.

Lauer, D. T., S. A. Morain, and V. V. Salomonson. 1997. The Landsat Program: Its Origins, Evolution, and Impacts. *Photogrammetric Engineering and Remote Sensing*, Vol. 63, pp. 831–838.

Li, R. 1998. Potential of High-Resolution Satellite Imagery for National Mapping Products. *Photogrammetric Engineering and Remote Sensing*, Vol. 64, pp. 1165–1169.

MacDonald, R. A. 1995. CORONA: Success for Space Reconnaissance, a Look into the Cold War, and a Revolution for Intelligence. *Photogrammetric Engineering and Remote Sensing*, Vol. 61, pp. 689–719.

Mack, P. 1990. *Viewing the Earth: The Social Constitution of the Landsat Satellite System*. Cambridge, MA: MIT Press, 270 pp.

Markham, B. L. 2004. Landsat Sensor Performance: History and Current Status. *IEEE Transactions on Geoscience and Remote Sensing*, Vol. 42, pp. 2691–2694.

Mather, P. M., and I. A. Newman. 1995. U.K. Global Change Federal Metadata Network. Chapter 9 in TERRA-2: Understanding the Terrestrial Environment (P. M. Mather, ed.). New York: Wiley, pp. 103–111.

Morain, S. A., and A. M. Budge. 1995. *Earth Observing Platforms and Sensors CD-ROM*. Bethesda, MD: American Society for Photogrammetry and Remote Sensing.

Ruffner, K. C. (ed.). 1995. *CORONA: America's First Satellite Program*. Washington, DC: Center for the Study of Intelligence, 360 pp.

Salomonson, V. V. (ed.). 1997. The 25th Anniversary of Landsat-1 (Special issue). *Photogrammetric Engineering and Remote Sensing*, Vol. 63, No. 7.

Salomonson, V. V., J. R. Irons, and D. L. Williams. 1996. The Future of Landsat: Implications for Commercial Development. In *American Institute for Physics, Proceedings 325* (M. El-Genk and R. P. Whitten, eds.). College Park, MD: American Institute of Physics, pp. 353–359.

Slater, P. N. 1980. *Remote Sensing: Optics and Optical Systems*. Reading, MA: Addison-Wesley, 575 pp.

Stoney, W. E. 2008. *ASPRS Guide to Land Imaging Satellites*. Bethesda, MD: American Society for Photogrammetry and Remote Sensing, *www.asprs.org/news/satellites/ASPRS_DATABASE_021208.pdf*.

Surazakov, Arzhan, and V. Aizen. 2010. Positional Accuracy Evaluation of Declassified Hexagon KH-9 Mapping Camera Imagery. *Photogrammetric Engineering and Remote Sensing*, Vol. 76, pp. 603–608.

Taranick, J. V. 1978. *Characteristics of the Landsat Multispectral Data System*. USGS Open File Report 78-187. Reston, VA: U.S. Geological Survey, 76 pp.

Williams, D. L., S. N. Goward, and T. L. Arvidson. 2006. Landsat: Yesterday, Today, and Tomorrow. *Photogrammetric Engineering and Remote Sensing*, Vol. 72, pp. 1171–1178.

Wulder, M. A., J. C. White, S. N. Goward, J. G. Masek, J. R. Irons, M. Herold, W. B. Cohen, T. R. Loveland, and C. E. Woodcock. 2008. Landsat Continuity: Issues and Opportunities for Land Cover Monitoring. *Remote Sensing of Environment*, Vol. 112, pp. 955–969.

Zhou, G., and K. Jezek. 2002. Orthorectifying 960's Declassified Intelligence Satellite Photography DISP of Greenland. *IEEE Geoscience and Remote Sensing*, Vol. 40, pp. 1247–1259.

CORONA

CORONA is the project designation for the satellite reconnaissance system operated by the United States during the interval 1960–1972. CORONA gathered photographic imagery that provided strategic intelligence on the activities of Soviet industry and strategic forces. For many years this imagery and details of the CORONA system were closely guarded as national security secrets. In 1995, when the Soviet threat was no longer present and CORONA had been replaced by more advanced systems, President Clinton announced that CORONA imagery would be declassified and released to the public. MacDonald (1995) and Ruffner (1995) provide detailed descriptions of the system. Curran (1985) provides an overview based on open sources available prior to the declassification.

Historical Context

When Dwight Eisenhower became president of the United States in 1953, he was distressed to learn of the rudimentary character of the nation's intelligence estimates concerning the military stature of the Soviet Union. Within the Soviet sphere, closed societies and rigid security systems denied Western nations the information they required to prepare reliable estimates of Soviet and allied nations' military capabilities. As a result, Eisenhower feared the United States faced the danger of developing policies based on speculation or political dogma rather than reliable estimates of the actual situation.

His concern led to a priority program to build the U-2 system, a high-altitude reconnaissance aircraft designed to carry a camera system of unprecedented capabilities. The first flight occurred in the summer of 1956. Although U-2 flights imaged only a small portion of the Soviet Union, they provided information that greatly improved U.S. estimates of Soviet capabilities. The U-2 program was intended only as a stopgap measure; it was anticipated that the Soviet Union would develop countermeasures that would prevent long-term use of the system. In fact, the flights continued for about 4 years, ending in 1960 when a U-2 plane was shot down near Sverdlovsk in the Soviet Union. This event ended use of the U-2 over the Soviet Union, although it continued to be a valuable military asset for observing other regions of the world. Later it was employed in the civilian sphere for collecting imagery for environmental analyses.

At the time of the U-2 incident, work was already under way to design a satellite system that would provide photographic imagery from orbital altitudes, thereby avoiding risks to pilots and the controversy arising from U.S. overflights of another nation's territory. Although CORONA had been conceived beforehand, the Soviet Union's launch of the first artificial Earth-orbiting satellite in October 1957 increased the urgency of the effort.

The CORONA satellite was designed and constructed as a joint effort of the U.S. Air Force and the Central Intelligence Agency (CIA), in collaboration with private contractors who designed and manufactured launch systems, satellites, cameras, and films. Virtually every element of the system extended technical capabilities beyond their known limits. Each major component encountered difficulties and failures that were, in time, overcome to eventually produce an effective, reliable system. During the interval June 1959–December 1960, the system experienced a succession of failures, first with the launch system, then with the satellite, next the recovery system, and finally with the cameras and films. Each problem was identified and solved within a remarkably short time, so that CORONA was in effect operational by August 1959, only 3 months after the end of the U-2 flights over the Soviet Union. Today CORONA's capabilities may seem commonplace, as we are familiar with artificial satellites, satellite imagery, and high-resolution films. However, at the time CORONA extended reconnaissance capabilities far beyond what even technical experts thought might be feasible.

As a result, the existence of the program was a closely held secret for many years. Although the Soviet Union knew in a general way that a reconnaissance satellite effort was under development, details of the system, the launch schedule, the nature of the imagery, and the ability of interpreters to derive information from the imagery were closely guarded secrets. Secrecy denied the Soviets knowledge of the capabilities of the system, which might have permitted development of effective measures for deception. In time, the existence of the program became more widely known in the United States, although details of the system's capabilities were still secret.

CORONA is the name given to the satellite system, whereas the camera systems carried KEYHOLE (KH) designations that were familiar to those who used the imagery. The designations KH-1 through KH-5 refer to different models of the cameras used for the CORONA program. Most of the imagery was acquired with the KH-4 and KH-5 systems (including KH-4A and KH-4B).

Satellite and Orbit

The reconnaissance satellites were placed into near-polar orbits, with, for example, an inclination of 77°, apogee of 502 mi., and perigee of 116 mi. Initially missions would last only 1 day; by the end of the program, missions extended for 16 days. Unlike the other satellite systems described here, images were returned to Earth by a capsule (the satellite recovery vehicle) ejected from the principal satellite at the conclusion of each mission. The recovery vehicle was designed to withstand the heat of reentry into the Earth's atmosphere and to deploy a parachute at an altitude of about 60,000 ft. The capsule was then recovered in the air by specially designed aircraft. Recoveries were planned for the Pacific Ocean near Hawaii. Capsules were designed to sink if not recovered within a few days to prevent recovery by other nations in the event of malfunction. Several capsules were, in fact, lost when this system failed. Later models of the satellite were designed with two capsules, which extended the length of CORONA missions to as long as 16 days.

Cameras

Camera designs varied as problems were solved and new capabilities were added. MacDonald (1995) and Ruffner (1995) provide details of the evolution of the camera systems; here the description focuses on the main feature of the later models that acquired much of the imagery in the CORONA archive. The basic camera design, the KH-3, manufactured by Itek Corporation, was a vertically oriented panoramic camera with a 24-in. focal length. The camera's 70° panoramic view was acquired by the mechanical motion of the system at right angles to the line of flight. Image motion compensation (see Chapter 3) was employed to correct the effects of satellite motion relative to the Earth's surface during exposure.

The KH-4 camera acquired most of the CORONA imagery (Figure 6.22). The KH-4B (sometimes known as the MURAL system) consisted of two KH-3 cameras oriented to observe the same area from different perspectives, thereby providing stereo capability (Figure 6.23). One pointed 15° forward along the flight path; the other was aimed 15° aft. A small-scale index image provided the context for proper orientation of the panoramic imagery, stellar cameras viewed the pattern of stars, and horizon cameras viewed the Earth's horizon to assure correct orientation of the spacecraft and the panoramic cameras. A related system, named ARGON, acquired photographic imagery to compile accurate maps of the Soviet Union, using cameras with 3-in. and 1.5-in. focal lengths.

Imagery

The CORONA archive consists of about 866,000 images, totaling about 400 mi. of film, acquired during the interval August 1960–May 1972. Although film format varied, most

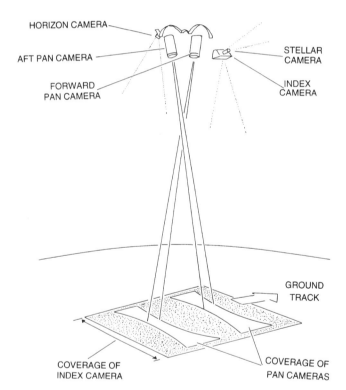

FIGURE 6.22. Coverage of a KH-4B camera system. The two stereo panoramic cameras point for and aft along the ground track, so a given area can be photographed first by the aft camera (pointing forward), then by the forward camera (pointing aft) (see Figure 6.23). Each image from the panoramic camera represents an area about 134.8 mi. (216.8 km) in length and about 9.9 mi. (15.9 km) wide at the image's greatest width. The index camera provides a broad-scale overview of the coverage of the region; the stellar and horizon cameras provide information to maintain satellite stability and orientation. Based on MacDonald (1995).

FILM SUPPLY
CASSETTE

FILM PATH

TAKE-UP
CASSETTES

INDEX
CAMERA

STEREO PANORAMIC CAMERAS

FIGURE 6.23. Line drawing of major components of the KH-4B camera (based on MacDonald, 1995). The two panoramic cameras are pointed forward and aft along the ground track to permit acquisition of stereo coverage as shown in Figure 6.22. Each of the two take-up cassettes could be ejected sequentially to return exposed film to Earth. The index camera, as shown in Figure 6.22, provided a broad-scale overview to assist in establishing the context for coverage from the panoramic cameras.

imagery was recorded on 70-mm film in strips about 25 or 30 in. long. Actual image width is typically either 4.4 in. or 2.5 in. wide. Almost all CORONA imagery was recorded on black-and-white panchromatic film, although a small portion used a color infrared emulsion, and some was in a natural color emulsion. The earliest imagery is said to have a resolution of about 40 ft.; by the end of the program in 1972, the resolution was as fine as 6 ft., although detail varied greatly depending on atmospheric effects, illumination, and the nature of the target (Figure 6.24). The imagery is indexed to show coverage and dates. About 50% is said to be obscured by clouds, although the index does not record which scenes are cloud covered. Most of the early coverage was directed at the Soviet Union (Figure 6.25), although later coverage shows other areas, often focused on areas of current or potential international concern.

National Photographic Interpretation Center

Interpretation of CORONA imagery was largely centralized at the National Photographic Interpretation Center (NPIC), maintained by the CIA in a building in the navy yard in southeastern Washington, D.C. Here teams of photointerpreters from the CIA, armed forces, and allied governments were organized to examine imagery to derive intelligence pertaining to the strategic capabilities of the Soviet Union and its satellite nations. Film from each mission was immediately interpreted to provide immediate reports on topics of current significance.

Then imagery was reexamined to provide more detailed analyses of less urgent developments. In time, the system accumulated an archive that provided a retrospective record of the development of installations, testing of weapons systems, and operations of units in the field. By this means, image analysts could trace the development of individual military and indus-

FIGURE 6.24. Severodvinsk Shipyard (on the White Sea coastline, near Archangel, Russia, formerly USSR) as imaged by KH-4B camera, 10 February 1969. This image is much enlarged from the original image. From CIA.

trial installations, recognize unusual activities, and develop an understanding of the strategic infrastructure of the Soviet Union. Although the priority focus was on understanding the strategic capabilities of Soviet military forces, CORONA imagery was also used to examine agricultural and industrial production, environmental problems, mineral resources, population patterns, and other facets of the Soviet Union's economic infrastructure.

Reports from NPIC photointerpreters formed only one element of the information considered by analysts as they prepared intelligence estimates; they also studied other sources, such as electronic signals, reports from observers, and press reports. However, CORONA photographs had a central role in this analysis because of their ready availability, reliability, and timeliness.

Image Availability

Until 1995, the CIA maintained the CORONA archive. It was released to the National Archives and Records Administration and to the USGS after declassification. The National Archive retains the original negatives and provides copies for purchase by the public through the services of contractors. The USGS stores duplicate negatives to provide imagery to the public and maintains a digital index as part of the Earth Explorer online product search index:

http://earthexplorer.usgs.gov

Other organizations may decide to purchase imagery for resale to the public, possibly in digitized form, or to sell selected scenes of special interest.

Uses

CORONA provided the first satellite imagery of the Earth's surface and thereby extends the historical record of satellite imagery into the late 1950s, about 10 years before Landsat. Therefore, it may be able to assist in assessment of environmental change, trends in human use of the landscape, and similar phenomena (Plate 8). Further, CORONA imagery records many of the pivotal events of the cold war and may thereby form an important source for historians who wish to examine and reevaluate these events more closely than was possible before imagery was released. Ruffner (1995) provides examples that show archaeological and geologic applications of CORONA imagery. There may well be other applications not yet identified.

However, today's interpreters of this imagery may face some difficulties. Many of the original interpretations depended not only on the imagery itself but also on the skill and experience of the interpreters (who often were specialized in identification of specific kinds of equipment), access to collateral information, and availability of specialized equipment. Further, many critical interpretations depended on the experience and expertise of interpreters with long experience in analysis of images of specific weapons systems—experience that is not readily available to today's analysts. Therefore, it may be difficult to reconstruct or reevaluate interpretations of an earlier era.

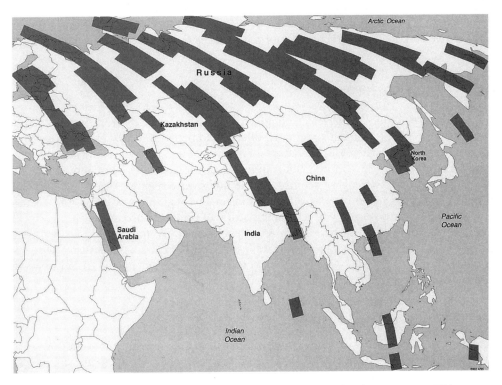

FIGURE 6.25. Typical coverage of a KH-4B camera, Eurasian land mass. From CIA.

CHAPTER SEVEN

Active Microwave

7.1. Introduction

This chapter describes active microwave sensor systems. Active microwave sensors form an example of an *active sensor*—a sensor that broadcasts a directed pattern of energy to illuminate a portion of the Earth's surface, then receives the portion scattered back to the instrument. This energy forms the basis for the imagery we interpret. Because passive sensors (e.g., photography) are sensitive to variations in solar illumination, their use is constrained by time of day and weather. In contrast, active sensors generate their own energy, so their use is subject to fewer constraints, and they can be used under a wider range of operational conditions. Further, because active sensors use energy generated by the sensor itself, its properties are known in detail. Therefore, it is possible to compare transmitted energy with received energy to judge with more precision than is possible with passive sensors the characteristics of the surfaces that have scattered the energy.

7.2. Active Microwave

The microwave region of the electromagnetic spectrum extends from wavelengths of about 1 mm to about 1 m. This region is, of course, far removed from those in and near the visible spectrum, where our direct sensory experience can assist in our interpretation of images and data. Thus formal understanding of the concepts of remote sensing are vital to understanding imagery acquired in the microwave region. As a result, the study of microwave imagery is often a difficult subject for beginning students and requires more attention than is usually necessary for study of other regions of the spectrum. The family of sensors discussed here (Figure 7.1) includes active microwave sensors (imaging radars carried by either aircraft or satellites). Another class of microwave sensors—*passive microwave sensors*—are sensitive to microwave energy emitted from the Earth's surface, to be discussed in Chapter 9.

Active Microwave Sensors

Active microwave sensors are *radar* devices, instruments that transmit a microwave signal, then receive its reflection as the basis for forming images of the Earth's surface. The rudimentary components of an imaging radar system include a transmitter, a receiver, an antenna array, and a recorder. A *transmitter* is designed to transmit repetitive pulses of

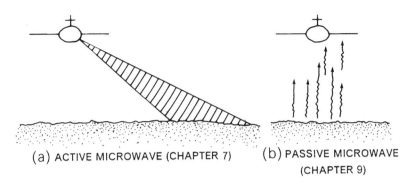

(a) ACTIVE MICROWAVE (CHAPTER 7)   (b) PASSIVE MICROWAVE
(CHAPTER 9)

FIGURE 7.1. Active and passive microwave remote sensing. (a) Active microwave sensing, using energy generated by the sensor, as described in this chapter. (b) Passive microwave sensing, which detects energy emitted by the Earth's surface, described in Chapter 9.

microwave energy at a given frequency. A *receiver* accepts the reflected signal as received by the antenna, then filters and amplifies it as required. An *antenna array* transmits a narrow beam of microwave energy. Such an array is composed of *waveguides*, devices that control the propagation of an electromagnetic wave such that waves follow a path defined by the physical structure of the guide. (A simple waveguide might be formed from a hollow metal tube.) Usually the same antenna is used both to transmit the radar signal and to receive its echo from the terrain. Finally, a *recorder* records and/or displays the signal as an image. Numerous refinements and variations of these basic components are possible; a few are described below in greater detail.

Side-Looking Airborne Radar

Radar is an acronym for "radio detection and ranging." The "ranging" capability is achieved by measuring the time delay between the time a signal is transmitted toward the terrain and the time its echo is received. Through its ranging capability, possible only with active sensors, radar can accurately measure the distance from the antenna to features on the ground. A second unique capability, also a result of radar's status as an active sensor, is its ability to detect frequency and polarization shifts. Because the sensor transmits a signal of known wavelength, it is possible to compare the received signal with the transmitted signal. From such comparisons imaging radars detect changes in frequency that give them capabilities not possible with other sensors.

Side-looking airborne radar (SLAR) imagery is acquired by an antenna array aimed to the side of the aircraft, so that it forms an image of a strip of land parallel to, and at some distance from, the ground track of the aircraft. As explained below, the designation SAR (synthetic aperture radar) also describes radar imagery. The resulting image geometry differs greatly from that of other remotely sensed images. One of SLAR's most unique and useful characteristics is its ability to function during inclement weather. SLAR is often said to possess an "all-weather" capability, meaning that it can acquire imagery in all but the most severe weather conditions. The microwave energy used for SLAR imagery is characterized by wavelengths long enough to escape interference from clouds and light rain (although not necessarily from heavy rainstorms). Because SLAR systems are independent of solar illumination, missions using SLAR can be scheduled at night or during early morning or evening hours when solar illumination might be unsatisfactory for acquiring aerial photography. This advantage is especially important for imaging radars carried by the Earth-orbiting satellites described later in this chapter.

Radar images typically provide crisp, clear representations of topography and drainage (Figure 7.2). Despite the presence of the geometric errors described below, radar images typically provide good positional accuracy; in some areas they have provided the data for small-scale base maps depicting major drainage and terrain features. Analysts have registered TM and SPOT data to radar images to form composites of the two quite different kinds of information. Some of the most successful operational applications of SLAR imagery have occurred in tropical climates, where persistent cloud cover has prevented acquisition of aerial photography and where shortages of accurate maps create a context in which radar images can provide cartographic information superior to that on existing conventional maps. Another important characteristic of SLAR imagery is its synoptic view of the landscape. SLAR's ability to clearly represent the major topographic and drainage features within relatively large regions at moderate image scales makes it a valuable addition to our repertoire of remote sensing imagery. Furthermore, because it acquires images in the microwave spectrum, SLAR may show detail and information that differ greatly from those shown by sensors operating in the visible and near infrared spectra.

Origins and History

The foundations for imaging radars were laid by scientists who first investigated the nature and properties of microwave and radio energy. James Clerk Maxwell (1831–1879) first defined essential characteristics of electromagnetic radiation; his mathematical descrip-

FIGURE 7.2. Radar image of a region near Chattanooga, Tennessee, September 1985 (X-band, HH polarization). This image has been processed to produce pixels of about 11.5 m in size. From USGS.

tions of the properties of magnetic and electrical fields prepared the way for further theoretical and practical work. In Germany, Heinrich R. Hertz (1857–1894) confirmed much of Maxwell's work and further studied properties and propagation of electromagnetic energy in microwave and radio portions of the spectrum. The *hertz*, the unit for designation of frequencies (Chapter 2), is named in his honor. Hertz was among the first to demonstrate the reflection of radio waves from metallic surfaces and thereby to begin research that led to development of modern radios and radars.

In Italy, Guglielmo M. Marconi (1874–1937) continued the work of Hertz and other scientists, in part by devising a practical antenna suitable for transmitting and receiving radio signals. In 1895 he demonstrated the practicability of the wireless telegraph. After numerous experiments over shorter distances, he demonstrated in 1901 the feasibility of long-range communications by sending signals across the Atlantic; in 1909 he shared the Nobel Prize in Physics. Later he proposed that ships could be detected by using the reflection of radio waves, but there is no evidence that his suggestion influenced the work of other scientists.

The formal beginnings of radar date from 1922, when A. H. Taylor and L. C. Young, civilian scientists working for the U.S. Navy, were conducting experiments with high-frequency radio transmissions. Their equipment was positioned near the Anacostia River near Washington, D.C., with the transmitter on one bank of the river and the receiver on the other. They observed that the passage of a river steamer between the transmitter and the receiver interrupted the signal in a manner that clearly revealed the potential of radio signals as a means for detecting the presence of large objects (Figure 7.3). Taylor and Young recognized the significance of their discovery for marine navigation in darkness and inclement weather and, in a military context, its potential for detection of enemy vessels. Initial efforts to implement this idea depended on the placement of transmitters and receivers at separate locations, so that a continuous microwave signal was reflected from an object, then recorded by a receiver placed some distance away. Designs evolved so that a single instrument contained both the transmitter and the receiver at a single location, integrated in a manner that permitted use of a pulsed signal that could be reflected from the target back to the same antenna that transmitted the signal. Such instruments were first devised during the years 1933–1935 more or less simultaneously in the United States (by Young and Taylor), Great Britain, and Germany. A British inventor, Sir Robert Watson-Watt, is sometimes given credit as the inventor of the first radar system, although Young and Taylor have been credited by other historians.

Subsequent improvements were based mainly on refinements in the electronics required to produce high-power transmissions over narrow wavelength intervals, to carefully time short pulses of energy, and to amplify the reflected signal. These and other developments led to rapid evolution of radar systems in the years prior to World War II. Due to the profound military significance of radar technology, World War II prompted radar's sophistication. The postwar era perfected imaging radars. Experience with conventional radars during World War II revealed that radar reflection from ground surfaces (*ground clutter*) varied greatly according to terrain, season, settlement patterns, and so

FIGURE 7.3. Beginnings of radar. Schematic diagram of the situation that led to the experiments by Young and Taylor.

on and that reflection from ocean surfaces was modified by winds and waves. Ground clutter was undesirable for radars designed to detect aircraft and ships, but later systems were designed specifically to record the differing patterns of reflection from ground and ocean surfaces. The side-looking characteristic of SLAR was desirable because of the experiences of reconnaissance pilots during the war, who often were required to fly low-level missions along predictable flight paths in lightly armed aircraft to acquire the aerial photography required for battlefield intelligence. The side-looking capability provided a means for acquiring information at a distance using an aircraft flying over friendly territory. Later, the situation posed by the emergence of the cold war in Europe, with clearly defined frontiers, and strategic requirements for information within otherwise closed borders also provided an incentive for development of sensors with SLAR's capabilities. The experience of the war years also must have provided an intense interest in development of SLAR's all-weather capabilities, because Allied intelligence efforts were several times severely restricted by the absence of aerial reconnaissance during inclement weather. Thus the development of imaging radars is linked to military and strategic reconnaissance, even though many current applications focus on civilian requirements.

7.3. Geometry of the Radar Image

The basics of the geometry of a SLAR image are illustrated in Figure 7.4. Here the aircraft is viewed head-on, with the radar beam represented in vertical cross section as the fan-shaped figure at the side of the aircraft. The upper edge of the beam forms an angle with a horizontal line extended from the aircraft; this angle is designated as the *depression angle* of the far edge of the image. Upper and lower edges of the beam, as they intersect

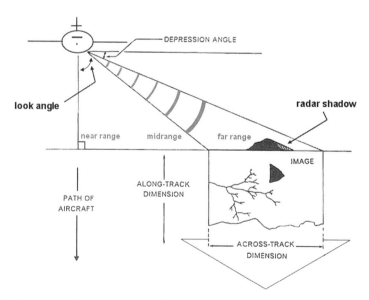

FIGURE 7.4. Geometry of an imaging radar system. The radar beam illuminates a strip of ground parallel to the flight path of the aircraft; the reflection and scattering of the microwave signal from the ground forms the basis for the image.

with the ground surface, define the edges of the radar image; the forward motion of the aircraft (toward the reader, out of the plane of the illustration) forms what is usually the "long" dimension of the strip of radar imagery.

The smallest depression angle forms the *far-range* side of the image. The near-range region is the edge nearest to the aircraft. Intermediate regions between the two edges are sometimes referred to as *midrange* portions of the image. Steep terrain may hide areas of the imaged region from illumination by the radar beam, causing *radar shadow*. Note that radar shadow depends on topographic relief and the direction of the flight path in relation to topography. Within an image, radar shadow depends also on depression angle, so that (given equivalent topographic relief) radar shadow will be more severe in the far-range portion of the image (where depression angles are smallest) or for those radar systems that use shallow depression angles. (A specific radar system is usually characterized by a fixed range of depression angles.)

Radar systems measure distance to a target by timing the delay between a transmitted signal and its return to the antenna. Because the speed of electromagnetic energy is a known constant, the measure of time translates directly to a measure of distance from the antenna. Microwave energy travels in a straight path from the aircraft to the ground—a path that defines the *slant-range* distance, as if one were to stretch a length of string from the aircraft to a specific point on the ground as a measure of distance. Image interpreters prefer images to be presented in *ground-range* format, with distances portrayed in their correct relative positions on the Earth's surface. Because radars collect all information in the slant-range domain, radar images inherently contain geometric artifacts, even though the image display may ostensibly appear to match a ground-range presentation.

One such error is *radar layover* (Figure 7.5). At near range, the top of a tall object is closer to the antenna than is its base. As a result, the echo from the top of the object reaches the antenna before the echo from the base. Because radar measures all distances with respect to time elapsed between transmission of a signal and the reception of its echo, the top of the object appears (i.e., in the slant-range domain) to be closer to the antenna than does its base. Indeed, it is closer, if only the slant-range domain is considered. However, in the ground-range domain (the context for correct positional representation and for accurate measurement), both the top and the base of the object both occupy the same geographic position. In the slant-range domain of the radar image, they occupy different image positions—a geometric error analogous perhaps to relief displacement in the context of aerial photography.

Radar layover is depicted in Figure 7.5. Here the topographic feature *ABC* is shown with *AB* = *BC* in the ground-range representation. However, because the radar can position *A*, *B*, and *C* only by the time delay with relation to the antenna, it must perceive the relationships between *A*, *B*, and *C* as shown in the slant range (image plane). Here *A* and *B* are reversed from their ground-range relationships, so that *ABC* is now *bac*, due to the fact that the echo from *B* must be received before the echo from *A*.

A second form of geometric error, *radar foreshortening*, occurs in terrain of modest to high relief depicted in the mid- to far-range portion of an image (Figure 7.6). Here the slant-range representation depicts *ABC* in their correct relationships *abc*, but the distances between them are not accurately shown. Whereas *AB* = *BC* in the ground-range domain, *ab* < *bc* when they are projected into the slant range. Radar foreshortening tends to cause images of a given terrain feature to appear to have steeper slopes than they do in nature on the near-range side of the image and to have shallower slopes than they do

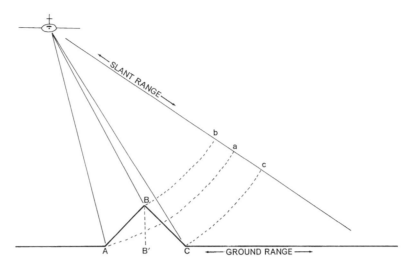

FIGURE 7.5. Radar layover. In the ground-range domain, *AB* and *BC* are equal. Because the radar can measure only slant-range distances, *AB* and *BC* are projected onto the slant-range domain, represented by the line *bac*. The three points are not shown in their correct relationship because the slant-range distance from the antenna to the points does not match to their ground-range distances. Point *B* is closer to the antenna than is point *A*, so it is depicted on the image as closer to the edge of the image.

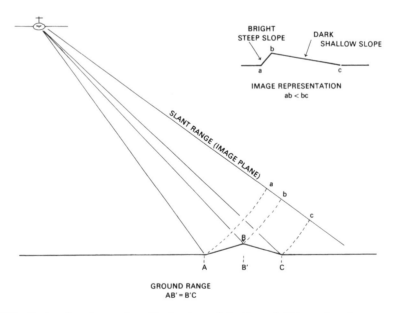

FIGURE 7.6. Radar foreshortening. Projection of *A*, *B*, and *C* into the slant-range domain distorts the representations of *AB* and *BC*, so that *ab* appears shorter, steeper, and brighter than it should be in a faithful rendition, and *bc* appears longer, shallower in slope, and darker than it should be.

in nature on the far-range side of the feature (Figure 7.7). Thus a terrain feature with equal fore and back slopes may be imaged to have shorter, steeper, and brighter slopes than it would in a correct representation, and the image of the back slope would appear to be longer, shallower, and darker than it would in a correct representation. Because depression angle varies with position on the image, the amount of radar foreshortening in the image of a terrain feature depends not only on the steepness of its slopes but also on its position on the radar image. As a result, apparent terrain slope and shape on radar images are not necessarily accurate representations of their correct character in nature. Thus care should be taken when interpreting these features.

FIGURE 7.7. Image illustrating radar fore-shortening, Death Valley, California. The unnatural appearance of the steep terrain illustrates the effect of radar foreshortening, when a radar system observes high, steep topography at steep depression angles. The radar observes this terrain from the right—radar foreshortening creates the compressed appearance of the mountainous terrain in this scene. From NASA-JPL; SIR-C/X-SAR October 1994. P43883.

7.4. Wavelength

Imaging radars normally operate within a small range of wavelengths with the rather broad interval defined at the beginning of this chapter. Table 7.1 lists primary subdivisions of the active microwave region, as commonly defined in the United States. These divisions and their designations have an arbitrary, illogical flavor that is the consequence of their origin during the development of military radars, when it was important to conceal the use of specific frequencies for given purposes. To preserve military security, the designations were intended as much to confuse unauthorized parties as to provide convenience for authorized personnel. Eventually these designations became established in everyday usage, and they continue to be used even though there is no longer a requirement for military secrecy.

Although experimental radars can often change frequency, or sometimes even use several frequencies (for a kind of "multispectral radar"), operational systems are generally designed to use a single wavelength band. Airborne imaging radars have frequently used C-, K-, and X-bands. As described elsewhere in this chapter, imaging radars used for satellite observations often use L-band frequencies.

The choice of a specific microwave band has several implications for the nature of the radar image. For real aperture imaging radars, spatial resolution improves as wavelength becomes shorter with respect to antenna length (i.e., for a given antenna length, resolution is finer with use of a shorter wavelength). Penetration of the signal into the soil is in part a function of wavelength—for given moisture conditions, penetration is greatest at longer wavelengths. The longer wavelengths of microwave radiation (e.g., relative to visible radiation) mean that imaging radars are insensitive to the usual problems of atmospheric attenuation—usually only very heavy rain will interfere with transmission of microwave energy.

7.5. Penetration of the Radar Signal

In principle, radar signals are capable of penetrating what would normally be considered solid features, including vegetative cover and the soil surface. In practice, it is very difficult to assess the existence or amount of radar penetration in the interpretation of specific images. Penetration is assessed by specifying the *skin depth*, the depth to which the strength of a signal is reduced to $1/e$ of its surface magnitude, or about 37%. Separate

TABLE 7.1. Radar Frequency Designations

Band	Wavelengths
P-band	107–77 cm
UHF	100–30 cm
L-band	30–15 cm
S-band	15–7.5 cm
C-band	7.5–3.75 cm
X-band	3.75–2.40 cm
Ku-band	2.40–1.67 cm
K-band	1.67–1.18 cm
Ka-band	1.18–0.75 cm
VHF	1–10 m
UHF	10 cm–1 m

features are subject to differing degrees of penetration; specification of the skin depth, measured in standard units of length, provides a means of designating variations in the ability of radar signals to penetrate various substances.

In the absence of moisture, skin depth increases with increasing wavelength. Thus optimum conditions for observing high penetration would be in arid regions, using long-wavelength radar systems. Penetration is also related to surface roughness and to incidence angle; penetration is greater at steeper angles and decreases as incidence angle increases. We should therefore expect maximum penetration at the near-range edge of the image and minimum penetration at the far-range portion of the image.

The difficulties encountered in the interpretation of an image that might record penetration of the radar signal would probably prevent practical use of any information that might be conveyed to the interpreter. There is no clearly defined means by which an interpreter might be able to recognize the existence of penetration or to separate its effects from the many other variables that contribute to radar backscatter. For radar systems operating near the X-band and the K-band, empirical evidence suggests that the radar signal is generally scattered from the first surface it strikes, probably foliage in most instances. Even at the L-band, which is theoretically capable of a much higher degree of penetration, the signal is apparently scattered from the surface foliage in densely vegetated regions, as reported by Sabins (1983) after examining L-band imagery of forested regions in Indonesia.

Figure 7.8 illustrates these effects with X- and P-band SAR images of a heavily forested region of Colombia. The X-band image, collected at wavelengths of 2–4 cm, records features in a view that resembles an aerial photograph. The longer wavelengths of the P-band image reveal features below the vegetation canopy and, in some instances, below the soil surface, depicting terrain and land use features.

FIGURE 7.8. X- and P-band SAR images, Colombia. The X-band image, collected at wavelengths of 2–4 cm, records features in a view that resembles a view similar to that of an aerial photograph. Longer wavelengths of the P-band image reveals features below the vegetation canopy and, in some instances, below the soil surface, depicting terrain and land use features. From FUGRO-EarthData.

7.6. Polarization

The *polarization* of a radar signal denotes the orientation of the field of electromagnetic energy emitted and received by the antenna. Radar systems can be configured to transmit either horizontally or vertically polarized energy and to receive either horizontally or vertically polarized energy as it is scattered from the ground. Unless otherwise specified, an imaging radar usually transmits horizontally polarized energy and receives a horizontally polarized echo from the terrain. However, some radars are designed to transmit horizontally polarized signals but to separately receive the horizontally and vertically polarized reflections from the landscape. Such systems produce two images of the same landscape (Figure 7.9). One is the image formed by the transmission of a horizontally polarized signal and the reception of a horizontally polarized return signal. This is often referred to as the *HH image*, or the *like-polarized* mode. A second image is formed by the transmission of a horizontally polarized signal and the reception of the vertically polarized return; this is the *HV image*, or the *cross-polarized* mode.

By comparing the two images, the interpreter can identify features and areas that represent regions on the landscape that tend to depolarize the signal. Such areas will reflect the incident horizontally polarized signal back to the antenna as vertically polarized energy—that is, they change the polarization of the incident microwave energy. Such areas can be identified as bright regions on the HV image and as dark or dark gray regions on the corresponding HH image. Their appearance on the HV image is much brighter due to the effect of depolarization; the polarization of the energy that would have contributed to the brightness of the HH image has been changed, so it creates instead a bright area on the HV image. Comparison of the two images therefore permits detection of those areas that are good depolarizers.

TRANSMIT   RECEIVE

LIKE-POLARIZED IMAGE (HH or VV)

TRANSMIT   RECIEVE

CROSS-POLARIZED IMAGE (HV or VH)

FIGURE 7.9. Radar polarization. Many imaging radars can transmit and receive signals in both horizontally and vertically polarized modes. By comparing the like-polarized and cross-polarized images, analysts can learn about characteristics of the terrain surface. From NASA-JPL. P45541, SIR C/X SAR, October 1994.

This same information can be restated in a different way. A surface that is an ineffective depolarizer will tend to scatter energy in the same polarization in which it was transmitted; such areas will appear bright on the HH image and dark on the HV image. In contrast, a surface that is a "good" depolarizer will tend to scatter energy in a polarization different from that of the incident signal; such areas will appear dark on the HH image and bright on the HV image. Causes of depolarization are related to physical and electrical properties of the ground surface. A rough surface (with respect to the wavelength of the signal) may depolarize the signal.

Another cause of depolarization is volume scattering from an inhomogeneous medium; such scatter might occur if the radar signal is capable of penetrating beneath the soil surface (as might conceivably be possible in some desert areas where vegetation is sparse and the soil is dry enough for significant penetration to occur), where it might encounter subsurface inhomogenities, such as buried rocks or indurated horizons.

7.7. Look Direction and Look Angle

Look Direction

Look direction, the direction at which the radar signal strikes the landscape, is important in both natural and man-made landscapes. In natural landscapes, look direction is especially important when terrain features display a preferential alignment. Look directions perpendicular to topographic alignment will tend to maximize radar shadow, whereas look directions parallel to topographic orientation will tend to minimize radar shadow. In regions of small or modest topographic relief, radar shadow may be desirable as a means of enhancing microtopography or revealing the fundamental structure of the regional terrain. The extent of radar shadow depends not only on local relief but also on orientations of features relative to the flight path; those features positioned in the near-range portion (other factors being equal) will have the smallest shadows, whereas those at the far-range edge of the image will cast larger shadows (Figure 7.10). In areas of high relief, radar shadow is usually undesirable, as it masks large areas from observation.

In landscapes that have been heavily altered by human activities, the orientation of structures and land-use patterns are often a significant influence on the character of the radar return, and therefore on the manner in which given landscapes appear on radar imagery. For instance, if an urban area is viewed at a look direction that maximizes the scattering of the radar signal from structures aligned along a specific axis, it will have an appearance quite different from that of an image acquired at a look direction that tends to minimize reflection from such features.

Look Angle

Look angle, the depression angle of the radar, varies across an image, from relatively steep at the near-range side of the image to relatively shallow at the far-range side (Figure 7.11). The exact values of the look angle vary with the design of specific radar systems, but some broad generalizations are possible concerning the effects of varied look angles. First, the basic geometry of a radar image ensures that the resolution of the image must vary with look angle; at steeper depression angles, a radar signal illuminates a smaller

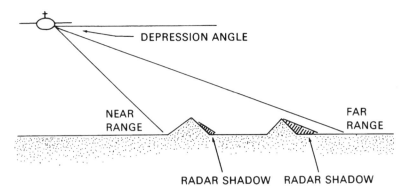

FIGURE 7.10. Radar shadow. Radar shadow increases as terrain relief increases and depression angle decreases.

area than does the same signal at shallow depression angles. Therefore, the spatial resolution, at least in the across-track direction, varies with respect to depression angle.

It has been shown that the sensitivity of the signal to ground moisture is increased as depression angle becomes steeper. Furthermore, the slant-range geometry of a radar image means that all landscapes are viewed (by the radar) at oblique angles. As a result, the image tends to record reflections from the sides of features. The obliqueness, and therefore the degree to which we view sides, rather than tops, of features, varies with look angle. In some landscapes, the oblique view may be very different from the overhead view to which we are accustomed in the use of other remotely sensed imagery. Such variations in viewing angle may contribute to variations in the appearance on radar imagery of otherwise similar landscapes.

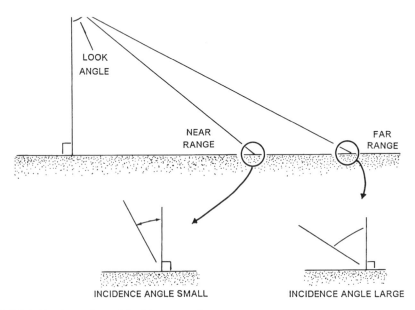

FIGURE 7.11. Look angle and incidence angle.

7.8. Real Aperture Systems

Real aperture SLAR systems (sometimes referred to as *brute force* systems), one of the two strategies for acquiring radar imagery, are the oldest, simplest, and least expensive of imaging radar systems. They follow the general model described earlier for the basic configuration of a SLAR system (Figure 7.4). The transmitter generates a signal at a specified wavelength and of a specified duration. The antenna directs this signal toward the ground and then receives its reflection. The reflected signal is amplified and filtered, then displayed on a cathode ray tube, where a moving film records the radar image line by line as it is formed by the forward motion of the aircraft.

The resolution of such systems is controlled by several variables. One objective is to focus the transmitted signal to illuminate as small an area as possible on the ground, as it is the size of this area that determines the spatial detail recorded on the image. If the area illuminated is large, then reflections from diverse features may be averaged together to form a single graytone value on the image, and their distinctiveness is lost. If the area is small, individual features are recorded as separate features on the image, and their identities are preserved.

The size of the area illuminated is controlled by several variables. One is antenna length in relation to wavelength. A long antenna length permits the system to focus energy on a small ground area. Thus real aperture systems require long antennas in order to achieve fine detail; limits on the ability of an aircraft to carry long antennas forms a practical limit to the resolution of radar images. This restriction on antenna length forms a barrier for use of real aperture systems on spacecraft: Small antennae would provide very coarse resolution from spacecraft altitudes, but practical limitations prevent use of large antennae.

The area illuminated by a real aperture SLAR system can be considered analogous to the spot illuminated by a flashlight aimed at the floor. As the flashlight is aimed straight down, the spot of light it creates is small and nearly circular in shape. As the flashlight beam is pointed toward the floor at increasingly further distances, the spot of light becomes larger, dimmer, and assumes a more irregular shape. By analogy, the near-range portions of a radar image will have finer resolution than the far-range portions. Thus antenna length in relation to wavelength determines the angular resolution of a real aperture system—the ability of the system to separate two objects in the along-track dimension of the image (Figure 7.12).

The relationship between resolution and antenna length and wavelength is given by the equation

$$\beta = \lambda/A \qquad\qquad \text{(Eq. 7.1)}$$

where β is the beamwidth, λ is the wavelength, and A is the antenna length. Real aperture systems can be designed to attain finer along-track resolution by increasing the length of the antenna and by decreasing wavelength. Therefore, these qualities form the design limits on the resolution of real aperture systems—antenna length is constrained by practical limits of aeronautical design and operation.

Radar systems have another, unique means of defining spatial resolution. The length of the radar pulse determines the ability of the system to resolve the distinction between two objects in the cross-track axis of the image (Figure 7.13). If long pulses strike two

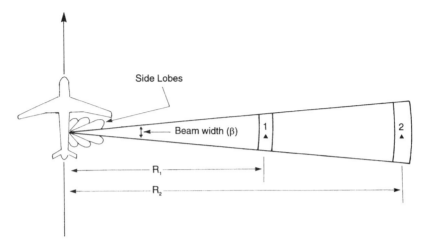

FIGURE 7.12. Azimuth resolution. For real aperture radar, the ability of the system to acquire fine detail in the along-track axis derives from its ability to focus the radar beam to illuminate a small area. A long antenna, relative to wavelength, permits the system to focus energy on a small strip of ground, improving detail recorded in the along-track dimension of the image. The *beam width* (β) measures this quality of an imaging radar. Beam width, in relation to range (R), determines detail—region 1 at range R_1 will be imaged in greater detail than region 2 at greater range R_2. Also illustrated here are *side lobes*, smaller beams of microwave energy created because the antenna cannot be perfectly effective in transmitting a single beam of energy.

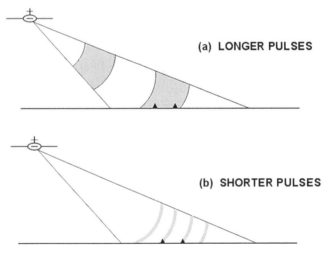

FIGURE 7.13. Effect of pulse length. (a) Longer pulse length means that the two objects shown here are illuminated by a single burst of energy, creating a single echo that cannot reveal the presence of two separate objects. (b) Shorter pulse length illuminates the two objects with separate pulses, creating separate echoes for each object. Pulse length determines resolution in the cross-track dimension of the image.

nearby features at the same time, they will record the two objects as a single reflection and therefore as a single feature on the image. In contrast, shorter pulses are each reflected separately from adjacent features and can record the distinctive identities and locations of the two objects.

7.9. Synthetic Aperture Systems

The alternative design is the *synthetic aperture radar* (SAR), which is based on principles and technology differing greatly from those of real aperture radars. Although SAR systems have greater complexity and are more expensive to manufacture and operate than are real aperture systems, they can overcome some of the limitations inherent to real aperture systems, and therefore they can be applied in a wider variety of applications, including observation from Earth-orbiting satellites.

Consider a SAR that images the landscape as depicted in Figure 7.14. At 1, the aircraft is positioned so that a specific region of the landscape is just barely outside the region illuminated by the SAR. At 2, it is fully within the area of illumination. At 3, it is just at the trailing edge of the illuminated area. Finally, at 4, the aircraft moves so that the region falls just outside the area illuminated by the radar beam. A synthetic aperture radar operates on the principle that objects within a scene are illuminated by the radar over an interval of time, as the aircraft moves along its flight path. A SAR system receives the signal scattered from the landscape during this interval and saves the complete history

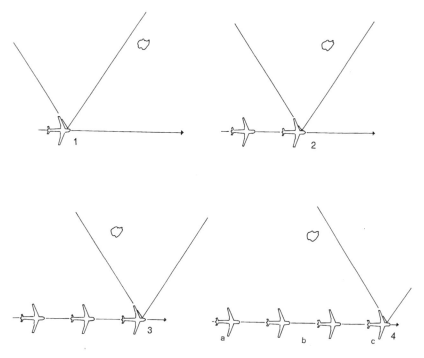

FIGURE 7.14. Synthetic aperture imaging radar. Synthetic aperture systems accumulate a history of backscattered signals from the landscape as the antenna moves along path *abc*.

of reflections from each object. Knowledge of this history permits later reconstruction of the reflected signals as though they were received by a single antenna occupying physical space *abc*, even though they were in fact received by a much shorter antenna that was moved in a path along distance 1-2-3-4. (Thus the term *synthetic aperture* denotes the artificial length of the antenna, in contrast to the *real* aperture based on the actual physical length of the antenna used with real aperture systems.)

In order to implement this strategy, it is necessary to define a practical means of assigning separate components of the reflected signal to their correct positions as the spatial representation of the landscape is re-created on the image. This process is, of course, extraordinarily complicated if each such assignment must be considered an individual problem in unraveling the complex history of the radar signal at each of a multitude of antenna positions.

Fortunately, this problem can be solved in a practical manner because of the systematic changes in frequency experienced by the radar signal as it is scattered from the landscape. Objects within the landscape experience different frequency shifts in relation to their distances from the aircraft track. At a given instant, objects at the leading edge of the beam reflect a pulse with an increase in frequency (relative to the transmitted frequency) due to their position ahead of the aircraft, and those at the trailing edge of the antenna experience a decrease in frequency (Figure 7.15). This is the *Doppler effect*, often explained by analogy to the change in pitch of a train whistle heard by a stationary observer as a train passes by at high speed. As the train approaches, the pitch appears higher than that of a stationary whistle due to the increase in frequency of sound waves. As the train passes the observer, then recedes into the distance, the pitch appears lower due to the decrease in frequency. Radar, as an active remote sensing system, is operated with full knowledge of the frequency of the transmitted signal. As a result, it is possible to compare the frequencies of transmitted and reflected signals to determine the nature and amount of frequency shift. Knowledge of frequency shift permits the system to assign

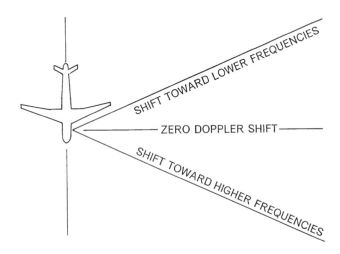

FIGURE 7.15. Frequency shifts experienced by features within the field of view of the radar system.

reflections to their correct positions on the image and to synthesize the effect of a long antenna.

7.10. Interpreting Brightness Values

Each radar image is composed of many image elements of varying brightness (Figure 7.16). Variations in image brightness correspond, at least in part, to place-to-place changes within the landscape; through knowledge of this correspondence, the image interpreter has a basis for making predictions, or inferences, concerning landscape properties. Unlike passive remote sensing systems, active systems illuminate the land with radiation of known and carefully controlled properties. Therefore, in principle, the interpreter should have a firm foundation for deciphering the meaning of the image because the only "unknowns" of the many variables that influence image appearance are the ground conditions—the object of study.

However, in practice, the interpreter of a radar image faces many difficult obstacles in making a rigorous interpretation of a radar image. First, most imaging radars are uncalibrated in the sense that image brightness values on an image cannot be quantitatively matched to backscattering values in the landscape. Typically, returned signals from a terrain span a very broad range of magnitudes. Furthermore, the features that compose even the simplest landscapes have complex shapes and arrangements and are formed from diverse materials of contrasting electrical properties. As a result, there are often few detailed models of the kinds of backscattering that should in principle be expected from separate classes of surface materials.

Direct experience and intuition are not always reliable guides to interpretation of images acquired outside the visible spectrum. In addition, many SLAR images observe

FIGURE 7.16. Two examples of radar images illustrating their ability to convey detailed information about quite different landscapes. Left: agricultural fields of the Maricopa Agricultural Experiment Station, Phoenix, Arizona (Ku-band, spatial resolution, about 1 m resolution); varied tones and textures distinguish separate crops and growth states. Right: structures near athletic fields, University of New Mexico, Albuquerque (Ku-band, spatial resolution, about 1 m.). Varied brightnesses and tones convey information about diffuse surfaces, specular reflection, and corner reflectors, which each carry specific meaning within the context of different landscapes. From Sandia National Laboratories. Reproduced by permission.

the landscape at very shallow depression angles. Because interpreters gain experience from their observations at ground level or from studying overhead aerial views, they may find the oblique radar view from only a few degrees above the horizon difficult to interpret.

Speckle

SAR images are subject to fine-textured effects that can create a grainy salt-and-pepper appearance when viewed in detail (such as that visible within the agricultural fields at the left of Figure 7.16). Speckle is created by radar illumination of separate scatterers that are too small to be individually resolved (i.e., small relative to wavelength). Because the radar signal is coherent (transmitted at a very narrow range of wavelengths), the energy scattered by small, adjacent, features tend either to be reinforced (*constructive interference*), or to be suppressed (*destructive interference*) (see later discussion pertaining to Figure 7.25). When high amplitude peaks of the waveform coincide, they create bright returns; alternatively, when the high amplitude peaks match to the low amplitude values of the waveforms, they tend to cancel each other, creating a dark return. Because speckle constitutes a form of noise (it does not convey useful information), it is usually processed either by *multilook* processes that illuminate the scene with slightly differing frequencies that produce independent returns—which then can be averaged to reduce the effect of the speckle—or by local averaging that smoothes the speckled effect. These two alternative filtering strategies are characterized either as *nonadaptive* or *adaptive*. Nonadaptive filters require less computation—they apply a single filter to the entire image uniformly, whereas the more complex adaptive filters adjust to match local properties of the terrain, thereby preserving natural edges and boundaries. All strategies, although intended to extract accurate estimates of backscatter, run the risk of eliminating genuine high-frequency information within the scene, so one must always balance benefits and losses within each image.

The Radar Equation

The fundamental variables influencing the brightness of a region on a radar image are formally given by the *radar equation*:

$$P_r = \frac{\sigma G^2 P_t \lambda^2}{(4\pi)^3 R^4} \qquad \text{(Eq. 7.2)}$$

Here P_r designates the power returned to the antenna from the ground surface; R specifies the range to the target from the antenna; P_t is the transmitted power; λ is the wavelength of the energy; and G is the antenna gain (a measure of the system's ability to focus the transmitted energy). All of these variables are determined by the design of the radar system and are therefore known or controlled quantities. The one variable in the equation not thus far identified is σ, the backscattering coefficient; σ is, of course, not controlled by the radar system but by the specific characteristics of the terrain surface represented by a specific region on the image. Whereas σ is often an incidental factor for the radar engineer, it is the primary focus of study for the image interpreter, as it is this quantity that carries information about the landscape.

The value of σ conveys information concerning the amount of energy scattered from a specific region on the landscape as measured by σ°, the *radar cross section*. It specifies the corresponding area of an isotropic scatterer that would return the same power as does the observed signal. The backscattering coefficient (σ°) expresses the observed scattering from a large surface area as a dimensionless ratio between two areal surfaces; it measures the average radar cross section per unit area. σ° varies over such wide values that it must be expressed as a ratio rather than as an absolute value.

Ideally, radar images should be interpreted with the objective of relating observed σ° (varied brightnesses) to properties within the landscape. It is known that backscattering is related to specific *system variables*, including wavelength, polarization, and azimuth, in relation to landscape orientation and depression angle. In addition, *landscape parameters* are important, including surface roughness, soil moisture, vegetative cover, and microtopography. Because so many of these characteristics are interrelated, making detailed interpretations of individual variables is usually very difficult, in part due to the extreme complexity of landscapes, which normally are intricate compositions of diverse natural and man-made features. Often many of the most useful landscape interpretations of radar images have attempted to recognize integrated units defined by assemblages of several variables rather than to separate individual components. The notion of "spectral signatures" is very difficult to apply in the context of radar imagery because of the high degree of variation in image tone as incidence angle and look direction change.

Moisture

Moisture in the landscape influences the backscattering coefficient through changes in the dielectric constant of landscape materials. (The *dielectric constant* is a measure of the ability of a substance to conduct electrical energy—an important variable determining the response of a substance that is illuminated with microwave energy.) Although natural soils and minerals vary in their ability to conduct electrical energy, these properties are difficult to exploit as the basis for remote sensing because the differences between dielectric properties of separate rocks and minerals in the landscape are overshadowed by the effects of even very small amounts of moisture, which greatly change the dielectric constant. As a result, the radar signal is sensitive to the presence of moisture both in the soil and in vegetative tissue; this sensitivity appears to be greatest at steep depression angles.

The presence of moisture also influences effective skin depth; as the moisture content of surface soil increases, the signal tends to scatter from the surface. As moisture content decreases, skin depth increases, and the signal may be scattered from a greater thickness of soil.

Roughness

A radar signal that strikes a surface will be reflected in a manner that depends both on characteristics of the surface and properties of the radar wave, as determined by the radar system and the conditions under which it is operated. The *incidence angle* (θ) is defined as the angle between the axis of the incident radar signal and a perpendicular to the surface that the signal strikes (Figure 7.17). If the surface is homogeneous with respect to its electrical properties and "smooth" with respect to the wavelength of the signal, then the reflected signal will be reflected at an angle equal to the incidence angle, with most of the energy directed in a single direction (i.e., specular reflection).

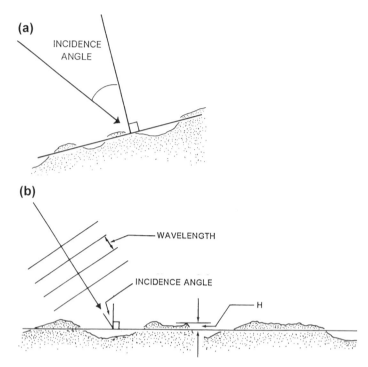

FIGURE 7.17. Measurement of incidence angle (a) and surface roughness (b).

For "rough" surfaces, reflection will not depend as much on incidence angle, and the signal will be scattered more or less equally in all directions (i.e., diffuse, or isotropic, scattering). For radar systems, the notion of a rough surface is defined in a manner considerably more complex than that familiar from everyday experience, as roughness depends not only on the physical configuration of the surface but also on the wavelength of the signal and its incidence angle (Table 7.2). Consider the physical configuration of the surface to be expressed by the standard deviation of the heights of individual facets (Figure 7.17). Although definitions of surface roughness vary, one common definition defines a rough surface as one in which the standard deviation of surface height (S_h) exceeds one-eighth of the wavelength (λ) divided by the cosine of the incidence angle (cos θ):

$$S_h > \lambda/(8 \cos \theta) \tag{Eq. 7.3}$$

where h is the average height of the irregularities. In practice, this definition means that a given surface appears rougher as wavelengths become shorter. Also, for a given wave-

TABLE 7.2. Surface Roughness Defined for Several Wavelengths

Roughness category	K-band (λ = 0.86 cm)	X-band (λ = 3 cm)	L-band (λ = 25 cm)
Smooth	h < 0.05 cm	h < 0.17 cm	h < 1.41 cm
Intermediate	h = 0.05–0.28 cm	h = 0.17–0.96 cm	h = 1.41–8.04 cm

Note. Data from Jet Propulsion Laboratory (1982).

length, surfaces will act as smooth scatterers as incidence angle becomes greater (i.e., equal terrain slopes will appear as smooth surfaces as depression angle becomes smaller, as occurs in the far-range portions of radar images).

Corner Reflectors

The return of the radar signal to the antenna can be influenced not only by moisture and roughness but also by the broader geometric configuration of targets. Objects that have complex geometric shapes, such as those encountered in an urban landscape, can create radar returns that are much brighter than would be expected based on size alone. This effect is caused by the complex reflection of the radar signal directly back to the antenna in a manner analogous to a ball that bounces from the corner of a pool table directly back to the player. This behavior is caused by objects classified as *corner reflectors*, which often are, in fact, corner-shaped features (such as the corners of buildings and the alleyways between them in a dense urban landscape) but that are also formed by other objects of complex shape. Corner reflectors are common in urban areas due to the abundance of concrete, masonry, and metal surfaces constructed in complex angular shapes (Figure 7.18). Corner reflectors can also be found in rural areas, formed sometimes by natural surfaces, but more commonly by metallic roofs of farm buildings, agricultural equipment, and items such as power line pylons and guardrails along divided highways.

Corner reflectors are important in interpretation of the radar image. They form a characteristic feature of the radar signatures of urban regions and identify other features, such as power lines, highways, and railroads (Figure 7.19). It is important to remember that the image of a corner reflector is not shown in proportion to its actual size: The returned energy forms a star-like burst of brightness that is proportionately much larger than the size of the object that caused it. Thus they can convey important information but do not appear on the image in their correct relative sizes.

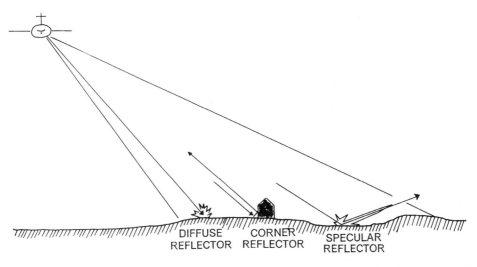

DIFFUSE   CORNER   SPECULAR
REFLECTOR   REFLECTOR   REFLECTOR

FIGURE 7.18. Three classes of features important for interpretation of radar imagery. See Figure 2.22.

FIGURE 7.19. SAR image of Los Angeles, California, acquired at L-band. This image illustrates the classes of features represented on radar imagery, including diffuse, specular, and corner reflectors. This image depicts the internal structure of the built-up urban region, including transportation, roads, and highways. Radar foreshortening is visible in its representation of the mountainous topography. From NASA-JPL; SIR-C/X-SAR, October 1994. PIA1738.

7.11. Satellite Imaging Radars

Scientists working with radar remote sensing have been interested for years in the possibility of observing the Earth by means of imaging radars carried by Earth satellites. Whereas real aperture systems cannot be operated at satellite altitudes without unacceptably coarse spatial resolution (or use of impractically large antennas), the synthetic aperture principle permits compact radar systems to acquire imagery of fine spatial detail at high altitudes. This capability, combined with the ability of imaging radars to acquire imagery in darkness, through cloud cover, and during inclement weather, provides the opportunity for development of a powerful remote sensing capability with potential to observe large parts of the Earth's ocean and land areas that might otherwise be unobservable because of remoteness and atmospheric conditions.

The following sections describe SAR satellite systems, beginning with accounts of experimental systems that evaluated concepts in advance of operational systems being designed. Table 7.3 lists several of the operational SAR satellite systems that have been deployed for specific application missions.

TABLE 7.3. Summary of Some SAR Satellite Systems

System	Dates	Bands	Polarizations	Lead organization
ERS-1	1991–2000	C-band	VV	European Space Agency
JERS-1	1992–1998	L-band	HH	Japan
SIR-C	1994	X/C/L-bands	full	USA
RADARSAT	1995–present	C-band	HH	Canada
ERS-2	1995–present	C-band	VV	European Space Agency
ENVISAT	2002–present	C-band	dual	European Space Agency
ALOS	2006–present	L-band	full	Japan
RADARSAT-2	2007–present	C-band	full	Canada
TerraSAR-X	2007–present	X-band	full	Germany
COSMO-SkyMed Constellation	2007–present	X-band		Italy
TanDEM-X	2010–present	X-band	full	Germany

Seasat Synthetic Aperture Radar

Seasat (Figure 7.20) was specifically tailored to observe the Earth's oceans by means of several sensors designed to monitor winds, waves, temperature, and topography. Many of the sensors detected active and passive microwave radiation, although one radiometer operated in the visible and near infrared spectra. (Seasat carried three microwave radiometers, an imaging radar, and a radiometer that operated both in the visible and in the infrared. Radiometers are described in Chapter 9.) Some sensors were capable of observing 95% of the Earth's oceans every 36 hours. Specifications for the Seasat SAR are as follows:

FIGURE 7.20. Seasat SAR geometry.

- *Launch*: 28 June 1978
- *Electrical system failure*: 10 October 1978
- *Orbit*: Nearly circular at 108° inclination; 14 orbits each day
- *Frequency*: L-band (1.275 GHz)
- *Wavelength*: 23 cm
- *Look direction*: Looks to starboard side of track
- *Swath width*: 100 km, centered 20° off-nadir
- *Ground resolution*: 25 m × 25 m
- *Polarization*: HH

Our primary interest here is the SAR, designed to observe ocean waves, sea ice, and coastlines. The satellite orbit was designed to provide optimum coverage of oceans, but its track did cross land areas, thereby offering the opportunity to acquire radar imagery of the Earth's surface from satellite altitudes. The SAR, of course, had the capability to operate during both daylight and darkness and during inclement weather, so (unlike most other satellite sensors) the Seasat SAR could acquire data on both ascending and descending passes. The high transmission rates required to convey data of such fine resolution meant that Seasat SAR data could not be recorded onboard for later transmission to ground stations; instead, the data were immediately transmitted to a ground station. Therefore, data could be acquired only when the satellite was within line of sight of one of the five ground stations equipped to receive Seasat data; these were located in California, Alaska, Florida, Newfoundland, and England.

The SAR was first turned on in early July 1978 (Day 10); it remained in operation for 98 days. During this time, it acquired some 500 passes of data: the longest SAR track covered about 4,000 km in ground distance. Data are available in digital form, as either digitally or optically processed imagery. All land areas covered are in the northern hemisphere, including portions of North America and western Europe.

Seasat SAR data have been used for important oceanographic studies. In addition, Seasat data have provided a foundation for the study of radar applications within the Earth sciences and for the examination of settlement and land-use patterns. Because Seasat data were acquired on both ascending and descending passes, the same ground features can be observed from differing look angles under conditions that hold most other factors constant, or almost constant.

Shuttle Imaging Radar-A

The Shuttle Imaging Radar (SIR) is a synthetic aperture imaging radar carried by the Shuttle Transportation System. (NASA's space shuttle orbiter has the ability to carry a variety of scientific experiments; the SIR is one of several.) SIR-A, the first scientific payload carried on board the shuttle, was operated for about 54 hours during the flight of *Columbia* in November 1981. Although the mission was reduced in length from original plans, SIR-A was able to acquire images of almost 4,000,000 mi^2 of the Earth's surface. Additional details of SIR-A are given below:

- *Launch*: 12 November 1981
- *Land*: 14 November 1981
- *Altitude*: 259 km

- *Frequency*: 1.278 GHz (L-band)
- *Wavelength*: 23.5 cm
- *Depression angle*: 40°
- *Swath width*: 50 km
- *Polarization*: HH
- *Ground resolution*: about 40 m × 40 m

SIR-A's geometric qualities were fixed, with no provision for changing depression angle. All data were recorded on magnetic tape and signal film on board the shuttle; after the shuttle landed, it was physically carried to ground facilities for processing into image products. First-generation film images were typically 12 cm (5 in.) in width at a scale of about 1:5,250,000 (Figure 7.21). About 1,400 ft. of film were processed; most has been judged to be of high quality. Portions of all continents (except Antarctica) were imaged to provide images of a wide variety of environments differing with respect to climate, vegetation, geology, land use, and other qualities.

Shuttle Imaging Radar-B

The Shuttle Imaging Radar-B (SIR-B) was the second imaging radar experiment for the space shuttle. It was similar in design to SIR-A, except that it provided greater flexibility in acquiring imagery at varied depression angles, as shown below:

- *Orbital altitude*: 225 km
- *Orbital inclination*: 57°
- *Frequency*: 1.28 GHz
- *Wavelength*: 23 cm

FIGURE 7.21. SIR-C image, northeastern China, northeast of Beijing, 1994. Light-colored dots indicate locations of villages sited on an agricultural plain, bordered by rough uplands, with reservoirs visible in the upper portion of the image. From JPL/NASA and the USDS EROS Data Center.

- *Resolution*: about 25 m × 17 m (at 60° depression) or about 25 m × 58 m (at 15° depression)
- *Swath width*: 40–50 km

SIR-B used horizontal polarization at L-band (23 cm). Azimuth resolution was about 25 m; range resolution varied from about 17 m at an angle of 60° to 58 m at 15°. The shuttle orbited at 225 km; the orbital track drifted eastward 86 km each day (at 45° latitude). In contrast to the fixed geometric configuration of the Seasat SAR and SIR-A, the radar pallet onboard the shuttle permitted control of the angle of observation, which thereby enabled operation of the SIR in several modes. A given image swath of 40–50 km could be imaged repeatedly, at differing depression angles, by changing the orientation of the antenna as spacecraft position changed with each orbit. This capability made it possible to acquire stereo imagery.

Also, the antenna could be oriented to image successive swaths and thereby build up coverage of a larger region than could be covered in any single pass. A mosaic composed of such images would be acquired at a consistent range of depression angles, although angles would, of course, vary within individual images.

The shuttle orbit also provided the opportunity to examine a single study area at several look directions. Varied illumination directions and angles permit imaging of a single geographic area at varied depression angles and azimuths and thereby enable (in concept) derivation of information concerning the influence of surface roughness and moisture content on the radar image and information concerning the interactions of these variables with look direction and look angle.

Thus an important function of SIR-B was its ability to advance understanding of the radar image itself and its role in remote sensing of the Earth's landscapes. The significance of SIR imagery arises in part from the repetitive nature of the orbital coverage, with an accompanying opportunity to examine temporal changes in such phenomena as soil moisture, crop growth, land use, and the like. Because of the all-weather capability of SAR, the opportunities to examine temporal variation with radar imagery may be even more significant than they were with Landsat imagery. Second, the flexibility of the shuttle platform and the SIR experiment permit examination of a wide range of configurations for satellite radars, over a broad range of geographic regions, by many subject areas in the Earth sciences. Therefore, a primary role for SIR-B was to investigate specific configurations and applications for the design of other satellite systems tailored to acquire radar imagery of the Earth's surface.

Shuttle Imaging Radar-C/X-Synthetic Aperture Radar System

In August 1994 the space shuttle conducted a third imaging radar experiment. The Shuttle Imaging Radar-C (SIR-C) is an SAR operating at both L-band (23 cm) and C-band (6 cm), with the capability to transmit and receive both horizontally and vertically polarized radiation. SIR-C, designed and manufactured by the Jet Propulsion Laboratory (Pasadena, California) and Ball Communications Systems Division, is one of the largest and most complex items ever built for flight on the shuttle. In addition to the use of two microwave frequencies and its dual polarization, the antenna has the ability to electronically aim the radar beam to supplement the capability of the shuttle to aim the antenna by maneuvering the spacecraft. Data from SIR-C can be recorded onboard using tape storage, or it can be transmitted by microwave to the Tracking and Data Relay Satellite System (TDRSS) link to ground stations (Chapter 6).

The X-synthetic aperture radar (X-SAR) was designed and built in Europe as a joint German–Italian project for flight on the shuttle, to be used independently or in coordination with SIR-C. X-SAR was an L-band SAR (VV polarization) with the ability to create highly focused radar beams. It was mounted in a manner that permits it to be aligned with the L- and C-band beams of the SIR-C.

Together, the two systems had the ability to gather data at three frequencies and two polarizations. The SIR-C/X-SAR experiment was coordinated with field experiments that collected ground data at specific sites devoted to studies of ecology, geology, hydrology, agriculture, oceanography, and other topics. These instruments continue the effort to develop a clearer understanding of the capabilities of SAR data to monitor key processes at the Earth's surface (Plate 9).

X-SAR characteristics can be briefly summarized:

- *Frequencies*:
  - X-band (3 cm)
  - C-band (6 cm)
  - L-band (23 cm)
- *Ground swath*: 15–90 km, depending on orientation of the antenna
- *Resolution*: 10–200 m

European Resource Satellite Synthetic Aperture Radar

The European Space Agency (ESA), a joint organization of several European nations, designed a remote sensing satellite with several sensors configured to conduct both basic and applied research. Here our primary interest is the SAR for European Resource Satellite-1 (ERS-1) (launched in 1991, in service till 2000) and ERS-2 (launched in 1995). One of the satellite's primary missions is to use several of its sensors to derive wind and wave information from 5 km × 5 km SAR scenes positioned within the SAR's 80-km swath width. Nonetheless, the satellite has acquired a library of images of varied land and maritime scenes (Figure 7.22).

Because of the SAR's extremely high data rate, SAR data cannot be stored onboard, so it must be acquired within range of a ground station. Although the primary ground control and receiving stations are located in Europe, receiving stations throughout the world are equipped to receive ERS data. ERS has a sun-synchronous, nearly polar orbit that crosses the equator at about 10:30 a.m. local sun time. Other characteristics include:

- *Frequency*: 5.3 GHz
- *Wavelength*: C-band (6 cm)
- *Incidence angle*: 23° at midrange (20° at near range; 26° at far range)
- *Polarization*: VV
- *Altitude*: 785 km
- *Spatial resolution*: 30 m
- *Swath width*: 100 km

ERS data are available from

RADARSAT International/MDA
Tel: 604-278-3411

FIGURE 7.22. ERS-1 SAR image, Sault St. Marie, Michigan–Ontario, showing forested areas as bright tones, rough texture. Within water bodies, dark areas are calm surfaces; lighter tones are rough surfaces caused by winds and currents. Copyright 1991 by ESA. Received and processed by Canada Centre for Remote Sensing; distributed by RADARSAT International.

Website: *www.rsi.ca*
NASA Alaska SAR Facility
Tel: 907-474-6166
Website: *www.asf.alaska.edu/dataset_documents/ers1_and_ers2_sar_images.html*

RADARSAT Synthetic Aperture Radar

RADARSAT is a joint project of the Canadian federal and provincial governments, the United States government, and private corporations. The United States, through NASA, provides launch facilities and the services of a receiving station in Alaska. Canada has special interests in use of radar sensors because of its large territory, the poor illumination prevailing during much of the year at high latitudes, unfavorable weather conditions, interest in monitoring sea ice in shipping lanes, and many other issues arising from assessing natural resources, especially forests and mineral deposits. Radar sensors, particularly those carried by satellites, provide capabilities tailored to address many of these concerns.

The RADARSAT's C-band radar was launched on 4 November 1995 into a sunsynchronous orbit at 98.6° inclination, with equatorial crossings at 6:00 a.m. and 6:00

p.m. An important feature of the RADARSAT SAR is the flexibility it offers to select among a wide range of trade-offs between area covered and spatial resolution and to use a wide variety of incidence angles (Figure 7.23). This flexibility increases opportunities for practical applications. Critical specifications include:

- *Frequency*: 5.3 GHz
- *Wavelength*: 5.6 cm (C-band)
- *Incidence angle*: varies, 10° to 60°
- *Polarization*: HH
- *Altitude*: 793–821 km
- *Spatial resolution*: varies, 100 × 100 m to 9 × 9 m
- *Swath width*: varies, 45–510 km
- *Repeat cycle*: 24 days

Data are available from:

RADARSAT International/MDA
13800 Commerce Parkway
Richmond V6V 2J3, BC, Canada
Tel: 604-278-3411
Fax: 604-231-2751
E-mail: *info@mdacorporation.com*
Website: *www.mdacorporation.com/index.shtml*

RADARSAT 2, launch in December, 2007, acquires imagery at spatial resolutions of 3 to 100 m, depending on the selection of imaging mode. Because it can acquire imagery looking laterally in either direction from the satellite track, it offers the capability to observe a specified region at frequent intervals.

Applications of RADARSAT SAR imagery include geological exploration and mapping, sea ice mapping, rice crop monitoring, flood delineation and mapping, coastal zone mapping, and oil spill/seep detection. Because superstructures of ships form good corner

FIGURE 7.23. RADARSAT SAR geometry.

reflectors, SAR data are effective in detecting ships at sea, and RADARSAT imagery has been used in the monitoring of shipping lanes. Figure 7.24 shows a RADARSAT SAR image of Cape Breton Island, Nova Scotia, Canada, on 28 November 1995; this image represents a region extending about 132 km east–west and 156 km north–south. At the time that RADARSAT acquired this image, this region was in darkness and was experiencing high winds and inclement weather, so this image illustrates radar's ability to operate under conditions of poor illumination and unfavorable weather. Sea conditions near Cape Breton Island's Cape North and Aspy Bay are recorded as variations in image tone. Ocean surfaces appearing black or dark are calmer waters sheltered from winds and currents by the configuration of the coastline or the islands themselves. Brighter surfaces are rougher seas, usually on the windward coastlines, formed as winds and current interact with subsurface topography. The urban area of Sydney is visible as the bright area in the lower center of the image.

FIGURE 7.24. RADARSAT SAR image, Cape Breton Island, Canada. RADARSAT's first SAR image, acquired 28 November 1995. Image copyright 1995 by Canadian Space Agency. Received by the Canada Center for Remote Sensing; processed and distributed by RADARSAT International.

Japanese Earth Resources Satellite-1

The Japanese Earth Resources Satellite-1 (JERS-1) is a satellite launched in February 1992 by the National Space Development Agency of Japan and continued in service till 1998. It carried a C-band SAR and an optical CCD sensor sensitive in seven spectral regions from 0.52 to 2.40 μm, with resolution of about 18 m. The JERS-1 SAR had the following characteristics:

- *Altitude*: 568 km
- *Orbit*: sun-synchronous
- *Wavelength*: 23 cm (L-band) (1.3 GHz)
- *Incidence angle*: 35°
- *Swath width*: 75 km
- *Spatial resolution*: 18 m
- *Polarization*: HH

The URLs for JERS-1 are:

www.eorc.nasda.go.jp/JERS-1

http://southport.jpl.nasa.gov/polar/jers1.html

COSMO–SkyMed

COSMO–SkyMed (Constellation of Small Satellites for Mediterranean Basin Observation) is a program of the Italian Space Agency, operated in collaboration with other agencies of the Italian government, for a system of SAR satellites. The system consists of a constellation of four satellites, launched in June and December of 2007, in October 2008, and in 2010, each with SAR systems. The first satellite was equipped with an X-band SAR and later satellites with more advanced multimode X-, C- L- and P-band instruments. The full constellation of four satellites offers a revisit time of a few hours on a global scale, which can be operated as individual systems or as pairs in an interferometric configuration (as outlined below) to acquire 3D SAR imagery by combining radar measurements from two satellites of the same target from separate incidence angles. COSMO–SkyMed is designed to meet both defense and civil applications, such as fires, urban planning, landslides, droughts, floods, Earthquakes, and management of natural resources in agriculture and forestry. For further information, see

www.telespazio.it/cosmo.html

www.astronautix.com/craft/coskymed.htm

Envisat

In 2002, the European Space Agency launched Envisat, designed to monitor a broad suite of environmental variables using ten different instruments. The sensor of primary interest here is the ASAR—the Advanced Synthetic Aperture Radar—an advanced synthetic aperture imaging radar based on experience gained from ERS-1 and ERS-2. Envisat completes 14 orbits each day, in a sun-synchronous orbit. ASAR operates at C-band, with the ability to employ a variety of polarizations, beam widths, and resolutions. Specifics are available at:

http://envisat.esa.int/m-s

7.12. Radar Interferometry

Interferometric SAR (*inSAR*, or sometimes, *ifSAR*)) uses two SAR images of the same region acquired from two different positions. In a very rough sense, SAR interferometry is comparable to the use of stereo photography to determine topography of a region by observation from two different perspectives. However, SAR interferometry is applied not in the optical domain of photogrammetry but in the realm of radar geometry, to exploit radar's status as an active sensor.

Envision two SAR images of the same region acquired simultaneously from slightly different flight (or orbital) tracks. This situation establishes a *spatial baseline*, in which two images are acquired simultaneously from separate tracks. Because SAR is an active sensor, the characteristics of the transmitted signal are known in detail. Also, because SAR illuminates the terrain with coherent radiation of known wavelength, it is possible to evaluate not only variations in brightnesses of the returned signals but also variations in the *phases* of the scattered signal—that is, the extent to which the peaks of the transmitted waveforms align with those of the scattered signal from the terrain (Figure 7.25). The composite formed by the interaction between phases of the transmitted and scattered signals is referred as an *interferogram*, which shows differences in phase between the transmitted and scattered signals (Figure 7.26). Because the differing positions of the two antennae are known, differences in phase can be translated to differences in terrain elevation (Figure 7.27).

Thus the interferogram can be processed to reveal differences in topographic elevation that have generated the phase differences, thereby providing an accurate representation of the terrain. This analytical process, known as *phase unwrapping*, generates elevation differences expressed in multiples of the SAR wavelength displayed as contour-like fringes, known as *wrapped color*. Plates 10 and 11 illustrate elevation data acquired using this technique. The most notable example of use of the across-track configuration for the two antennas was the 2000 Shuttle Radar Topography Mission (SRTM), in which a second antenna was extended from the U.S. space shuttle orbiter to complement the primary antenna in the shuttle cargo bay (Figure 7.28). This system permitted accurate mapping of the Earth's terrain between 60° N and 56° S latitude.

FIGURE 7.25. Phase. Waveforms that match with respect to corresponding points on their cycles are said to be *in phase*. Otherwise, they are *out of phase*. Phase is measured by the *phase angle* (θ), measured in degrees or radians; θ varies between 0 and 2π radians.

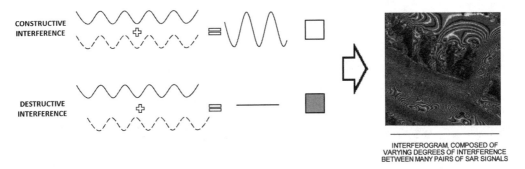

INTERFEROGRAM, COMPOSED OF
VARYING DEGREES OF INTERFERENCE
BETWEEN MANY PAIRS OF SAR SIGNALS

FIGURE 7.26. Constructive and destructive interference. Two waveforms that are in phase will combine to create a more intense brightness, indicated by the white pixel. Two waveforms that are out of phase will cancel each other out, creating a weak signal, signified by the dark pixel. InSAR signals received from the ground by two antennas at different locations will produce an array of many observations of varying degrees constructive and destructive interference, forming the kind of interference pattern illustrated at the right.

Other variations of this strategy can yield other kinds of information. If the two images are collected at different times, such as different orbital passes (*repeat-pass interferometry*), the system establishes a *temporal baseline*, which provides an image pair that can reveal changes that occurred during the interval between the acquisition of the two images. In a scientific context, the motion might record ocean currents, ice flow in glaciers, or ice floes in polar oceans. The sensitivity of such analyses depends on the nature of the temporal baseline, so that very short temporal baselines can record rather rapid motion (such as vehicular motion), whereas longer baselines can detect slower speeds, such as movement of surface ice in glaciers. Paired antennas operated from a single platform are sensitive to velocities of centimeters per second (e.g., suitable for observing moving vehicles or ocean waves). Longer temporal baselines (e.g., separate passes within the

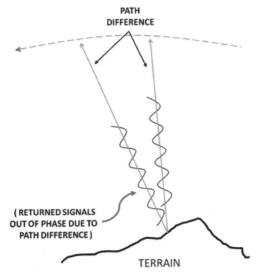

FIGURE 7.27. InSAR can measure terrain elevation because returned signals received at different orbital positions are out of phase due to differing path lengths from the antenna to the terrain.

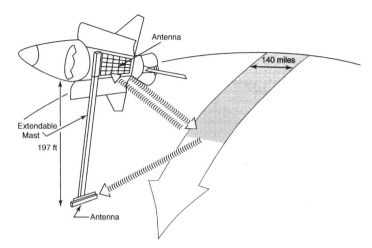

FIGURE 7.28. SRTM interferometry.

same orbital path) are effective in recording slower speeds, perhaps centimeters per day, such as the motion of glacial ice. Under these conditions, phase differences can reveal tectonic uplift or subsistence, movement of glaciers, or changes in vegetative cover, for example. These capabilities of acquiring reliable topographic data and of recording landscape changes have established interferometric SAR as an important tool in geophysics and Earth sciences.

Shuttle Radar Topography Mission

The most striking application of SAR interferometry is the Shuttle Radar Topography Mission (SRTM) (STS-99, Shuttle *Endeavour*), launched 11 February 2000, returned 22 February 2000, for a duration of about 11 days, 5½ hours. More information is available at

   www.jpl.nasa.gov/srtm

SRTM was designed to use C-band and X-band interferometric SAR to acquire topographic data over 80% of Earth's land mass (between 60° N and 56° S). One antenna was mounted, as for previous shuttle SAR missions, in the shuttle's payload bay; another antenna was mounted on a 60-m (200-ft.) mast extended after the shuttle attained orbit (see Figure 7.28). The configuration was designed for single-pass interferometry using C-band and X-band SAR imagery. The interferometric data have been used to produce digital topographic data for very large portions of the world at unprecedented levels of detail and accuracy. SRTM provided, for the first time, consistent, high-quality, cartographic coverage of most of the Earth's land areas (Figure 7.29).

SRTM data have been used for a broad range of practical applications in geology and geophysics, hydrologic modeling, aircraft navigation, transportation engineering, land-use planning, siting of communications facilities, and military training and logistical planning. Furthermore, they will support other remote sensing applications by facilitating rectification of remotely sensed data and registration of remotely acquired image data.

FIGURE 7.29. Sample SRTM data, Massanutten Mountain and the Shenandoah Valley of northern Virginia. Within the U.S., SRTM data are distributed at 30 m; elsewhere, at 90 m. From Jet Propulsion Laboratory, Pasadena, California, and the EROS Data Center, Sioux Falls, South Dakota. February 2000, PIA03382BW.

(SRTM data are consistent with U.S. National Map Accuracy Standards.) The SRTM was sponsored by the U.S. National Imagery and Mapping Agency (NIMA), NASA, the German aerospace center (Deutsches Zentrum fur Luft- und Raumfart [DLR]), and the Italian space agency (Agenzia Spaziale Italiana [ASI]) and managed by the Jet Propulsion Laboratory, Pasadena, California. Data are distributed by the EROS Data Center, Sioux Falls, South Dakota.

7.13. Summary

Radar imagery is especially useful because it complements characteristics of images acquired in other portions of the spectrum. Aerial photography, for example, provides excellent information concerning the distribution and status of the Earth's vegetation cover. The information it conveys is derived from biologic components of plant tissues. However, from aerial photography we learn little direct information about the physical structure of the vegetation. In contrast, although active microwave imagery provides no data about the biologic component of the plant cover, it does provide detailed information concerning the physical structure of plant communities. We would not expect to replace aerial photography or optical satellite data with radar imagery, but we could expect to be able to combine information from microwave imagery to acquire a more complete understanding of the character of the vegetation cover. Thus the value of any sensor must be assessed not only in the context of its specific capabilities but also in the context of its characteristics relative to other sensors.

7.14. Some Teaching and Learning Resources

- Free online software resources
  - ESA NEST: *www.array.ca/nest/tiki-index.php*
  - ROI_PAC: *www.asf.alaska.edu/softwaretools*

- ASF: *http://enterprise.lr.tudelft.nl.dors*
- DORIS: *www.hi.is/~ahoopper/stamps/index.html*
- IDIOT: *http://srv-43-200.bv.tu-berlin.de/idiot/*
- ESA: *http://envisat.esa.int/resources/softwaretools*

- Free online tutorials
  - ESA: *http://earth.esa.int/polsarpro/tutorial.html*
  - CCRS: *www.ccrs.nrcan.gc.ca/sarrso/index_e.php*
  - JPL: *http://southport.jpl.nasa.gov/scienceapps/dixon/index.html*
  - ENVI: *http://geology.isu.edu/dml/ENVI_Tutorials/SAR_Process.pdf*
  - NPA: *www.npagroup.com/insar/whatisinsar/insar_simple.htm#interferometry*

- Videos
  - Radar Satellite Mapping Floodings, University of Tartu
    www.youtube.com/watch?v=9fiR4RGah8U
  - InSAR—Interferometric Synthetic Aperture Radar
    www.youtube.com/watch?v=0SIhWWgzE1w&feature=related
  - Satellite Reconnaissance—SAR Lupe Germany [German narration, English subtitles]
    www.youtube.com/watch?v=roPlHnq7cFc
  - TerraSAR-X [German narration]
    www.youtube.com/watch?v=SPZV2xzU5kA&feature=related
  - TerraSAR-X–Deutschlands Radarauge im All
    www.youtube.com/watch?v=HxcOmldQ_78&feature=related
  - Synthetic Aperture Radar Video
    www.youtube.com/watch?v=P1k2RzHlHik
  - Comparison of SRTM and Intermap DTM
    www.youtube.com/watch?v=YNArJBJ—4U

Review Questions

1. List advantages for the use of radar images relative to images from aerial photography and Landsat TM. Can you identify disadvantages?

2. Imaging radars may not be equally useful in all regions of the Earth. Can you suggest certain geographic regions where they might be most effective? Are there other geographic zones where imaging radars might be less effective?

3. Radar imagery has been combined with data from other imaging systems, such as the Landsat TM, to produce composite images. Because these composites are formed from data from two widely separated portions of the spectrum, together they convey much more information than can either image alone. Perhaps you can suggest (from information already given in Chapters 3 and 6) some of the problems encountered in forming and interpreting such composites.

4. Why might radar images be more useful in many less developed nations than in industrialized nations? Can you think of situations in which radar images might be especially useful in the industrialized regions of the world?

5. A given object or feature will not necessarily have the same appearance on all radar images. List some of the factors that will determine the texture and tone of an object as it is represented on a radar image.

6. Seasat was, of course, designed for observation of Earth's oceans. Why are the steep depression angles of the Seasat SAR inappropriate for many land areas? Can you think of advantages for use of steep depression angles in some regions?

7. What problems would you expect to encounter if you attempted to prepare a mosaic from several radar images?

8. Why are synthetic aperture radars required for radar observation of the Earth by satellite?

9. Why was the shuttle imaging radar so important in developing a more complete understanding of interpretation of radar imagery?

References

Brown, W. M., and L. J. Porcello. 1969. An Introduction to Synthetic Aperture Radar. *IEEE Spectrum*, Vol. 6, No. 9, pp. 52–62.

Cheney, M. 2001. A Mathematical Tutorial on Synthetic Aperture Radar. *SIAM Review*, Vol. 43, pp. 301–312.

Drunpob, A., N. B. Chang, M. Beaman, C. Wyatt, and C. Slater. 2005. Seasonal Soil Moisture Variation Analysis Using RADARSAT-1 Satellite Imagery in a Semi-Arid Coastal Watershed. *IEEE International Workshop on the Analysis of Multi-Temporal Remote Sensing Images*, 186–190.

Elachi, C., et al. 1982a. Shuttle Imaging Radar Experiment. *Science*, Vol. 218, pp. 996–1003.

Elachi, C., et al. 1982b. Subsurface Valleys and Geoarcheology of the Eastern Sahara. *Science*, Vol. 218, pp. 1004–1007.

Evans, D. L., J. J. Plant, and E. R. Stofan. 1997. Overview of the Spaceborne Imaging Radar-C/Xband Synthetic Aperture Radar (SIR-C/X-SAR) Missions. *Remote Sensing of Environment*, Vol. 59, pp. 135–140.

Ford, J. P., J. B. Cimino, and C. Elachi. 1983. *Space Shuttle Columbia Views the World with Imaging Radar: The SIR-A Experiment* (JPL Publication 82-95). Pasadena, CA: Jet Propulsion Laboratory, 179 pp.

Gens, R. 1999. Quality Assessment of Interferometrically Derived Digital Elevation Models. *International Journal of Applied Earth Observation and Geoinformation*, Vol. 1, pp. 102–108.

Gens, R., and J. L. Vangenderen. 1996. SAR Interferometry: Issues, Techniques, Applications. *International Journal of Remote Sensing*, Vol. 17, pp. 1803–1835.

Henderson, F. M., and A. J. Lewis (eds.). 1998. *Principles and Applications of Imaging Radars* (*Manual of Remote Sensing*, Vol. 2, R. A. Ryerson, editor-in-chief, American Society for Photogrammetry and Remote Sensing). New York: Wiley, 866 pp.

Jensen, H., L. C. Graham, L. J. Porcello, and E. M. Leith. 1977. Side-Looking Airborne Radar. *Scientific American*, Vol. 237, pp. 84–95.

Jet Propulsion Laboratory. 1982. *The SIR-B Science Plan* (JPL Publication 82-78). Pasadena, CA: Jet Propulsion Laboratory, 90 pp.

Jordan, R. L., B. L. Honeycutt, and M. Werner. 1991. The SIR-C/X-SAR Synthetic Aperture Radar System. *Proceedings of IEEE*, Vol. 79, pp. 827–838.

Kimura, H., and Y. Yamaguchi. 2000. Detection of Landslide Areas Using Satellite Radar Technology. *Photogrammetric Engineering and Remote Sensing*, Vol. 66, pp. 337–344.

Lasne, Y. P., T. Paillou, T. August-Bernex, G. Ruffié, and G. Grandjean. 2004. A Phase Signature for Detecting Wet Subsurface Structures using Polarimetric L-Band SAR. *IEEE Transactions on Geoscience and Remote Sensing*, pp. 1683–1694.

Raney, R. K., A. P. Luscombe, E. J. Langham, and S. Ahmed. 1991. RADASAT. *Proceedings of IEEE*, Vol. 79, pp. 839–849.

Sabins, F. F. 1983. Geologic Interpretation of Space Shuttle Radar Images of Indonesia. *AAPG Bulletin*, Vol. 67, pp. 2076–2099.

Sahebi, M., R. J. Angles, and F. Bonn, 2002. A Comparison of Multi-Polarization and Multi-Angular Approaches for Estimating Bare Soil Surface Roughness from Spaceborne Radar Data. *Canadian Journal of Remote Sensing*, Vol. 28, pp. 641–652.

Simpson, R. B. 1966. Radar, Geographic Tool. *Annals, Association of American Geographers*, Vol. 56, pp. 80–96.

Ulaby, F., P. C. Dubois, and J. van Zyl. 1996. Radar Mapping of Surface Soil Moisture, *Journal of Hydrology*, Vol. 184, pp. 54–84.

CHAPTER EIGHT

Lidar

8.1. Introduction

Lidar—an acronym for "light detection and ranging"—can be considered analogous to radar imagery, in the sense that both families of sensors are designed to transmit energy in a narrow range of frequencies, then receive the backscattered energy to form an image of the Earth's surface. Both families are active sensors; they provide their own sources of energy, which means they are independent of solar illumination. More important, they can compare the characteristics of the transmitted and returned energy—the timing of pulses, the wavelengths, and the angles—so they can assess not only the brightness of the backscatter but also its angular position, changes in frequency, and the timing of reflected pulses. Knowledge of these characteristics means that lidar data, much like data acquired by active microwave sensors, can be analyzed to extract information describing the structure of terrain and vegetation features not conveyed by conventional optical sensors.

Because lidars are based on an application of *lasers*, they use a form of *coherent* light—light that is composed of a very narrow band of wavelengths—very "pure" with respect to color. Whereas ordinary light, even if it is dominated by a specific color, is composed of many wavelengths, with a diverse assemblage of waveforms, a laser produces light that is in phase ("coherent"), and composed of a narrow range of wavelengths ("monochromatic") (Figure 8.1.). Such light can be transmitted over large distances as a narrow beam that will diverge only slightly, in contrast with most light in our everyday experience.

The laser—an acronym for "light amplification by stimulated emission of radiation"—is an instrument that applies a strong electrical current to a "lasable" material, usually crystals or gases, such as rubies, CO_2, helium–neon, argon, and many other less familiar materials. Such *lasable* materials have atoms, molecules, or ions that emit light as they return to a normal ground state after excitement by a stimulus such as electricity or light. The emitted light forms the coherent beam described above. Each separate material provides a specific laser with its distinctive characteristics with respect to wavelength.

The laser provides an intense beam that does not diverge as it travels from the transmitter, a property that can favor applications involving heating, cutting (including surgery), etching, or illumination. Laser pointers, laser printers, CD players, scanners, bar code readers, and many other everyday consumer items are based on laser technology. Although imaging lasers, of course, do not use intense beams, they do exploit the focused, coherent nature of the beam to produce very detailed imagery. A laser uses mirrored sur-

FIGURE 8.1. Normal (top) and coherent (bottom) light.

faces to accumulate many pulses to increase the intensity of the light before it leaves the laser (Figure 8.2).

8.2. Profiling Lasers

Lasers were invented in the late 1950s. Initially they were used for scientific inquiry and industrial applications. The first environmental applications of lidars were used principally for *atmospheric profiling*: Static lasers can be mounted to point upward into the atmosphere to assess atmospheric aerosols. Solid particles suspended in the atmosphere direct a portion of the laser beam back to the ground, where it is measured to indicate the abundance of atmospheric particles. Because lasers can measure the time delay of the backscatter, they can assess the clarity of the atmosphere over a depth of several kilometers, providing data concerning the altitudes of the layers they detect.

The first airborne lasers were designed as *profiling lasers*—lasers aimed directly

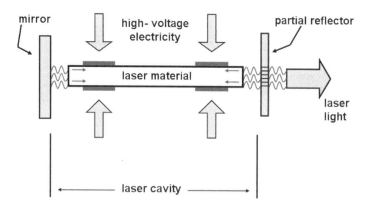

FIGURE 8.2. Schematic diagram of a simple laser. Energy, such as electricity, is applied to a substance, such as lasable gases (e.g., nitrogen, helium–neon) or materials (e.g., ruby crystals). When the materials return to their normal state, they emit coherent light, which is intensified before release by multiple reflections between the mirrored surfaces. Intensified light then can then pass through the semitransparent mirror to form the beam of coherent light that is emitted by the instrument.

beneath the aircraft in the nadir to illuminate a single region in the nadir position. (When used primarily to acquire topographic data, such instruments are known as *airborne laser altimeters*.) The forward motion of the aircraft carries the illuminated region forward to view a single track directly beneath the aircraft. Echoes from repetitive lidar pulses provide an elevation profile of the narrow region immediately beneath the aircraft (Figure 8.3). Although lidar profilers do not provide the image formats that we now expect, they provide a very high density of observations and are used as investigative tools for researchers investigating topography, vegetation structure, hydrography, and atmospheric studies, to list only a few of many applications.

8.3. Imaging Lidars

It is only relatively recently that lidars could be considered remote sensing instruments that collect images of the Earth's surface. By the late 1980s, several technologies matured and converged to create the context for the development of precision scanning lidar systems that we now know. *Inertial measurement units* (IMUs) enabled precise control and recording of orientation of aircraft (roll, pitch, and yaw). GPS could provide accurate records of geographic location of an aircraft as it acquired data. And development of highly accurate clocks permitted precise timing of lidar pulses required to create high-performance lidar scanning systems.

A lidar scanner can transmit up to 300,000 pulses each second, depending on the specific design and application. A scanning mirror directs the pulses back and forth across the image swath beneath the aircraft. The width of the swath is determined by the instrument's design and the operating conditions of the aircraft. Most imaging lidars

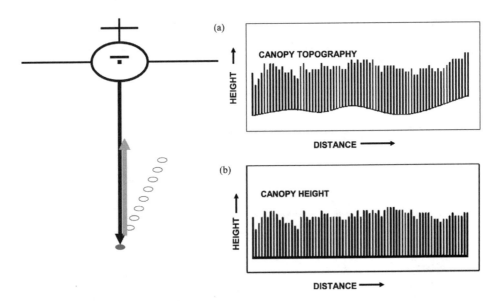

FIGURE 8.3. Schematic representation of an airborne laser profiler. (a) Acquisition of laser profiles; (b) Sample data gathered by a laser profiler, illustrating extraction of canopy height from the raw profile data.

use wavelengths in the visible (e.g., 0.532 µm, green, for penetration of water bodies) or near infrared (e.g., 1.64 µm for sensitivity to vegetation, ability to detect open water, and freedom from atmospheric scattering) regions of the spectrum.

Several alternative designs for imaging lidar instruments are in use (Habib, 2010). Figure 8.4 presents a schematic representation of a typical lidar system: (1) the system's laser (coordinated by the electronic component) generates a beam of coherent light, transmitted by a fiber optic cable to (2) a rotating mirror, offset to provide a scanning motion. The laser light is directed to a bundle of fiber optic cables that can be twisted to transmit the light as a linear beam. The oscillating motion of the mirror scans the laser beam side to side along the cross-track axis of the image, recording many thousands of returns each second. Because a lidar scanner is well integrated with GPS, IMU, and timing systems, these pulses can be associated with specific points on the Earth's surface.

As the reflected portion of the laser beam reaches the lidar aperture, it is received by another system of lenses and directed through fiber optic cables to another scanning lens (5), then directed through an optical system to filter the light before it is directed (6) to a receiving system to accept and direct the signal to the electronics component. The electronics coordinate timing of the pulses and permit matching of the signal with data from the inertial navigation system, and GPS. Together these components permit the system to accurately place each returned signal in its correct geographic position. Typically, two fiberglass bundles are configured to view the ground along a linear path. One transmits laser pulses, and an identical bundle receives the echoes. The system operates at such high speed that a high density of pulses are received from each square meter on the terrain. The timing capability of the lidar permits accurate assessment of distance and elevation, which permits formation of an image with detailed and accurate representation of elevations in the scene.

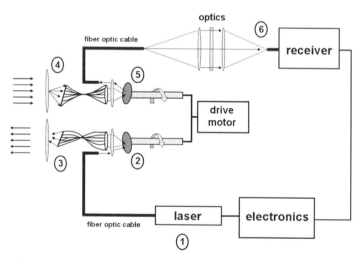

FIGURE 8.4. Schematic diagram of a lidar scanner. (1) The system's laser (coordinated by the electronic component) generates a beam of coherent light, transmitted by a fiber optic cable to (2) a rotating mirror, offset to provide a scanning motion. The laser light is directed to a bundle of fiber optic cables that are twisted to provide a linear beam and then directed through a system of lenses toward the ground. The energy received back from the terrain is received by another system of lenses and processed to form an image.

FIGURE 8.5. Examples of lidar flight lines. Upper image: raw lidar data, with brightness indicating increasing relative elevation. Lower image: lidar data processed with a hill-shading algorithm, revealing terrain texture; Wytheville, VA. From Virginia Department of Transportation. Copyright 2003, Commonwealth of Virginia. Reproduced by permission.

8.4. Lidar Imagery

Lidar imagery is acquired in parallel strips that match to form a continuous image of a region (Figure 8.5). In Figure 8.5 the upper image shows the raw lidar data, with each pixel representing the elevation of a specific point on the ground. Light tones represent higher elevations, darker tones represent lower elevations. The lower image is formed from the same data but is represented using a hill-shading technique that assumes that each pixel is illuminated from the upper left corner of the image. This effect creates image texture reminiscent of an aerial photograph, usually easier for casual interpretation.

Figures 8.6 and 8.7 show enlargements of portions of the same image, selected to show some of the distinctive qualities of lidar imagery. Figure 8.6 depicts an interstate interchange, with structures, forests, pastureland, and cropland, represented with precision and detail. Figure 8.7 shows a nearby region. A deep quarry is visible at the center. Open land and forest are again visible. The parallel strips visible near the upper right of this image depict mature corn fields—an indication of the detail recorded by these images.

8.5. Types of Imaging Lidars

Lidar instruments differ in their design. Most of the kind that are used as altimeters use *pulsed lasers* that generate very carefully timed bursts of light. These generate range information by measuring the time delay (travel time) for emitted and received pulses. Because light travels at a known constant velocity, the time for a pulse to return to the sensor translates directly to a distance, or range, between the aircraft and the target. For a given emitted pulse there is a corresponding waveform of return energy (or intensity) per unit of time. Some lidar sensors digitize this whole waveform in a discretized fashion, and are thus called *waveform lidars*. Others, known as *discrete return lidars*, only record the time and intensity of four to five returns for each emitted pulse. The means by which these discrete returns are selected from the return waveform depends on the sensor, but thresholding the intensity is common. It should be noted that the emitted pulse length

FIGURE 8.6. Section of a lidar image enlarged to depict detail. Upper image: raw lidar data, with brightness indicating increasing relative elevation. Lower image: lidar data processed with a hill-shading algorithm, revealing terrain texture; Wytheville, VA. From Virginia Department of Transportation. Copyright 2003, Commonwealth of Virginia. Reproduced by permission.

and the detector response time, along with other system engineering constraints, impose a minimum time (known as the dead time) between successive discrete returns. One practical outcome of this design is that a discrete return from understory vegetation that is near the ground surface often prevents accurate detection of ground surface elevation. This effect causes a systematic overestimate of ground elevation (and thus underestimates derived tree or building heights) in such areas.

The resolution of specific lidar systems can differ widely. Until recently, *large foot-print* systems were primarily waveform lidars and *small-footprint* systems were primarily discrete return lidars. This bifurcation based on footprint size is rapidly changing, however, with the advent of commercially available waveform digitizers for small-footprint systems. Flying height and beam divergence, along with other system design parameters, affect footprint size.

8.6. Processing Lidar Image Data

Organizations that collect lidar data usually use their own in-house processing systems specifically designed to manipulate the very large data volumes associated with each lidar mission, to perform the special operations tailored for each specific instrument, and to meet the requirements for each project. Filtering, detection of anomalies, and verification require reference to existing maps digital elevation models (see below), satellite imagery, and aerial photography. Often lidar missions are completed using ancillary photography gathered coincident with the lidar data.

As outlined below, a typical lidar project requires production of (1) a surface elevation model (SEM) representing the first surface intercepted by the lidar pulse, (2) a bare-earth digital elevation model (DEM) representing the terrain surface after removal of vegetation or structures, and (3) a canopy layer, representing the height of the canopy above the terrain surface. Some lidar data may be further processed to identify, isolate, and extract features such as structures and vegetation cover or to be fused with other forms of remotely sensed data, such as CIR aerial photography. Such products are usually provided in formats suitable for standard applications software, so the results can be further processed by end users.

Because the instrument can accurately record the angle of individual pulses, the attitude of the aircraft (using inertial navigational systems), and the position of the aircraft (using GPS), a lidar system can create an image-like array representing variation of elevation within the areas under observation (i.e., each pulse can be associated with a scan angle and data describing the position and orientation of the aircraft) (Figure 8.8.). Depending on the design of a specific lidar system, the positions of individual returns (known as *postings*) are irregularly spaced in their original, unprocessed form. These irregular positions are later interpolated to form a regular grid that provides a more

FIGURE 8.7. Section of a lidar image enlarged to depict detail. Left: raw lidar data, with brightness indicating increasing relative elevation. Right: Lidar data processed with a hillshading algorithm, revealing terrain texture; Wytheville, VA. From Virginia Department of Transportation. Copyright 2003, Commonwealth of Virginia. Reproduced by permission.

AS COLLECTED

AFTER INTERPOLATION

FIGURE 8.8. Acquisition of lidar data. Lidar systems acquire data by scanning in the pattern suggested by the top diagram; details vary according to specific systems. The pattern of returns is then interpolated to generate the regular array that forms the lidar image. Examples of actual scan pattern and interpolated data are depicted in Plate 12.

coherent representation of the terrain. Plate 12 shows this effect rather well: the upper portion of the illustration shows the original postings, whereas the lower half shows the systematic grid formed by the interpolation process.

Thus each return from the ground surface can be precisely positioned in xyz space to provide an array that records both position and elevation. For small-footprint lidars, horizontal accuracy might be in the range of 20–30 cm and vertical accuracy in the range of 15–20 cm. Therefore, the array, or image, forms a detailed DEM. With the addition of ground control (points of known location that can be accurately located within the imaged region, often found using GPS), lidar can provide data comparable in detail and positional accuracy to those acquired by photogrammetric analysis of aerial photographs. A lidar can record different kinds of returns from the terrain. Some returns, known as *primary returns*, originate from the first objects a lidar pulse encounters—often the upper surface of a vegetation canopy (Figure 8.9). In addition, portions of a pulse pass through gaps in the canopy into the interior structure of leaves and branches to lower vegetation layers and the ground surface itself. This energy creates echoes known as *secondary*, or *partial, returns*. Therefore, for complex surfaces such as forests with multiple canopies, some portions of a pulse might be reflected from upper and middle portions of the canopy and other portions from the ground surface at the base (Figure 8.10).

The total collection of lidar returns for a region can be examined to separate those returns that originated above a specified level from those that originated below that level.

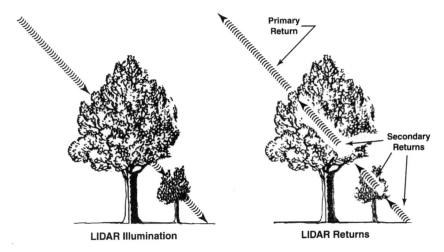

FIGURE 8.9. Schematic diagram of primary and secondary lidar returns.

This kind of approach can then form a kind of filtering process to separate ground returns from nonground returns and, with additional analysis, can thereby separate the terrain surface from overlying vegetation cover and structures (Figure 8.11 and Plate 13). Lidars are distinctive as one of the few sensors that can reliably differentiate between multiple imaged layers.

Lidar data may not accurately represent shorelines, stream channels, and ridges. Contours derived from lidar data may not form hydrographically coherent surfaces com-

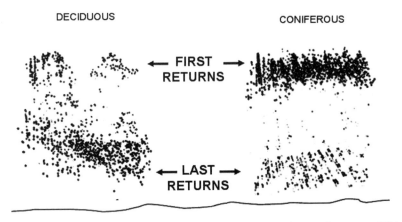

FIGURE 8.10. Primary and secondary lidar returns from two forested regions. This illustration represents lidar return from two separate forested areas, shown in profile. The dots near the top of the diagram represent the returns that are received first (primary returns), and the dots at the lower and central portions of the diagram represent the return received later (secondary returns). Note the contrast between the dome-shaped canopy formed by the crowns of the deciduous forest (left) and the peaked crowns of the coniferous canopy (right). The coniferous forest has only sparse undergrowth, while the deciduous forest is characterized by abundant undergrowth. From Peter Sforza and Sorin Popescu.

FIGURE 8.11. Bare-earth lidar surface with its first reflective surface shown above it. The data are from a multiple-return lidar system and show the wealth of data that can be extracted from lidar data. These data were collected at an altitude of 12,000 ft., with a nominal post spacing of 15 ft. and a vertical accuracy of about 14 in. RMSE. From EarthData.

parable to those represented on the usual contour maps. For terrain analysis, it is common for the analyst to insert breaklines, usually by manual inspection of the data in digital format, to separate the data array into discrete units that can be treated individually. *Breaklines* are interconnected points that define abrupt changes in terrain, such as edges of roads, drainage ditches, and ridgelines. Other boundaries outline exclusion areas, such as water bodies and dense forest, that are to be excluded from contouring. The analyst can digitize geomorphic features such as drainageways, road edges, sides, and ditches as "hard breaklines." More subtle variations in topography can be mapped as "soft breaklines." These can be inserted manually, then subsequently be considered while the data are used to generate a digital terrain model.

Often data are processed and organized in units that correspond to flightlines (Figure 8.5, for example). Such units are arbitrary, so some prefer to prepare units that correspond to USGS topographic quadrangles to facilitate storage and manipulation, especially if the lidar data are to be used in coordination with other data. In contrast, analysts who use the lidar data for modeling may prefer to organize the data in geographic units corresponding to drainage basins or to place the edges between units at ridgelines, where the seams may be less likely to influence the analysis.

Lidar imagery is increasingly finding a role as a source of data for DEMs. The lidar bare earth surface isolates topographic information from the effects of dense vegetation, which can be significant in other sources of elevation data. The lidar data can be interpolated to form a DEM, which often displays high accuracy and detail. Photogrammetric methods often encounter difficulties in representing topography covered by dense vegetation and the complex terrain near streams and drainageways, which in many applications may be regions where high accuracy is most valued. Although lidar data may be a superior source of elevation data in many situations, many applications require removal of buildings and other structures from the data and careful consideration of interpolation procedures.

8.7. Summary

Lidar data permit reliable, if perhaps imperfect, separation of vegetation from terrain—a capability unique among competing remote sensing instruments. Lidar provides a highly accurate, detailed representation of terrain. Its status as an active sensor permits convenience and flexibility in flight planning due to its insensitivity to variations in weather and solar illumination—both important constraints on aerial photography.

Lidar data provide detailed, spatial data of high accuracy and precision. Lidar data provide direct measurement of surface elevation, with detail and accuracy usually associated only with photogrammetric surveys. Some applications replace photogrammetric applications of aerial photography. Many of the best-known applications have focused on urban regions, which experience continuing needs for detailed information concerning building densities, urban structures, and building footprints. Lidar data are used for highway planning, pipeline routing planning and route planning and for designing wireless communication systems in urban regions (Figure 8.12 and Plate 14). However, several states have completed or are planning to acquire state-wide lidar coverage to support floodplain mapping programs and other efforts with broader geographic reach. Lidar

FIGURE 8.12. Old Yankee Stadium, New York City, as represented by lidar data. This lidar image was collected in the summer of 2000 as part of a citywide project to acquire data to identify line-of-sight obstructions for the telecommunications industry. The lidar data were collected from an altitude of 2,500 m, with a vertical accuracy of 1 m RMSE and a nominal post spacing of 4 m. From EarthData.

data have also been used to study forest structure, as the detailed and accurate information describing canopy configuration and structure may permit accurate mapping of timber volume. It clearly will find a broad range of environmental applications, in which its detailed representations of terrain will open new avenues of inquiry not practical with coarser data. As lidar archives acquire increasing geographic scope and temporal depth, the field will be able to expand its reach to examine sequential changes in vegetation cover, land use, and geomorphology (for example) with a precision and accuracy not previously feasible. However, practical applications of lidar data continue to face challenges, especially in defining practical processing protocols that can find wide acceptance across a range of user communities and application fields.

8.8. Some Teaching and Learning Resources

- Lidar: Light Detection and Ranging
  www.youtube.com/user/ASPRS#p/u/1/hxiRkTtBQp8
  and
  www.youtube.com/watch?v=hxiRkTtBQp8&fmt=22

- Lidar surface shadow model for Boston's Back Bay
  www.youtube.com/watch?v=s4OhzaIXMhg&NR=1

- Pylon Lidar Survey
  www.youtube.com/watch?v=Dv6a0KgTbiw

- Terrapoint Aerial Services—LiDAR Flight Simulation
  www.youtube.com/watch?v=GSPcyhSAgTQ&NR=1

- LandXplorer: LiDAR Scan of London
  www.youtube.com/watch?v=F2xy-US46PQ&NR=1

- Airborne 1 Hoover Dam LiDAR Fly-Through
  www.youtube.com/watch?v=JvauCmPAjuI

- Lidar Survey
  www.youtube.com/watch?v=f1P42oQHN_M&feature=related

- eCognition Image Analysis: Extracting Tree Canopy from Lidar
  www.youtube.com/watch?v=OR1Se18Zd4E

- 3D Map of Bournemouth
  www.youtube.com/watch?v=jANDq2Ad5H4&feature=related

Review Questions

1. Review some of the strengths of lidar data relative to other forms of remotely sensed data discussed thus far.

2. Many observers believe that the increasing availability of lidar will displace the current role of aerial photography for many applications. What are some of the reasons that might lead people to believe that lidar could replace many of the remote sensing tasks now fulfilled by aerial photography?

3. Can you identify reasons that aerial photography might yet retain a role, even in competition with strengths of lidar data?

4. If aerial photography is largely replaced by lidar, do you believe that there will still be a role for teaching aerial photography as a topic in a university remote sensing course? Explain.

5. Identify some reasons that lidar might be effectively used in combination with other data. What might be some of the difficulties encountered in bringing lidar data and (for example) fine-resolution optical satellite imagery together?

6. Assume for the moment that lidar data become much cheaper, easier to use, and, in general, more widely available to remote sensing practitioners. What kinds of new remote sensing analyses might become possible, or what existing analyses might become more widespread?

7. The text discusses how lidar imagery is based on the convergence of several technologies. Review your notes to list these technologies. Think about the technologic, scientific, social and economic contexts that foster the merging of these separate capabilities. How do you think we can prepare now to encourage future convergences of other technologies (now unknown) that might lead to advances in remote sensing instruments?

8. Lidar imagery many not be equally useful in all regions of the Earth. Can you suggest certain geographic regions or environments in which lidar data might not be effective?

9. Discuss some of the considerations that might be significant in deciding in what season to acquire lidar data of your region.

10. Identify some of the special considerations that might be significant in planning acquisition of lidar data in urban regions.

References

Anon. 2010. Airborne Lidar Processing Software. *GIM International* (February), pp. 14–15. (See also February 2007.)

Blair, J. B., D. L. Rabine, and M. A. Hofton. 1999. The Laser Vegetation Imaging Sensor: A Medium-Altitude, Digitization-Only, Airborne Laser Altimeter for Mapping Vegetation and Topography. *ISPRS Journal of Photogrammetry and Remote Sensing*, Vol. 54, pp. 115–122.

DeLoach, S. R., and J. Leonard. 2000. Making Photogrammetric History. *Professional Surveyor*, Vol. 20, No. 4, pp. 6–11.

Flood, M. 2001. Laser altimetry: From Science to Commercial LiDAR Mapping. *Photogrammetric Engineering and Remote Sensing*, Vol. 67, pp. 1209–1217.

Flood, M. 2002. Product Definitions and Guidelines for Use in Specifying LiDAR Deliverables. *Photogrammetric Engineering and Remote Sensing*, Vol. 67, pp. 1209–1217.

Fowler, R. A. 2000, March. The Lowdown on Lidar. *Earth Observation Magazine*, pp. 27–30.

Habib, Ayman F. 2010. Airborne LIDAR Mapping. Chapter 23 in *Manual of Geospatial Science and Technology*. 2nd Ed.) (J. D. Bossler, ed.) London: Taylor & Francis. pp. 439-465.

Hill, J. M., L. A. Graham, and R. J. Henry. 2000. Wide-Area Topographic Mapping Using Airborne Light Detection and Ranging (LIDAR) Technology. *Photogrammetric Engineering and Remote Sensing*, Vol. 66, pp. 908–914, 927, 960.

Hyyppä, J., W. Wagner, M. Hollaus, and H. Hyyppä. 2009. Airborne Laser Mapping. Chapter 14 in the *Sage Handbook of Remote Sensing* (T. A. Warner, M. Duane Nellis, and G. M. Foody, eds.). London: Sage, pp. 199-211.

Lefsky, M. A., D. Harding, W. B. Cohen, G. Parker, and H. H. Shugart. 1999. Surface Lidar Remote Sensing of Basal Area and Biomass in Deciduous Forests of Eastern Maryland, USA. *Remote Sensing of Environment*, Vol. 67, pp. 83–98.

Nelson, R., W. Krabill, and J. Tonelli. 1988a. Estimating Forest Biomass and Volume Using Airborne Laser Data. *Remote Sensing of Environment*, Vol. 24, pp. 247–267.

Nelson, R., R. Swift, and W. Krabill, W. 1988. Using Airborne Lasers to Estimate Forest Canopy and Stand Characteristics. *Journal of Forestry*, Vol. 86, pp. 31–38.

Nilsson, M. 1996. Estimation of Tree Heights and Stand Volume using an Airborne Lidar System. *Remote Sensing of Environment*, Vol. 56, pp. 1–7.

Popescu, S. C. 2002. *Estimating Plot-Level Forest Biophysical Parameters Using Small-Footprint Airborne Lidar Measurements*, PhD Dissertation, Virginia Tech, Blacksburg, VA, 144 pp.

Romano, M. E. 2004. Innovation in Lidar Processing. *Photogrammetric Engineering and Remote Sensing*, Vol. 70, pp. 1201–1206.

Sapeta, K. 2000. Have You Seen the Light? LIDAR Technology Is Creating Believers. *GEO-World*, Vol. 13, No. 10, pp. 32–36.

Wehr, A., and U. Lohr (eds.). 1999. Airborne Laser Scanning (Theme Issue). *ISPRS Journal of Photogrammetry and Remote Sensing*, Vol. 54, Nos. 2–3.

Wher, A., and U. Lohr. 1999. Airborne Laser Scanning: An Introduction and Overview. *ISPRS Journal of Photogrammetry and Remote Sensing*, Vol. 54, pp. 68–92.

White, S. A., and Y. Wang. 2003. Utilizing DEMs Derived from LIDAR Data to Analyze Morphologic Change in the North Carolina Coastline. *Remote Sensing of Environment*, Vol. 85., pp. 39–47.

Zhang, K., and D. Whitman. 2005. Comparison of Three Algorithms for Filtering Airborne Lidar Data. *Photogrammetric Engineering and Remote Sensing*, Vol. 71, pp. 313–324.

PLATE 1. Color and color infrared aerial photograph, Torch Lake, Michigan. Top: natural color aerial photograph. Bottom: color infrared aerial photograph of approximately the same region. Photographs courtesy of U.S. Environmental Protection Agency.

PLATE 2. High altitude aerial photograph: Corpus Christi, Texas, 31 January 1995. Image provided by U.S. Geological Survey.

PLATE 3. Density slicing. Density slicing is a form of image enhancement that assigns colors to brightness ranges within a black-and-white image. Colors do not add new information, but can assist in enhancing information present within the original image. Left: the original black-and-white image, with a histogram representing the frequencies of brightness values. Right: the same image after application of the density slice, with a histogram colored to represent approximate assignment of colors. The analyst can adjust the levels represented by colors to emphasize those features of greatest interest.

PLATE 4. Thematic Mapper Color Composite (Bands 1, 3, and 4). This image forms a color composite providing spectral information approximately equivalent to a color infrared aerial photograph (compare with Plates 1 and 2). This image shows the same region depicted by Figures 6.13 and 6.14: New Orleans, Louisiana, 16 September 1982. Scene ID: 40062-15591.

PLATE 5. SPOT color composite. SPOT 3 HRV image of the Strait of Bonifacio, separating Corsica (on the north) from Sardinia (on the south), 27 September 1993. CNES/SPOT Image. Reproduced by permission.

PLATE 6. SeaWifS image, 26 February 2000. This image shows a broad-scale view of the western coast of Africa. The tan-colored swirl is dust carried in winds blowing to the west from eastern Africa. Dust is lifted to altitudes as high as 15,000 ft.; some of it will be blown to the west, causing air quality alerts in Caribbean cities and contributing to sedimentation that threatens the health of coral reefs. © 2000 Orbital Imaging Corporation. All rights reserved.

PLATE 7. Canberra, Australia, as imaged by IKONOS multispectral imagery. This image illustrates high-resolution optical imagery, with a nominal resolution of about 1 m, that conveys spatial detail comparable to some aerial photographs (compare with Plate 2). Satellite image by GeoEye.

PLATE 8. Aral Sea shrinkage, 1962–1994. The left image is a mosaic of CORONA images from 1962, revealing the extent of the Aral Sea. The right image, acquired by AVHRR in 1994, has been prepared to the same scale and orientation as the CORONA image. The comparison reveals the dramatic decrease in limits of the Aral Sea during this interval, due mainly to large-scale diversion for irrigation of waters feeding the lake. Images prepared by U.S. Central Intelligence Agency and U.S. Geological Survey.

PLATE 9. Multifrequency SAR image, Barstow, California. This spaceborne SAR image depicts the Mojave Desert near Barstow, California (an area 46 by 30 mi (75 by 48 km) using several frequencies and polarizations. It was acquired by Spaceborne Imaging Radar-C/X-Band Synthetic Aperture Radar (SIR-C/X-SAR) on board the space shuttle *Endeavour*, April 1994. Colors represent frequencies and polarizations as follows: red is C-band, HV; green is L-band, HH; and blue is the ratio of C-band to L-band, both HH.

The city of Barstow is the pinkish area in the lower left center, on the banks of the Mojave River, visible as the diagonal light blue sinuous line. The orange circular and rectangular patches are irrigated agricultural fields. Radar data have been used to evaluate effects of irrigation on soil stability and land degradation. Sparsely vegetated areas of sand and fine gravel appear blue, and rocky hills and rougher gravel deposits appear mostly in shades of orange and brown. Jet Propulsion Laboratory, Pasadena California. P-4747308.

PLATE 10. Radar interferometry, Honolulu, Hawaii, 18 February 2000. The Shuttle Radar Topography Mission (SRTM) uses two radar antennae to observe the same topography from separate positions. Processing of the two sets of microwave returns generates topographic data, shown here using the "wrapped color" bands to signify elevation classes. SRTM data have been processed to provide elevation data for a very large proportion of the earth's surface. Image provided by Jet Propulsion Laboratory, Pasadena, California (PIA02720).

PLATE 11. Radar interferometry, Missouri River floodplain. This is a combined radar and topography image of an area along the Missouri River that experienced severe flooding and levee failure in the summer of 1993. The image represents an area of about 5 km × 10 km (3.1 × 6.2 mi.) centered south of the town of Glasgow in central Missouri.

The meandering course of the Missouri River is seen as the dark curving band on the left side of the image. The predominantly blue area on the left half of the image is the river's floodplain, which was completely inundated during the 1993 flood. The colors represent elevations, with low areas shown in purple, intermediate elevations in blue, green, and yellow, and the highest areas shown as orange. The total elevation range is 85 m (279 ft.). The higher yellow-and-orange area on the right side of the image shows the topography and drainage patterns typical of upland regions of this part of the midwestern United States. Dark streaks in the floodplain are agricultural areas that were severely damaged by levee failures during the flooding. The region enclosed by the C-shaped bend in the river in the upper part of the image is Lisbon Bottoms. A powerful outburst of water from a failed levee on the north side of Lisbon Bottoms scoured a deep channel across the fields, which shows as a purple band. As the flood waters receded, deposits of sand and silt were left behind, which now appear as dark, smooth streaks in the image. The yellow areas within the blue, near the river, are clumps of trees sitting on slightly higher ground within the floodplain. Microwave energy is sensitive to the rough, irregular form of the treetops, so they are represented as higher (yellow) than the fields.

The image was acquired by the NASA/JPL Topographic Synthetic Aperture Radar system (TOPSAR) that flew over the area aboard a DC-8 aircraft in August 1994. The elevations are obtained by radar interferometry, in which the radar signals are transmitted by one antenna and echoes are received by two antennas aboard the aircraft. The two sets of received signals are combined using computer processing to produce a topographic map. The brightness of the image represents the radar backscatter at C-band in the vertically transmitted and received polarization. Radar and interferometry processing for this image was performed at the Jet Propulsion Laboratory, Pasadena, California; image generation was performed at Washington University, St. Louis (P-46311).

PLATE 12. Interpolation of lidar returns. Most lidar systems do not collect data in a regular grid. In addition, the amount and detail of the lidar data is overwhelming for many applications. In the top half of this image, the irregular spacing of the lidar ground points can be seen, as well as the jerky, irregular nature of contours created directly from the lidar. In the bottom half, a regular grid has been interpolated from the lidar data, generalizing and smoothing the surface. The improvement in the appearance of the contour lines is apparent. These data were collected from an altitude of 12,000 ft., at a nominal post spacing of 15 ft., and a vertical accuracy of about 14 in. RMSE. The contours are at a 5-ft. interval. Image courtesy of Fugro EarthData, Inc.

PLATE 13. Lidar data used to model the built environment. Data from a multiple-return lidar system can be used to develop advanced surface models. The first reflective surface is used to determine the height of the buildings, while the last returns are used to construct a ground elevation model. Building footprints from the existing GIS are incorporated to define feature edges and enhance the final 3D surface. These data were collected from an altitude of 12,000 ft., at a nominal post spacing of 15 ft., and a vertical accuracy of about 14 in. RMSE. Images courtesy of Fugro EarthData, Inc.

PLATE 14. Glideslope surface estimated from lidar data. This image shows a rapidly growing use for lidar data. Unlike traditional photogrammetric techniques, lidar captures the first reflective surface by design. This allows quick and easy assessment of a variety of obstructions that cannot be collected as easily from stereo photography. In this example, the imaginary floor of a runway approach corridor is depicted in 3D as a flat blue surface. First return lidar data are rendered in their true location, and the penetrations of the trees through the "glideslope" can be identified. GIS can be used to map these obstructions for removal. These data were collected from an altitude of 12,000 ft., at a nominal post spacing of 15 ft., and a vertical accuracy of about 14 in. RMSE. Image courtesy of Fugro EarthData, Inc.

PLATE 15. Thermal images of residential structures showing thermal properties of separate elements of wooden structures. (March 1979, 11 P.M.; air temperature = 2°C.) Left: image shows the original thermal image (white = hot; dark = cool). Right: a color-coded version of the same data shown at the top, with each change in color representing a change in temperature of about 1.5°C, over a range from 2°C to 11°C. Colors represent increasing temperatures in the sequence black, magenta, blue, cyan, green, yellow, red, and white. Thermal images provided by Daedalus Enterprises, Inc., Ann Arbor, Michigan.

PLATE 16. Landscape near Erfurt, Germany, as observed by a thermal scanner in the region between 8.5 and 12 μm. The thermal data have been geometrically and radiometrically processed, then superimposed over digital elevation data, with reds and yellows assigned to represent warmer temperatures and blues and greens to represent cooler temperatures. See Chapter 9 for extended discussion. Images courtesy of EUROSENSE-BELFOTOP N.W. and Daedalus Enterprises Inc., Ann Arbor Michigan.

PLATE 17. Image cubes. Two examples of AVIRIS images displayed in image cube format. The top portion shows the image as recorded at the shortest wavelength, in conventional format with a perspective view. Sides of the cubes represent edges of each image as recorded in 220 channels, with the longest wavelength at the bottom (see Figure 15.4). Changes in colors along the sides of the cube represent changes in the spectral properties of surfaces as wavelength changes; the horizontal black stripes represent wavelengths of strong atmospheric absorption. Left: Moffet Field, CA. Right: Rogers Dry Lake, California. Images courtesy of NASA-Jet Propulsion Laboratory.

PLATE 18. Change image depicting the Stoughton WI F3 tornado of 18 August 2005. Red pixels indicate tornado damage as revealed by spectral differences between the two images. For such images to be of value in the response and assessment effort, there must be a short time difference between the before and after images, clear weather for collection of the revisit image, and collection of a pair of images (in this instance, TM and ASTER) with comparable spectral and spatial resolution. Courtesy of Environmental Remote Sensing at the Space Science and Engineering Center, University of Wisconsin-Madison, *http://ersc.ssec. wisc.edu/tornadoes.php*

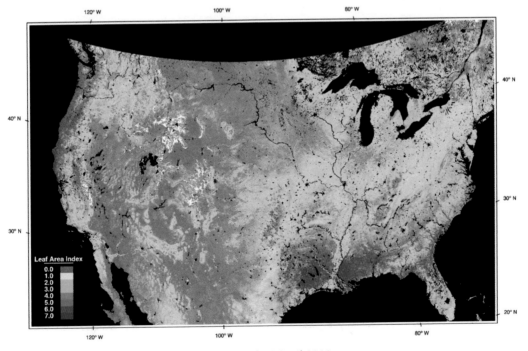

PLATE 19. MODIS leaf area index (LAI), 24 March–8 April 2000.

PLATE 20. Example of a ratio image. The left image shows a portion of a SPOT HRV image of a landscape in Rondonia, Brazil. The long axis of the rectangle marked on the image represents 2 km on the ground. Red represents forested regions; blue, dry grasses of open pasture; and lighter reds and oranges, brushy regrowth in open land. The right image shows the same image as represented using the normalized difference vegetation index, as discussed in the text. White represents pixels dominated by healthy vegetation. Black represents very sparsely vegetated pixels or, in this region, grassy regions with seasonally dry vegetation. Gray regions are pixels with varying amounts of brushy regrowth. Image © CNES/SPOT Image. Reproduced by permission.

PLATE 21. Time-integrated NDVI (TIN) estimates a daily (interpolated) NDVI above a baseline summed for the entire duration of the growing season and interpreted as growing season canopy photosynthetic activity (related to primary productivity). This image shows TIN for 2008. Darker blues and greens represent the higher values; beiges and browns, the lowest values. Shows a pattern related to vegetation cover, moisture, and insolation. USGS Remote Sensing Phenology. Acquired using AVHRR satellite sensors. USGS EROS Center (*http://phenology.cr.usgs.gov/*).

PLATE 22. Band ratios used to study lithologic differences, Cuprite, Nevada, mining district. Variations in color represent variations in ratios between spectral bands, as discussed in the text. Image courtesy of Geo-Sectra Corporation. From Vincent et al. (1984, p. 226).

PLATE 23. Alacranes Atoll, Gulf of Mexico. This image shows the coral reef system of the Alacranes Reef, 65 mi. due north of Progreso, a port town on the northern tip of the Yucatan Peninsula. It is a large reef system, 17 mi. long and 13 mi. wide, with five islands and atolls. In 1994, the UN decreed the reef a protected Reserve of the Biosphere. IKONOS 0.82-m resolution, 4 September 2008. Satellite image by GeoEye.

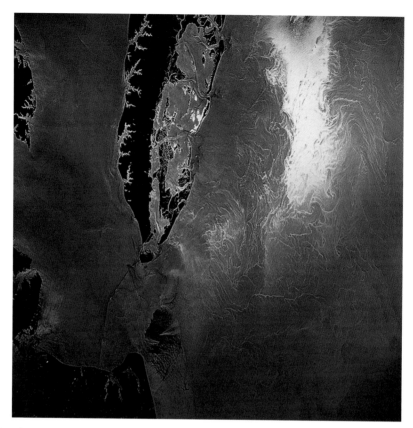

PLATE 24. Chesapeake Bay, as photographed 8 June 1991 from the space shuttle. Land areas appear in silhouette, having been underexposed in order to record hydrographic features. As a result, the land–water interface is very sharply delineated. Note for example, the coastal marshes discussed in Chapter 19 (section 19.10, Figures 19.19–19.22). The Chesapeake Bay occupies the left-hand portion of the image; Norfolk, Virginia, and associated naval facilities occupy the lower left corner but are not visible due to the underexposure. Just to the right of the Chesapeake Bay Bridge-Tunnel a curved front visible as the dark line at the mouth of the bay separates the waters of the bay from those of the Atlantic. The coastal waters of the Virginia coast are visually distinct from those of the bay, due to currents, surface oils, temperature, and wind and wave patterns. Near this front, wakes of several ships are visible. Photo courtesy of NASA (STS-40-614-047).

PLATE 25. Belgian port of Zeebrugge. The port and nearby waters were observed by the Daedalus digital multispectral scanner using 12 channels in the visible, near infrared, and thermal regions. Image data have been processed to provide correct geometry and radiometry, and, with the use of onsite observations, analyzed to reveal sediment content within the upper 1 m of the water column. Reds and yellows indicate high sediment content; and blues and greens represent clearer water. The left image shows conditions at low tide; the right image, high tide. Images courtesy of EUROSENSE-BELFOTOP N.W. and Daedalus Enterprises, Inc., Ann Arbor Michigan.

PLATE 26. Atlantic Ocean's Gulf Stream, 2 May 2001, as imaged by the MODIS spectroradiometer. The red pixels in this image show the warmer areas, greens are intermediate values, and blues are relatively low values. See Chapter 19 for a more complete discussion. These images were produced from data collected and processed by the University of Wisconsin—Madison's MODIS direct broadcast receiving station. Images courtesy of Liam Gumley, MODIS Atmosphere Team, University of Wisconsin—Madison Cooperative Institute for Meteorological Satellite Studies. NASA Visible Earth, EOS Project Science Office, NASA Goddard Space Flight Center.

PLATE 27. Landsat TM quarter scene depicting Santa Rosa del Palmar, Bolivia (northwest of Santa Bruise), July 1992, TM bands 3, 4, and 5. Here four distinct land-use patterns are visible. The landscape in the southeastern portion of the scene is dominated by broad-scale mechanized agriculture practiced by a Mennonite community. The upper right (northeastern) region of the image is a mountainous area occupied by a diminishing population of Indians, who practice a form of slash-and-burn agriculture, visible here as the dispersed patches of light green. In the image's northwest and central regions, the national government has encouraged broad-scale clearing of forest for agriculture following the construction of highways into the forest. In the upper left, these clearings appear as light green spots aligned northwest to southeast; at the center of each patch is a central facility providing colonists with fertilizers, pesticides, and staples. In the image's southwestern corner, the complex field pattern reflects an established agricultural landscape occupied by Japanese immigrants. GEOPIC image courtesy of Earth Satellite Corporation, Rockville, Maryland. Used by permission.

PLATE 28. Images depicting global remote sensing data. Top: AVHRR data representing sea surface temperature compiled from data collected during July 1984. The warmest surfaces are shown as red; then, in decreasing sequence, oranges, yellows, greens, and finally blues, representing the coolest temperatures. Bottom: composite data from two separate NASA satellites combined to represent the earth's biosphere. In the oceans, colors represent phytoplankton concentrations (highest values in browns and yellows, lower values in purples and dark blues). On land, greens represent dense vegetation, yellows depict sparsely vegetated regions. Images prepared by NASA Goddard Space Flight Center.

PLATE 29. Deepwater Horizon oil release, 2010. This annotated MODIS/Aqua satellite image, 14 July 2010, shows that the area of oil slicks and sheen appears greatly reduced: slicks cover approximately 3,786 sq. mi. (9,805 km^2) on this image. Radar satellite images taken on July 11 and July 12 confirm this smaller slick area. Persistent, moderately strong winds over the previous few days (ranging from 7 to 20 mi. per hr.) may have dispersed thinner portions of the slick over much of the region. Image courtesy of SkyTruth: *www.skytruth.org*

CHAPTER NINE

Thermal Imagery

9.1. Introduction

The literal meaning of *infrared* is "below the red," indicating that its frequencies are lower than those in the red portion of the visible spectrum. With respect to wavelength, however, the infrared spectrum is "beyond the red," having wavelengths longer than those of red radiation. The far infrared is often defined as having wavelengths from about 0.76 μm to about 1,000 μm (1 mm); there are great differences between properties of radiation within this range, as noted below.

The infrared spectrum was discovered in 1800 by Sir William Herschel (1738–1822), a British astronomer who was searching for the relationship between heat sources and visible radiation. Later, in 1847, two Frenchmen, A. H. L. Fizeau (1819–1896) and J. B. L. Foucault (1819–1868), demonstrated that infrared radiation has optical properties similar to those of visible light with respect to reflection, refraction, and interference patterns.

The infrared portion of the spectrum extends beyond the visible region to wavelengths about 1 mm (Figure 9.1). The shorter wavelengths of the infrared spectrum, near the visible, behave in a manner analogous to visible radiation. This region forms the reflective infrared spectrum (sometimes known as the "near infrared"), extending from about 0.7 μm to 3.0 μm. Many of the same kinds of films, filters, lenses, and cameras that we use in the visible portion of the spectrum can also be used, with minor variations, for imaging in the near infrared. The very longest infrared wavelengths are in some respects similar to the shorter wavelengths of the microwave region.

This chapter discusses use of radiation from about 7 to 18 μm for remote sensing of landscapes. This spectral region is often referred to as the *emissive infrared* or the *thermal infrared*. Of course, emission of thermal energy occurs over many wavelengths, so some prefer to use the term *far infrared* for this spectral region. In addition, this chapter also considers passive remote sensing of shorter microwave radiation, which resembles thermal remote sensing in some respects.

Remote sensing in the mid and far infrared is based on a family of imaging devices that differ greatly from the cameras and films used in the visible and near infrared. The interaction of the mid and far infrared with the atmosphere is also quite different from that of shorter wavelengths. The far infrared regions are free from the scattering that is so important in the ultraviolet and visible regions, but absorption by atmospheric gases restricts uses of the mid and far infrared spectrum to specific atmospheric windows. Also, the kinds of information acquired by sensing the far infrared differ from those acquired

FIGURE 9.1. Infrared spectrum. This schematic view of the infrared spectrum identifies principal designations within the infrared region, and approximate definitions, with some illustrative examples. Regions are not always contiguous because some portions of the infrared spectrum because some portions are unavailable for remote sensing due to atmospheric effects.

in the visible and near infrared. Variations in emitted energy in the far infrared provide information concerning surface temperature and thermal properties of soils, rocks, vegetation, and man-made structures. Inferences based on thermal properties lead to inferences about the identities of surface materials.

9.2. Thermal Detectors

Before the 1940s, the absence of suitable instruments limited use of thermal infrared radiation for aerial reconnaissance. Aerial mapping of thermal energy depends on use of a sensor that is sufficiently sensitive to thermal radiation that variations in apparent temperature can be detected by an aircraft moving at considerable speed high above the ground. Early instruments for thermographic measurements examined differences in electrical resistance caused by changes in temperature. But such instruments could function only when in close proximity to the objects of interest. Although such instruments could be useful in an industrial or laboratory setting, they are not sufficiently sensitive for use in the context of remote sensing; they respond slowly to changes in temperature and cannot be used at distances required for remote sensing applications.

During the late 1800s and early 1900s, the development of *photon detectors* (sometimes called, *thermal photon detectors*) provided a practical technology for use of the thermal portion of the spectrum in remote sensing. Such detectors are capable of responding directly to incident photons by reacting to changes in electrical resistance, providing a sensitivity and speed of response suitable for use in reconnaissance instruments. By the 1940s, a family of photon detectors had been developed to provide the basis for electro-optical instruments used in several portions of the thermal infrared spectrum.

Detectors are devices formed from substances known to respond to energy over a defined wavelength interval, generating a weak electrical signal with a strength related to the radiances of the features in the field of view of the sensor. (Often, the sensitivity of such materials increases to practical levels when the substances are cooled to very low temperatures to increase sensitivity and reduce noise.) The electrical current is amplified, then used to generate a digital signal that can be used to form a pictorial image, roughly similar in overall form to an aerial photograph (Figure 9.2).

Detectors have been designed with sensitivities for many of the spectral intervals of interest in remote sensing, including regions of the visible, near infrared, and ultraviolet spectra. Detectors sensitive in the thermal portion of the spectrum are formed from rather exotic materials, such as *indium antimonide* (InSb) and mercury-doped germanium (Ge:Hg). InSb has a peak sensitivity near 5 μm in the mid-infrared spectrum, and Ge:Hg has a peak sensitivity near 10 μm in the far infrared spectrum (Figure 9.3). Mercury cadmium telluride (MCT) is sensitive over the range 8–14 μm. To maintain maximum sensitivity, such detectors must be cooled to very low temperatures (–196°C or –243°C) using liquid nitrogen or liquid helium.

The sensitivity of the detector is a significant variable in the design and operation of the system. Low sensitivity means that only large differences in brightness are recorded ("coarse radiometric resolution") and most of the finer detail in the scene is lost. High sensitivity means that finer differences in scene brightness are recorded ("fine radiometric resolution"). The *signal-to-noise ratio* (SNR or S/N ratio) expresses this concept (Chapter 4). The "signal" in this context refers to differences in image brightness caused by actual variations in scene brightness. "Noise" designates variations unrelated to scene brightness. Such variations may be the result of unpredictable variations in the performance of the system. (There may also be random elements contributed by the landscape and the atmosphere, but here "noise" refers specifically to that contributed by the sensor.) If noise is large relative to the signal, the image does not provide a reliable representation of the feature of interest. Clearly, high noise levels will prevent imaging of subtle features. Even if noise levels are low, there must be minimum contrast between a feature and its

FIGURE 9.2. Use of thermal detectors.

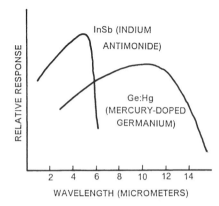

FIGURE 9.3. Sensitivity of some common thermal detectors.

background (i.e., a minimum magnitude for the signal) for the feature to be imaged. Also, note that increasing fineness of spatial resolution decreases the energy incident on the detector, with the effect of decreasing the strength of the signal. For many detectors, noise levels may remain constant even though the level of incident radiation decreases; if so, the increase in spatial resolution may be accompanied by decreases in radiometric resolution, as suggested in Chapter 4.

9.3. Thermal Radiometry

A *radiometer* is a sensor that measures the intensity of radiation received within a specified wavelength interval and within a specific field of view. Figure 9.4 gives a schematic view of a radiometer. A lens or mirror gathers radiation from the ground, then focuses it on a detector positioned in the focal plane. A field stop may restrict the field of view, and filters may be used to restrict the wavelength interval that reaches the detector. A charac-

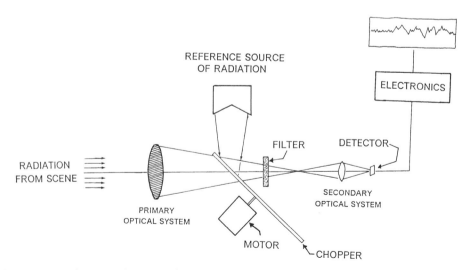

FIGURE 9.4. Schematic diagram of a radiometer.

teristic feature of radiometers is that radiation received from the ground is compared with a reference source of known radiometric qualities.

A device known as a *chopper* is capable of interrupting the radiation that reaches the detector. The chopper consists of a slotted disk or similar device rotated by an electrical motor so that, as the disk rotates, it causes the detector to alternately view the target, then the reference source of radiation. Because the chopper rotates very fast, the signal from the detector consists of a stream of data that alternately measures the radiance of the reference source, then radiation from the ground. The amplitude of this signal can be used to determine the radiance difference between the reference and the target. Because the reference source has known radiance, radiance of the target can then be estimated.

Although there are many variations on this design, this description identifies the most important components of radiometers. Related instruments include *photometers*, which operate at shorter wavelengths and often lack the internal reference source, and *spectrometers*, which examine radiance over a range of wavelengths. Radiometers can be designed to operate at different wavelength intervals, including portions of the infrared and ultraviolet spectra. By carefully tailoring the sensitivity of radiometers, scientists have been able to design instruments that are very useful in studying atmospheric gases and cloud temperatures. Radiometers used for Earth resource study are often configured to view only a single trace along the flight path; the output signal then consists of a single stream of data that varies in response to differences in radiances of features along the flight line. A scanning radiometer can gather data from a corridor beneath the aircraft; output from such a system resembles those from some of the scanning sensors discussed in earlier chapters.

Spatial resolution of a radiometer is determined by an *instantaneous field of view* (IFOV), which is in turn controlled by the sensor's optical system, the detector, and flying altitude. Radiometers often have relatively coarse spatial resolution—for example, satellite-borne radiometers may have spatial resolution of 60 m to 100 km or more—in part because of the desirability of maintaining high radiometric resolution. To ensure that the sensor receives enough energy to make reliable measurements of radiance, the IFOV is defined to be rather large; a smaller IFOV would mean that less energy would reach the detector, that the signal would be much too small with respect to system noise, and that the measure of radiance would be much less reliable.

The IFOV can be informally defined as the area viewed by the sensor if the motion of the instrument were to be suspended so that it records radiation from only a single patch of ground. The IFOV can be more formally expressed as the angular field of view (β) of the optical system (Figure 9.5). The projection of this field of view onto the ground surface defines the circular area that contributes radiance to the sensor. Usually for a particular sensor it is expressed in radians (r); to determine the IFOV for a particular image, it is necessary to know the flying altitude (H) and to calculate the size of the circular area viewed by the detector.

From elementary trigonometry it can be seen that the diameter of this area (D) is given as

$$D = H\beta \qquad\qquad \text{(Eq. 9.1)}$$

as illustrated in Figure 9.5. Thus, for example, if the angular field of view is 1.0 milliradians (mr) (1 mr = 0.001 r) and the flying altitude (H) is 400 m above the terrain, then:

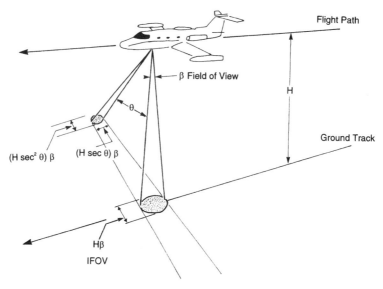

FIGURE 9.5. Instantaneous field of view (IFOV). At nadir, the diameter of the IFOV is given by flying altitude (H) and the instrument's field of view (θ). As the instrument scans to the side, it observes at angle β; in the off-nadir position, the IFOV becomes larger and unevenly shaped.

$$D = H\beta$$

$$D = 400 \times 1.0 \times 0.001$$

$$D = 0.40 \text{ m}$$

Because a thermal scanner views a landscape over a range of angles as it scans from side to side, the IFOV varies in size depending on the angle of observation (θ). Near the nadir (ground track of the aircraft), the IFOV is relatively small; near the edge of the image, the IFOV is large. This effect is beneficial in one sense because it compensates for the effects of the increased distance from the sensor to the landscape, thereby providing consistent radiometric sensitivity across the image.

Other effects are more troublesome. Equation 9.1 defines the IFOV at nadir as $H\beta$. Most thermal scanners scan side to side at a constant angular velocity, which means that, in a given interval of time, they scan a larger cross-track distance near the sides of the image than at the nadir. At angle θ, the IFOV measures $H \sec \theta \beta$ in the direction of flight and $H \sec^2 \theta \beta$ along the scan axis. Thus, near the nadir, the IFOV is small and symmetrical; near the edge of the image, it is larger and elongated in the direction of flight. The variation in the shape of the IFOV creates geometric errors in the representations of features—a problem discussed in subsequent sections. The variation in size means that radiance from the scene is averaged over a larger area and can be influenced by the presence of small features of contrasting temperature. Although small features can influence data for IFOVs of any size, the impact is more severe when the IFOV is large.

9.4. Microwave Radiometers

Microwave emissions from the Earth convey some of the same information carried by thermal (far infrared) radiation. Even though their wavelengths are much longer than those of thermal radiation, microwave emissions are related to temperature and emissivity in much the same manner as is thermal radiation. *Microwave radiometers* are very sensitive instruments tailored to receive and record radiation in the range from about 0.1 mm to 3 cm. Whereas the imaging radars discussed in Chapter 7 are active sensors that illuminate the terrain with their own energy, microwave radiometers are passive sensors that receive microwave radiation naturally emitted by the environment. The strength and wavelength of such radiation is largely a function of the temperature and emissivity of the target. Thus, although microwave radiometers, like radars, use the microwave region of the spectrum, they are functionally most closely related to the thermal sensors discussed in this chapter.

In the present context, we are concerned with microwave emissions from the Earth, which indirectly provide information pertaining to vegetation cover, soil moisture status, and surface materials. Other kinds of studies, peripheral to the field of remote sensing, derive information from microwave emissions from the Earth's atmosphere or from extraterrestrial objects. In fact, the field of microwave radiometry originated with radio astronomy, and some of its most dramatic achievements have been in the reconnaissance of extraterrestrial objects.

A *microwave radiometer* consists of a sensitive receiving instrument typically in the form of a horn- or dish-shaped antenna that observes a path directly beneath the aircraft or satellite; the signal gathered by the antenna is electronically filtered and amplified and then displayed as a stream of digital data, or, in the instance of scanning radiometers, as an image. As with thermal radiometers, microwave radiometers have a reference signal from an object of known temperature. The received signal is compared with the reference signal as a means of deriving the radiance of the target.

Examination of data from a microwave radiometer can be very complex due to many factors that contribute to a given observation. The component of primary interest is usually energy radiated by the features within the IFOV; of course, variations within the IFOV are lost, as the sensor can detect only the average radiance within this area. The atmosphere also radiates energy, so it contributes radiance, depending on moisture content and temperatures. In addition, solar radiation in the microwave radiation can be reflected from the surface to the antenna.

9.5. Thermal Scanners

The most widely used imaging sensors for thermal remote sensing are known as *thermal scanners*. Thermal scanners sense radiances of features beneath the aircraft flight path and produce digital and/or pictorial images of that terrain. There are several designs for thermal scanners. *Object-plane scanners* view the landscape by means of a moving mirror that oscillates at right angles to the flight path of the aircraft, generating a series of parallel (or perhaps overlapping) scan lines that together image a corridor directly beneath the aircraft. *Image-plane scanners* use a wider field of view to collect a more comprehensive image of the landscape; this image is then moved, by means of a moving mirror, relative to the detector. In either instance, the instrument is designed as a series of lenses and mirrors configured to acquire energy from the ground and to focus it on the detector.

An *infrared scanning system* consists of a scanning unit, with a gyroscopic roll connection unit, infrared detectors (connected to a liquid nitrogen cooling unit), and an amplification and control unit (Figure 9.6). A magnetic tape unit records data for later display as a video image; some systems may provide film recording of imagery as it is acquired. Together these units might weigh about 91 kg (200 lbs.), and they are usually mounted in an aircraft specially modified to permit the scanning unit to view the ground through an opening in the fuselage.

Infrared energy is collected by a scanning mirror that scans side to side across the flight path, in a manner similar to that described earlier for the Landsat MSS. The typical field of view might be as wide as 77°, so an aircraft flying at an altitude of 300 m (1,000 ft.) could record a strip as wide as 477 m (1,564 ft.). The forward motion of the aircraft generates the along-track dimension of the imagery. The mirror might make as many as 80 scans per second, with each scan representing a strip of ground about 46 cm (18 in.) in width (assuming the 300-m altitude mentioned above).

Energy collected by the scanning mirror is focused first on a parabolic mirror and then on a flat, stationary mirror that focuses energy on the infrared detector unit. The infrared detector unit consists of one of the detectors mentioned above, confined in a vacuum container, and cooled by liquid nitrogen (to reduce electronic noise and enhance the sensitivity of the detector). On some units, the detector can be easily changed by removing a small unit, then replacing it with another. The detector generates an electrical signal that varies in strength in proportion to the radiation received by the mirror and focused on the detector. The signal from the detector is very weak, however, so it must be

FIGURE 9.6. Thermal scanner.

amplified before it is recorded by the magnetic tape unit that is connected to the scanner. A *roll correction unit*, consisting in part of a gyroscope, senses side-to-side motion of the aircraft and sends a signal that permits the electronic control unit to correct the signal to reduce geometric errors caused by aircraft instability. After the aircraft has landed, the digital storage unit from the sensor is removed from the aircraft, to be prepared for analysis.

9.6. Thermal Properties of Objects

All objects at temperatures above absolute zero emit thermal radiation, although the intensity and peak wavelength of such radiation varies with the temperature of the object, as specified by the radiation laws outlined in Chapter 2. For remote sensing in the visible and near infrared, we examine contrasts in the abilities of objects to reflect direct solar radiation to the sensor. For remote sensing in the far infrared spectrum, we sense differences in the abilities of objects and landscape features to absorb shortwave visible and near infrared radiation, then to emit this energy as longer wavelengths in the far infrared region.

Thus, except for geothermal energy, man-made thermal sources, and range and forest fires, the immediate source of emitted thermal infrared radiation is shortwave solar energy. Direct solar radiation (with a peak at about 0.5 μm in the visible spectrum) is received and absorbed by the landscape (Chapter 2). The amount and spectral distribution of energy emitted by landscape features depend on the thermal properties of these features, as discussed below. The contrasts in thermal brightness, observed as varied gray tones on the image, are used as the basis for identification of features.

A *blackbody* is a theoretical object that acts as a perfect absorber and emitter of radiation; it absorbs and reemits all energy that it receives. Although the blackbody is a theoretical concept, it is useful in describing and modeling the thermal behavior of actual objects. Moreover, it is possible to approximate the behavior of blackbodies in laboratory experiments.

As explained in Chapter 2, as the temperature of a blackbody increases, the wavelength of peak emission decreases in accordance with Wien's displacement law (Eq. 2.5). The Stefan–Boltzmann law (Eq. 2.4) describes mathematically the increase in total radiation emitted (over a range of wavelengths) as the temperature of a blackbody increases.

Emissivity (ε_λ) is a ratio between emittance of an object in relation to emittance of a blackbody at the same temperature:

$$\varepsilon_\lambda = \frac{\text{Radiant emittance of an object}}{\text{Radiant emittance of a blackbody at the same temperature}} \quad \text{(Eq. 9.2)}$$

(See also Eq. 2.3.) The subscript (λ) sometimes used with ε signifies that λ has been measured for specific wavelengths. Emissivity therefore varies from 0 to 1, with 1 signifying a substance with a thermal behavior identical to that of a blackbody. Table 9.1 lists emissivities for some common materials. Note that many of the substances commonly present in the landscape (e.g., soil, water) have emissivities rather close to 1. Note, however, that emissivity can vary with temperature, wavelength, and angle of observation.

TABLE 9.1. Emissivities of Some Common Materials

Material	Temperature (°C)	Emissivity[a]
Polished copper	50–100	0.02
Polished brass	200	0.03
Polished silver	100	0.03
Steel alloy	500	0.35
Graphite	0–3,600	0.7–0.8
Lubricating oil (thick film on nickel base)	20	0.82
Snow	–10	0.85
Sand	20	0.90
Wood (planed oak)	20	0.90
Concrete	20	0.92
Dry soil	20	0.92
Brick (red common)	20	0.93
Glass (polished plate)	20	0.94
Wet soil (saturated)	20	0.95
Distilled water	20	0.96
Ice	–10	0.96
Carbon lamp black	20–400	0.96
Lacquer (matte black)	100	0.97

Note. Data from Hudson (1969) and Weast (1986).

[a]Measured at normal incidence over a range of wavelengths.

Graybodies

An object that has an emissivity less than 1.0 but has constant emissivity over all wavelengths is known as a *graybody* (Figure 9.7). A *selective radiator* is an object with an emissivity that varies with respect to wavelength. If two objects in the same setting are at the same temperature but have different emissivities, the one having the higher emissivity will radiate more strongly. Because the sensor detects radiant energy (apparent temperature) rather than the kinetic ("true") temperature, precise interpretation of an image requires knowledge of emissivities of features shown on the image.

Heat

Heat is the internal energy of a substance arising from the motion of its component atoms and molecules. *Temperature* measures the relative warmth or coolness of a substance. It

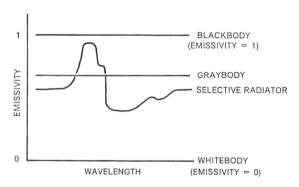

FIGURE 9.7. Blackbody, graybody, whitebody.

is the kinetic temperature or average thermal energy of molecules within a substance. *Kinetic temperature*, sometimes known as the *true temperature*, is measured using the usual temperature scales, most notably the Fahrenheit, Celsius (centigrade), and Kelvin (absolute) scales. *Radiant* (or *apparent*) *temperature* measures the emitted energy of an object. Photons from the radiant energy are detected by the thermal scanner.

Heat capacity is the ratio of the change in heat energy per unit mass to the corresponding change in temperature (at constant pressure). For example, we can measure the heat capacity of pure water to be 1 calorie (cal) per gram (g), meaning that 1 cal is required for each gram to raise its temperature by 1°C. The specific heat of a substance is the ratio of its heat capacity to that of a reference substance. Because the reference substance typically is pure water, specific heat is often numerically equal to its heat capacity. Because a calorie is defined as the amount of heat required to raise by 1°C the temperature of 1 g of pure water, use of water as the reference means that heat capacity and specific heat will be numerically equivalent. In this context, specific heat can be defined as the amount of heat (measured in calories) required to raise the temperature of 1 g of a substance 1°C.

Thermal conductivity is a measure of the rate at which a substance transfers heat. Conductivity is measured as calories per centimeter per second per degree Celsius, so it measures calories required to transfer a change in temperature over specified intervals of length and time.

Some of these variables can be integrated into a single measure, called *thermal inertia* (*P*), defined as

$$P = \sqrt{KC\rho} \qquad \text{(Eq. 9.3)}$$

where K is the thermal conductivity (cal $\cdot$ cm^{-1} $\cdot$ sec^{-1} $\cdot$ °C^{-1}); C is the heat capacity (cal $\cdot$ g^{-1} $\cdot$ °C^{-1}); and ρ is the density (g $\cdot$ cm^{-3}). *Thermal inertia*, *P*, is then measured in cal $\cdot$ cm^{-2} $\cdot$ °C^{-1} $\cdot$ sec$^{-1/2}$. Thermal inertia measures the tendency of a substance to resist changes in temperature, or, more precisely, the rate of heat transfer at the contact between two substances.

In the context of remote sensing, thermal inertia indicates the ability of a surface to retain heat during the day and reradiate it at night, thereby signaling the nature of the varied thermal properties of the local terrain. The thermal inertia of specific surfaces (perhaps at the depths of several centimeters) will be determined by their physical characteristics, including mineralogy, particle size, compactness and lithification of mineral grains, and presence and depth of unconsolidated surface materials, such as sand, dust, and other loose sediments. Thus implementation of remote sensing in the context of thermal inertia observes landscapes within daily cycle of heating and cooling—although a single thermal image provides only an isolated snapshot of relative temperatures, a *pair* of carefully timed thermal snapshots permits observation of temperature *changes* between the warmest and coolest portions of the day, and therefore offers an opportunity to observe differences in the thermal properties of the materials at the Earth's surface.

Temperatures of materials with low thermal inertia change significantly during the daily heating–cooling cycle, whereas temperatures of materials with high thermal inertias will respond more slowly. Thermal inertia characterizes a material's ability to conduct and store heat and therefore its ability to retain heat during the day, then reradiate it at night. For remote sensing, thermal inertia represents a complex composite of factors such as particle size, soil cover, moisture, bedrock, and related terrain features. Relative thermal inertia can sometimes be approximated by assessing the amplitude of the diurnal temperature curve (i.e., the difference between daily maximum and minimum surface temperatures). Assessing differences in thermal inertia of the materials at the Earth's surface, in the context of other characteristics, can help to characterize these surface materials and their properties.

9.7. Geometry of Thermal Images

Thermal scanners, like all remote sensing systems, generate geometric errors as they gather data. These errors mean that representations of positions and shapes of features depicted on thermal imagery do not match to their correct planimetric forms. Therefore, images cannot be directly used as the basis for accurate measurements. Some errors are caused by aircraft or spacecraft instability. As the aircraft rolls and pitches, the scan lines lose their correct positional relationships, and, of course, the features they portray are not accurately represented in the image.

Thermal imagery also exhibits relief displacement analogous to that encountered in aerial photography (Figure 9.8). Thermal imagery, however, does not have the single central perspective of an aerial photograph, but rather a separate nadir for each scan line. Thus the focal point for relief displacement is the nadir for each scan line, or, in effect, the trace of the flight path on the ground. Thus relief displacement is projected from a line that follows the center of the long axis of the image. At the center of the image, the sensor views objects from directly overhead, and planimetric positions are correct. However, as distance from the centerline increases, the sensor tends to view the sides rather than only

FIGURE 9.8. Relief displacement and tangential scale distortion.

the tops of features, and relief displacement increases. These effects are visible in Figure 9.9; the tanker and the tanks appear to lean outward from a line that passes through the center of the image. The effect increases toward the edges of the image.

Figure 9.9 also illustrates other geometric qualities of thermal line scan imagery. Although the scanning mirror rotates at a constant speed, the projection of the IFOV onto the ground surface does not move (relative to the ground) at equal speed because of the varied distance from the aircraft to the ground. At nadir, the sensor is closer to the ground than it is at the edge of the image; in a given interval of time, the sensor scans a shorter distance at nadir than it does at the edge of the image. Therefore, the scanner produces a geometric error that tends to compress features along an axis oriented perpendicular to the flight line and parallel to the scan lines. In Figure 9.9 this effect, known as *tangential scale distortion*, is visible in the shapes of the cylindrical storage tanks. The images of those nearest the flight line are more circular, whereas the shapes of those furthest from the flight line (nearest the edge of the image) are compressed along an axis perpendicular to the flight line. Sometimes the worst effects can be removed by corrections applied as the film image is generated, although it is often necessary to avoid use of the extreme edges of the image.

9.8. The Thermal Image and Its Interpretation

The image generated by a thermal scanner is a strip of black-and-white film depicting thermal contrasts in the landscape as variations in gray tones (e.g., Figures 9.9 and 9.10). Usually, brighter tones (whites and light grays) represent warmer features; darker tones (dark grays and blacks) represent cooler features. In some applications the black-and-white image may be subjected to level slicing or other enhancements that assign distinc-

FIGURE 9.9. Thermal image of an oil tanker and petroleum storage facilities near the Delaware River, 19 December 1979. This image, acquired at 11:43 p.m., shows discharge of warm water into the Delaware River and thermal patterns related to operation of a large petrochemical facility. Thermal image from Daedulus Enterprises, Inc., Ann Arbor, Michigan.

FIGURE 9.10. This thermal image shows another section of the same facility imaged in Figure 9.9. Thermal image from Daedulus Enterprises, Inc., Ann Arbor, Michigan.

tive hues to specific gray tones as an aid for manual interpretation. Often it is easier for the eye to separate subtle shades of color than the variations in gray on the original image. Such enhancements are simply manipulations of the basic infrared image; they do not represent differences in means of acquisition or in the quality of the basic information available for interpretation.

For any thermal infrared image, the interpreter must always determine (1) whether the image at hand is a positive or a negative image and (2) the time of day that the image was acquired. Sometimes it may not be possible to determine the correct time of day from information within the image itself; misinterpretation can alter the meaning of gray tones on the image and render the resulting interpretation useless.

As the sensor views objects near the edge of the image, the distance from the sensor to the ground increases. This relationship means that the IFOV is larger nearer the edges of the image than it is near the flight line.

Thermal scanners are generally uncalibrated, so they show relative radiances rather than absolute measurements of radiances. However, some thermal scanners do include reference sources that are viewed by the scanning system at the beginning and end of each scan. The reference sources can be set at specific temperatures that are related to those expected to be encountered in the scene. Thus each scan line includes values of known temperature that permit the analyst to estimate temperatures of objects within the image.

In addition, errors caused by the atmosphere and by the system itself prevent precise interpretation of thermal imagery. Typical system errors might include recording noise, variations in reference temperatures, and detector errors. Full correction for atmospheric conditions requires information not usually available in detail, so often it is necessary to use approximations, or value-based samples acquired at a few selected times and places, then extrapolated to estimate values elsewhere. Also, the *atmospheric path traveled* by radiation reaching the sensor varies with angle of observation, which changes as the instrument scans the ground surface. These variations in angle lead to errors in observed values in the image.

Even when accurate measures of radiances are available, it is difficult to derive data for kinetic temperatures from the apparent temperature information within the image. Derivation of kinetic temperatures requires knowledge of emissivities of the materials. In some instances, such knowledge may be available, as the survey may be focused on a known area that must be repeatedly imaged to monitor changes over time (e.g., as moisture conditions change). But many other surveys examine areas not previously studied in detail, and information regarding surface materials and their emissivities may not be known.

Emissivity is a measure of the effectiveness of an object in translating temperature into emitted radiation (and in converting absorbed radiation into a change in observed temperature). Because objects differ with respect to emissivity, observed differences in emitted infrared energy do not translate directly into corresponding differences in temperature. As a result, it is necessary to apply knowledge of surface temperature or of emissivity variations to accurately study surface temperature patterns from thermal imagery. Because knowledge of these characteristics assumes a very detailed prior knowledge of the landscape, such interpretations should be considered as appropriate for examination of a distribution known already in some detail rather than for reconnaissance of an unknown pattern (e.g., one might already know the patterns of soils and crops at an agricultural experiment station but may wish to use the imagery to monitor temperature patterns). Often estimated values for emissivity are used, or assumed values are applied to areas of unknown emissivity.

Also, it should be recognized that the sensor records radiances of the surfaces of objects. Because radiances may be determined at the surface of an object by a layer perhaps as thin as 50 μm, a sensor may record conditions that are not characteristic of the subsurface mass, which is probably the object of the study. For example, evaporation from a water body or a moist soil surface may cool the thin layer of moisture at the contact point with the atmosphere. Because the sensor detects radiation emitted at this surface layer, the observed temperature may differ considerably from that of the remaining mass of the soil or water body. Leckie (1982) estimates that calibration error and other instrument errors are generally rather small, although they may be important in some instances. Errors in estimating emissivity and in attempts to correct for atmospheric effects are likely to be the most important sources of error in quantitative studies of thermal imagery.

In many instances, a thermal image must be interpreted to yield qualitative rather than quantitative information. Although some applications do require interpretations of quantitative information, there are many others for which qualitative interpretation is completely satisfactory. An interpreter who is well informed about the landscape represented on the image, the imaging system, the thermal behavior of various materials, and the timing of the flight is prepared to derive considerable information from an image, even though it may not be possible to derive precise temperatures from the image.

The *thermal landscape* is a composite of the familiar elements of surface material, topography, vegetation cover, and moisture. Various rocks, soils, and other surface materials respond differently to solar heating. Thus in some instances the differences in thermal properties listed in Table 9.1 can be observed in thermal imagery. However, the thermal behavior of surface materials is also influenced by other factors. For example, slopes that face the sun will tend to receive more solar radiation than slopes that are shadowed by topography. Such differences are, of course, combined with those arising from different surface materials. Also, the presence and nature of vegetation alters the thermal behavior of the landscape. Vegetation tends to heat rather rapidly but can also shade areas, creating patterns of warm and cool.

Water tends to retain heat, to cool slowly at night, and to warm slowly during the daytime. In contrast, many soils and rocks (if dry) tend to release heat rapidly at night and to absorb heat quickly during the daytime. Even small or modest amounts of moisture can greatly alter the thermal properties of soil and rock. Therefore, thermal sensors can be very effective in monitoring the presence and movement of moisture in the environment. In any given image, the influences of surface materials, topography, vegetation, and moisture can combine to cause very complex image patterns. However, often it is possible to isolate the effect of some of these variables and therefore to derive useful information concerning, for example, movement of moisture or the patterns of differing surface materials.

Timing of acquisition of thermal imagery is very important. The optimum times vary according to the purpose and subject of the study, so it is not possible to specify universally applicable rules. Because the greatest thermal contrast tends to occur during the daylight hours, sometimes thermal images are acquired in the early afternoon to capture the differences in thermal properties of landscape features. However, in the 3–6 μm range, the sensor may record reflected as well as emitted thermal radiation, so daytime missions in this region may not be optimum for thermal information. Also during daytime, the sensor may record thermal patterns caused by topographic or cloud shadowing; although shadows may sometimes be useful in interpretation, they are more likely to complicate analysis of a thermal image, so it is usually best to avoid acquiring heavily shadowed images. In a daytime image, water bodies typically appear as cool relative to land, and bare soil, meadow, and wooded areas appear as warm features.

Some of the problems arising from daytime images are avoided by planning missions just before dawn. Shadows are absent, and sunlight, of course, cannot cause reflection (at shorter wavelengths) or shadows. However, thermal contrast is lower, so it may be more difficult to distinguish between broad classes of surfaces based on differences in thermal behavior. On such an image, water bodies would appear as warm relative to land. Forested areas may also appear to be warm. Open meadows and dry, bare soil are likely to appear as cool features.

The thermal images of petroleum storage facilities (Figures 9.9 and 9.10) show thermal contrasts that are especially interesting. A prominent feature on Figure 9.9 is the bright thermal plume discharged by the tributary to the Delaware River. The image clearly shows the sharp contrast in temperature as the warm water flows into the main channel, then disperses and cools as it is carried downstream. Note the contrast between the full and partially full tanks and the warm temperatures of the pipelines that connect the tanker with the storage tanks. Many of the same features are also visible in Figure 9.10, which also shows a partially loaded tanker with clear delineation of the separate storage tanks in the ship.

Thermal imagery has obvious significance for studies of heat loss, thermal efficiency, and effectiveness of insulation in residential and commercial structures. Plate 15 shows thermal images of two residential structures observed at night during the winter season. Windows and walls form principal avenues for escape of heat, and the chimney shows as an especially bright feature. Note that the walkways, paved roads, and parked automobiles show as cool features.

In Figure 9.11, two thermal images depict a portion of the Cornell University cam-

FIGURE 9.11. Two thermal images of a portion of the Cornell University campus in Ithaca, New York. Thermal images from Daedulus Enterprises, Inc., Ann Arbor, Michigan.

pus in Ithaca, New York, acquired in January (left) and again the following November (right). Campus buildings are clearly visible, as are losses of heat through vents in the roofs of buildings and at manholes where steampipes for the campus heating system join or change direction. The left-hand image shows a substantial leak in a steam pipe as it passes over the bridge in the right center of the image. On the right, a later image of the same region shows clearly the effects of repair of the defective section

Figure 9.12 shows Painted Rock Dam, Arizona, as depicted by both an aerial photograph and a thermal infrared image. The aerial photograph (top) was taken at about

FIGURE 9.12. Painted Rock Dam, Arizona, 28 January 1979. Aerial photograph (top) and thermal image (bottom). Thermal image from Daedulus Enterprises, Inc., Ann Arbor, Michigan.

10:30 a.m.; the thermal image was acquired at about 7:00 a.m. the same day. The prominent linear feature is a large earthen dam, with the spillway visible at the lower left. On the thermal image, the open water upstream from the dam appears as a uniformly white (warm) region, whereas land areas are dark (cool)—a typical situation for early morning hours, before solar radiation has warmed the Earth. On the downstream side of the dam, the white (warm) regions reveal areas of open water or saturated soil. The open water in the spillway is, of course, expected, but the other white areas indicate places where there may be seepage and potentially weak points in the dam structure.

Figure 9.13 shows thermal images of a power plant acquired at four different stages of the tidal cycle. The discharge of warm water is visible as the bright plume in the upper left of each image. At the top, at low tide (5:59 a.m.), the warm water is carried downstream toward the ocean. The second image, acquired at flood tide (8:00 a.m.), shows the deflection of the plume by rising water. In the third image, taken at high tide (10:59 a.m.), the plume extends downstream for a considerable distance. Finally, in the bottom image, acquired at ebb tide (2:20 p.m.), the shape of the plume reflects the reversal of tidal flow once again.

If imagery or data for two separate times are available, it may be possible to employ knowledge of thermal inertia as a means of studying the pattern of different materials at the Earth's surface. Figure 9.14 illustrates the principles involved—it represents two images acquired at times that permit observation of extremes of temperature, perhaps near noontime and again just before dawn. These two sets of data permit estimation of the ranges of temperature variation for each region on the image. Because these varia-

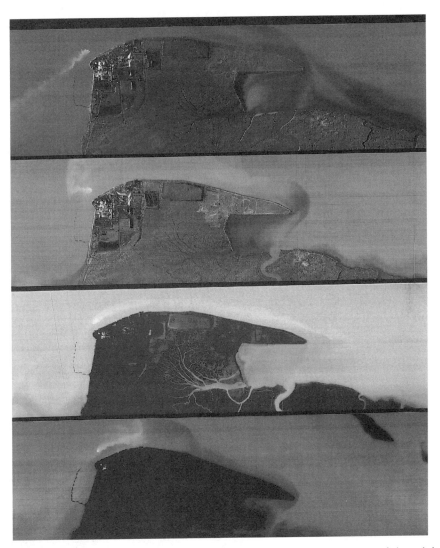

FIGURE 9.13. Thermal images of a power plant acquired at different states of the tidal cycle. Thermal images from Daedulus Enterprises, Inc., Ann Arbor, Michigan.

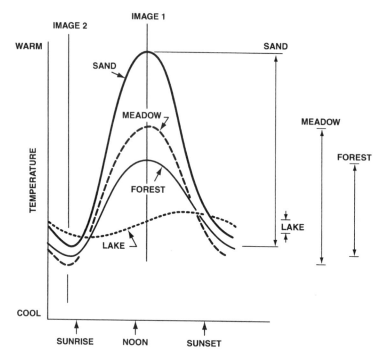

FIGURE 9.14. Schematic illustration of diurnal temperature variation of several broad classes of land cover.

tions are determined by the thermal inertias of the substances, they permit interpretation of features represented by the images. Leckie (1982) notes that misregistration can be a source of error in comparisons of day and night images, although such errors are thought to be small relative to other errors.

In Figure 9.14 images are acquired at noon, when temperatures are high, and again just before dawn, when temperatures are lowest. By observing the differences in temperature, it is possible to estimate thermal inertia. Thus the lake, composed of a material that resists changes in temperature, shows rather small changes, whereas the surface of open sand displays little resistance to thermal changes and exhibits a wider range of temperatures.

Plate 16 shows another example of diurnal temperature changes. These two images show the landscape near Erfurt, Germany, as observed by a thermal scanner in the region between 8.5 and 12 µm. The thermal data have been geometrically and radiometrically processed, then superimposed over digital elevation data, with reds and yellows assigned to represent warmer temperatures and blues and greens to represent cooler temperatures. The city of Erfurt is positioned at the edge of a limestone plateau, which is visible as the irregular topography in the foreground of the image. The valley of the river Gera is visible in the lower left, extending across the image to the upper center region of the image.

The top image represents this landscape as observed just after sunset. The urbanized area and much of the forested topography south of the city show as warm reds and yellows. The bottom image shows the same area observed just before sunrise, when temperatures are at their coolest. As the open land of the rural landscape and areas at the periphery of the urbanized areas have cooled considerably, the forested region on the plateau south of the city is now the warmest surface, due to the thermal effects of the forest canopy.

Figure 9.15 depicts a pair of thermal images that illustrates the same principle applied to seasonal, rather than diurnal, temperature contrasts. At left is an ASTER image (discussed in Chapter 21) that represents thermal variation within the terrain near Richmond, Virginia, as acquired in March, when the landscape is cool relative to the water bodies in this region. In contrast, a thermal image of the same region acquired in October shows a reversal of the relative brightnesses within the terrain—the land surface is now warm relative to the water bodies.

9.9. Heat Capacity Mapping Mission

The Heat Capacity Mapping Mission (HCMM) (in service from April 1978 to September 1980) was a satellite system specifically designed to evaluate the concept that orbital observations of temperature differences at the Earth's surface at different points in the daily heating–cooling cycle might provide a basis for estimation of thermal inertia and other thermal properties of surface materials. Although HCMM was designed as an experiment and is long out of service, it deserves mention as an innovative attempt to exploit the thermal region of the spectrum using a satellite platform.

The satellite was in a sun-synchronous orbit, at an altitude of 620 km, bringing

MARCH   **OCTOBER**

FIGURE 9.15. ASTER thermal images illustrating seasonal temperature differences. The March and October images illustrate differences in the responses of land and water to seasonal temperature variations.

it over the equator at 2:00 p.m. local sun time. At 40°N latitude the satellite passed overhead at 1:30 p.m., then again about 12 hours later at 2:30 a.m. These times provided observations at two points on the diurnal heating–cooling cycle (Figure 9.16). (It would have been desirable to use a time later in the morning just before sunrise, in order to provide higher thermal contrast, but orbital constraints dictated the time differences between passes). The repeat cycle varied with latitude—at midlatitude locations the cycle was 5 days. The 12-hour repeat coverage was available for some locations at 16-day intervals, but other locations received only 36-hour coverage. The HCMM radiometer used two channels. One, in the reflective portion of the spectrum (0.5–1.1 μm), had a spatial resolution of about 500 m × 500 m. A second channel was available in the thermal infrared region, 10.5–12.5 μm, and had a spatial resolution of 600 m × 600 m. The image swath was about 716 km wide.

Estimation of thermal inertia requires that two HCMM images be registered to observe apparent temperatures at different points on the diurnal heating–cooling cycle. Because the day and night images must be acquired on different passes with different inclinations (Figure 9.17), the registration of the two images is often much more difficult than might be the case with similar images. The rather coarse resolution means that normally distinct landmarks are not visible or are more difficult to recognize. Watson et al. (1982) have discussed some of the problems encountered in registering HCMM images.

NASA's Goddard Space Flight Center received and processed HCMM data, calibrating the thermal data and performing geometric registration, when possible, for the two contrasting thermal images. Ideally, these two images permit reconstruction of a thermal inertia map of the imaged area. In fact, atmospheric conditions varied during the intervals between passes, and the measured temperatures were influenced by variations in cloud cover, wind, evaporation at the ground surface, and atmospheric water vapor. As a result, the measured differences are *apparent thermal inertia* (ATI) rather than a true measure of thermal inertia. Due to the number of approximations employed and scaling of the values, the values for ATI are best considered to be measures of relative thermal inertia.

As an experimental prototype, HCMM was in service for a relatively short interval and acquired data for rather restricted regions of the Earth. Some of the concepts developed and tested by HCMM have been employed in the design of other programs, most notably the NASA Earth Observing System (Chapter 21).

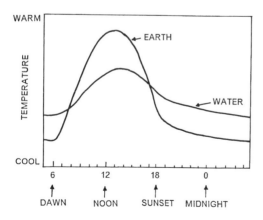

FIGURE 9.16. Diurnal temperature cycle.

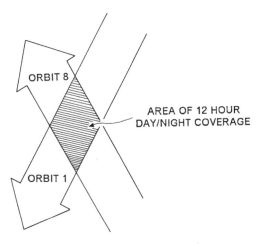

FIGURE 9.17. Overlap of two HCCM passes at 40° latitude.

Data are available in both film and digital formats; HCMM scenes show areas about 700 km × 700 km (1,127 mi. × 1,127 mi.) in size. Nonetheless, the HCMM archive, maintained by the National Space Science Data Center (World Data Center-A, Code 601, NASA Goddard Space Flight Center, Greenbelt, Maryland 20771) provides coverage for many midlatitude regions (including North America, Australia, Europe, and northern Africa).

Price (1985) cautioned that the HCMM data product (ATI) can be unreliable in some agricultural regions due to effects of evaporation of surface moisture, which can artificially reduce the diurnal amplitude of soil heat flux relative to the amplitude in areas in which the surface is dry. Thus apparent thermal inertia should not be used in regions having such variability in surface moisture.

9.10. Landsat Multispectral Scanner and Thematic Mapper Thermal Data

Chapter 6 mentioned, but did not discuss in any detail, the thermal data collected by the TM and by the Landsat 3 MSS. The Landsat 3 MSS, unlike those of earlier Landsats, included a channel (MSS band 5, formerly designated as band 8) sensitive to thermal data in the region 10.4–12.6 µm. Due to decreased sensitivity of the thermal detectors, spatial resolution of this band was significantly coarser than that of the visible and near infrared channels (237 m × 237 m vs. 79 m × 79 m). Each of the MSS band-5 pixels therefore corresponds to about nine pixels from one of the other four bands. Landsat 3 band 5 was designed to detect apparent temperature differences of about 15°C. Although initial images were reported to meet this standard, progressive degradation of image quality was observed as the thermal sensors deteriorated over time. In addition, the thermal band was subject to other technical problems. Because of these problems, the MSS thermal imaging system was turned off about 1 year after Landsat 3 was launched.

As a result, relatively few images were acquired using the MSS thermal band, and relatively little analysis has been attempted. Lougeay (1982) examined a Landsat 3 thermal image of an alpine region in Alaska with the prospect that the extensive glaciation in this region could contribute to sufficient thermal contrast to permit interpretation of terrain features. The coarse spatial resolution, solar heating of south-facing slopes, and extensive shadowing due to rugged topography and low sun angle contributed to difficulties in interpretation. Although Lougeay was able to recognize major landscape features, his analysis is more of a suggestion of potential applications of the MSS thermal data than an illustration of their usefulness.

Price (1981) also studied the Landsat MSS thermal data. He concluded that the thermal band provided new information not conveyed by the other MSS channels. Rather broad landcover classes (including several classes of open water, urban and suburban land, vegetated regions, and barren areas) could readily be identified on the scene that he examined. The time of the Landsat overpass was too early to record maximum thermal contrast (which occurs in early afternoon), and analyses of temperatures are greatly complicated by topographic and energy balance effects at the ground surface and the influence of the intervening atmosphere. Despite the possibility of adding new information to the analysis, Price advised against routine use of the thermal channel in the usual analyses of Landsat MSS data. In general, the experience of other analysts has confirmed this conclusion, and these data have not found widespread use.

The Landsat TM includes a thermal band, usually designated as TM band 6, sensitive in the region 10.4–12.5 μm (Figure 9.18). It has lower radiometric sensitivity and coarser spatial resolution (about 120 m) relative to other TM bands. Some studies (e.g., Toll, 1985) suggest that the TM thermal band does not significantly add to the accuracy of the usual land-cover analysis, possibly due to some of the same influences noted in the Price study mentioned above. However, it seems likely that continued study of characteristics of these data will reveal more information about their characteristics and lead to further research.

9.11. Summary

Thermal imagery is a valuable asset for remote sensing because it conveys information not easily derived from other forms of imagery. The thermal behavior of different soils, rocks, and construction materials can permit derivation of information not present in other images. The thermal properties of water contrast with those of many other landscape materials, so that thermal images can be very sensitive to the presence of moisture in the environment. And the presence of moisture is itself often a clue to the differences between different classes of soil and rock.

Of course, use of data from the far infrared region can present its own problems. Like all images, thermal imagery has geometric errors. Moreover, the analyst cannot derive detailed quantitative interpretations of temperatures unless detailed knowledge of emissivity is at hand. Timing of image acquisition can be critical. Atmospheric effects can pose serious problems, especially from satellite altitudes. Because the thermal landscape differs so greatly from the visible landscape, it may often be necessary to use aerial photography to locate familiar landmarks while interpreting thermal images. Existing archives of thermal imagery are not comparable in scope to those for aerial photography or satellite data (such as those of Landsat or SPOT), so it may be difficult to acquire suitable thermal data unless it is feasible to purchase custom-flown imagery.

FIGURE 9.18. TM band 6, showing a thermal image of New Orleans, Louisiana, 16 September 1982. This image shows the same area and was acquired at the same time as Figures 6.13 and 6.14. Resolution of TM band 6 is coarser than the other bands, so detail is not as sharp. Here dark image tone represents relatively cool areas, and bright image tone represents relatively warm areas. The horizontal banding near the center of the image is caused by a defect in the operation of the TM, discussed in Chapter 10. From USGS image.

9.12. Some Teaching and Learning Resources

- Atlanta and Heat Island Effect
  www.youtube.com/watch?v=e62a1MYr3XY

- Urban Heat Islands on the Weather Channel
  www.youtube.com/watch?v=t-sXHl3l-rM&NR=1

- A Year of Thermal Pollution from Oyster Creek Nuclear Plant
  www.youtube.com/watch?v=HNVLr01dEeI

- Hurricane Flossie
  www.youtube.com/watch?v=o1q30qeiA_8

- Hurricane Flossie Updated
  www.youtube.com/watch?v=p_rxu0fJ2h8&NR=1

- Aerial Thermal Imaging Demo
  www.youtube.com/watch?v=GiQfZGjFfz4

- Passive Microwave Sensing of Sea Ice Extent
  www.youtube.com/watch?v=Q-d2-JOHCvc

Review Questions

1. Explain why choice of time of day is so important in planning acquisition of thermal imagery.

2. Would you expect season of the year to be important in acquiring thermal imagery of, for example, a region in Pennsylvania? Explain.

3. In your new job as an analyst for an institution that studies environmental problems in coastal areas, it is necessary for you to prepare a plan to acquire thermal imagery of a tidal marsh. List the important factors you must consider as you plan the mission.

4. Many beginning students are surprised to find that geothermal heat and man-made heat are of such modest significance in the Earth's energy balance and in determining information presented on thermal imagery. Explain, then, what it is that thermal imagery does depict, and why thermal imagery is so useful in so many disciplines.

5. Can you understand why thermal infrared imagery is considered so useful to so many scientists even though it does not usually provide measurements of actual temperatures? Can you identify situations in which it might be important to be able to determine actual temperatures from imagery?

6. Fagerlund et al. (1982) report that it is possible to judge from thermal imagery whether storage tanks for gasoline and oil are empty, full, or partially full. Examine Figures 9.9 and 9.10 to find examples of each. How can you confirm from evidence on the image itself that full tanks are warm (bright) and empty tanks are cool (dark)?

7. Examine Figure 9.9; is the tanker empty, partially full, or full? Examine Figure 9.10; are these tankers empty, partially full, or full? From your inspection of the imagery can you determine something about the construction of tankers and the procedures used to empty or fill tankers?

8. What information would be necessary to plan an aircraft mission to acquire thermal imagery to study heat loss from a residential area in the northeastern United States?

9. In what ways would thermal imagery be important in agricultural research?

10. Outline ways in which thermal imagery would be especially useful in studies of the urban landscape.

References

Abrams, H. 2000. The Advanced Spaceborne Thermal Emission and Reflection Radiometer (ASTER): Data Products for the High Spatial Resolution Imager on NASA's Terra Platform. *International Journal of Remote Sensing*, Vol. 21, pp. 847–859.

Colcord, J. E. 1981. Thermal Imagery Energy Surveys. *Photogrammetric Engineering and Remote Sensing*, Vol. 47, pp. 237–240.

Dennison, P. E., K. Charoensir, D. A. Roberts, S. H. Petersen, and R. O. Green. 2006. Wildfire Temperature and Land Cover Modeling using Hyperspectral Data. *Remote Sensing of Environment*, Vol. 100, pp. 212–222.

Fagerlund, E., B. Kleman, L. Sellin, and H. Svenson. 1970. Physical Studies of Nature by Thermal Mapping. *Earth Science Reviews*, Vol. 6, pp. 169–180.

Fraser, R., and J. J. Kay. 2004. Energy Analysis of Ecosystems: Establishing a Role for Thermal Remote Sensing. Chapter 8 in *Thermal Remote Sensing in Land Surface Processes* (D. A. Quattrochi and J. C. Luvall, eds.). Boca Raton, FL: CRC Press, pp. 283–360.

Gillespie, A. R., and A. B. Kahle. 1978. Construction and Interpretation of a Digital Thermal Inertia Image. *Photogrammetric Engineering and Remote Sensing*, Vol. 43, pp. 983–1000.

Gluch, R., D. A. Quattrochi, and J. C. Luvall. 2006. A Multiscale Approach to Urban Thermal Analysis. *Remote Sensing of Environment*, Vol. 104, pp. 123–131.

Goddard Space Flight Center. 1978. *Data Users Handbook, Heat Capacity Mapping Mission (HCMM), for Applications Explorer Mission-A (AEM))*. Greenbelt, MD: NASA.

Hudson, R. D. 1969. Infrared Systems Engineering. New York: Wiley, 642 pp.

Jakosky, B. M., M. T. Mellon, H. H. Kieffer, and P. R. Christensen. 2000. The Thermal Inertia of Mars from the Mars Global Surveyor Thermal Emission Spectrometer. *Journal of Geophysical Research*, Vol. 105, pp. 9643–9652.

Kahle, A. B. 1977. A Simple Model of the Earth's Surface for Geologic Mapping by Remote Sensing. *Journal of Remote Sensing*, Vol. 82, pp. 1673–1680.

Leckie, D. G. 1982. An Error Analysis of Thermal Infrared Line-Scan Data for Quantitative Studies. *Photogrammetric Engineering and Remote Sensing*, Vol. 48, pp. 945–954.

McVicar, T. R., and D. L. B. Jupp. 1999. Estimating One-Time-of-Day Meteorological Data from Standard Daily Data as Inputs to Thermal Remote Sensing-Based Energy Balance Models. *Agricultural and Forest Meteorology*, Vol. 96, pp. 219–238.

Mellon, M. T., B. M. Jakosky, H. H. Kieffer, and P. R. Christensen. 2000. High-Resolution Thermal-Inertia Mapping from the Mars Global Surveyor Thermal Emission Spectrometer. *Icarus*, Vol. 148, pp. 437–455.

Moore, R. K. (ed.). 1975. Microwave Remote Sensors. Chapter 9 in *Manual of Remote Sensing* (R. G. Reeves, ed.). Falls Church, VA: American Society of Photogrammetry, pp. 399–538.

Norman, J. R., and F. Becker. 1995. Terminology in Thermal Infrared Remote Sensing of Natural Surfaces. *Remote Sensing Reviews*, Vol. 12, pp. 159–173.

Prata, A. J., V. Caselles, C. Coll, J. A. Sobrino, and C. Ottlé. 1995. Thermal Remote Sensing of Land Surface Temperature from Satellites: Current Status and Future Prospects. *Remote Sensing Reviews*, Vol. 12, pp. 175–224.

Pratt, D. A., and C. D. Ellyett. 1979. The Thermal Inertia Approach to Mapping of Soil Moisture and Geology. *Remote Sensing of Environment*, Vol. 8, pp. 151–168.

Price, J. C. 1978. Thermal Inertia Mapping: A New View of the Earth. *Journal of Geophysical Research*, Vol. 82, pp. 2582–2590.

Price, J. C. 1981. The Contribution of Thermal Data in Landsat Multispectral Classification Photogrammetric Engineering and Remote Sensing, Vol. 47, pp. 229–236.

Price, J. C. 1985. On the Analysis of Thermal Infrared Imagery: The Limited Utility of Apparent Thermal Inertia. *Remote Sensing of Environment*, Vol. 18, pp. 59–73.

Quattrochi, D. A., and J. C. Luvall. 1999. Thermal Infrared Remote Sensing for Analysis of Landscape Ecological Processes: Methods and Applications. *Landscape Ecology*, Vol. 14, pp. 577–598.

Quattrochi, D. A., and J. C. Luvall. 2004. *Thermal Remote Sensing in Land Surface Processing*. Boca Raton, FL: CRC Press, 142 pp.

Quattrochi, D. A., and J. C. Luvall. 2009. Thermal Remote Sensing in Earth Science Research. Chapter 5 in *The Sage Handbook of Remote Sensing* (T. A. Warner, M. D. Nellis, and G. M. Foody, eds.). London: Sage, pp. 64–78.

Ramsey, M. S., and J. H. Fink. 1999. Estimating Silicic Lava Vesicularity with Thermal Remote Sensing: A New Technique for Volcanic Mapping and Monitoring. *Bulletin of Volcanology*, Vol. 61, pp. 32–39.

Sabins, F. 1969. Thermal Infrared Imaging and Its Application to Structural Mapping, Southern California. *Geological Society of America Bulletin*, Vol. 80, pp. 397–404.

Schott, J. R., and W. J. Volchok. 1985. Thematic Mapper Infrared Calibration. *Photogrammetric Engineering and Remote Sensing*, Vol. 51, pp. 1351–1358.

Short, N. M., and L. M. Stuart. 1982. *The Heat Capacity Mapping Mission (HCMM) Anthology* (NASA Special Publication 465). Washington, DC: U.S. Government Printing Office, p. 264.

Toll, D. L. 1985. Landsat-4 Thematic Mapper Scene Characteristics of a Suburban and Rural Area. *Photogrammetric Engineering and Remote Sensing*, Vol. 51, pp. 1471–1482.

Watson, K., S. Hummer-Miller, and D. L. Sawatzky. 1982. Registration of Heat Capacity Mapping Mission Day and Night Images. *Photogrammetric Engineering and Remote Sensing*, Vol. 48, pp. 263–268.

Weast, R. C. (ed.). 1986. *CRC Handbook of Chemistry and Physics*. Boca Raton, FL: CRC Press.

Xue, Y., and A. P. Cracknell. 1995. Advanced Thermal Inertia Modeling. *International Journal of Remote Sensing*, Vol. 16, pp. 431–446.

Image Resolution

10.1. Introduction and Definitions

In very broad terms, *resolution* refers to the ability of a remote sensing system to record and display fine spatial, spectral, and radiometric detail. A working knowledge of resolution is essential for understanding both practical and conceptual aspects of remote sensing. Our understanding or lack of understanding of resolution may be the limiting factor in our efforts to use remotely sensed data, especially at coarse spatial resolutions.

For scientists with an interest in instrument design and performance, measurement of resolution is of primary significance in determining the optimum design and configuration of individual elements (e.g., specific lenses, detectors, or photographic emulsions) of a remote sensing system. Here our interest focuses on understanding image resolution in terms of the entire remote sensing system, regardless of our interests in specific elements of the landscape. Whether our focus concerns soil patterns, geology, water quality, land use, or vegetation distributions, knowledge of image resolution is a prerequisite for understanding the information recorded on the images we examine.

The purpose of this chapter is to discuss image resolution as a separate concept in recognition of its significance throughout the field of remote sensing. It attempts to outline generally applicable concepts without ignoring special and unique factors that apply in specific instances.

Estes and Simonett (1975) define resolution as "the ability of an imaging system . . . to record fine detail in a distinguishable manner" (p. 879). This definition includes several key concepts. The emphasis on the imaging *system* is significant because in most practical situations it makes little sense to focus attention on the resolving power of a single element of the system (e.g., the detector array) if another element (e.g., the camera lens) limits the resolution of the final image. "Fine detail" is, of course, a relative concept, as is the specification that detail be recorded in a "distinguishable" manner. These aspects of the definition emphasize that resolution can be clearly defined only by operational definitions applicable under specified conditions.

For the present, it is sufficient to note that there is a practical limit to the level of detail that can be acquired from a given aerial or satellite image. This limit we define informally as the *resolution* of the remote sensing system, although it must be recognized that image detail also depends on the character of the scene that has been imaged, atmospheric conditions, illumination, and the experience and ability of the image interpreter.

Most individuals think of resolution as *spatial resolution*, the fineness of the spatial detail visible in an image. "Fine detail" in this sense means that small objects can be identified on an image. But other forms of resolution are equally important. *Radiometric resolution* can be defined as the ability of an imaging system to record many levels of brightness. *Coarse* radiometric resolution would record a scene using only a few brightness levels or a few bits (i.e., at very high contrast), whereas *fine* radiometric resolution would record the same scene using many levels of brightness. *Spectral resolution* denotes the ability of a sensor to define fine wavelength intervals. Hyperspectral sensors (Chapter 15) generate images composed of 200 or more narrowly defined spectral regions. These data represent an extreme of spectral resolution relative to the TM or Landsat MSS images, which convey spectral information in only a few rather broad spectral regions.

Finally, *temporal resolution* is an important consideration in many applications. Remote sensing has the ability to record sequences of images, thereby representing changes in landscape patterns over time. The ability of a remote sensing system to record such a sequence at relatively close intervals generates a dataset with *fine* temporal resolution. In contrast, systems that can record images of a given region only at infrequent intervals produce data at *coarse* temporal resolution. In some applications, such as flood or disaster mapping, temporal resolution is a critical characteristic that might override other desirable qualities. Clearly, those applications that attempt to monitor dynamic phenomena, such as news events, range fires, land-use changes, traffic flows, or weather-related events, will have an interest in temporal resolution.

In many situations, there are clear trade-offs between different forms of resolution. For example, in traditional photographic emulsions, increases in spatial resolving power are based on decreased size of film grain, which produces accompanying decreases in radiometric resolution (i.e., the decreased sizes of grains in the emulsion portray a lower range of brightnesses). In other systems there are similar trade-offs. Increasing spatial detail requires, in scanning systems, a smaller instantaneous field of view (i.e., energy reaching the sensor has been reflected from a smaller ground area). If all other variables have been held constant, this must translate into decreased energy reaching the sensor; lower levels of energy mean that the sensor may record less "signal" and more "noise," thereby reducing the usefulness of the data. This effect can be compensated for by broadening the spectral window to pass more energy (i.e., decreasing spectral resolution) or by dividing the energy into fewer brightness levels (i.e., decreasing radiometric resolution). Of course, overall improvements can be achieved by improved instrumentation or by altering operating conditions (e.g., flying at a lower altitude). The general situation, however, seems to demand costs in one form of resolution for benefits achieved in another.

10.2. Target Variables

Observed spatial resolution in a specific image depends greatly on the character of the scene that has been imaged. In complex natural landscapes, identification of the essential variables influencing detail observed in the image may be difficult, although many of the key factors can be enumerated. Contrast is clearly one of the most important influences on spatial and radiometric resolution. *Contrast* can be defined as the difference in brightness between an object and its background. If other factors are held constant, high contrast favors recording of fine spatial detail; low contrast produces coarser detail. A black automobile imaged against a black asphalt background will be more difficult to observe than a white vehicle observed under the same conditions.

The significance of contrast as an influence on spatial resolution illustrates the inter-relationships between the various forms of resolution and emphasizes the reality that no single element of system resolution can be considered in isolation from the others. It is equally important to distinguish between contrast in the original scene and that recorded on the image of that scene; the two may be related, but not necessarily in a direct fashion (see Figure 4.6). Also, it should be noted that contrast in the original scene is a dynamic quality that, for a given landscape, varies greatly from season to season (with changes in vegetation, snow cover, etc.) and within a single day (as angle and intensity of illumination change).

The shape of an object or feature is significant. *Aspect ratio* refers to the length of a feature in relation to its width. Usually long, thin features, such as highways, railways, and rivers, tend to be visible on aerial imagery, even in circumstances in which their widths are much less than the nominal spatial resolution of the imagery. *Regularity of shape* favors recording of fine detail. Features with regular shapes, such as cropped agricultural fields, tend to be recorded in fine detail, whereas complex shapes will be imaged in coarser detail.

The *number* of objects in a pattern also influences the level of detail recorded by a sensor. For example, the pattern formed by the number and regular arrangement of tree crowns in an orchard favors the imaging of the entire pattern in fine detail. Under similar circumstances, the crown of a single isolated tree might not be visible on the imagery.

Extent and uniformity of background contributes to resolution of fine detail in many distributions. For example, a single automobile in a large, uniform parking area or a single tree positioned in a large cropped field will be imaged in detail not achieved under other conditions.

10.3. System Variables

Remember that the resolution of individual sensors depends in part on the design of that sensor and in part on its operation at a given time. For example, resolution of an aerial photograph (Chapter 3) is determined by the quality of the camera lens, flying altitude, scale, and the design of the camera. For scanning systems such as the Landsat MSS/TM (Chapter 6) or thermal scanners (Chapter 9), the IFOV determines many of the qualities of image resolution. The IFOV depends, of course, on the optical system (the angular field of view) and operating altitude. Speed of the scanning motion and movement of the vehicle that carries the sensor will also have their effects on image quality. For active microwave sensors (Chapter 7), image resolution is determined by beamwidth (antenna gain), angle of observation, wavelength, and other factors discussed previously.

10.4. Operating Conditions

For all remote sensing systems, the operating conditions, including flying altitude and ground speed, are important elements influencing the level of detail in the imagery. Atmospheric conditions can be included as important variables, especially for satellite and high-altitude imagery.

10.5. Measurement of Resolution

Ground Resolved Distance

Perhaps the simplest measure of spatial resolution is *ground resolved distance* (GRD), defined simply as the dimensions of the smallest objects recorded on an image. One might speak of the resolution of an aerial photograph as being "2 m," meaning that objects of that size and larger could be detected and interpreted from the image in question. Smaller objects presumably would not be resolved and, therefore, would not be interpretable.

Such measures of resolution may have utility as a rather rough suggestion of usable detail, but must be recognized as having only a very subjective meaning. The objects and features that compose the landscape vary greatly in size, shape, contrast with background, and pattern. Usually we have no means of relating a given estimate of GRD to a specific problem of interest. For example, the spatial resolution of U.S. Department of Agriculture (USDA) 1:20,000 black-and-white aerial photography is often said to be "about 1 m," yet typically one can easily detect on these photographs the painted white lines in parking lots and highways; these lines may be as narrow as 6–9 in. Does this mean that the resolution of this photography should be assessed as 6 in. rather than 1 m? Only if we are interested in the interpretation of long, thin features that exhibit high contrast with their background could we accept such an estimate as useful. Similarly, the estimate of 1 m may be inappropriate for many applications.

Line Pairs per Millimeter

Line pairs per millimeter (LPM) is a means of standardizing the characteristics of targets used to assess image resolution. Essentially, it is a means of quantifying, under controlled conditions, the estimate of GRD by using a standard target, positioned on the ground, which is imaged by the remote sensing system under specified operating conditions.

Although many targets have been used, the resolution target designed by the U.S. Air Force (USAF) has been a standard for a variety of studies (Figure 10.1). This target consists of parallel black lines positioned against a white background. The width of spaces between lines is equal to that of the lines themselves; their length is five times their width. As a result, a block of three lines and the two white spaces that separate them form a square. This square pattern is reproduced at varied sizes to form an array consisting of

FIGURE 10.1. Bar target used in resolution studies.

bars of differing widths and spacings. Sizes are controlled to produce changes in spacing of the bars (spatial frequency) of 12%. Repetition of the pattern at differing scales ensures that the image of the pattern will include at least one pattern so small that individual lines and their spaces will not be fully resolved.

If images of two objects are visually separated, they are said to be "spatially resolved." Images of the USAF resolution target are examined by an interpreter to find that smallest set of lines in which the individual lines are all completely separated along their entire length. The analyst measures the width of the image representation of one "line pair" (i.e., the width of the image of one line and its adjacent white space) (Figure 10.2). This measurement provides the basis for the calculation of the number of line pairs per millimeter (or any other length we may choose; LPM is standard for many applications). For example, in Figure 10.2 the width of a line and its adjacent gap is measured to be 0.04 mm. From 1 line pair/0.04 mm we find a resolution of 25 LPM.

For aerial photography, this measure of resolution can be translated into GRD by the relationship

$$GRD = \frac{H}{(f)(R)} \qquad\qquad \text{(Eq. 10.1)}$$

where GRD is ground resolved distance in meters; H is the flying altitude above the terrain in meters; f is the focal length in millimeters; and R is the system resolution in line pairs per millimeter.

Such measures have little predictable relationship to the actual size of landscape features that might be interpreted in practical situations, because seldom will the features of interest have the same regularity of size, shape, and arrangement and the high contrast of the resolution target used to derive the measures. They are, of course, valuable as comparative measures for assessing the performance of separate systems under the same operating conditions or of a single system under different conditions.

Although the USAF target has been widely used, other resolution targets have been developed. For example, a colored target has been used to assess the spectral fidelity of color films (Brooke, 1974), and bar targets have been constructed with contrast ratios somewhat closer to conditions observed during actual applications. For many years, a USGS target formed by a large array painted on the roof of the USGS National Center in Reston, Virginia, was a means of assessing aerial imagery under operational conditions

0.02 mm

FIGURE 10.2. Use of the bar target to find LPM.

from high altitudes. Currently the USGS maintains fly-over target ranges near Sioux Falls, South Dakota, for evaluating performance of both aircraft and satellite sensors.

Modulation Transfer Function

The *modulation transfer function* (MTF) records system response to a target array with elements of varying spatial frequency (i.e., unlike the bar targets described above, targets used to find MTFs are spaced at varied intervals). Often the target array is formed from bars of equal length spaced against a white background at intervals that produce a sinusoidal variation in image density along the axis of the target.

Modulation refers to changes in the widths and spacings of the target. *Transfer* denotes the ability of the imaging system to record these changes on the image—that is, to "transfer" these changes from the target to the image. Because the target is explicitly designed with spatial frequencies too fine to be recorded on the image, some frequencies (the high frequencies at the closest spacings) cannot be imaged. The "function" then shows the degree to which the image records specified frequencies (Figure 10.3).

Although the MTF is probably the "best" measure of the ability of an imaging system as a whole or of a single component of that system to record spatial detail, the complexity of the method prevents routine use in many situations. The MTF can be estimated using simpler and more readily available targets, including the USAF target described above (Welch, 1971).

10.6. Mixed Pixels

As spatial resolution interacts with the fabric of the landscape, a special problem is created in digital imagery by those pixels that are not completely occupied by a single, homogeneous category. The subdivision of a scene into discrete pixels acts to average brightnesses over the entire pixel area. If a uniform or relatively uniform land area occupies the pixel, then similar brightnesses are averaged, and the resulting digital value forms a reasonable representation of the brightnesses within the pixel. That is, the average value does not differ greatly from the values that contribute to the average. However, when a

FIGURE 10.3. Modulation transfer function. The inset illustrates an example of a target used in estimating the modulation transfer function. The solid line illustrates a hypothecal modulation transfer for an image that records the pattern with high fidelity; the dashed line represents an image that cannot portray the high-frequency variation at the right side of the target.

pixel area is composed of two or more areas that differ greatly with respect to brightness, then the average is composed of several very different values, and the single digital value that represents the pixel may not accurately represent any of the categories present (Figure 10.4).

An important consequence of the occurrence of mixed pixels is that pure spectral responses of specific features are mixed together with the pure responses of other features. The mixed response sometimes known as a *composite signature* does not match the *pure signatures* that we wish to use to map the landscape. Note, however, that sometimes composite signatures can be useful because they permit us to map features that are too complex to resolve individually.

Nonetheless, mixed pixels are also a source of error and confusion. In some instances, the digital values from mixed pixels may not resemble any of the several categories in the scene; in other instances, the value formed by a mixed pixel may resemble those from other categories in the scene but not actually present within the pixel—an especially misleading kind of error.

Mixed pixels occur often at the edges of large parcels or along long linear features, such as rivers or highways, where contrasting brightnesses are immediately adjacent to one another (Figure 10.5). The edge, or border, pixels then form opportunities for errors in digital classification. Scattered occurrences of small parcels (such as farm ponds observed at the resolution of the Landsat MSS) may produce special problems because they may be represented *only* by mixed pixels, and the image analyst may not be aware of the presence of the small areas of high contrast because they occur at subpixel sizes. An especially difficult situation can be created by landscapes composed of many parcels that are small relative to the spatial resolution of the sensor. A mosaic of such parcels will create an array of digital values, *all* formed by mixed pixels (Figure 10.6).

It is interesting to examine the relationships between the numbers of mixed pixels in a given scene and the spatial resolution of the sensor. Studies have documented the increase in numbers of mixed pixels that occurs as spatial resolution decreases. Because the numbers, sizes, and shapes of landscape parcels vary greatly with season and geographic setting, there can be no generally applicable conclusions regarding this problem.

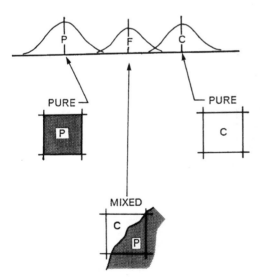

FIGURE 10.4. False resemblance of mixed pixels to a third category.

FIGURE 10.5. Edge pixels.

Yet examination of a few simple examples may help us understand the general character of the problem.

Consider the same contrived scene that is examined at several different spatial resolutions (Figure 10.7). This scene consists of two contrasting categories, with two parcels of one superimposed against the more extensive background of the other. This image is then examined at four levels of spatial resolution; for each level of detail, pixels are categorized as "background," "interior," or "border." (Background and interior pixels consist only of a single category; border pixels are those composed of two categories.) A tabulation of proportions of the total in each category reveals a consistent pattern (Table 10.1). As resolution becomes coarser, the number of mixed pixels increases (naturally) at the expense of the number of pure background and pure interior pixels. In this example, interior pixels experience the larger loss, but this result is the consequence of the specific circumstances of this example and is unlikely to reveal any generally applicable conclusions.

If other factors could be held constant, it would seem that fine spatial resolution would offer many practical advantages, including capture of fine detail. Note, however, the substantial increases in the total numbers of pixels required to achieve this advantage; note, too, that increases in the numbers of pixels produce compensating disadvantages, including increased costs. Also, this example does not consider another important effect often encountered as spatial resolution is increased: The finer detail may resolve features

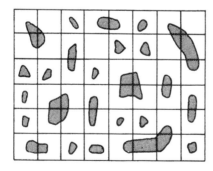

"MOSAIC" PIXELS
ALL PIXELS ARE MIXED PIXELS

WORST CASE SITUATION:
FEATURE REPRESENTED
ONLY BY MIXED PIXELS

FIGURE 10.6. Mixed pixels generated by an image of a landscape composed of small pixels.

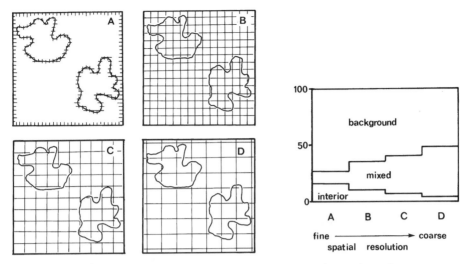

FIGURE 10.7. Influence of spatial resolution on proportions of mixed pixels.

not recorded at coarser detail, thereby increasing, rather than decreasing, the proportions of mixed pixels. This effect may explain some of the results observed by Sadowski and Sarno (1976), who found that classification accuracy decreased as spatial resolution became finer.

Marsh et al. (1980) have reviewed strategies for resolving the percentages of components that compose the ground areas with mixed pixels. The measured digital value for each pixel is determined by the brightnesses of distinct categories within that pixel area projected on the ground, as integrated by the sensor over the area of the pixel. For example, the projection of a pixel on the Earth's surface may encompass areas of open water (W) and forest (F). Assume that we know that (1) the digital value for such a pixel is "mixed," not "pure"; (2) the mean digital value for water in all bands is (W_i); (3) the mean digital value for forest in all bands is (F_i); and (4) the observed value of the mixed pixel in all spectral bands is (M_i). We wish, then, to find the areal percentages PW and PF that contribute to the observed value M_i.

Marsh et al. outline several strategies for estimating PW and PF under these conditions; the simplest, if not the most accurate, is the weighted average method:

$$PW = (M_i - F_i)/(W_i - F_i) \qquad \text{(Eq. 10.2)}$$

An example can be shown using the following data:

TABLE 10.1. Summary of Data Derived from Figure 10.7

Spatial resolution		Total	Mixed	Interior	Background
Fine	A	900	109	143	648
	B	225	59	25	141
	C	100	34	6	60
Coarse	D	49	23	1	25

	Band			
	1	2	3	4
Means for the mixed pixel (M_i)	16	12	16	18
Means for forest (F_i)	23	16	32	35
Means for water (W_i)	9	8	0	1

Using Equation 10.2, the areal proportion of the mixed pixel composed of the water category can be estimated as follows:

Band 1: $P_W = (16 - 23)/(9 - 23) = -7/-14 = 0.50$

Band 2: $P_W = (12 - 16)/(8 - 16) = -4/-8 = 0.50$

Band 3: $P_W = (16 - 32)/(0 - 32) = -16/-32 = 0.50$

Band 4: $P_W = (18 - 35)/(1 - 35) = -17/-34 = 0.50$

Thus the mixed pixel is apparently composed of about 50% water and 50% forest. Note that in practice we may not know which pixels are mixed and may not know the categories that might contribute to the mixture. Note also that this procedure may yield different estimates for each band. Other procedures, too lengthy for concise description here, may give more suitable results in some instances (Marsh et al., 1980).

10.7. Spatial and Radiometric Resolution: Simple Examples

Some of these effects can be illustrated by contrived examples prepared by manipulation of a black-and-white image of a simple agricultural scene (Figures 10.8 and 10.9). In each case, the upper left panel shows the original image, followed by successive images contrived to show effects of coarser spatial resolution or coarser radiometric resolution.

Figure 10.8 shows a sequence in which the radiometric resolution is held constant, but the spatial resolution has been degraded by averaging brightnesses over increasingly large blocks of adjacent pixels, thereby decreasing the spatial resolution by factors of 5, 10, 25, and so on. At the image size shown here, the effects of lower resolutions are not visually obvious until rather substantial decreases have been made. As pixels are averaged over blocks of 25 pixels, the loss in detail is visible, and at levels of 50 and 100 pixels, the image has lost the detail that permits recognition and interpretation of the meaning of the scene. Note also the effects of mixed pixels at high-contrast edges, where the sharp changes in brightness blur into broad transitional zones at the coarser levels of detail. At the coarse level of detail of the final image in Figure 10.8, only the vaguest suggestions of the original pattern are visible.

For Figure 10.9, the spatial resolution has been held constant while the radiometric resolution has been reduced from 8 bits (256 brightness levels) to 1 bit (2 brightness levels). As is the case for the previous example, changes are visually evident only after the image has been subjected to major reductions in detail. At higher radiometric resolutions, the scene has a broad range of brightness levels, allowing the eye to see subtle features in the landscape. At lower resolutions, the brightnesses have been represented at only a few levels (ultimately, only black and white), depicting only the coarsest outline of the scene.

FIGURE 10.8. Spatial resolution. A contrived example in which spatial resolution has been degraded by averaging brightnesses over increasingly larger blocks of adjacent pixels.

Keep these examples in mind as you examine images from varied sensors. Any sensor is designed to record specific levels of spatial and radiometric resolution; these qualities determine its effectiveness in portraying features on the landscape. Broader levels of resolution may be adequate for rather coarse-textured landscapes. For example, even very low levels of resolution may be effective for certain well-defined tasks, such as separating

FIGURE 10.9. Radiometric resolution. A contrived example in which spatial resolution has been held constant while radiometric resolution has been reduced from 8 bits (256 brightness levels) to 1 bit (2 brightness levels).

open water from land, detecting cloud patterns, and the like. Note also that finer resolution permits more subtle distinctions but also records detail that may not be relevant to the task at hand and may tend to complicate analysis.

10.8. Interactions with the Landscape

Although most discussions of image resolution tend to focus on sensor characteristics, understanding the significance of image resolution in the application sciences requires assessment of the effect of specific resolutions on images of specific landscapes or classes of landscapes. For example, relatively low resolution may be sufficient for recording the essential features of landscapes with rather coarse fabrics (e.g., the broadscale patterns of the agricultural fields of the North American Great Plains) but inadequate for imaging complex landscapes composed of many small parcels with low contrast.

Podwysocki's studies (1976a, 1976b) of field sizes in the major grain-producing regions of the world is an excellent example of the systematic investigation of this topic. His research can be placed in the context of the widespread interest in accurate forecasts of world wheat production in the years that followed large international wheat purchases by the Soviet Union in 1972. Computer models of biophysical processes of crop growth and maturation could provide accurate estimates of yields (given suitable climatological data), but estimates of total production also require accurate estimates of planted acreage. Satellite imagery would seem to provide the capability to derive the required estimates of area plowed and planted. Podwysocki attempted to define the extent to which the spatial resolution of the Landsat MSS would be capable of providing the detail necessary to provide the required estimates.

He examined Landsat MSS scenes of the United States, China, the Soviet Union, Argentina, and other wheat-producing regions, sampling fields for measurements of length, width, and area. His data are summarized by frequency distributions of field sizes for samples of each of the world's major wheat-producing regions. (He used his samples to find the Gaussian distributions for each of his samples, so he was able to extrapolate the frequency distributions to estimate frequencies at sizes smaller than the resolution of the MSS data.) Cumulative frequency distributions for his normalized data reveal the percentages of each sample that equal or exceed specific areas (Figure 10.10). For example, the curve for India reveals that 99% (or more) of this sample were at least 1 ha in size, that all were smaller than about 100 ha (247 acres), and that we can expect the Indian wheat fields to be smaller than those in Kansas. These data and others presented in his study provide the basis for evaluating the effectiveness of a given resolution in monitoring features of specified sizes. This example is especially instructive because it emphasizes not only the differences in average field size in the different regions (shown in Figure 10.10 by the point at which each curve crosses the 50% line) but also the differences in variation of field size between the varied regions (shown in Figure 10.10 by the slopes of the curves).

In a different analysis of relationships between sensor resolution and landscape detail, Simonett and Coiner (1971) examined 106 sites in the United States, each selected to represent a major land-use region. Their study was conducted prior to the launch of Landsat 1 with the objective of assessing the effectiveness of MSS spatial resolution in recording differences between major land-use regions in the United States. Considered

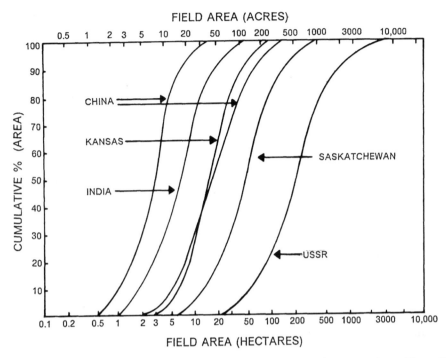

FIGURE 10.10. Field size distributions for selected wheat-producing regions. From Podwysocki, 1976a.

as a whole, their sites represent a broad range of physical and cultural patterns in the 48 coterminous states.

For each site they simulated the effects of imaging with low-resolution imagery by superimposing grids over aerial photographs, with grid dimensions corresponding to ground distances of 800, 400, 200, and 100 ft. Samples were randomly selected within each site. Each sample consisted of the *number* of land-use categories within cells of each size and thereby formed a measure of landscape diversity, as considered at several spatial resolutions. For example, those landscapes that show only a single land-use category at the 800-ft. resolution have a very coarse fabric and would be effectively imaged at the low resolution of satellite sensors. Those landscapes that have many categories within the 100-ft. grid are so complex that very fine spatial resolution would be required to record the pattern of landscape variation. Their analysis grouped sites according to their behavior at various resolutions. They reported that natural landscapes appeared to be more susceptible than man-made landscapes to analysis at the relatively coarse resolutions of the Landsat MSS.

Welch and Pannell (1982) examined Landsat MSS (bands 2 and 4) and Landsat 3 RBV images (in both pictorial and digital formats) to evaluate their suitability as sources of landscape information at levels of detail consistent with a map scale of 1:250,000. Images of three study areas in China provided a variety of urban and agricultural landscapes for study, representing a range of spatial detail and a number of geographical settings. Their analysis of modulation transfer functions reveals that the RBV imagery represents an improvement in spatial resolution of about 1.7 over the MSS imagery and that Landsat 4 TM provided an improvement of about 1.4 over the RBV (for target:background contrasts of about 1.6:1).

Features appearing on each image were evaluated with corresponding representations on 1:250,000 maps in respect to size, shape, and contrast. A numerical rating system provided scores for each image based on the numbers of features represented and the quality of the representations on each form of imagery. MSS images portrayed about 40–50% of the features shown on the usual 1:250,000 topographic maps. MSS band 2 was the most effective for identification of airfields; band 4 performed very well for identification and delineation of water bodies. Overall, the MSS images achieved scores of about 40–50%. RBV images attained higher overall scores (50–80%), providing considerable improvement in representation of high-contrast targets but little improvement in imaging of detail in fine-textured urban landscapes. The authors concluded that spatial resolutions of MSS and RBV images were inadequate for compilation of usual map detail at 1:250,000.

10.9. Summary

This chapter highlights the significance of image resolution as a concept that extends across many aspects of remote sensing. Although the special and unique elements of any image must always be recognized and understood, many of the general aspects of image resolution can assist us in understanding how to interpret remotely sensed images. Although there has long been an intense interest in measuring image resolution, especially in photographic systems, it is clear that much of our more profound understanding has been developed through work with satellite scanning systems such as the Landsat MSS. Such data were of much coarser spatial resolution than any studied previously. As more and more attention was focused on their analysis and interpretation (Chapters 11 and 12), it was necessary to develop a better understanding of image resolution and its significance for specific tasks. Now much finer resolution data are available, but we can continue to develop and apply our knowledge of image resolution to maximize our ability to understand and interpret these images.

Review Questions

1. Most individuals are quick to appreciate the advantages of fine resolution. However, there may well be *disadvantages* to fine-resolution data relative to data of coarser spatial, spectral, and radiometric detail. Suggest what some of these disadvantages might be.

2. Imagine that the spatial resolution of the digital remote sensing system is increased from about 80 m to 40 m. List some of the consequences, assuming that image coverage remains the same. What would be some of the consequences of *decreasing* detail from 80 m to 160 m?

3. You examine an image of the U.S. Air Force resolution target and determine that the image distance between the bars in the smallest pair of lines is 0.01 mm. Find the LPM for this image. Find the LPM for an image in which you measure the distance to be 0.04 mm. Which image has finer resolution?

4. For each object or feature listed below, discuss the characteristics that will be signifi-

cant in our ability to resolve the object on a remotely sensed image. Categorize each as "easy" or "difficult" to resolve clearly. Explain.

   a. A white car parked alone in an asphalt parking lot.

   b. A single tree in a pasture.

   c. An orchard.

   d. A black cat in a snow-covered field.

   e. Painted white lines on a crosswalk across an asphalt highway.

   f. Painted white lines on a crosswalk across a concrete highway.

   g. A pond.

   h. A stream.

5. Write a short essay describing how spatial resolution, spectral resolution, and radiometric resolution are interrelated. Is it possible to increase one kind of resolution without influencing the others?

6. Review Chapters 1–9 to identify the major features that influence spatial resolution of images collected by the several kinds of sensors described. Prepare a table to list these factors in summary form.

7. Explain why some objects might be resolved clearly in one part of the spectrum but poorly in another portion of the spectrum.

8. Although the U.S. Air Force resolution target is very useful for evaluating some aspects of remotely sensed images, it is not necessarily a good indication of the ability of a remote sensing system to record patterns that are significant for environmental studies. List some of the reasons this might be true.

9. Describe ideal conditions for achieving maximum spatial resolution.

References

Badhwar, G. B. 1984. Automatic Corn–Soybean Classification Using Landsat MSS Data: II. Early Season Crop Proportion Estimation. *Remote Sensing of Environment*, Vol. 14, pp. 31–37.

Badhwar, G. D. 1984. Use of Landsat-Derived Profile Features for Spring Small-Grains Classification. *International Journal of Remote Sensing*, Vol. 5, pp. 783–797.

Badhwar, G. D., J. G. Carnes, and W. W. Austen. 1982. Use of Landsat-Derived Temporal Profiles for Corn–Soybean Feature Extraction and Classification. *Remote Sensing of Environment*, Vol. 12, pp. 57–79.

Badhwar, G. D., C. E. Garganti, and F. V. Redondo. 1987. Landsat Classification of Argentina Summer Crops. *Remote Sensing of Environment*, Vol. 21, pp. 111–117.

Brooke, R. K. 1974. *Spectral/Spatial Resolution Targets for Aerial Imagery*. (Technical Report ETL-TR-74-3). Ft. Belvoir, VA: U.S. Army Engineer Topographic Laboratories, 20 pp.

Chhikara, R. S. 1984. Effect of Mixed (Boundary) Pixels on Crop Proportion Estimation. *Remote Sensing of Environment*, Vol. 14, pp. 207–218.

Crapper, P. F. 1980. Errors Incurred in Estimating an Area of Uniform Land Cover Using Landsat. *Photogrammetric Engineering and Remote Sensing*, Vol. 46, pp. 1295–1301.

Estes, J. E., and D. S. Simonett. 1975. Fundamentals of Image Interpretation. Chapter 14 in *Manual of Remote Sensing* (R. G. Reeves, ed.). Bethesda, MD: American Society for Photogrammetry and Remote Sensing, pp. 869–1076.

Ferguson, M. C., G. D. Badhwar, R. S. Chhikara, and D. E. Pitts. 1986. Field Size Distributions for Selected Agricultural Crops in the United States and Canada. *Remote Sensing of Environment*, Vol. 19, pp. 25–45.

Hall, F. G., and G. D. Badhwar. 1987. Signature Extendable Technology: Global Space-Based Crop Recognition. *IEEE Transactions on Geoscience and Remote Sensing*, Vol. GE-25, pp. 93–103.

Hallum, C. R., and C. R. Perry. 1984. Estimating Optimal Sampling Unit Sizes for Satellite Surveys. *Remote Sensing of Environment*, Vol. 14, pp. 183–196.

Hyde, R. F., and N. J. Vesper. 1983. Some Effects of Resolution Cell Size on Image Quality. *Landsat Data Users Notes*, Issue 29, pp. 9–12.

Irons, J. R., and D. L. Williams. 1982. Summary of Research Addressing the Potential Utility of Thematic Mapper Data for Renewable Resource Applications. *Harvard Computer Graphics Week*. Cambridge, MA: Graduate School of Design, Harvard University.

Latty, R. S., and R. M. Hoffer. 1981. Computer-Based Classification Accuracy Due to the Spatial Resolution Using Per-Point versus Per-Field Classification Techniques. In *Proceedings, 7th International Symposium on Machine Processing of Remotely Sensed Data*. West Lafayette, IN: Laboratory for Applications of Remote Sensing, pp. 384–393.

MacDonald, D. E. 1958. Resolution as a Measure of Interpretability. *Photogrammetric Engineering*, Vol. 24, No. 1, pp. 58–62.

Markham, B. L., and J. R. G. Townshend. 1981. Land Cover Classification Accuracy as a Function of Sensor Spatial Resolution. In *Proceedings, 15th International Symposium on Remote Sensing of Environment*. Ann Arbor: University of Michigan Press, pp. 1075–1090.

Marsh, S. E., P. Switzer, and R. J. P. Lyon. 1980. Resolving the Percentage of Component Terrains within Single Resolution Elements. *Photogrammetric Engineering and Remote Sensing*, Vol. 46, pp. 1079–1086.

Pitts, D. E., and G. Badhwar. 1980. Field Size, Length, and Width Distributions Based on LACIE Ground Truth Data. *Remote Sensing of Environment*, Vol. 10, pp. 201–213.

Podwysocki, M. H. 1976a. *An Estimate of Field Size Distribution for Selected Sites in the Major Grain Producing Countries* (Publication No. X-923-76-93). Greenbelt, MD: Goddard Space Flight Center, 34 pp.

Podwysocki, M. H. 1976b. *Analysis of Field Size Distributions: LACIE Test Sites 5029, 5033, 5039, Anwhei Province, People's Republic of China* (Publication No. X-923-76-145). Greenbelt, MD: Goddard Space Flight Center, 8 pp.

Potdar, M. B. 1993. Sorghum Yield Modeling Based on Crop Growth Parameters Determined from Visible and Near-IR Channel NOAA AVHRR Data. *International Journal of Remote Sensing*, Vol. 14, pp. 895–905.

Sadowski, F., and J. Sarno. 1976. *Forest Classification Accuracy as Influenced by Multispectral Scanner Spatial Resolution* (Report for Contract NAS9-14123:NASA). Houston, TX: Lyndon Baines Johnson Space Center.

Salmonowicz, P. H. 1982. USGS Aerial Resolution Targets. *Photogrammetric Engineering and Remote Sensing*, Vol. 48, pp. 1469–1473.

Simonett, D. S., and J. C. Coiner. 1971. Susceptibility of Environments to Low Resolution Imaging for Land Use Mapping. In *Proceedings, 7th International Symposium on Remote Sensing of Environment*. Ann Arbor: University of Michigan Press, pp. 373–394.

Tucker, C. J. 1980. Radiometric Resolution for Monitoring Vegetation: How Many Bits Are Needed? *International Journal of Remote Sensing*, Vol. 1, pp. 241–254.

Wehde, M. E. 1979. Spatial Quantification of Maps or Images: Cell Size or Pixel Size Implication. In *Joint Proceedings, American Society of Photogrammetry and American Congress of Surveying and Mapping.* Bethesda, MD: American Society for Photogrammetry and Remote Sensing, pp. 45–65.

Welch, R. 1971. Modulation Transfer Functions. *Photogrammetric Engineering*, Vol. 47, pp. 247–259.

Welch, R., and C. W. Pannell. 1982. Comparative Resolution of Landsat 3 MSS and RBV Images of China. *Photogrammetric Record*, Vol. 10, pp. 575–586.

PART THREE

ANALYSIS

CHAPTER ELEVEN

Preprocessing

11.1. Introduction

In the context of digital analysis of remotely sensed data, *preprocessing* refers to those operations that are preliminary to the principal analysis. Typical preprocessing operations could include (1) radiometric preprocessing to adjust digital values for effects of a hazy atmosphere and/or (2) geometric preprocessing to bring an image into registration with a map or another image. Once corrections have been made, the data can then be subjected to the primary analyses described in subsequent chapters. Thus preprocessing forms a preparatory phase that, in principle, improves image quality as the basis for later analyses that will extract information from the image.

It should be emphasized that, although certain preprocessing procedures are frequently used, there can be no definitive list of "standard" preprocessing steps, because each project requires individual attention, and some preprocessing decisions may be a matter of personal preference. Furthermore, the quality of image data varies greatly, so some data may not require the preprocessing that would be necessary in other instances. Also, preprocessing alters image data. Although we may assume that such changes are beneficial, the analyst should remember that preprocessing may create artifacts that are not immediately obvious. As a result, the analyst should tailor preprocessing to the data at hand and the needs of specific projects, using only those preprocessing operations essential to obtain a specific result.

11.2. Radiometric Preprocessing

Radiometric preprocessing influences the brightness values of an image to correct for sensor malfunctions or to adjust the values to compensate for atmospheric degradation. Any sensor that observes the Earth's surface using visible or near visible radiation will record a mixture of two kinds of brightnesses. One is the brightness derived from the Earth's surface—that is, the brightnesses that are of interest for remote sensing. But the sensor also observes the brightness of the atmosphere itself—the effects of atmospheric scattering (Chapter 2). Thus an observed digital brightness value (e.g., "56") might be in part the result of surface reflectance (e.g., "45") and in part the result of atmospheric scattering (e.g., "11"). Of course we cannot immediately distinguish between the two brightnesses, so one objective of atmospheric correction is to identify and separate these two components so that the main analysis can focus on examination of correct surface brightness (the "45" in this example). Ideally, atmospheric correction should find a separate correction for each pixel in the scene; in practice, we may apply the same correction to an entire band or apply a single factor to a local region within the image.

Preprocessing operations to correct for atmospheric degradation fall into three rather broad categories. First are those procedures known as radiative transfer code (RTC) computer models, which model the physical behavior of solar radiation as it passes through the atmosphere. Application of such models permits observed brightnesses to be adjusted to approximate true values that might be observed under a clear atmosphere, thereby improving image quality and accuracies of analyses. Because RTC models the physical process of scattering at the level of individual particles and molecules, this approach has important advantages with respect to rigor, accuracy, and applicability to a wide variety of circumstances. But they also have significant disadvantages. Often they are complex, usually requiring detailed in situ data acquired simultaneously with the image and/or satellite data describing the atmospheric column at the time and place of acquisition of an image. Such data may be difficult to obtain in the necessary detail and may apply only to a few points within a scene. Although meteorological satellites, as well as a growing number of remote sensing systems, collect atmospheric data that can contribute to atmospheric corrections of imagery, procedures for everyday applications of such methods are not now at hand for most analysis.

A second approach to atmospheric correction of remotely sensed imagery is based on examination of spectra of objects of known or assumed brightness recorded by multispectral imagery. From basic principles of atmospheric scattering, we know that scattering is related to wavelength, sizes of atmospheric particles, and their abundance. If a known target is observed using a set of multispectral measurements, the relationships between values in the separate bands can help assess atmospheric effects. This approach is often known as "image-based atmospheric correction" because it aspires to adjust for atmospheric effect solely, or mainly, from evidence available within the image itself.

Ideally, the target consists of a natural or man-made feature that can be observed with airborne or ground-based instruments at the time of image acquisition, so the analyst could learn from measurements independent of the image the true brightness of the object when the image was acquired. However, in practice we seldom have such measurements, and therefore we must look for features of known brightness that commonly, or fortuitously, appear within an image. In its simplest form, this strategy can be implemented by identifying a very dark object or feature within the scene. Such an object might be a large water body or possibly shadows cast by clouds or by large topographic features. In the infrared portion of the spectrum, both water bodies and shadows should have brightness at or very near zero, because clear water absorbs strongly in the near infrared spectrum and because very little infrared energy is scattered to the sensor from shadowed pixels. Analysts who examine such areas, or the histograms of the digital values for a scene, can observe that the lowest values (for dark areas, such as clear water bodies) are not zero but some larger value. Typically, this value will differ from one band to the next, so, for example, for Landsat band 1, the value might be 12; for band 2, the value might be 7; for band 3, 2; and for band 4, 2 (Figure 11.1). These values, assumed to represent the value contributed by atmospheric scattering for each band, are then subtracted from all digital values for that scene and that band. Thus the lowest value in each band is set to zero, the dark black color assumed to be the correct tone for a dark object in the absence of atmospheric scattering. Such dark objects form one of the most common examples of *pseudo-invariant objects*, which are taken as approximations of features with spectral properties

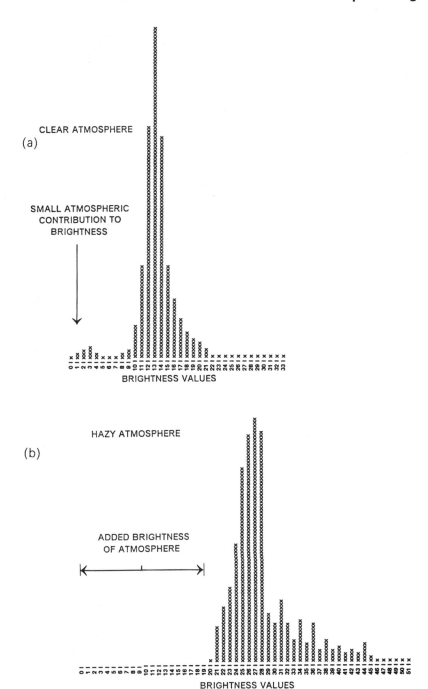

FIGURE 11.1. Histogram minimum method for correction of atmospheric effects. The lowest brightness value in a given band is taken to indicate the added brightness of the atmosphere to that band and is then subtracted from all pixels in that band. (a) Histogram for an image acquired under clear atmospheric conditions; the darkest pixel is near zero brightness. (b) Histogram for an image acquired under hazy atmospheric conditions; the darkest pixels are relatively bright, due to the added brightness of the atmosphere.

that do not change significantly over time, so their spectral characteristics sometimes can form the basis for normalization of multispectral scenes to a standard reference. This strategy forms one of the simplest, most direct methods for adjusting digital values for atmospheric degradation (Chavez, 1975), known sometimes as the *histogram minimum method (HMM)* or the *dark object subtraction (DOS) technique.*

This procedure has the advantages of simplicity, directness, and almost universal applicability, as it exploits information present within the image itself. Yet it must be considered as an approximation; atmospheric effects change not only the position of the histogram on the axis but also its shape (i.e., not all brightnesses are affected equally) (Figure 11.2). (Chapter 2 explained that the atmosphere can cause dark pixels to become brighter and bright pixels to become darker, so application of a single correction to all pixels will provide only a rough adjustment for atmospheric effects. Thus, it is said that the DOS technique is capable of correction for additive effects of atmospheric scattering, but not for multiplicative effects.) Although Moran et al. (1992) found that DOS was the least accurate of the methods evaluated in their study, they recognized that its simplicity and the prospect of pairing it with other strategies might form the basis for a practical technique for operational use. Because the subtraction of a constant from all pixels in a scene will have a larger proportional impact on the spectra of dark pixels than on brighter pixels, users should apply caution in using it for scenes in which spectral characteristics of dark features, such as water bodies or perhaps coniferous forests, for example, might form important dimensions of a study. Despite such concerns, the technique, both in its basic form and in the various modifications described below, has been found to be satisfactory for a variety of remote sensing applications (e.g., Baugh and Groeneveld, 2006; Nield et al., 2007).

A more sophisticated approach retains the idea of examining brightness of objects within each scene but attempts to exploit knowledge of interrelationships between separate spectral bands. Chavez (1975) devised a procedure that paired values from each band with values from a near infrared spectral channel. The Y intercept of the regression line is then taken as the correction value for the specific band in question. Whereas the HMM procedure is applied to entire scenes or to large areas, the regression technique can be applied to local areas (of possibly only 100–500 pixels each), ensuring that the adjustment is tailored to conditions important within specific regions. An extension of the regression technique is to examine the *variance–covariance matrix*, the set of variances and covariances between all band pairs on the data. (This is the *covariance matrix method* [CMM] described by Switzer et al., 1981.) Both procedures assume that within a specified image region, variations in image brightness are due to topographic irregularities and reflectivity is constant (i.e., land-cover reflectivity in several bands is uniform for the region in question). Therefore, variations in brightness are caused by small-scale topographic shadowing, and the dark regions reveal the contributions of scattering to each band. Although these assumptions may not always be strictly met, the procedure, if applied with care and with knowledge of the local geographic setting, seems to be robust and often satisfactory.

11.3. Some More Advanced Atmospheric Correction Tools

Further Refinements of DOS

The basic DOS strategy has been evaluated and refined by further research. Chavez (1988) proposed a modification of the basic DOS technique, in which an analyst selects a relative atmospheric scattering model that estimates scattering effects across all the spectral

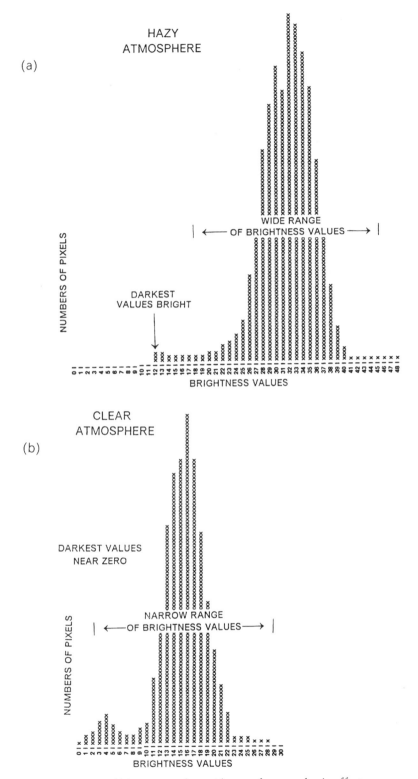

FIGURE 11.2. Inspection of histograms for evidence of atmospheric effects.

bands, given an estimate of the value of an initial band. Gilabert et al. (1994) proposed a modified DOS method for adjusting Landsat TM data for atmospheric scattering by identifying dark surfaces in TM bands 1 (blue) and 3 (red) and applying an algorithm to estimate atmospheric aerosols in the two TM bands. Teillet and Fedosejevs (1995) outline a dark target algorithm applicable to Landsat TM images by extraction of digital counts for specified dark objects, conversion to reflectance values, and determination of aerosol optical depth. The key uncertainties in the approach are the assumed values of the dark target surface reflectance in the relevant spectral bands and the radiometric sensor calibration. Chavez (1996) proposed use of the cosine of the solar zenith angle (cosine estimation of atmospheric transmittance [COST]) to supplement the DOS estimate of atmospheric transmittance. Song et al. (2001) propose a method that adds the effect of Rayleigh scattering to conventional DOS.

MODTRAN

MODTRAN (MODerate resolution atmospheric TRANsmission) is a computer model for estimating atmospheric transmission of electromagnetic radiation under specified conditions. MODTRAN was developed by the U.S. Air Force and the Spectral Science Corporation, which have patented some aspects of the model. (The most recent version is MODTRAN 5; an earlier system, LOWTRAN, is now considered obsolete). MODTRAN estimates atmospheric emission, thermal scattering, and solar scattering (including Rayleigh, Mie, single, and multiple scattering), incorporating effects of molecular absorbers and scatterers, aerosols, and clouds for wavelengths from the ultraviolet region to the far infrared. It uses various standard atmospheric models based on common geographic locations and also permits the user to define an atmospheric profile with any specified set of parameters. The model offers several options for specifying prevailing aerosols, based on common aerosol mixtures encountered in terrestrial conditions (e.g., rural, urban, maritime). Within MODTRAN, the estimate of visibility serves as an estimate of the magnitude of atmospheric aerosols. See

http://modtran.org

www.kirtland.af.mil/library/factsheets/factsheet.asp?id=7915

MODTRAN has been validated by extensive use in varied applications and serves also as a component of other models.

ATCOR

ATCOR is a proprietary system for implementing atmospheric correction based on the MODTRAN 4 model, developed by the German Aerospace Center (DLR) and marketed under license by ReSe Applications Schläpfer (*www.rese.ch/atcor/*). It provides separate models for satellite sensors (ATCOR 2/3), with small or moderate fields of view sensors and low relief terrain, whereas ATCOR 4 is designed for aircraft systems (including both optical and thermal instruments) and accommodates more rugged terrain. It features options specifically tailored for many of the more common satellite and aircraft sensors, extraction of specific band ratios (see Chapter 17), and several measures of net radiation, surface flux, albedo, and reflectance.

6S

6S (Second Simulation of the Satellite Signal in the Solar Spectrum; Vermote et al., 1997) simulates the signal observed by a satellite sensor for a Lambertian target at mean sea level. Developed at both the University of Maryland and the Laboratoire d'Optique Atmospherique, University of Lille, France, the code is widely used in a variety of atmospheric correction algorithms, including that developed for the MODIS surface reflectance products and the atmosphere removal algorithm (ATREM) developed at the University of Colorado at Boulder (Gao and Goetz 1990). 6S and MODTRAN5 (described above) are among the most widely used radiative transfer models in remote sensing.

Radiative transfer is the physical phenomenon of energy transfer through a medium as electromagnetic radiation. As the electromagnetic radiation travels through the medium (which, for this discussion, is the atmosphere), it is affected by three wavelength dependent processes, as follows: absorption (energy loss), scattering (energy redistribution), and emission (energy gain) (introduced in Chapter 1). The equation of radiative transfer, although not presented here because of its complexity, relates all these processes. Solutions to the radiative transfer equation are myriad, with scores of models available for various applications. Key to each solution is the means by which coefficients for emission, scattering, and absorption are estimated. 6S presents a robust and vetted solution to the radiative transfer equation. Among particularly important features of the model are its ability to take into account (1) target altitude, (2) polarization by molecules and aerosols, (3) nonuniform targets, and (4) the interaction of the atmosphere and the BRDF of the target. Further information can be obtained from Vermote et al. (1997) and the user manual, available online at *http://6s.ltdri.org*.

11.4. Calculating Radiances from DNs

As explained in Chapter 4, digital data formatted for distribution to the user community present pixel values as digital numbers (DNs), expressed as integer values to facilitate computation and transmission and to scale brightnesses for convenient display. Although DNs have practical value, they do not present brightnesses in the physical units (Watts per square meter per micrometer per steradian) (cf. Chapter 2) necessary to understand the optical processes that underlie the phenomena that generated the observed brightnesses. When image brightnesses are expressed as DNs, each image becomes an individual entity, without a defined relationship to other images or to features on the ground. DNs express accurate relative brightnesses within an image but cannot be used to examine brightnesses over time (from one date to another), to compare brightnesses from one instrument to another, to match one scene with another, nor to prepare mosaics of large regions. Further, DNs cannot serve as input for models of physical processes in (for example) agriculture, forestry, or hydrology. Therefore, conversion to radiances forms an important transformation to prepare remotely sensed imagery for subsequent analyses.

DNs can be converted to radiances using data derived from the instrument calibration provided by the instrument's manufacturer. Because calibration specifications can drift over time, recurrent calibration of aircraft sensors is a normal component of operation and maintenance and, for some instruments, may form a standard step in preflight preparation. However, satellite sensors, once launched, are unavailable for recalibration in the laboratory, although their performance can be evaluated by directing the sensor to view onboard calibration targets or by imaging a flat, uniform landscape, such as carefully selected desert terrain. As examples, calibration procedures and specifications for Landsat sensors can be found in Markham (1986), Markham et al. (2003), and Chander et al. (2009).

For a given sensor, spectral channel, and DN, the corresponding radiance value (L) can be calculated as

$$L = [(L_{max} - L_{min})/(Q_{calmax} - Q_{calmin})] \times (Q_{cal} - Q_{calmin}) + L_{min}) \qquad \text{(Eq. 11.1)}$$

where L_λ is the spectral radiance at the sensor's aperture [W/(m^2 sr µm)]; Q_{cal} is the quantized calibrated pixel value (DN); Q_{calmin} is the minimum quantized calibrated pixel value corresponding to L_{min_λ} (DN); Q_{calmax} is the maximum quantized calibrated pixel value corresponding to L_{max_λ} (DN); L_{min} is the spectral at-sensor radiance that is scaled to Q_{calmin} [W/(m^2 sr µm)]; and L_{min} is the spectral at-sensor radiance that is scaled to Q_{calmax} [W/(m^2 sr µm)]. Calibration information for specific Landsat MSS and TM sensors has been presented in system-specific review articles (e.g., Chander et al., 2007; Markham and Barker, 1986; Markham and Chander, 2003). Chander et al. (2009) provide, in a single document, a summary and update of calibration data for Landsat instruments, as well as related U.S. systems. For commercial satellite systems, essential calibration data are typically provided in header information accompanying image data or at websites describing specific systems and their data.

11.5. Estimation of Top of Atmosphere Reflectance

Accurate measurement of brightness, whether measured as radiances or as DNs, is not optimal because such values are subject to modification by differences in sun angle, atmospheric effects, angle of observation, and other effects that introduce brightness errors unrelated to the characteristics we wish to observe. It is much more useful to observe the *proportion* of radiation reflected from varied objects relative to the amount of the same wavelengths incident upon the object. This proportion, known as *reflectance*, is useful for defining the distinctive spectral characteristics of objects.

Reflectance (R_{rs}) has already been introduced in Chapter 2 as the relative brightness of a surface, as measured for a specific wavelength interval:

$$\text{Reflectance} = \frac{\text{Observed brightness}}{\text{Irradiance}} \qquad \text{(Eq. 11.2)}$$

As a ratio, reflectance is a dimensionless number (varying between 0 and 1), but it is commonly expressed as a percentage. In the usual practice of remote sensing, R_{rs} is not directly measurable, because normally we record only the observed brightness and must estimate the brightness incident upon the object. Precise estimation of reflectance would require detailed in situ measurement of wavelength, angles of illumination and observation, and atmospheric conditions at the time of observation. Because such measurements are impractical on a routine basis, we must approximate the necessary values.

Usually, for the practice of remote sensing, the analyst has a measurement of the brightness of a specific pixel at the lens of a sensor, sometimes known as *at-aperture, in-band* radiance, signifying that it represents radiance measured at the sensor (i.e., not at ground level) for a specific spectral channel (W/m² × sr × μm). To estimate the reflectance of the pixel, it is necessary to assess how much radiation was incident upon the object before it was reflected to the sensor. In the ideal, the analyst could measure this brightness at ground level just before it was redirected to the sensor. Although analysts sometimes collect this information in the field using specialized instruments known as spectroradiometers (Chapter 13), typically the incident radiation must be estimated using a calculation of exoatmospheric radiation (ESUN$_\lambda$) for a specific time, date, and place.

$$Q_Q = (\pi \times L_\lambda \times d^2)/(ESUN_\lambda \times \cos\theta) \qquad \text{(Eq. 11.3)}$$

where Q_Q is the at-sensor, in-band reflectance; L_λ is the at-sensor, in-band radiance; ESUN$_\lambda$ is the band-specific, mean solar exoatmospheric irradiance; θ_s is the solar zenith angle; and d is the Earth–Sun distance (expressed in astronomical units) for the date in question (see *http://sunearth.gsfc.nasa.gov*). To find the appropriate Earth–Sun distance, navigate to *http://ssd.jpl.nasa.gov/cgi-bin/eph* to generate an ephemeris for the overpass time indicated in the header information for the scene in question. The value of d (delta in the ephemeris output) gives the Earth–Sun distance in astronomical units (AU), between 0.9 and 1.1, with several decimal places. The time, date, solar zenith angle, and position of the scene center can usually be found in the header information for the scene.

Such estimation of exoatmospheric radiation does not, by definition, allow for atmospheric effects that will alter brightness as it travels from the outer edge of the atmosphere to the object at the surface of the Earth, so it forms only a rough approximation of the true value. However, for many purposes it forms a serviceable approximation that permits estimation of relative reflectances within a scene.

11.6. Destriping and Related Issues

Optical/mechanical scanners are subject to a kind of radiometric error known as *striping*, or *dropped scan lines*, which appear as a horizontal banding caused by small differences in the sensitivities of detectors within the sensor (or sometimes the complete failure of a detector). Within a given band, such differences appear on images as banding where individual scan lines exhibit unusually brighter or darker brightness values that contrast with the background brightnesses of the "normal" detectors. Landsat MSS (and less so for the TM) was subject to this phenomenon. Within the Landsat MSS this error is known as *sixth-line striping* (Figure 11.3) due to the design of the instrument. Because MSS detectors are positioned in arrays of six (Chapter 6), an anomalous detector response appears as linear banding at intervals of six lines. Striping may appear on only one or two bands of a multispectral image or may be severe for only a portion of a band. Other forms of digital imagery sometimes exhibit similar effects, all caused by difficulties in maintaining consistent calibration of detectors within a sensor. Campbell and Liu (1995) found that striping and related defects in digital data imagery seemed to have minimal impact on the character of the data—much less than might be suggested by the appearance of the imagery.

Although sixth-line striping is often clearly visible as an obvious banding (Figure

FIGURE 11.3. Sixth-line striping in Landsat MSS data.

11.3), it may also be present in a more subtle form that may escape casual visual inspection. *Destriping* refers to the application of algorithms to adjust incorrect brightness values to values thought to be near the correct values. Some image processing software provides special algorithms to detect striping. Such procedures search through an image line by line to look for systematic differences in average brightnesses of lines spaced at intervals of six; examination of the results permits the analyst to have objective evidence of the existence or absence of sixth-line striping (Table 11.1). If striping is present, the analyst must make a decision. If no correction is applied, the analysis must proceed with brightness values that are known to be incorrect. Conversely, efforts to correct for such serious errors may yield rather rudimentary approximations of the (unknown) true values. Often striping may be so severe that it is obvious the bad lines must be adjusted.

A variety of destriping algorithms have been devised. All identify the values generated by the defective detectors by searching for lines that are noticeably brighter or darker than the lines in the remainder of the scene. These lines are presumably the bad

TABLE 11.1. Results of a Sixth-Line Striping Analysis

	Mean	Standard deviation
Entire image:	19.98	7.06
Detector 1:	19.07	6.53
Detector 2:	19.50	6.97
Detector 3:	19.00	6.78
Detector 4:	22.03	7.60
Detector 5:	21.09	7.34
Detector 6:	19.17	6.51

lines caused by the defective detectors (especially if they occur at intervals of six lines). Then the destriping procedure estimates corrected values for the bad lines. There are many different estimation procedures; most belong to one of two groups. One approach is to replace bad pixels with values based on the average of adjacent pixels not influenced by striping; this approach is based on the notion that the missing value is probably quite similar to the pixels that are nearby (Figure 11.4a). A second strategy is to replace bad pixels with new values based on the mean and standard deviation of the band in question or on statistics developed for each detector (Figure 11.4b). This second approach is based on the assumption that the overall statistics for the missing data must, because there are so many pixels in the scene, resemble those from the good detectors.

The algorithm described by Rohde et al. (1978) combines elements of both strategies. Their procedure attempts to bring all values in a band to a normalized mean and variance, based on overall statistics for the entire band. Because brightnesses of individual regions within the scene may vary considerably from these overall values, a second algorithm can be applied to perform a local averaging to remove the remaining influences of striping. Some destriping algorithms depend entirely on local averaging; because they

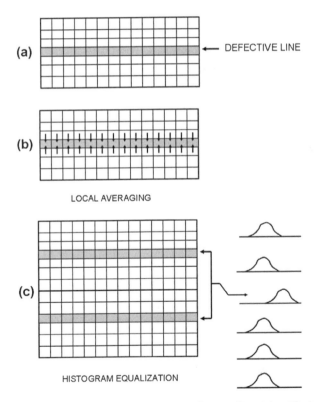

FIGURE 11.4. Two strategies for destriping. (a) Defective line identified with an array of pixels. (b) Local averaging—pixels in the defective line are replaced with an average of values of neighboring pixels in adjacent lines. (c) Histogram normalization—data from all lines are accumulated at intervals of six lines (for the Landsat MSS); the histogram for defective detectors displays an average different from the others. A correction shifts the values for the defective lines to match the positions for the other lines within the image.

tend to degrade image resolution and introduce statistical dependencies between adjacent brightness values, it is probably best to be cautious in their application if alternatives are available.

Table 11.1 shows the results for an analysis of striping in a Landsat MSS scene, using the histogram normalization approach. The tabulation provides means and standard deviations of lines organized such that all lines collected by detector 1 are grouped together, all those collected by detector 2 are grouped together, and so on. All the lines collected by the first detector are found by starting at the first line, then skipping to line 7, then to line 13, and so on. All the lines collected by the second detector are found by starting at line 2, then skipping to line 8, then to line 14, and so on. For the image reported in Table 11.1, it is clear that detectors 4 and 5 are producing results that differ from the others; the means and standard deviations of the pixels collected by these detectors are higher than those of pixels collected by other detectors. The differences seem rather small—only two or three brightness values—yet the effect is usually quite obvious to the eye (Figure 11.3), probably because the brightness differences are organized in distinct linear patterns that attract the attention of the observer's visual system.

11.7. Identification of Image Features

In the context of image processing, the terms *feature extraction* and *feature selection* have specialized meanings. "Features" are not geographical features visible on an image, but are rather "statistical" characteristics of image data, including, for example, individual bands or combinations of bands that carry the most potent information within the scene. *Feature extraction* usually identifies specific bands or channels that are of greatest value for any analysis, whereas *feature selection* indicates selection of image information that is specifically tailored for a particular application.

Thus these processes could also be known as "information extraction"—the isolation of those components that are most useful in portraying the essential elements of an image. In theory, discarded data contain noise and errors present in original data. Thus feature extraction may increase accuracy. In addition, these processes reduce the number of spectral channels, or bands, that must be analyzed, thereby reducing computational demands; the analyst can work with fewer but more potent channels. The reduced data set may convey almost as much information as does the complete data set.

Multispectral data, by their nature, consist of several channels of data. Although some images may have as few as 3, 4, or 7 channels (Chapter 6), other image data may have many more, possibly 200 or more channels (Chapter 15). With so much data, processing of even modest-sized images requires considerable time. In this context, feature selection assumes considerable practical significance, as image analysts wish to reduce the amount of data while retaining effectiveness and/or accuracy.

Our examples here are based on TM data, which provide enough channels to illustrate the concept but are compact enough to be reasonably concise (Table 11.2). A variance–covariance matrix shows interrelationships between pairs of bands. Some pairs show rather strong correlations—for example, bands 1 and 3 and 2 and 3 both show correlations above 0.9. High correlation between pairs of bands means that the values in the two channels are closely related. Thus, as values in channel 2 rise or fall, so do those in channel 3; one channel tends to duplicate information in the other. Feature selection attempts to identify, then remove, such duplication so that the dataset can include maximum information using the minimum number of channels.

For example, for data represented by Table 11.2, bands 3, 5, and 6 might include almost as much information as the entire set of seven channels, because band 3 is closely related to bands 1 and 2, band 5 is closely related to bands 4 and 7, and band 6 carries information largely unrelated to any others. Therefore, the discarded channels (1, 2, 4, and 7) each resemble one of the channels that have been retained. So a simple approach to feature selection discards unneeded bands, thereby reducing the number of channels. Although this kind of selection can be used as a sort of rudimentary feature extraction, typically feature selection is a more complex process based on statistical interrelationships between channels.

A more powerful approach to feature selection applies a method of data analysis called *principal components analysis* (PCA). This presentation offers only a superficial description of PCA, as more complete explanation requires the level of detail provided by Davis (2002) and others. In essence, PCA identifies the optimum linear combinations of the original channels that can account for variation of pixel values within an image. Linear combinations are of the form

$$A = C_1X_1 + C_2X_2 + C_3X_3 + C_4X_4 \qquad \text{(Eq. 11.4)}$$

where X_1, X_2, X_3, and X_4 are pixel values in four spectral channels and C_1, C_2, C_3, and C_4 are coefficients applied individually to the values in the respective channels. A represents a transformed value for the pixel. Assume, as an example, that $C_1 = 0.35$, $C2 = -0.08$, $C_3 = 0.36$, and $C_4 = 0.86$. For a pixel with $X_1 = 28$, $X_2 = 29$, $X_3 = 21$, $X_4 = 54$, the transformation assumes a value of 61.48. Optimum values for coefficients are calculated by a procedure that ensures that the values they produce account for maximum variation within the entire dataset. Thus this set of coefficients provides the maximum information that can be conveyed by any single channel formed by a linear combination of the original channels. If we make an image from all the values formed by applying this procedure to an entire image, we generate a single band of data that provides an optimum depiction of the information present within the four channels of the original scene.

The effectiveness of this procedure depends, of course, on calculation of the optimum coefficients. Here our description must be, by intention, abbreviated, because calculation of the coefficients is accomplished by methods described by upper-level statistics texts or discussions such as those of Davis (2002) and Gould (1967). For the present, the

TABLE 11.2. Correlation Matrix for Seven Bands of a TM Scene

	1	2	3	4	5	6	7
1.	1.00						
2.	0.92	1.00					
3.	0.90	0.94	1.00				
4.	0.39	0.59	0.48	1.00			
5.	0.49	0.66	0.67	0.82	1.00		
6.	0.03	0.08	0.02	0.18	0.12	1.00	
7.	0.67	0.76	0.82	0.60	0.90	0.02	1.00

important point is that PCA permits identification of a set of coefficients that concentrates maximum information in a single band.

The same procedure also yields a second set of coefficients that will yield a second set of values (we could represent this as the *B* set, or *B* image) that will be a less effective conveyor of information but will still represent variation of pixels within the image. In all, the procedure will yield seven sets of coefficients (one set for each band in the original image) and therefore will produce seven sets of values, or bands (here denoted as *A, B, C, D, E, F,* and *G*), each in sequence conveying less information than the preceding band. Thus in Table 11.3 transformed channels I and II (each formed from linear combinations of the seven original channels) together account for about 93% of the total variation in the data, whereas channels III–VII together account for only about 7% of the total variance. The analyst may be willing to discard the variables that convey 7% of the variance as a means of reducing the number of channels. The analyst still retains 93% of the original information in a much more concise form. Thus feature selection reduces the size of the dataset by eliminating replication of information.

The effect is easily seen in Figure 11.5, which shows transformed data for a subset of a Landsat TM scene. The first principal-component (PC) images, PC I and PC II, are the most potent; PC III, PC IV, PC V, and PC VI show the decline in information content, such that the images for the higher PC (such as PCs V and VI in Figure 11.5) record artifacts of system noise, atmospheric scatter, topographic shadowing, and other undesirable contributions to image brightness. If such components are excluded from subsequent analysis, it is likely that accuracy can be retained (relative to the entire set of seven channels) while also reducing time and cost devoted to the analysis. A color presentation of the first three components, assigning each to one of the additive primaries (red, green, and blue), is usually effective in presenting a concise, potent portrayal of the information con-

TABLE 11.3. Results of Principal Components Analysis of Data in Table 11.2

	Component						
	I	II	III	IV	V	VI	VII
				Eigenvectors			
% var.:	82.5%	10.2%	5.3%	1.3%	0.4%	0.3%	0.1%
EV:	848.44	104.72	54.72	13.55	4.05	2.78	0.77
	0.14	0.35	0.60	0.07	−0.14	−0.66	−0.20
	0.11	0.16	0.32	0.03	−0.07	−0.15	−0.90
	0.37	0.35	0.39	−0.04	−0.22	0.71	−0.36
	0.56	−0.71	0.37	−0.09	−0.18	0.03	−0.64
	0.74	0.21	−0.50	0.06	−0.39	−0.10	0.03
	0.01	−0.05	0.02	0.99	0.12	0.08	−0.04
	0.29	0.42	−0.08	−0.09	0.85	0.02	−0.02
				Loadings			
Band 1	0.562	0.519	0.629	0.037	−0.040	−0.160	−0.245
Band 2	0.729	0.369	0.529	0.027	−0.307	−0.576	−0.177
Band 3	0.707	0.528	0.419	−0.022	−0.659	−0.179	−0.046
Band 4	0.903	−0.401	0.150	−0.017	0.020	0.003	−0.003
Band 5	0.980	0.098	−0.166	0.011	−0.035	−0.008	−0.001
Band 6	0.144	−0.150	0.039	0.969	0.063	0.038	−0.010
Band 7	0.873	0.448	−0.062	−0.033	0.180	0.004	−0.002

FIGURE 11.5. Feature selection by principal components analysis. These images depict six of the seven principal components for the image described by Tables 11.2 and 11.3. The first principal component image (PC I), formed from a linear combination of data from all seven original bands, accounts for over 80% of the total variation of the image data. PC II and PC III present about 10% and 5% of the total variation, respectively. The higher components (e.g., PC V and PC VI) account for very low proportions of the total variation and convey mainly noise and error, as is clear by the image patterns they show.

veyed by a multichannel multispectral image. Note, however, that because each band is a linear combination of the original channels, the analyst must be prepared to interpret the meaning of the new channels. In some instances this task is relatively straightforward; in other instances, it can be very difficult to unravel the meaning of a PCA image. Further, the student should remember that the PCA transformation applies only to the specific image at hand and that each new image requires a new recalculation of the PCA. For some applications, this constraint limits the effectiveness of the technique.

11.8. Subsets

Because of the very large sizes of many remotely sensed images, analysts typically work with those segments of full images that specifically pertain to the task at hand. Therefore, to minimize computer storage and the analyst's time and effort, one of the first tasks in each project is to prepare *subsets*, portions of larger images selected to show only the region of interest.

Although selecting subsets would not appear to be one of the most challenging tasks in remote sensing, it can be more difficult than one might first suppose. Often subsets must be "registered" (matched) to other data or to other projects, so it is necessary to find distinctive landmarks in both sets of data to ensure that coverages coincide spatially. Second, because time and computational effort devoted to matching images to maps or other images (as described below) increase with large images, it is often convenient to prepare subsets before registration. Yet if the subset is too small, then it may be difficult to identify sufficient landmarks for efficient registration. Therefore, it may be useful to prepare a preliminary subset large enough to conduct the image registration effectively before selecting the final, smaller subset for analytical use (Figure 11.6).

The same kinds of considerations apply in other steps of an analysis. Subsets should be large enough to provide context required for the specific analysis at hand. For example, it may be important to prepare subsets large enough to provide sufficient numbers of training fields for image classification (Chapter 12) or a sufficient set of sites for accuracy assessment (Chapter 14).

FIGURE 11.6. Subsets. Sometimes a subset of a particular area is too small to encompass sufficient points to allow the subset to be accurately matched to an accurate map (discussed in Section 11.8). Selection of an intermediate temporary subset permits accurate registration using an adequate number of control points. After the temporary subset has been matched to the map, the study area can be selected more precisely without concern for the distribution of control points.

11.9. Geometric Correction by Resampling

A critical consideration in the application of remote sensing is preparation of planimetrically correct versions of aerial and satellite images so that they will match to other imagery and to maps and will provide the basis for accurate measurements of distance and area. Previously, in Chapter 3, we learned that aerial imagery has inherent positional errors that prevent use of the raw imagery for positional measurements. Chapter 3 also introduced the orthophoto, which is a planimetrically correct version of an aerial (or satellite) image created by analysis of stereo imagery through removal of positional effects of topographic relief to produce a positionally accurate image. Similar products can be prepared by applying a precise knowledge of the internal geometry of the instrument and of the sensor's position in space relative the terrain to derive planimetrically correct versions of an aerial or satellite image. Because such images are prepared by applying basic optical principles and details of the instrument calibration, they form the preferred standard for positionally correct images.

A second approach to image registration, known as *image resampling*, approaches the problem in a completely different manner. No effort is made to apply our knowledge of system geometry; instead, the images are treated simply as an array of values that must be manipulated to create another array with the desired geometry. Resampling scales, rotates, translates, and performs related manipulations as necessary to bring the geometry of an image to match a particular reference image of desired properties. Such operations can be seen essentially as an interpolation problem similar to those routinely considered in cartography and related disciplines. Such processes constitute the practice of *image resampling*—the application of interpolation to bring an image into registration with another image or a planimetrically correct map. Image resampling forms a convenient alternative to the analytical approach, as it does not require the detailed data describing the instrument and its operation (which may not be at hand). Although resampling may provide useful representations of images, users should recognize that resampled images are not equivalent to orthographic representations but are produced by arbitrary transformations that bring a given image into registration with another map or image. Although resampling can apply to vector data, our discussion here refers to raster images.

Resampling is related to but distinct from *georeferencing*. Georeferencing indicates that resampling of an image requires matching not only to a reference image but also to reference points that correspond to specific known locations on the ground. Georeferenced images are presented in a specific geographic projection so that the image is presented in a defined projection and coordinate system. Georeferencing is, of course, important for images that are to be used as locational references or to be matched to other maps and images. Although our discussion here focuses on the resampling process, in most image processing systems resampling and georeferencing are usually part of a single process.

In Figure 11.7, the input image is represented as an array of open dots, each representing the center of a pixel in the uncorrected image. Superimposed over this image is a second array, symbolized by the solid dots, which shows the centers of pixels in the image transformed (as described below) to have the desired geometric properties (the "output" image).

The locations of the output pixels are derived from locational information provided by *ground control points* (GCPs), places on the input image that can be located with precision on the ground and on planimetrically correct maps. (If two images are to be registered,

FIGURE 11.7. Resampling. Open circles (O) represent the reference grid of known values in the input image. Black dots (•) represent the regular grid of points to be estimated to form the output image. Each resampling method employs a different strategy to estimate values at the output grid, given known values for the input grid.

GCPs must be easily recognized on both images.) The locations of these points establish the geometry of the output image and its relationship to the input image. Thus this first step establishes the framework of pixel positions for the output image using the GCPs.

The next step is to decide how to best estimate the values of pixels in the corrected image, based on information in the uncorrected image. The simplest strategy from a computational perspective is simply to assign each "corrected" pixel the value from the nearest "uncorrected" pixel. This is the *nearest-neighbor* approach to resampling (Figure 11.8). It has the advantages of simplicity and the ability to preserve the original values of the unaltered scene—an advantage that may be critical in some applications. The nearest-neighbor method is considered the most computationally efficient of the methods usually applied for resampling. On the other hand, it may create noticeable positional errors, which may be severe in linear features where the realignment of pixels may be noticeable, and may exhibit similar artifacts when applied to imagery of uneven terrain acquired by a pushbroom scanner. Its principal advantage is that it makes few, if any, alterations to pixel values—an advantage in applications where even minor changes may be considered significant.

A second, more complex, approach to resampling is *bilinear interpolation* (Figure 11.9). Bilinear interpolation calculates a value for each output pixel based on a weighted

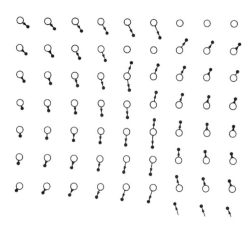

FIGURE 11.8. Nearest-neighbor resampling. Each estimated value (•) receives its value from the nearest point on the reference grid (O).

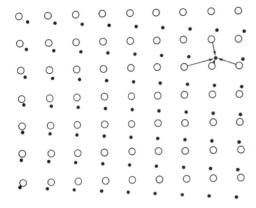

FIGURE 11.9. Bilinear interpolation. Each estimated value (•) in the output image is formed by calculating a weighted average of the values of the four nearest neighbors in the input image (O). Each estimated value is weighted according to its distance from the known values in the input image.

average of the four nearest input pixels. In this context, "weighted" means that nearer pixel values are given greater influence in calculating output values than are more distant pixels. Because each output value is based on several input values, the output image will not have the unnaturally blocky appearance of some nearest-neighbor images. The image therefore has a more "natural" look. Yet there are important changes. First, because bilinear interpolation creates new pixel values, the brightness values in the input image are lost. The analyst may find that the range of brightness values in the output image differs from those in the input image. Such changes to digital brightness values may be significant in later processing steps. Second, because the resampling is conducted by averaging over areas (i.e., blocks of pixels), it decreases spatial resolution by a kind of "smearing" caused by averaging small features with adjacent background pixels.

Finally, the most sophisticated, most complex, and (possibly) most widely used resampling method is *cubic convolution* (Figure 11.10). Cubic convolution uses a weighted average of values within a neighborhood that extends about two pixels in each direction, usually encompassing 16 adjacent pixels. Typically, the images produced by cubic convolution resampling are much more attractive than those of other procedures, but the data are altered more than are those of nearest-neighbor or bilinear interpolation, the computations are more intensive, and the minimum number of GCPs is larger. Figure 11.11

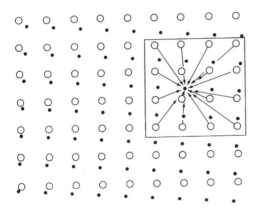

FIGURE 11.10. Cubic convolution. Each estimated value in the output matrix (•) is found by assessing values within a neighborhood of 16 pixels in the input image (O).

CUBIC CONVOLUTION   **NEAREST NEIGHBOR**

FIGURE 11.11. Illustration of image resampling. The image at the left has been resampled using both cubic convolution and nearest-neighbor strategies. The arrows mark areas illustrating the failure of the nearest-neighbor algorithm to record the continuity of linear features.

illustrates the contrast between effects of applying cubic convolution resampling and the nearest-neighbor algorithm to the same image—the nearest-neighbor strategy can create a blocky, angular appearance when applied to linear features, whereas the cubic convolution presents a much more natural appearance. Although both bilinear interpolation and cubic convolution alter pixel values as they interpolate to register images, in most applications these changes are probably insignificant in the analytical context.

Identification of GCPs

A practical problem in applying image registration procedures is the selection of control points (Figure 11.12). GCPs are features that can be located with precision and accuracy on accurate maps yet that are also easily located on digital images. Ideally, GCPs could be as small as a single pixel, if one could be easily identified against its background. In practice, most GCPs are likely to be spectrally distinct areas as small as a few pixels. Examples might include intersections of major highways, distinctive water bodies, edges of land-cover parcels, stream junctions, and similar features (Figure 11.13). Although identification of such points may seem to be an easy task, in fact, difficulties that might emerge during this step can form a serious roadblock to the entire analytical process, as another procedure may depend on completion of an accurate registration.

ROAD INTERSECTIONS   RIVERS AND WATER BODIES

FIGURE 11.12. Selection of distinctive ground control points.

FIGURE 11.13. Examples of ground control points. GCPs must be identifiable both on the image and on a planimetrically correct reference map.

Typically, it is relatively easy to find a rather small or modest-sized set of control points. However, in some scenes, the analyst finds it increasingly difficult to expand this set, as one has less and less confidence in each new point added to the set of GCPs. Thus there may be a rather small set of "good" GCPs, points that the analyst can locate with confidence and precision both on the image and on an accurate map of the region.

The locations may also be a problem. In principle, GCPs should be dispersed throughout the image, with good coverage near edges. Obviously, there is little to be gained from having a large number of GCPs if they are all concentrated in a few regions of the image. Analysts who attempt to expand areal coverage to ensure good dispersion are forced to consider points in which it is difficult to locate GCPs with confidence. Therefore the desires to select "good" GCPs and to achieve good dispersion may work against each other such that the analyst finds it difficult to select a judicious balance. Analysts should anticipate difficulties in selecting GCPs as they prepare subsets early in the analytical process. If subsets are too small, or if they do not encompass important landmarks, the analysts may later find that the subset region of the image does not permit selection of a sufficient number of high-quality GCPs.

Bernstein et al. (1983) present information that shows how registration error decreases as the number of GCPs is increased. Obviously, it is better to have more rather than fewer GCPs. But, as explained above, the quality of GCP accuracy may decrease as their number increases because the analyst usually picks the best points first. They recommend that 16 GCPs may be a reasonable number if each can be located with an accuracy of one-third of a pixel. This number may not be sufficient if the GCPs are poorly distributed or if the nature of the landscape prevents accurate placement.

Many image-processing programs permit the analyst to anticipate the accuracy of the registration by reporting errors observed at each GCP if a specific registration has been applied. The standard measure of the location error is the *root mean square* (rms) error, which is the standard deviation of the difference between actual positions of GCPs and their calculated positions (i.e., after registration). These differences are known as the *residuals.* Usually rms is reported in units of image pixels for both north–south and east–west directions (Table 11.4). Note that this practice reports locational errors among the GCPs, which may not always reflect the character of errors encountered at other pixels. Nonetheless, it is helpful to be able to select the most useful GCPs. If analysts wish to assess the overall accuracy of the registration, some of the GCPs should be withheld from the registration procedure and then used to evaluate its success.

Some images distributed by commercial enterprises or governmental agencies have been georeferenced by standardized processing algorithms to meet specified standards, so some users may find that the positional accuracy of such images is satisfactory for their needs. Such images are designated by specific processing levels, as discussed in Section 11.12. For example, Landsat TM data are terrain corrected using cubic convolution, high quality digital elevation models, and an established archive of GCPs (see Gutman et al., 2008, and *http://landsat.usgs.gov/Landsat_Processing_Details.php*). Other image providers employ similar procedures.

11.10. Data Fusion

Data fusion refers to processes that bring images of varied resolution in to a single image that incorporates, for example, the high spatial resolution of a panchromatic image with the multispectral content of a multiband image at coarser resolution. Because of the broad range of characteristics of imagery collected by varied sensors described in previous chapters and the routine availability of digital data, there has been increasing incentive to bring data from varied systems together into a single image. Such products are often valuable because they can integrate several independent sources of information into a single image. (Because image fusion prepares data for visual interpretation, discussed in Chapter 5, rather than for digital analysis, it differs from the other preprocessing techniques discussed in this chapter. However, placement of this topic in this unit permits development of concepts in a more progressive sequence than would otherwise be possible.)

Although data fusion can be applied to a varied selection of different forms of remotely sensed data (e.g., merging multispectral data with radar imagery or multispectral images with digital elevation data), the classic example of image fusion involves the merging of a multispectral image of relatively coarse spatial resolution with another image of the same region acquired at finer spatial resolution. Examples might include fusing multispectral SPOT data (20-m spatial resolution) with the corresponding panchromatic SPOT scene (at 10-m spatial resolution) of the same region or multispectral TM data with corresponding high-altitude aerial photography.

The technique assumes that the images are compatible in the sense that they were acquired on or about the same date and that the images register to each other. The analyst must rectify the images to be sure that they both share the same geometry and register them to be sure that they match spatially. In this context, the basic task of image fusion is to substitute the spatial detail of the fine-resolution image detail to replace one of the multispectral bands, then apply a technique to restore the lost multispectral content of the discarded band from the coarse-resolution multispectral image. Specifics of these procedures are beyond the scope of this discussion. Chavez et al. (1991), Wald et al. (1997), and Carter (1998) are among the authors who have described these procedures in some detail. Briefly stated, the techniques fall into three classes.

Spectral domain procedures project the multispectral bands into spectral data space,

TABLE 11.4. Sample Tabulation of Data for GCPs

Point no.	Image X pixel	X pixel residual	Image Y pixel	Y pixel residual
1	1269.75	−0.2471E+00	1247.59	0.1359E+02
2	867.91	−0.6093E+01	1303.90	0.8904E+01
3	467.79	−0.1121E+02	1360.51	0.5514E+01
4	150.52	0.6752E+02	1413.42	−0.8580E+01
5	82.20	−0.3796E+01	163.19	0.6189E+01
6	260.89	0.2890E+01	134.23	0.5234E+01
7	680.59	0.3595E+01	70.16	0.9162E+01
8	919.18	0.1518E+02	33.74	0.1074E+02
9	1191.71	0.6705E+01	689.27	0.1127E+02
10	1031.18	0.4180E+01	553.89	0.1189E+02
11	622.44	−0.6564E+01	1029.43	0.8427E+01
12	376.04	−0.5964E+01	737.76	0.6761E+01
13	162.56	−0.7443E+01	725.63	0.8627E+01
14	284.05	−0.1495E+02	1503.73	0.1573E+02
15	119.67	−0.8329E+01	461.59	0.4594E+01
16	529.78	−0.2243E+00	419.11	0.5112E+01
17	210.42	−0.1558E+02	1040.89	−0.1107E+01
18	781.85	−0.2915E+02	714.94	−0.1521E+03
19	1051.54	−0.4590E+00	1148.97	0.1697E+02
20	1105.95	0.9946E+01	117.04	0.1304E+02

Note. X rms error = 18.26133
Y rms error = 35.33221
Total rms error = 39.77237

Point no.	Error	Error contribution by point
1	13.5913	0.3417
2	10.7890	0.2713
3	12.4971	0.3142
4	68.0670	1.7114
5	7.2608	0.1826
6	5.9790	0.1503
7	9.8416	0.2474
8	18.5911	0.4674
9	13.1155	0.3298
10	12.6024	0.3169
11	10.6815	0.2686
12	9.0161	0.2267
13	11.3944	0.2865
14	21.6990	0.5456
15	9.5121	0.2392
16	5.1174	0.1287
17	15.6177	0.3927
18	154.8258	3.8928
19	16.9715	0.4267
20	16.3982	0.4123

then find the new (transformed) band most closely correlated with the panchromatic image. That spectral content can then be assigned to the high-resolution panchromatic image.

The *intensity–hue–saturation (IHS) technique* (Carper et al., 1990) is one of the spectral domain procedures. IHS refers to the three dimensions of multispectral data that we know in the everyday context as "color." *Intensity* is equivalent to brightness, *hue* refers to dominant wavelength (what we regard as "color"), and *saturation* specifies purity, the degree to which a specific color is dominated by a single wavelength. For the IHS fusion technique, the three bands of the lower resolution image are transformed from normal red–green–blue (RGB) space into IHS space. The high-resolution image is then stretched to approximate the mean and variance of the intensity component. Next, the stretched high-resolution image is substituted for the intensity component of the original image, and then the image is projected back into RGB data space. The logic of this strategy lies in the substitution of the intensity component by the high-resolution image, which is characterized by an equivalent range of intensity but superior spatial detail.

The *principal components transformation* (PCT) conducts a PCA of the raw low-resolution image. The high-resolution image is then stretched to approximate the mean and variance of the first PC. Next, the stretched high-resolution image is substituted for the first PC of the low-resolution image, and then the image is reconstructed back into its usual appearance using the substituted high-resolution image as the first PC. The logic of this PCT approach rests on the fact that the first PC of a multispectral image often conveys most of the brightness information in the original image.

Spatial domain procedures extract the high-frequency variation of a fine-resolution image and then insert it into the multispectral framework of a corresponding coarse-resolution image. The *high-pass filter* (HPF) *technique* is an example. An HPF is applied to the fine-resolution image to isolate and extract the high-frequency component of the image. It is this high-frequency information that conveys the scene's fine spatial detail. This high-frequency element is then introduced into the low-resolution image (with compensation to preserve the original brightness of the scene) to synthesize the fine detail of the high-resolution image. The ARSIS technique proposed by Ranchin and Wald (2000) applies the wavelet transform as a method of simulating the high-resolution detail within the coarse-resolution image.

Algebraic procedures operate on images at the level of the individual pixel to proportion spectral information among the three bands of the multispectral image, so that the replacement (high-resolution) image used as a substitute for one of the bands can be assigned correct spectral brightness. The *Brovey transform* finds proportional brightnesses conveyed by the band to be replaced and assigns it to the substitute. Its objective of preserving the spectral integrity of the original multispectral image is attained if the panchromatic image has a spectral range equivalent to the combined range of the three multispectral bands. Because this assumption is often not correct for many of the usual multispectral images, this procedure does not always preserve the spectral content of the original image.

The *multiplicative model* (MLT) multiplies each multispectral pixel by the corresponding pixel in the high-resolution image. To scale the brightnesses back to an approximation of their original ranges, the square root of the combined brightnesses is calculated. The result is a combined brightness that requires some sort of weighting to restore some approximation of the relative brightnesses of the individual bands. Although these weights are selected arbitrarily, many have found the procedure to produce satisfactory results.

Composite images created by image fusion are intended for visual interpretation. Comparative evaluations of the different approaches have examined different combinations of fused data using different criteria, so it is difficult to derive definitive reports. Chavez et al. (1991) found that the HPF technique was superior to the IHS and PCT methods for fused TM and high-altitude photographic data. Carter (1998) found that the IHS and HPF methods produced some of the best results. Wald et al. (1997) proposed a framework for evaluating the quality of fused images and emphasized that the character of the scene recorded by the imagery is an important consideration. Because fused images are composites derived from arbitrary manipulations of data, the combined images are suited for visual interpretation but not for further digital classification or analysis.

11.11. Image Data Processing Standards

Digital data collected by remote sensing instruments must be processed to adjust for artifacts introduced in recording and transmission and to format for convenient use. In the ideal, scientists prefer data that originate close to the original values collected by the instrument. However, all data require at least minimal processing to remove the most severe errors introduced by the instrument, transmission, the atmosphere, and processing algorithms. As users address applications that are less basic in nature and have an increasingly applied character, they will have requirements focused increasingly on data characteristics that match to their needs for interpretation and on standardized products that will be consistent with equivalent data used by other analysts. Therefore, among the community of users of image data, there will be some who prefer data that have higher levels of processing that prepare them for convenient use—for example, to correct for atmospheric effects or to project them into useful geographic projections. Image providers apply such operations through batch processing, applying the same algorithms to all imagery. It is important to implement such processing operations in a systematic manner, documented so that users understand what operations have been performed and exactly how they were implemented.

However, processing choices that may be satisfactory for the majority of users may not be optimal for users with more specialized requirements, so image providers offer a hierarchy of processing choices. The *NASA Earth Science Reference Handbook* (Parkinson et al., 2006) (Table 11.5) provides an example of how different levels of processing are defined and designated. Thus, for example, level 0 data will be processed only to remove or correct engineering artifacts, whereas higher level processing tailors data for convenient use by applications scientists and, in doing so, simultaneously introduces artifacts not present in the level 0 versions. Scientists conducting basic research, or who have defined their own processing algorithms tailored to specific problems, often prefer to acquire level 0 data so they can carefully control preprocessing of the data and avoid artifacts introduced by the standardized batch processes. In contrast, those conducting applied research may prefer standardized processing to ensure consistency with results produced by other analysts. Individual agencies and commercial enterprises will each have their own protocols for processing image data, so although Table 11.5 provides an introduction to basic issues, it does not apply specifically to data beyond those addressed in the NASA publication. Other organizations may use similar designations for their processing levels but may have different specifications for each designation. Each data provider, whether a governmental agency or a commercial enterprise, will provide its own definitions and designations, usually through a user's manual or Web page. No specific processing level is appropriate for all applications, so users should select the processing level appropriate for the project at hand.

TABLE 11.5. Image Data Processing Levels as Defined by NASA (2006)

Level	Characteristics
0	Reconstructed, unprocessed instrument and payload data at full resolution, with any and all communications artifacts (e.g., synchronization frames, communications headers, duplicate data) removed.
1a	Reconstructed, unprocessed instrument data at full resolution, time referenced, and annotated with ancillary information, including radiometric and geometric calibration coefficients and georeferencing parameters (e.g., platform ephemeris) computed and appended but not applied to the Level 0 data (or if applied, in a manner that Level 0 is fully recoverable from Level 1a data).
1b	Level 1a data that have been processed to sensor units (e.g., radar backscatter cross section, brightness temperature, etc.); not all instruments have Level 1b data; Level 0 data is not recoverable from Level 1b data.
2	Derived geophysical variables (e.g., ocean wave height, soil moisture, ice concentration) at the same resolution and location as Level 1 source data.
3	Variables mapped on uniform space–time grid scales, usually with some completeness and consistency (e.g., missing points interpolated, complete regions mosaicked together from multiple orbits, etc.).
4	Model output or results from analyses of lower-level data (i.e., variables that were not measured by the instruments but instead are derived from these measurements).

Note. Based on Parkinson et al. (2006, p. 31).

11.12. Summary

It is important to recognize that many of the preprocessing operations used today have been introduced into the field of remote sensing from the related fields of pattern recognition and image processing. In such disciplines, the emphasis is usually on detection or recognition of objects as portrayed on digital images. In this context, the digital values have much different significance than they do in remote sensing. Often, the analysis requires recognition simply of contrasts between different objects or the study of objects against their backgrounds, detection of edges, and reconstruction of shapes from the configuration of edges and lines. Digital values can be manipulated freely to change image geometry or to enhance images without concern that their fundamental information content will be altered.

However, in remote sensing we usually are concerned with much more subtle variations in digital values, and we are concerned when preprocessing operations alter the digital values. Such changes may alter spectral signatures, contrasts between categories, or variances and covariances of spectral bands.

Review Questions

1. How can an analyst know whether preprocessing is advisable? Suggest how you might make this determination.

2. How can an analyst determine whether specific preprocessing procedures have been effective?

3. Can you identify situations in which application of preprocessing might be inappropriate? Explain.

4. Discuss the merits of preprocessing techniques that improve the visual appearance of an image but do not alter its basic statistical properties. Are visual qualities important in the context of image analysis?

5. Examine images and maps of your region to identify prospective GCPs. Evaluate the pattern of GCPs; is the pattern even, or is it necessary to select questionable points to attain an even distribution?

6. Are optimum decisions regarding preprocessing likely to vary according to the subject of the investigation? For example, would optimum preprocessing decisions for a land-cover analysis differ from those for a hydrologic or geologic analysis?

7. Assume for the moment that sixth-line striping in MSS data has a purely visual impact, with no effect on the underlying statistical qualities of the image. In your judgment, should preprocessing procedures be applied? Why or why not?

8. Can you identify analogies for preprocessing in other contexts?

9. Suppose an enterprise offers to sell images with preprocessing already completed. Would such a product be attractive to you? Why or why not?

References

Baugh, W. M., and D. P. Groeneveld. 2006. Broadband Vegetation Index Performance Evaluated for a Low-Cover Environment. *International Journal of Remote Sensing*, Vol. 27, pp. 4715–4730.

Berstein, R. 1983. Image Geometry and Rectification. Chapter 21 in Manual of Remote Sensing. Chapter 21 in Manual of Remote Sensing, 2nd Ed. R. N. Colwell, ed. Falls Church, VA. American Society of Photogrammetry, pp. 873–922.

Davis, J. C. 2002. *Statistics and Data Analysis in Geology*. New York: Wiley, 638 pp.

Franklin, S. E., and P. T. Giles. 1995. Radiometric Processing of Aerial Imagery and Satellite Remote Sensing Imagery. *Computers and Geosciences*, Vol. 21, pp. 413–423.

Gould, P. 1967. On the Geographical Interpretation of Eigenvalues. *Transactions, Institute of British Geographers*, Vol. 42, pp. 53–86.

Gutman, G., Byrnes, R., Masek, J., Covington, S., Justice, C., Franks, S., and R. Headley. 2008. Towards Monitoring Land Cover and Land-Use Changes at a Global Scale: 2008. The Global Land Survey 2005. *Photogrammetric Engineering and Remote Sensing*, Vol. 74, pp. 6–10.

Holben, B., E. Vermote, Y. J. Kaufman, D. Tarè, and V. Kalb. 1992. Aerosol Retrieval over Land from AVHRR Data. *IEEE Transactions on Geoscience and Remote Sensing*, Vol. 30, pp. 212–222.

Mausel, P. W., W. J. Kramber, and J. Lee. 1990. Optimum Band Selection for Supervised Classification of Multispectral Data. *Photogrammetric Engineering and Remote Sensing*, Vol. 56, pp. 55–60.

Nield, S. J., J. L. Boettinger, and R. D. Ramsey. 2007. Digitally Mapping Gypsic and Natric Soil Areas Using Landsat ETM Data. *Soil Science Society of America Journal*, Vol. 71, pp. 245–252.

Parkinson, C. L., A. Ward, and M. D. King (eds.). 2006. *Earth Science Reference Handbook—A Guide to NASA's Earth Science Program and Earth Observing Satellite Missions*. Washington, DC: NASA, 277 pp., *http://eospso.gsfc.nasa.gov/ftp_docs/2006ReferenceHandbook.pdf*.

Pitts, D. E., W. E. McAllum, and A. E. Dillinger. 1974. The Effect of Atmospheric Water Vapor on Automatic Classification of ERTS Data. In *Proceedings of the Ninth International Symposium on Remote Sensing of Environment*. Ann Arbor: Institute of Science and Technology, University of Michigan, pp. 483–497.

Rohde, W. G., J. K. Lo, and R. A. Pohl. 1978. EROS Data Center Landsat Digital Enhancement Techniques and Imagery Availability, 1977. *Canadian Journal of Remote Sensing*, Vol. 4, pp. 63–76.

Westin, T. 1990. Precision Rectification of SPOT Imagery. *Photogrammetric Engineering and Remote Sensing*, Vol. 56, pp. 247–253.

Data Fusion

Carper, W. J., T. M. Lillesand, and R. W. Kiefer. 1990. The Use of Intensity Hue Saturation Transformations for Merging SPOT Panchromatic and Multispectral Image Data. *Photogrammetric Engineering and Remote Sensing*, Vol. 56, pp. 459–467.

Carter, D. B. 1998. *Analysis of Multiresolution Data Fusion Techniques*. MS thesis, Virginia Polytechnic Institute and State University, Blacksburg, 61 pp.

Chavez, P. S., Jr., S. C. Sides, and J. A. Anderson. 1991. Comparison of Three Different Methods to Merge Multiresolution and Multispectral Data: Landsat TM and SPOT Panchromatic. *Photogrammetric Engineering and Remote Sensing*, Vol. 57, pp. 265–303.

Pohl, C., and J. L. van Genderen. 1988. Multisensor Image Fusion in Remote Sensing: Concepts, Methods, and Applications. *International Journal of Remote Sensing*, Vol. 19, pp. 823–854.

Ranchin, T., and L. Wald. 2000. Fusion of High Spatial and Spectral Resolution Images: The ARSIS Concept and Its Implementation. *Photogrammetric Engineering and Remote Sensing*, Vol. 66, pp. 49–61.

Schowengerdt, R. A. 1980. Reconstruction of Multispatial Multispectral, Image Data Using Spatial Frequency Contents. *Photogrammetric Engineering and Remote Sensing*, Vol. 46, pp. 1325–1334.

Wald, L., T. Ranchin, and M. Mangolini. 1997. Fusion of Satellite Images of Different Spatial Resolutions: Assessing the Quality of Resulting Images. *Photogrammetric Engineering and Remote Sensing*, Vol. 63, pp. 691–699.

Reflectance

Gilabert, M. A., C. Conese, and F. Maselli. 1994. An Atmospheric Correction Method for the Automatic Retrieval of Surface Reflectances from TM images. *International Journal of Remote Sensing*, Vol. 15, no. 10, pp. 2065–2086.

Moran, M. S., R. D. Jackson, P. N. Slater, and P. M. Teillet. 1992. Evaluation of Simplified Procedures for Retrieval of Land Surface Reflectance Factors from Satellite Sensor Output. *Remote Sensing of Environment*, Vol. 41, pp. 169–184.

Song, C., C. E. Woodcock, K. C. Seto, M. P. Lenny, and S. A. Macomber. 2001. Classification and Change Detection using Landsat TM: When and How to Correct for Atmospheric Effects. *Remote Sensing of Environment*, Vol. 75, pp. 230–244.

Atmospheric Correction

Air Force Research Laboratory. 1998. *Modtran Users Manual, Versions 3.7 and 4.0*. Hanscom Air Force Base, MA.

Anderson, G. P., A. Berk, P. K. Acharya, M. W. Matthew, L. S. Bernstein, J. H. Chetwynd, et al. 2000. MODTRAN4: Radiative Transfer Modeling for Remote Sensing. In *Algo-*

rithms for Multispectral, Hyperspectral, and Ultraspectral Imagery: VI (S. S. Chen and M. R. Descour eds.). *Proceedings of SPIE*, Vol. 4049, pp. 176–183.

Berk, A., L. S. Bernstein, and D. C. Robertson. 1989. *MODTRAN: A Moderate Resolution Model for LOWTRAN 7.* Hanscom Air Force Base, MA: U.S. Air Force Geophysics Laboratory, 38 pp.

Brach, E. J., A. R. Mack, and V. R. Rao. 1979. Normalization of Radiance Data for Studying Crop Spectra over Time with a Mobile Field Spectro-Radiometer. *Canadian Journal of Remote Sensing*, Vol. 5, pp. 33–42.

Campbell, J. B. 1993. Evaluation of the Dark-Object Subtraction Method of Adjusting Digital Remote Sensing Data for Atmospheric Effects. Digital Image Processing and Visual Communications Technologies in the Earth and Atmospheric Sciences: II (M. J. Carlotto, ed.), *SPIE Proceedings*, Vol. 1819. pp. 176–188.

Campbell, J. B., R. M. Haralick, and S. Wang. 1984. Interpretation of Topographic Relief from Digital Multispectral Imagery. In *Remote Sensing* (P. N. Slater, ed.), *SPIE Proceedings*, Vol. 475, pp. 98–116.

Campbell, J. B., and X. Liu. 1994. Application of Dark Object Subtraction to Multispectral Data. In *Proceedings, International Symposium on Spectral Sensing Research (ISSSR '94)*. Alexandria, VA: U.S. Army Corps of Engineers Topographic Engineering Center, pp. 375–386.

Campbell, J. B., and X. Liu. 1995. Chromaticity Analysis in Support of Multispectral Remote Sensing. In *Proceedings, ACSM/ASPRS Annual Convention and Exposition*. Bethesda, MD: American Society for Photogrammetry and Remote Sensing, pp. 724–932.

Campbell, J. B., and L. Ran. 1993. CHROM: A C Program to Evaluate the Application of the Dark Object Subtraction Technique to Digital Remote Sensing Data. *Computers and Geosciences*, Vol. 19, pp. 1475–1499.

Chavez, P. S. 1975. Atmospheric, Solar, and M.T.F. Corrections for ERTS Digital Imagery. *Proceedings, American Society of Photogrammetry*. Bethesda, MD: American Society for Photogrammetry and Remote Sensing, pp. 69–69a.

Chavez, P. 1988. An Improved Dark-Object Subtraction Technique for Atmospheric Scattering Correction of Multispectral Data. *Remote Sensing of Environment*, Vol. 24, pp. 459–479.

Chavez, P. 1996. Image-based Atmospheric Corrections—Revisited and Improved. *Photogrammetric Engineering and Remote Sensing*, Vol. 62, pp. 1025–1036.

Gao, B.-C., and A. F. H. Goetz. 1990. Column Atmospheric Water Vapor and Vegetation Liquid Water Retrievals from Airbourne Mapping Spectrometer Data. *Journal of Geophysical Research–Atmosphere*, Vol. 95(D4), pp. 3564–3594.

Gao, B.-C., K. H. Heidebrecht, and A. F. H. Goetz. 1993. Derivation of Scaled Surface Reflectances from AVIRIS Data. *Remote Sensing of Environment*, Vol. 44, pp. 165–178.

Lambeck, P. F., and J. F. Potter. 1979. Compensation for Atmospheric Effects in Landsat Data. In *The LACIE Symposium: Proceedings of Technical Sessions: Vol. 2*. Houston: NASA-JSC, pp. 723–738.

Potter, J. F. 1984. The Channel Correlation Method for Estimating Aerosol Levels from Multispectral Scanner Data. *Photogrammetric Engineering and Remote Sensing*, Vol. 50, pp. 43–52.

Song, C., C. E. Woodcock, K. C. Seto, M. P. Lenny, and S. A. Macomber. 2001. Classification and Change Detection Using Landsat TM Data: When and How to Correct Atmospheric Effects? *Remote Sensing of Environment*, Vol. 75, pp. 230–244.

Switzer, P., W. S. Kowalick, and R. J. P. Lyon. 1981. Estimation of Atmospheric Path-Radiance by the Covariance Matrix Method. *Photogrammetric Engineering and Remote Sensing*, Vol. 47, pp. 1469–1476.

Teillet, P. M., and G. Fedosejevs. 1995. On the Dark Target Approach to Atmospheric Correction of Remotely Sensed Data. *Canadian Journal of Remote Sensing*, Vol. 21, No. 4, pp. 374–387.

Vermote, E. F., D. Tanré, J. L. Deuzé, M. Herman, and J.-J. Morcrette. 1997. Second Simulation of the Satellite Signal in the Solar Spectrum, 6S. *IEEE Transactions on Geoscience and Remote Sensing*, Vol. 35, pp. 675–686.

Wu, J., D. Wang, and M. E. Bauer. 2005. Image-Based Atmospheric Correction of Quickbird Imagery of Minnesota Cropland. *Remote Sensing of Environment*, Vol. 99, pp. 315–325.

Calculation of Radiance

Chander, G., B. L. Markham, and J. A. Barsi. 2007. Revised Landsat 5 Thematic Mapper Radiometric Calibration. *IEEE Transactions on Geoscience and Remote Sensing*, Vol. 4, no. 3. pp. 490–494.

Chander, G., B. L. Markham, and D. L. Helder. 2009. Summary of Current Radiometric Calibration Coefficients for Landsat MSS, TM, ETM+, and EO-1 ALI Sensors. *Remote Sensing of Environment*, Vol. 113, pp. 893–903.

Markham, B. L., and J. L. Barker. 1986. Landsat MSS and TM Post-Calibration Dynamic Ranges, Exoatmospheric Reflectances and At-Satellite Temperatures. *EOSAT Landsat Technical Notes*, No. 1.

Markham, B. L., J. L. Barker, E. Kaita, J. Seiferth, and R. Morfitt. 2003. On-Orbit Performance of the Landsat-7 ETM+ Radiometric Calibrators. *International Journal of Remote Sensing*, Vol. 24, pp. 265–285.

Radiometric Normalization

Furby, S. L., and N. A. Campbell. 2001. Calibrating Images from Different Dates to "Like-Value" Digital Counts. *Remote Sensing of Environment*, Vol. 77, pp. 186–196.

Hadjimitsis, D. G., C. R. I. Clayton, and V. S. Hope. 2004. An Assessment of the Effectiveness of Atmospheric Correction Algorithms through the Remote Sensing of Some Reservoirs. *International Journal of Remote Sensing*, Vol. 25, pp. 3651–3674.

Heo, J. and T. W. FitzHugh. 2000. A Standardized Radiometric Normalization Method for Change Detection using Remotely Sensed Imagery. *Photogrammetric Engineering and Remote Sensing*, Vol. 66, pp. 173–181.

Huang, C., L. Yang, C. Homer, B. Wylie, J. Vogelman, and T. DeFelice. (n.d.) At-Satellite Reflectance: A First-Order Normalization of Landsat 7 ETM+ Images. USGS. Available at *landcover.usgs.gov/pdf/huang2.pdf*.

Masek, J., C. Huang, R. Wolfe, W. Cohen, F. Hall, J. Kutler, et al. 2008. LEDAPS: North American Forest Disturbance Mapped from a Decadal Landsat Record. *Remote Sensing of Environment*, Vol. 112, pp. 2914–2926.

Schott, J. R., C. Salvaggio, and W. J. Volchock. 1988. Radiometric Scene Normalization Using Pseudoinvariant Features. *Remote Sensing of Environment*, Vol. 26, pp. 1–16.

CHAPTER TWELVE

Image Classification

12.1. Introduction

Digital image classification is the process of assigning pixels to classes. Usually each pixel is treated as an individual unit composed of values in several spectral bands. By comparing pixels to one another and to pixels of known identity, it is possible to assemble groups of similar pixels into classes that are associated with the informational categories of interest to users of remotely sensed data. These classes form regions on a map or an image, so that after classification the digital image is presented as a mosaic of uniform parcels, each identified by a color or symbol (Figure 12.1). These classes are, in theory, homogeneous: Pixels within classes are spectrally more similar to one another than they are to pixels in other classes. In practice, of course, each class will display some diversity, as each scene will exhibit some variability within classes.

Image classification is an important part of the fields of remote sensing, image analysis, and pattern recognition. In some instances, the classification itself may be the object of the analysis. For example, classification of land use from remotely sensed data (Chapter 20) produces a maplike image as the final product of the analysis. In other instances,

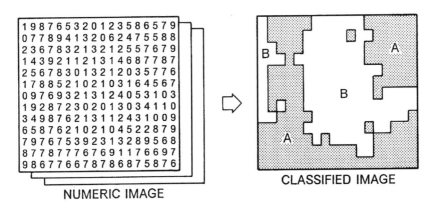

FIGURE 12.1. Numeric image and classified image. The classified image (right) is defined by examining the numeric image (left) and then grouping together those pixels that have similar spectral values. Here class "A" is defined by bright values (6, 7, 8, and 9). Class "B" is formed from dark pixels (0, 1, 2, and 3). Usually there are many more classes and at least three or four spectral bands.

the classification may be only an intermediate step in a more elaborate analysis, in which the classified data form one of several data layers in a GIS. For example, in a study of water quality (Chapter 19), an initial step may be to use image classification to identify wetlands and open water within a scene. Later steps may then focus on more detailed study of these areas to identify influences on water quality and to map variations in water quality. Image classification, therefore, forms an important tool for examination of digital images—sometimes to produce a final product, other times as one of several analytical procedures applied to derive information from an image.

The term *classifier* refers loosely to a computer program that implements a specific procedure for image classification. Over the years scientists have devised many classification strategies. The analyst must select a classification method that will best accomplish a specific task. At present it is not possible to state that a given classifier is "best" for all situations, because the characteristics of each image and the circumstances for each study vary so greatly. Therefore, it is essential that each analyst understand the alternative strategies for image classification so that he or she may be prepared to select the most appropriate classifier for the task at hand.

The simplest form of digital image classification is to consider each pixel individually, assigning it to a class based on its several values measured in separate spectral bands (Figure 12.2). Sometimes such classifiers are referred to as spectral or point classifiers because they consider each pixel as a "point" observation (i.e., as values isolated from their neighbors). Although point classifiers offer the benefits of simplicity and economy, they are not capable of exploiting the information contained in relationships between each pixel and those that neighbor it. Human interpreters, for example, could derive little information using the point-by-point approach, because humans derive less information from the brightnesses of individual pixels than they do from the context and the patterns of brightnesses, of groups of pixels, and from the sizes, shapes, and arrangements of parcels of adjacent pixels. These are the same qualities that we discussed in the context of manual image interpretation (Chapter 5).

As an alternative, more complex classification processes consider groups of pixels within their spatial setting within the image as a means of using the textural information so important for the human interpreter. These are *spatial*, or *neighborhood*, classifiers,

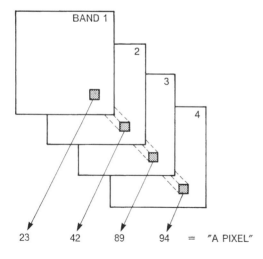

FIGURE 12.2. Point classifiers operate on each pixel as a single set of spectral values considered in isolation from its neighbors.

which examine small areas within the image using both spectral and textural information to classify the image (Figure 12.3). Spatial classifiers are typically more difficult to
program and much more expensive to use than point classifiers. In some situations spatial
classifiers have demonstrated improved accuracy, but few have found their way into routine use for remote sensing image classification.

Another kind of distinction in image classification separates supervised classification
from unsupervised classification. *Supervised classification* procedures require considerable interaction with the analyst, who must guide the classification by identifying areas
on the image that are known to belong to each category. *Unsupervised classification*, on
the other hand, proceeds with only minimal interaction with the analyst in a search for
natural groups of pixels present within the image. The distinction between supervised
and unsupervised classification is useful, especially for students who are first learning
about image classification. But the two strategies are not as clearly distinct as these definitions suggest, for some methods do not fit neatly into either category. These so-called
hybrid classifiers share characteristics of both supervised and unsupervised methods.

12.2. Informational Classes and Spectral Classes

Informational classes are the categories of interest to the users of the data. Informational
classes are, for example, the different kinds of geological units, different kinds of forest,
or the different kinds of land use that convey information to planners, managers, administrators, and scientists who use information derived from remotely sensed data. These
classes are the information that we wish to derive from the data—they are the object of
our analysis. Unfortunately, these classes are not directly recorded on remotely sensed
images; we can derive them only indirectly, using the evidence contained in brightnesses
recorded by each image. For example, the image cannot directly show geological units,
but rather only the differences in topography, vegetation, soil color, shadow, and other

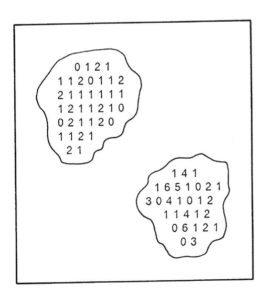

FIGURE 12.3. Image texture, the basis for
neighborhood classifiers. A neighborhood
classifier considers values within a region of
the image in defining class membership. Here
two regions within the image differ with
respect to average brightness and also with
respect to "texture"—the uniformity of pixels within a small neighborhood.

factors that lead the analyst to conclude that certain geological conditions exist in specific areas.

Spectral classes are groups of pixels that are uniform with respect to the brightnesses in their several spectral channels. The analyst can observe spectral classes within remotely sensed data; if it is possible to define links between the spectral classes on the image and the informational classes that are of primary interest, then the image forms a valuable source of information. Thus remote sensing classification proceeds by matching spectral categories to informational categories. If the match can be made with confidence, then the information is likely to be reliable. If spectral and informational categories do not correspond, then the image is unlikely to be a useful source for that particular form of information. Seldom can we expect to find exact one-to-one matches between informational and spectral classes. Any informational class includes spectral variations arising from natural variations within the class. For example, a region of the informational class "forest" is still "forest," even though it may display variations in age, species composition, density, and vigor, which all lead to differences in the spectral appearance of a single informational class. Furthermore, other factors, such as variations in illumination and shadowing, may produce additional variations even within otherwise spectrally uniform classes.

Thus informational classes are typically composed of numerous *spectral subclasses,* spectrally distinct groups of pixels that together may be assembled to form an informational class (Figure 12.4). In digital classification, we must often treat spectral subclasses as distinct units during classification but then display several spectral classes under a single symbol for the final image or map to be used by planners or administrators (who are, after all, interested only in the informational categories, not in the intermediate steps required to generate them).

In subsequent chapters we will be interested in several properties of spectral classes. For each band, each class is characterized by a mean, or average, value that of course represents the typical brightness of each class. In nature, all classes exhibit some variability around their mean values: Some pixels are darker than the average, others a bit brighter. These departures from the mean are measured as the *variance,* or sometimes by the *standard deviation* (the square root of the variance).

FIGURE 12.4. Spectral subclasses.

Often we wish to assess the distinctiveness of separate spectral classes, perhaps to determine whether they really are separate classes or whether they should be combined to form a single, larger class. A crude measure of the distinctiveness of two classes is simply the difference in the mean values; presumably classes that are very different should have big differences in average brightness, whereas classes that are similar to one another should have small differences in average brightness. This measure is a bit too simple, because it does not take into account the differences in variability between classes.

Another simple measure of distinctiveness is the *normalized difference* (ND), found by dividing the difference in class means by the sum of their standard deviations. For two classes A and B, with means A and B, and standard deviations s_a and s_b, the ND is defined as:

$$ND = \frac{\overline{A} - \overline{B}}{s_a - s_b}$$ (Eq. 12.1)

As an example, Table 12.1 summarizes properties of several classes, then uses these properties to show the normalized difference between these classes in several spectral channels. Note that some pairs of categories are very distinct relative to others. Also note that some spectral channels are much more effective than others in separating categories from one another.

12.3. Unsupervised Classification

Unsupervised classification can be defined as the identification of natural groups, or structures, within multispectral data. The notion of the existence of natural, inherent groupings of spectral values within a scene may not be intuitively obvious, but it can be demonstrated that remotely sensed images are usually composed of spectral classes that are reasonably uniform internally in respect to brightnesses in several spectral chan-

TABLE 12.1. Normalized Difference

	MSS band				MSS band			
	1	2	3	4	1	2	3	4
	Water				Forest			
x	37.5	31.9	22.8	6.3	26.9	16.6	55.7	32.5
s	0.67	2.77	2.44	0.82	1.21	1.49	3.97	3.12
	Crop				Pasture			
x	37.7	38.0	52.3	27.3	28.6	22.0	53.4	32.9
s	3.56	5.08	4.13	4.42	1.51	5.09	13.16	3.80

Water – forest (band 6): $\frac{55.7 - 22.8}{2.44 + 3.97} = \frac{32.9}{6.41} = 5.13$

Crop – pasture (band 6): $\frac{52.3 - 53.4}{4.13 + 13.16} = \frac{1.1}{17.29} = 0.06$

nels. Unsupervised classification is the definition, identification, labeling, and mapping of these natural classes.

Advantages

The advantages of unsupervised classification (relative to supervised classification) can be enumerated as follows:

• *No extensive prior knowledge of the region is required.* Or, more accurately, the nature of knowledge required for unsupervised classification differs from that required for supervised classification. To conduct supervised classification, detailed knowledge of the area to be examined is required to select representative examples of each class to be mapped. To conduct unsupervised classification, no detailed prior knowledge is required, but knowledge of the region is required to interpret the meaning of the results produced by the classification process.

• *Opportunity for human error is minimized.* To conduct unsupervised classification, the operator may perhaps specify only the number of categories desired (or, possibly, minimum and maximum limits on the number of categories) and sometimes constraints governing the distinctness and uniformity of groups. Many of the detailed decisions required for supervised classification are not required for unsupervised classification, so the analyst is presented with less opportunity for error. If the analyst has inaccurate preconceptions regarding the region, those preconceptions will have little opportunity to influence the classification. The classes defined by unsupervised classification are often much more uniform with respect to spectral composition than are those generated by supervised classification.

• *Unique classes are recognized as distinct units.* Such classes, perhaps of very small areal extent, may remain unrecognized in the process of supervised classification and could inadvertently be incorporated into other classes, generating error and imprecision throughout the entire classification.

Disadvantages and Limitations

The disadvantages and limitations of unsupervised classification arise primarily from reliance on "natural" groupings and difficulties in matching these groups to the informational categories that are of interest to the analyst.

• *Unsupervised classification identifies spectrally homogeneous classes within the data that do not necessarily correspond to the informational categories that are of interest to the analyst.* As a result, the analyst is faced with the problem of matching spectral classes generated by the classification to the informational classes that are required by the ultimate user of the information. Seldom is there a simple one-to-one correspondence between the two sets of classes.

• *The analyst has limited control over the menu of classes and their specific identities.* If it is necessary to generate a specific menu of informational classes (e.g., to match to other classifications for other dates or adjacent regions), the use of unsupervised classification may be unsatisfactory.

• *Spectral properties of specific informational classes will change over time* (on a seasonal basis, as well as over the years). As a result, relationships between informational classes and spectral classes are not constant, and relationships defined for one image cannot be extended to others.

Distance Measures

Some of the basic elements of unsupervised classification can be illustrated using data presented in Tables 12.2 and 12.3. These values can be plotted on simple diagrams constructed using brightnesses of two spectral bands as orthogonal axes (Figures 12.5 and 12.6). Plots of band pairs 4 and 5 and band pairs 5 and 6 are sufficient to illustrate the main points of significance here. These two-dimensional plots illustrate principles that can be extended to include additional variables (e.g., to analyze Landsat ETM+ bands 1–5 simultaneously). The additional variables create three- or four-dimensional plots of points in multidimensional data space, which are difficult to depict in a diagram but can be envisioned as groups of pixels represented by swarms of points with depth in several dimensions (Figure 12.7).

The general form for such diagrams is illustrated by Figure 12.7. A diagonal line of points extends upward from a point near the origin. This pattern occurs because a pixel that is dark in one band will often tend to be dark in another band, and as brightness increases in one spectral region, it tends to increase in others. (This relationship often holds for spectral measurements in the visible and near infrared; it is not necessarily observed in data from widely separated spectral regions.) Forty values are plotted in each of the diagrams in Figure 12.5. Specific groupings, or clusters, are evident; these clusters

TABLE 12.2. Landsat MSS Digital Values for February

	MSS band					MSS band			
	1	2	3	4		1	2	3	4
1.	19	15	22	11	21.	24	24	25	11
2.	21	15	22	12	22.	25	25	38	20
3.	19	13	25	14	23.	20	29	19	3
4.	28	27	41	21	24.	28	29	18	2
5	27	25	32	19	25.	25	26	42	21
6.	21	15	25	13	26.	24	23	41	22
7.	21	17	23	12	27.	21	18	12	12
8.	19	16	24	12	28.	25	21	31	15
9.	19	12	25	14	29.	22	22	31	15
10.	28	29	17	3	30.	26	24	43	21
11.	28	26	41	21	31.	19	16	24	12
12.	19	16	24	12	32.	30	31	18	3
13.	29	32	17	3	33.	28	27	44	24
14.	19	16	22	12	34.	22	22	28	15
15.	19	16	24	12	35.	30	31	18	2
16.	19	16	25	13	36.	19	16	22	12
17.	24	21	35	19	37.	30	31	18	2
18.	22	18	31	14	38.	27	23	34	20
19.	21	18	25	13	39.	21	16	22	12
20.	21	16	27	13	40.	23	22	26	16

Note. These are raw digital values for a forested area in central Virginia as acquired by the Landsat 1 MSS in February 1974. These values represent the same area as those in Table 12.3, although individual pixels do not correspond.

TABLE 12.3. Landsat MSS Digital Values for May

	MSS band						MSS band			
	1	2	3	4			1	2	3	4
1.	34	28	22	6		21.	26	16	52	29
2.	26	16	52	29		22.	30	18	57	35
3.	36	35	24	6		23.	30	18	62	28
4.	39	41	48	23		24.	35	30	18	6
5.	26	15	52	31		25.	36	33	24	7
6.	36	28	22	6		26.	27	16	57	32
7.	28	18	59	35		27.	26	15	57	34
8.	28	21	57	34		28.	26	15	50	29
9.	26	16	55	30		29.	26	33	24	27
10.	32	30	52	25		30.	36	36	27	8
11.	40	45	59	26		31.	40	43	51	27
12.	33	30	48	24		32.	30	18	62	38
13.	28	21	57	34		33.	28	18	62	38
14.	28	21	59	35		34.	36	33	22	6
15.	36	38	48	22		35.	35	36	56	33
16.	36	31	23	5		36.	42	42	53	26
17.	26	19	57	33		37.	26	16	50	30
18.	36	34	25	7		38.	42	38	58	33
19.	36	31	21	6		39.	30	22	59	37
20.	27	19	55	30		40.	27	16	56	34

Note. These are raw digital values for a forested area in central Virginia as acquired by the Landsat 1 MSS in May 1974. These values represent the same area as those in Table 12.2, although individual pixels do not correspond.

may correspond to informational categories of interest to the analyst. Unsupervised classification is the process of defining such clusters in multidimensional data space and (if possible) matching them to informational categories.

When we consider large numbers of pixels, the clusters are not usually as distinct, because pixels of intermediate values tend to fill in the gaps between groups (Figure 12.7). We must therefore apply a variety of methods to assist us in the identification of those groups that may be present within the data but may not be obvious to visual inspection. Over the years image scientists and statisticians have developed a wide variety of procedures for identifying such clusters—procedures that vary greatly in complexity and effec-

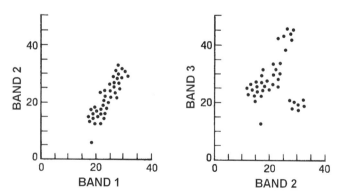

FIGURE 12.5. Two-dimensional scatter diagrams illustrating the grouping of pixels within multispectral data space.

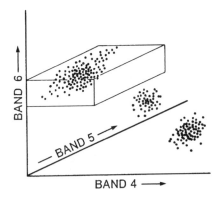

FIGURE 12.6. Sketch illustrating multidimensional scatter diagram. Here three bands of data are shown.

tiveness. The general model for unsupervised classification can, however, be illustrated using one of the simplest classification strategies, which can serve as the foundation for understanding more complex approaches.

Figure 12.8 shows two pixels, each with measurements in several spectral channels, plotted in multidimensional data space in the same manner as those illustrated in Figure 12.6. For ease of illustration, only two bands are shown here, although the principles illustrated extend to as many bands as may be available.

Unsupervised classification of an entire image must consider many thousands of pixels. But the classification process is always based on the answer to the same question: Do the two pixels belong to the same group? For this example, the question is, Should pixel C be grouped with A or with B? (see the lower part of Figure 12.8). This question can be answered by finding the distance between pairs of pixels. If the distance between A and C is greater than that between B and C, then B and C are said to belong to the same group, and A may be defined as a member of a separate class.

There are thousands of pixels in a remotely sensed image; if they are considered individually as prospective members of groups, the distances to other pixels can always be used to define group membership. How can such distances be calculated? A number of methods for finding distances in multidimensional data space are available. One of the simplest is *Euclidean distance*:

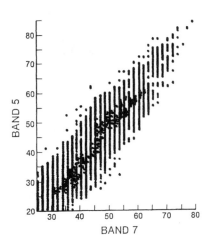

FIGURE 12.7. Scatter diagram. Data from two Landsat MSS bands illustrate the general form of the relationship between spectral measurements in neighboring regions of the spectrum. The diagram shows several hundred points; when so many values are shown, the distinct clusters visible in Figures 12.5 and 12.6 are often not visible. The groups may still be present, but they can be detected only with the aid of classification algorithms that can simultaneously consider values in several spectral bands. From Todd et al. (1980, p. 511). Copyright 1980 by the American Society for Photogrammetry and Remote Sensing. Reproduced by permission.

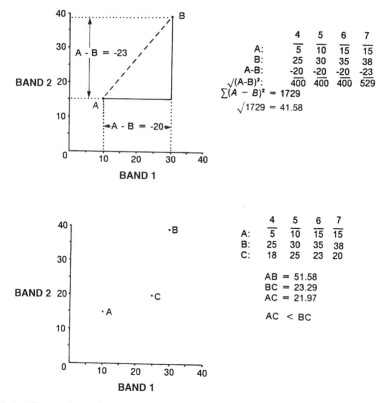

FIGURE 12.8. Illustration of Euclidean distance measure.

$$D_{ab} = \left[\sum_{i=1}^{n} (a_i - b_i)^2 \right]^{1/2} \qquad \text{(Eq. 12.2)}$$

where i is one of n spectral bands, a and b are pixels, and D_{ab} is the distance between the two pixels. The distance calculation is based on the Pythagorean theorem (Figure 12.9):

$$c = \sqrt{a^2 + b^2} \qquad \text{(Eq. 12.3)}$$

In this instance we are interested in distance c; a, b, and c are measured in units of the two spectral channels.

$$c = D_{ab} \qquad \text{(Eq. 12.4)}$$

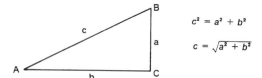

$$c^2 = a^2 + b^2$$
$$c = \sqrt{a^2 + b^2}$$

FIGURE 12.9. Definition of symbols used for explanation of Euclidean distance.

To find D_{ab}, we need to find distances a and b. Distance a is found by subtracting values of A and B in MSS channel 7 ($a = 38 - 15 = 23$). Distance b is found by finding the difference between A and B with respect to channel 6 ($b = 30 - 10 = 20$).

$$D_{ab} = c = \sqrt{20^2 + 23^2}$$

$$D_{ab} = \sqrt{400 + 529} = \sqrt{929}$$

$$D_{ab} = 30.47$$

This measure can be applied to as many dimensions (spectral channels) as might be available, by addition of distances. For example:

	Landsat MSS band			
	1	2	3	4
Pixel A	34	28	22	6
Pixel B	26	16	52	29
Difference	8	12	−30	−23
(Difference)^2	64	144	900	529

Total of (differences)^2 = 1,637

$\sqrt{\text{total}}$ = 40.5

The upper part of Figure 12.8 shows another worked example.

Thus the Euclidean distance between A and B is equal to 40.45 distance units. This value in itself has little significance, but in relation to other distances it forms a means of defining similarities between pixels. For example, if we find that distance $ab = 40.45$ and that distance $ac = 86.34$, then we know that pixel A is closer (i.e., more nearly similar) to B than it is to C and that we should form a group from A and B rather than A and C. Unsupervised classification proceeds by making thousands of distance calculations as a means of determining similarities for the many pixels and groups within an image. Usually the analyst does not actually know any of these many distances that must be calculated for unsupervised classification, as the computer presents only the final classified image without the intermediate steps necessary to derive the classification. Nonetheless, distance measures are the heart of unsupervised classification.

But not all distance measures are based on Euclidean distance. Another simple measure of determining distance is the L_1 *distance*, the sum of the absolute differences between values in individual bands (Swain and Davis, 1978). For the example given above, the L_1 distance is 73 (73 = 8 + 12 + 30 + 23). Other distance measures have been defined for unsupervised classification; many are rather complex methods of scaling distances to promote effective groupings of pixels.

Unsupervised classification often proceeds in an interactive fashion to search for an optimal allocation of pixels to categories, given the constraints specified by the analyst. A computer program for unsupervised classification includes an algorithm for calculation of distances as described above (sometimes the analyst might be able to select between several alternative distance measures) and a procedure for finding, testing, then revising classes according to limits defined by the analyst. The analyst may be required to specify limits on the number of clusters to be generated, to constrain the diversity of values within classes, or to require that classes exhibit a specified minimum degree of distinctness with respect to neighboring groups. Specific classification procedures may define distances differently; for example, it is possible to calculate distances to the centroid of each group, or to the closest member of each group, or perhaps to the most densely occupied region of a cluster. Such alternatives represent refinements of the basic strategy outlined here, and each may offer advantages in certain situations. Also, there are many variations in the details of how each classification program may operate; because so many distance measures must be calculated for classification of a remotely sensed image, most programs use variations of these basic procedures that accomplish the same objectives with improved computational efficiency.

Sequence for Unsupervised Classification

A typical sequence might begin with the analyst specifying minimum and maximum numbers of categories to be generated by the classification algorithm. These values might be based on the analyst's knowledge of the scene or on the user's requirements that the final classification display a certain number of classes. The classification starts with a set of arbitrarily selected pixels as cluster centers; often these are selected at random to ensure that the analyst cannot influence the classification and that the selected pixels are representative of values found throughout the scene. The classification algorithm then finds distances (as described above) between pixels and forms initial estimates of cluster centers as permitted by constraints specified by the analyst. The class can be represented by a single point, known as the "class centroid," which can be thought of as the center of the cluster of pixels for a given class, even though many classification procedures do not always define it as the exact center of the group. At this point, classes consist only of the arbitrarily selected pixels chosen as initial estimates of class centroids. In the next step, all the remaining pixels in the scene are assigned to the nearest class centroid. The entire scene has now been classified, but this classification forms only an estimate of the final result, as the classes formed by this initial attempt are unlikely to be the optimal set of classes and may not meet the constraints specified by the analyst.

To begin the next step, the algorithm finds new centroids for each class, as the addition of new pixels to the classification means that the initial centroids are no longer accurate. Then the entire scene is classified again, with each pixel assigned to the nearest centroid. And again new centroids are calculated; if the new centroids differ from those found in the preceding step, then the process repeats until there is no significant change detected in locations of class centroids and the classes meet all constraints required by the operator.

Throughout the process the analyst generally has no interaction with the classification, so it operates as an "objective" classification within the constraints provided by the analyst. Also, the unsupervised approach identifies the "natural" structure of the image in the sense that it finds uniform groupings of pixels that form distinct classes without the influence of preconceptions regarding their identities or distributions. The entire process, however, cannot be considered to be "objective," as the analyst has made decisions regarding the data to be examined, the algorithm to be used, the number of classes to be found, and (possibly) the uniformity and distinctness of classes. Each of these decisions influences the character and the accuracy of the final product, so it cannot be regarded as a result isolated from the context in which it was made.

Many different procedures for unsupervised classification are available; despite their diversity, most are based on the general strategy just described. Although some refinements are possible to improve computational speed and efficiency, this approach is in essence a kind of wearing down of the classification problem by repetitive application assignment and reassignment of pixels to groups. Key components of any unsupervised classification algorithm are effective methods of measuring distances in data space, identifying class centroids, and testing the distinctness of classes. There are many different strategies for accomplishing each of these tasks; an enumeration of even the most widely used methods is outside the scope of this text, but some are described in the articles listed in the references.

AMOEBA

A useful variation on the basic strategy for unsupervised classification has been described by Bryant (1979). The AMOEBA classification operates in the manner of usual unsupervised classification, with the addition of a contiguity constraint that considers the locations of values as spectral classes are formed. The analyst specifies a tolerance limit that governs the diversity permitted as classes are formed. As a class is formed by a group of neighboring pixels, adjacent pixels belonging to other classes are considered as prospective members of the class if it occurs as a small region within a larger, more homogeneous background (Figure 12.10). If the candidate pixel has values that fall within the tolerance limits specified by the analyst, the pixel is accepted as a member of the class despite the fact that it differs from other members. Thus locations as well as spectral properties of pixels form classification criteria for AMOEBA and similar algorithms.

Although this approach increases the spectral diversity of classes—normally an undesirable quality—the inclusion of small areas of foreign pixels as members of a more

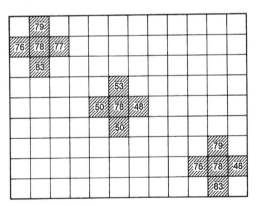

FIGURE 12.10. AMOEBA spatial operator. This contrived example shows AMOEBA as it considers classification of three pixels, all with the same digital value of "78." Upper left: The central pixel has a value similar to those of its neighbors; it is classified together with its neighbors. Center: The central pixel differs from its neighbors, but due to its position within a background of darker but homogeneous pixels, it would be classified with the darker pixels. Lower left: The central pixel would be classified with the darker pixels at the top, left, and bottom. In all three situations, the classification assignment is based on values at locations adjacent to the sample pixel.

extensive region may satisfy our notion of the proper character of a geographic region. Often, in the manual preparation of maps, the analyst will perform a similar form of generalization as a means of eliminating small inclusions within larger regions. This kind of generalization presents cartographic data in a form suitable for presentation at small scale. The AMOEBA classifier was designed for application to scenes composed primarily of large homogeneous regions, such as the agricultural landscapes of the North American prairies. For such scenes, it seems to work well, although it may not be as effective in more complex landscapes composed of smaller parcels (Story et al., 1984).

Assignment of Spectral Categories to Informational Categories

The classification results described thus far provide uniform groupings of pixels—classes that are uniform with respect to the spectral values that compose each pixel. These spectral classes are significant only to the extent that they can be matched to one or more informational classes that are of interest to the user of the final product. These spectral classes may sometimes correspond directly to informational categories. For example, in Figure 12.11 the pixels that compose the group closest to the origin correspond to "open water." This identification is more or less obvious from the spectral properties of the class; few other classes will exhibit such dark values in both spectral channels. However, seldom can we depend on a clear identification of spectral categories from spectral values alone. Often it is possible to match spectral and informational categories by examining patterns on the image; many informational categories are recognizable by the positions, sizes, and shapes of individual parcels and their spatial correspondence with areas of known identity.

Often, however, spectral classes do not match directly to informational classes. In some instances, informational classes may occur in complex mixtures and arrangements. Perhaps forested patches are scattered in small areas against a more extensive background of grassland. If these forested areas are small relative to the spatial resolution of the sensor, then the overall spectral response for such an area will differ from either "forest" or "grassland," due to the effect of mixed pixels on the spectral response. This region may be assigned, then, to a class separate from either the forest or the grassland classes (Chapter 10). It is the analyst's task to apply his or her knowledge of the region and the sensor to identify the character of the problem and to apply an appropriate remedy.

Or, because of spectral diversity, even nominally uniform informational classes may manifest themselves as a set of spectral classes. For example, a forested area may be

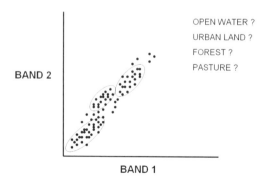

FIGURE 12.11. Assignment of spectral categories to image classes. Unsupervised classification defines the clusters defined schematically on the scatter diagram. The analyst must decide which, if any, match to the list of informational categories that form the objective of the analysis.

recorded as several spectral clusters, due perhaps to variations in density, age, aspect, shadowing, and other factors that alter the spectral properties of a forested region but do not alter the fact that the region belongs to the informational class "forest." The analyst must therefore examine the output to match spectral categories from the classification with the informational classes of significance to those who will use the results.

Thus a serious practical problem with unsupervised classification is that clear matches between spectral and informational classes are not always possible; some informational categories may not have direct spectral counterparts, and vice versa. Furthermore, the analyst does not have control over the nature of the categories generated by unsupervised classification. If a study is conducted to compare results with those from an adjacent region or from a different date, for example, it may be necessary to have the same set of informational categories on both maps or the comparison cannot be made. If unsupervised classification is to be used, it may be difficult to generate the same sets of informational categories on both images.

12.4. Supervised Classification

Supervised classification can be defined informally as the process of using samples of known identity (i.e., pixels already assigned to informational classes) to classify pixels of unknown identity (i.e., to assign unclassified pixels to one of several informational classes). Samples of known identity are those pixels located within *training areas*, or *training fields*. The analyst defines training areas by identifying regions on the image that can be clearly matched to areas of known identity on the image. Such areas should typify spectral properties of the categories they represent and, of course, must be homogeneous in respect to the informational category to be classified. That is, training areas should not include unusual regions, nor should they straddle boundaries between categories. Size, shape, and position must favor convenient identification both on the image and on the ground. Pixels located within these areas form the *training samples* used to guide the classification algorithm to assign specific spectral values to appropriate informational classes. Clearly, the selection of these training data is a key step in supervised classification.

Advantages

The advantages of supervised classification, relative to unsupervised classification, can be enumerated as follows. First, the analyst has control of a selected menu of informational categories tailored to a specific purpose and geographic region. This quality may be vitally important if it becomes necessary to generate a classification for the specific purpose of comparison with another classification of the same area at a different date or if the classification must be compatible with those of neighboring regions. Under such circumstances, the unpredictable (i.e., with respect to number, identity, size, and pattern) qualities of categories generated by unsupervised classification may be inconvenient or unsuitable. Second, supervised classification is tied to specific areas of known identity, determined through the process of selecting training areas. Third, the analyst using supervised classification is not faced with the problem of matching spectral categories on the final map with the informational categories of interest (this task has, in effect, been addressed during the process of selecting training data). Fourth, the operator may be able to detect serious errors in classification by examining training data to determine whether they have been correctly classified by the procedure—inaccurate classification of training data indicates serious problems in the classification or selection of training data, although correct classification of training data does not always indicate correct classification of other data.

Disadvantages and Limitations

The disadvantages of supervised classification are numerous. First, the analyst, in effect, imposes a classification structure on the data (recall that unsupervised classification searches for "natural" classes). These operator-defined classes may not match the natural classes that exist within the data and therefore may not be distinct or well defined in multidimensional data space. Second, training data are often defined primarily with reference to informational categories and only secondarily with reference to spectral properties. A training area that is "100% forest" may be accurate with respect to the "forest" designation but may still be very diverse with respect to density, age, shadowing, and the like, and therefore form a poor training area. Third, training data selected by the analyst may not be representative of conditions encountered throughout the image. This may be true despite the best efforts of the analyst, especially if the area to be classified is large, complex, or inaccessible.

Fourth, conscientious selection of training data can be a time-consuming, expensive, and tedious undertaking, even if ample resources are at hand. The analyst may experience problems in matching prospective training areas as defined on maps and aerial photographs to the image to be classified. Finally, supervised classification may not be able to recognize and represent special or unique categories not represented in the training data, possibly because they are not known to the analyst or because they occupy very small areas on the image.

Training Data

Training fields are areas of known identity delineated on the digital image, usually by specifying the corner points of a square or rectangular area using line and column numbers within the coordinate system of the digital image. The analyst must, of course, know the correct class for each area. Usually the analyst begins by assembling and studying maps and aerial photographs of the area to be classified and by investigating selected sites in the field. (Here we assume that the analyst has some field experience with the specific area to be studied, is familiar with the particular problem the study is to address, and has conducted the necessary field observations prior to initiating the actual selection of training data.) Specific training areas are identified for each informational category, following the guidelines outlined below. The objective is to identify a set of pixels that accurately represent spectral variation present within each informational region (Figure 12.12).

Key Characteristics of Training Areas

Numbers of Pixels

An important concern is the overall number of pixels selected for each category; as a general guideline, the operator should ensure that several individual training areas for each category provide a total of at least 100 pixels for each category.

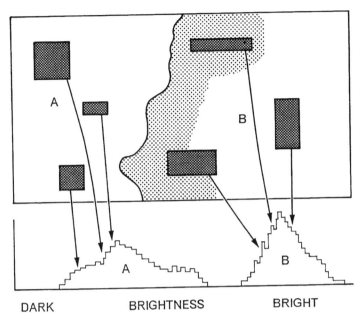

FIGURE 12.12. Training fields and training data. Training fields, each composed of many pixels, sample the spectral characteristics of informational categories. Here the shaded figures represent the training fields, each positioned carefully to estimate the spectral properties of each class, as depicted by the histograms. This information provides the basis for classification of the remaining pixels outside the training fields.

Size

Sizes of training areas are important. Each must be large enough to provide accurate estimates of the properties of each informational class. Therefore, they must as a group include enough pixels to form reliable estimates of the spectral characteristics of each class (hence, the minimum figure of 100 suggested above). Individual training fields should not, on the other hand, be too big, as large areas tend to include undesirable variation. (Because the total number of pixels in the training data for each class can be formed from many separate training fields, each individual area can be much smaller than the total number of pixels required for each class.) Joyce (1978) recommends that individual training areas be at least 4 ha (10 acres) in size at the absolute minimum and preferably include about 16 ha (40 acres). Small training fields are difficult to locate accurately on the image. To accumulate an adequate total number of training pixels, the analyst must devote more time to definition and analysis of the additional training fields. Conversely, use of large training fields increases the opportunity for inclusion of spectral inhomogeneities. Joyce (1978) suggests 65 ha (160 acres) as the maximum size for training fields.

Joyce specifically refers to Landsat MSS data. So in terms of numbers of pixels, he is recommending from 10 to about 40 pixels for each training field. For TM or SPOT data, of course, his recommendations would specify different numbers of pixels, as the resolutions of these sensors differ from that of the MSS. Also, since the optimum sizes of training fields vary according to the heterogeneity of each landscape and each class, each analyst should develop his or her own guidelines based on experience acquired in specific circumstances.

Shape

Shapes of training areas are not important, provided that shape does not prohibit accurate delineation and positioning of correct outlines of regions on digital images. Usually it is easiest to define square or rectangular areas; such shapes minimize the number of vertices that must be specified, usually the most bothersome task for the analyst.

Location

Location is important, as each informational category should be represented by several training areas positioned throughout the image. Training areas must be positioned in locations that favor accurate and convenient transfer of their outlines from maps and aerial photographs to the digital image. As the training data are intended to represent variation within the image, they must not be clustered in favored regions of the image that may not typify conditions encountered throughout the image as a whole. It is desirable for the analyst to use direct field observations in the selection of training data, but the requirement for an even distribution of training fields often conflicts with practical constraints, as it may not be practical to visit remote or inaccessible sites that may seem to form good areas for training data. Often aerial observation or use of good maps and aerial photographs can provide the basis for accurate delineation of training fields that cannot be inspected in the field. Although such practices are often sound, it is important to avoid development of a cavalier approach to selection of training data that depends completely on indirect evidence in situations in which direct observation is feasible.

Number

The optimum number of training areas depends on the number of categories to be mapped, their diversity, and the resources that can be devoted to delineating training areas. Ideally, each informational category or each spectral subclass should be represented by a number (5–10 at a minimum) of training areas to ensure that the spectral properties of each category are represented. Because informational classes are often spectrally diverse, it may be necessary to use several sets of training data for each informational category, due to the presence of spectral subclasses. Selection of multiple training areas is also desirable because later in the classification process it may be necessary to discard some training areas if they are discovered to be unsuitable. Experience indicates that it is usually better to define many small training areas than to use only a few large areas.

Placement

Placement of training areas may be important. Training areas should be placed within the image in a manner that permits convenient and accurate location with respect to distinctive features, such as water bodies, or boundaries between distinctive features on the image. They should be distributed throughout the image so that they provide a basis for representation of the diversity present within the scene. Boundaries of training fields should be placed well away from the edges of contrasting parcels so that they do not encompass edge pixels.

Uniformity

Perhaps the most important property of a good training area is its uniformity, or homogeneity (Figure 12.13). Data within each training area should exhibit a unimodal frequency distribution for each spectral band to be used (Figure 12.14a). Prospective training areas that exhibit bimodal histograms should be discarded if their boundaries cannot be adjusted to yield more uniformity. Training data provide values that estimate the means, variances, and covariances of spectral data measured in several spectral channels. For each class to be mapped, these estimates approximate the mean values for each band, the variability of each band, and the interrelationships between bands. As an ideal, these values should represent the conditions present within each class within the scene and thereby form the basis for classification of the vast majority of pixels within each scene that do not belong to training areas. In practice, of course, scenes vary greatly in complexity, and individual analysts differ in their knowledge of a region and in their ability to define training areas that accurately represent the spectral properties of informational classes. Moreover, some informational classes are not spectrally uniform and cannot be conveniently represented by a single set of training data.

Significance of Training Data

Scholz et al. (1979) and Hixson et al. (1980) discovered that selection of training data may be as important as or even more important than choice of classification algorithm in determining classification accuracies of agricultural areas in the central United States. They concluded that differences in the selection of training data were more important influences on accuracy than were differences among some five different classification procedures.

The results of their studies show little difference in the classification accuracies achieved by the five classification algorithms that were considered, if the same training statistics were used. However, in one part of the study, a classification algorithm given two alternative training methods for the same data produced significantly differ-

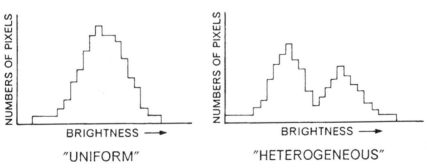

FIGURE 12.13. Uniform and heterogeneous training data. On the left, the frequency histogram of the training data has a single peak, indicating a degree of spectral homogeneity. Data from such fields form a suitable basis for image classification. On the right, a second example of training data displays a bimodal histogram that reveals that this area represents two, rather than one, spectral classes. These training data are not satisfactory for image classification and must be discarded or redefined.

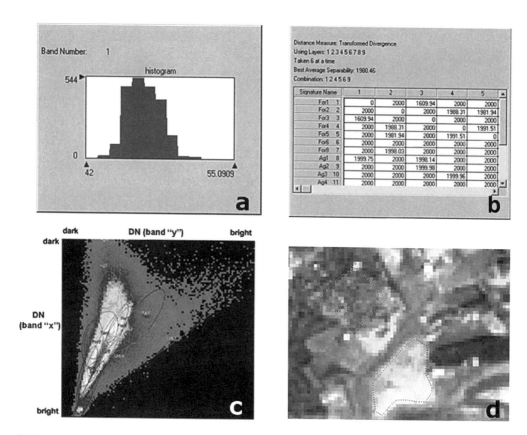

FIGURE 12.14. Examples of tools to assess training data. (a) Frequency histogram; (b) ram; (c) a point cloud of pixels in multispectral data space; (d) seed pixels, which allow the analyst to view regions of pixels with values similar to those of a selected pixel, to assist in defining the edges of training fields.

ent results. This finding suggests that the choice of training method, at least in some instances, is as important as the choice of classifier. Scholz et al. (1979, p. 4) concluded that the most important aspect of training fields is that all cover types in the scene must be adequately represented by a sufficient number of samples in each spectral subclass.

Campbell (1981) examined the character of the training data as it influences accuracy of the classification. His examples showed that adjacent pixels within training fields tended to have similar values; as a result, the samples that compose each training field may not be independent samples of the properties within a given category. Training samples collected in contiguous blocks may tend to underestimate the variability within each class and to overestimate the distinctness of categories. His examples also show that the degree of similarity varies between land-cover categories, from band to band, and from date to date. If training samples are selected randomly within classes, rather than as blocks of contiguous pixels, effects of high similarity are minimized, and classification accuracies improve. Also, his results suggest that it is probably better to use a large number of small training fields rather than a few large areas.

Idealized Sequence for Selecting Training Data

Specific circumstances for conducting supervised classification vary greatly, so it is not possible to discuss in detail the procedures to follow in selecting training data, which will be determined in part by the equipment and software available at a given facility. However, it is possible to outline an idealized sequence as a suggestion of the key steps in the selection and evaluation of training data.

1. *Assemble information,* including maps and aerial photographs of the region to be mapped.

2. *Conduct field studies,* to acquire firsthand information regarding the area to be studied. The amount of effort devoted to field studies varies depending on the analyst's familiarity with the region to be studied. If the analyst is intimately familiar with the region and has access to up-to-date maps and photographs, additional field observations may not be necessary.

3. *Carefully plan collection of field observations* and choose a route designed to observe all parts of the study region. Maps and images should be taken into the field in a form that permits the analyst to annotate them, possibly using overlays or photocopies of maps and images. It is important to observe all classes of terrain encountered within the study area, as well as all regions. The analyst should keep good notes, keyed to annotations on the image, and cross-referenced to photographs. If observations cannot be timed to coincide with image acquisition, they should match the season in which the remotely sensed images were acquired.

4. *Conduct a preliminary examination of the digital scene.* Determine landmarks that may be useful in positioning training fields. Assess image quality. Examine frequency histograms of the data.

5. *Identify prospective training areas,* using guidelines proposed by Joyce (1978) and outlined here. Sizes of prospective areas must be assessed in the light of scale differences between maps or photographs and the digital image. Locations of training areas must be defined with respect to features easily recognizable on the image and on the maps and photographs used as collateral information.

6. *Display the digital image, then locate and delineate training areas on the digital image.* Be sure to place training area boundaries well inside parcel boundaries to avoid including mixed pixels within training areas (Figure 12.14d). At the completion of this step, all training areas should be identified with respect to row and column coordinates within the image.

7. For each spectral class, *evaluate the training data,* using the tools available in the imaging-possessing system of choice (Figure 12.14c). Assess uniformity of the frequency histogram, class separability as revealed by the divergence matrix, the degree to which the training data occupy the available spectral data space, and their visual appearance on the image (Figure 12.14b).

8. *Edit training fields to correct problems identified in step 7.* Edit boundaries of training fields as needed. If necessary, discard those areas that are not suitable. Return to Step 1 to define new areas to replace those that have been eliminated.

9. *Incorporate training data information into a form suitable for use in the classification procedure* and proceed with the classification process as described in subsequent sections of this chapter.

Specific Methods for Supervised Classification

A variety of different methods have been devised to implement the basic strategy of supervised classification. All of them use information derived from the training data as a means of classifying those pixels not assigned to training fields. The following sections outline only a few of the many methods of supervised classification.

Parallelepiped Classification

Parallelepiped classification, sometimes also known as box decision rule, or level-slice procedures, are based on the ranges of values within the training data to define regions within a multidimensional data space. The spectral values of unclassified pixels are projected into data space; those that fall within the regions defined by the training data are assigned to the appropriate categories.

An example can be formed from data presented in Table 12.4. Here Landsat MSS bands 5 and 7 are selected from a larger data set to provide a concise, easily illustrated example. In practice, four or more bands can be used. The ranges of values with respect to band 5 can be plotted on the horizontal axis in Figure 12.15. The extremes of values in band 7 training data are plotted on the vertical axis, then projected to intersect with the ranges from band 5. The polygons thus defined (Figure 12.15) represent regions in data space that are assigned to categories in the classification. As pixels of unknown identity are considered for classification, those that fall within these regions are assigned to the category associated with each polygon, as derived from the training data. The procedure can be extended to as many bands, or as many categories, as necessary. In addition, the decision boundaries can be defined by the standard deviations of the values within the training areas rather than by their ranges. This kind of strategy is useful because fewer pixels will be placed in an "unclassified" category (a special problem for parallelepiped classification), but it also increases the opportunity for classes to overlap in spectral data space.

Although this procedure for classification has the advantages of accuracy, directness,

TABLE 12.4. Data for Example Shown in Figure 12.15

	Group A				Group B			
	Band				Band			
	1	2	3	4	1	2	3	4
	34	28	22	3	28	18	59	35
	36	35	24	6	28	21	57	34
	36	28	22	6	28	21	57	30
	36	31	23	5	28	14	59	35
	36	34	25	7	30	18	62	28
	36	31	21	6	30	18	62	38
	35	30	18	6	28	16	62	36
	36	33	24	2	30	22	59	37
	36	36	27	10	27	16	56	34
High	34	28	18	10	27	14	56	28
Low	36	36	27	3	30	22	62	38

Note. These data have been selected from a larger data set to illustrate parallelepiped classification.

FIGURE 12.15. Parallelepiped classification. Ranges of values within training data (Table 12.4) define decision boundaries. Here only two spectral bands are shown, but the method can be extended to several spectral channels. Other pixels, not included in the training data, are assigned to a given category if their positions fall within the polygons defined by the training data.

and simplicity, some of its disadvantages are obvious. Spectral regions for informational categories may intersect. Training data may underestimate actual ranges of classification and leave large areas in data space and on the image unassigned to informational categories. Also, the regions as defined in data space are not uniformly occupied by pixels in each category; those pixels near the edges of class boundaries may belong to other classes. Also, if training data do not encompass the complete range of values encountered in the image (as is frequently the case), large areas of the image remain unclassified, or the basic procedure described here must be modified to assign these pixels to logical classes. This strategy was among the first used in the classification of Landsat data and is still used, although it may not always be the most effective choice for image classification.

Minimum Distance Classification

Another approach to classification uses the central values of the spectral data that form the training data as a means of assigning pixels to informational categories. The spectral data from training fields can be plotted in multidimensional data space in the same manner illustrated previously for unsupervised classification. Values in several bands determine the positions of each pixel within the clusters that are formed by training data for each category (Figure 12.16). These clusters may appear to be the same as those defined earlier for unsupervised classification. However, in unsupervised classification, these clusters of pixels were defined according to the "natural" structure of the data. Now, for supervised classification, these groups are formed by values of pixels within the training fields defined by the analyst.

Each cluster can be represented by its centroid, often defined as its mean value. As unassigned pixels are considered for assignment to one of the several classes, the multidimensional distance to each cluster centroid is calculated, and the pixel is then assigned to the closest cluster. Thus the classification proceeds by always using the "minimum distance" from a given pixel to a cluster centroid defined by the training data as the spectral manifestation of an informational class.

Minimum distance classifiers are direct in concept and in implementation but are

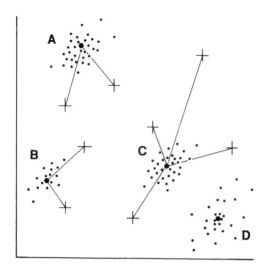

FIGURE 12.16. Minimum distance classifier. Here the small dots represent pixels from training fields and the crosses represent examples of large numbers of unassigned pixels from elsewhere on the image. Each of the pixels is assigned to the closest group, as measured from the centroids (represented by the larger dots) using the distance measures discussed in the text.

not widely used in remote sensing work. In its simplest form, minimum distance classification is not always accurate; there is no provision for accommodating differences in variability of classes, and some classes may overlap at their edges. It is possible to devise more sophisticated versions of the basic approach just outlined by using different distance measures and different methods of defining cluster centroids.

ISODATA

The ISODATA classifier (Duda and Hart, 1973) is a variation on the minimum distance method; however, it produces results that are often considered to be superior to those derived from the basic minimum distance approach. ISODATA is often seen as a form of supervised classification, although it differs appreciably from the classical model of supervised classification presented at the beginning of this chapter. It is a good example of a technique that shares characteristics of both supervised and unsupervised methods (i.e., "hybrid" classification) and provides evidence that the distinction between the two approaches is not as clear as idealized descriptions imply.

ISODATA starts with the training data selected, as previously described; these data can be envisioned as clusters in multidimensional data space. All unassigned pixels are then assigned to the nearest centroid. Thus far the approach is the same as described previously for the minimum distance classifier. Next, new centroids are found for each group, and the process of allocating pixels to the closest centroids is repeated if the centroid changes position. The process is repeated until there is no change, or only a small change, in class centroids from one iteration to the next. A step-by-step description is as follows:

1. Choose initial estimates of class means. These can be derived from training data; in this respect, ISODATA resembles supervised classification.

2. All other pixels in the scene are assigned to the class with the closest mean, as considered in multidimensional data space.

3. Class means are then recomputed to include effects of those pixels that may have been reassigned in Step 2.

4. If any class mean changes in value from Step 2 to Step 3, then the process returns to Step 2 and repeats the assignment of pixels to the closest centroid. Otherwise, the result at the end of Step 3 represents the final results.

Steps 2, 3, and 4 use the methodology of unsupervised classification, although the requirement for training data identifies the method essentially as supervised classification. It could probably best be considered as a hybrid rather than as a clear example of either approach.

Maximum Likelihood Classification

In nature the classes that we classify exhibit natural variation in their spectral patterns. Further variability is added by the effects of haze, topographic shadowing, system noise, and the effects of mixed pixels. As a result, remote sensing images seldom record spectrally pure classes; more typically, they display a range of brightnesses in each band. The classification strategies considered thus far do not consider variation that may be present within spectral categories and do not address problems that arise when frequency distributions of spectral values from separate categories overlap. For example, for application of a parallelepiped classifier, the overlap of classes is a serious problem because spectral data space cannot then be neatly divided into discrete units for classification. This kind of situation arises frequently because often our attention is focused on classifying those pixels that tend to be spectrally similar rather than those that are distinct enough to be easily and accurately classified by other classifiers.

As a result, the situation depicted in Figure 12.17 is common. Assume that we examine a digital image representing a region composed of three-fourths forested land and one-fourth cropland. The two classes "Forest" and "Cropland" are distinct with respect to average brightness, but extreme values (very bright forest pixels or very dark crop pixels) are similar in the region where the two frequency distributions overlap. (For clarity, Figure 12.17 shows data for only a single spectral band, although the principle extends to values observed in several bands and to more than the two classes shown here.) Brightness value "45" falls into the region of overlap, where we cannot make a clear assignment

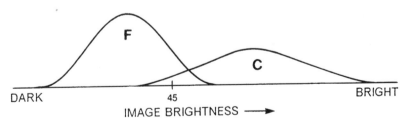

FIGURE 12.17. Maximum likelihood classification. These frequency distributions represent pixels from two training fields; the zone of overlap depicts pixel values common to both categories. The relation of the pixels within the region of overlap to the overall frequency distribution for each class defines the basis for assigning pixels to classes. Here, the relationship between the two histograms indicates that the pixel with the value "45" is more likely to belong to the Forest ("F") class rather than the Crop ("C") class.

to either "Forest" or "Cropland." Using the kinds of decision rules mentioned above, we cannot decide which group should receive these pixels unless we place the decision boundary arbitrarily.

In this situation, an effective classification would consider the relative likelihoods of "45 as a member of Forest" and "45 as a member of Cropland." We could then choose the class that would maximize the probability of a correct classification, given the information in the training data. This kind of strategy is known as *maximum likelihood* classification—it uses the training data as a means of estimating means and variances of the classes, which are then used to estimate the probabilities. Maximum likelihood classification considers not only the mean, or average, values in assigning classification but also the variability of brightness values in each class.

The maximum likelihood decision rule, implemented quantitatively to consider several classes and several spectral channels simultaneously, forms a powerful classification technique. It requires intensive calculations, so it has the disadvantage of requiring more computer resources than do most of the simpler techniques mentioned above. Also, it is sensitive to variations in the quality of training data—even more so than most other supervised techniques. Computation of the estimated probabilities is based on the assumption that both training data and the classes themselves display multivariate normal (Gaussian) frequency distributions. (This is one reason that training data should exhibit unimodal distributions, as discussed above.) Data from remotely sensed images often do not strictly adhere to this rule, although the departures are often small enough that the usefulness of the procedure is preserved. Nonetheless, training data that are not carefully selected may introduce error.

Bayes's Classification

The classification problem can be expressed more formally by stating that we wish to estimate the "probability of Forest, given that we have an observed digital value 45," and the "probability of Cropland, given that we have an observed digital value 45." These questions are a form of conditional probabilities, written as "$P(F|45)$," and "$P(C|45)$," and read as "The probability of encountering category Forest, given that digital value 45 has been observed at a pixel," and "The probability of encountering category Cropland, given that digital value 45 has been observed at a pixel." That is, they state the probability of one occurrence (finding a given category at a pixel), given that another event has already occurred (the observation of digital value 45 at that same pixel). Whereas estimation of the probabilities of encountering the two categories at random (without a conditional constraint) is straightforward (here $P[F] = 0.50$, and $P[C] = 0.50$, as mentioned above), conditional probabilities are based on two separate events. From our knowledge of the two categories as estimated from our training data, we can estimate $P(45|F)$ ("the probability of encountering digital value 45, given that we have category Forest") and $P(45|C)$ ("the probability of encountering digital value 45, given that we have category Cropland"). For this example, $P(45|F) = 0.75$, and $P(45|C) = 0.25$.

However, what we want to know are values for probabilities of "Forest, given that we observe digital value 45" [$P(F|45)$] and "Cropland, given that we observe digital value 45" [$P(C|45)$], so that we can compare them to choose the most likely class for the pixel. These probabilities cannot be found directly from the training data. From a purely intuitive examination of the problem, there would seem to be no way to estimate these probabilities.

But, in fact, there is a way to estimate $P(F|45)$ and $P(C|45)$ from the information at hand. Thomas Bayes (1702–1761) defined the relationship between the unknowns $P(F|45)$ and $P(C|45)$ and the known $P(F)$, $P(C)$, $P(45|F)$, and $P(45|C)$. His relationship, now known as *Bayes's theorem*, is expressed as follows for our example:

$$P(F|45) = \frac{P(F)P(45|F)}{P(F)P(45|F) + P(C)P(45|C)} \qquad \text{(Eq. 12.5)}$$

$$P(C|45) = \frac{P(C)P(45|C)}{P(C)P(45|C) + P(F)P(45|F)} \qquad \text{(Eq. 12.6)}$$

In a more general form, Bayes's theorem can be written:

$$P(b_1|a_1) = \frac{P(b_1)P(a_1|b_1)}{P(b_1)P(a_1|b_1) + P(b_2)P(a_1|b_2) + \ldots} \qquad \text{(Eq. 12.7)}$$

where a_1 and a_2 represent alternative results of the first stage of the experiment, and where b_1 and b_2 represent alternative results for the second stage.

For our example, Bayes's theorem can be applied as follows:

$$P(F|45) = \frac{P(F)P(45|F)}{P(F)P(45|F) + P(C)P(45|C)} \qquad \text{(Eq. 12.8)}$$

$$= \frac{\frac{1}{2} \times \frac{3}{4}}{(\frac{1}{2} \times \frac{3}{4}) + (\frac{1}{2} \times \frac{1}{4})} = \frac{\frac{3}{8}}{\frac{4}{8}} = \frac{3}{4}$$

$$P(C|45) = \frac{P(C)P(45|C)}{P(C)P(45|C) + P(F)P(45|F)}$$

$$= \frac{\frac{1}{2} \times \frac{1}{4}}{(\frac{1}{2} \times \frac{1}{4}) + (\frac{1}{2} \times \frac{3}{4})} = \frac{\frac{1}{8}}{\frac{4}{8}} = \frac{1}{4}$$

So we conclude that this pixel is more likely to be "Forest" than "Cropland." Usually data for several spectral channels are considered, and usually we wish to choose from more than two categories, so this example is greatly simplified. We can extend this procedure to as many bands or as many categories as may be necessary, although the expressions become more complex than can be discussed here.

For remote sensing classification, application of Bayes's theorem is especially effective when classes are indistinct or overlap in spectral data space. It can also form a convenient vehicle for incorporating ancillary data (Section 12.5) into the classification, as the added information can be expressed as a conditional probability. In addition, it can provide a means of introducing costs of misclassification into the analysis. (Perhaps an error in misassignment of a pixel to Forest is more serious than a misassignment to Cropland.) Furthermore, we can combine Bayes's theorem with other classification procedures, so, for example, most of the pixels can be assigned using a parallelepiped classifier, and then a Bayesian classifier can be used for those pixels that are not within the decision boundaries or within a region of overlap. Some studies have shown that such classifiers are very accurate (Story et al., 1984).

Thus Bayes's theorem is an extremely powerful means of using information at hand to estimate probabilities of outcomes related to the occurrence of preceding events. The weak point of the Bayesian approach to classification is the selection of the training data. If the probabilities are accurate, Bayes's strategy must give an effective assignment of observations to classes. Of course, from a purely computational point of view, the procedure will give an answer with any values. But to ensure accurate classification, the training data must have a sound relationship to the categories they represent. For the multidimensional case, with several spectral bands, it is necessary to estimate for each category a mean brightness for each band and a variance–covariance matrix to summarize the variability of each band and its relationships with other bands. From these data we extrapolate to estimate means, variances, and covariances of entire classes. Usually this extrapolation is made on the basis that the data are characterized by multivariate normal frequency distributions. If such assumptions are not justified, the classification results may not be accurate.

If classes and subclasses have been judiciously selected and if the training data are accurate, Bayes's approach to classification should be as effective as any that can be applied. If the classes are poorly defined and the training data are not representative of the classes to be mapped, then the results can be no better than those for other classifiers applied under similar circumstances. Use of Bayes's approach to classification forms a powerful strategy, as it includes information concerning the relative diversities of classes, as well as the means and ranges used in the previous classification strategies. The simple example used here is based on only a single channel of data and on a choice between only two classes. The same approach can be extended to consider several bands of data and several sets of categories.

This approach to classification is extremely useful and flexible and, under certain conditions, provides what is probably the most effective means of classification given the constraints of supervised classification. Note, however, that in most applications, this strategy is limited by the quality of the estimates of the probabilities required for the classification; if these are accurate, the results can provide optimal classification; if they are makeshift values conjured up simply to provide numbers for computation, the results may have serious defects.

k-Nearest Neighbors

k-nearest-neighbors classifier (KNN) is a simple but efficient application of the distance-based classification approach. It requires a set of training data representing classes present within the image. The KNN procedure examines each pixel to be classified (Figure 12.18), then identifies the *k*-nearest training samples, as measured in multispectral data space. Typically *k* is set to be a relatively small integer divisor. The candidate pixel is then assigned to the class that is represented by the most samples among the *k* neighbors.

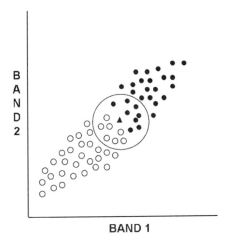

BAND 1

FIGURE 12.18. *k*-nearest-neighbors classi-fier. KNN assigns candidate pixels according to a "vote" of the *k* neighboring pixels, with *k* determined by the analyst.

Idealized Sequence for Conducting Supervised Classification

The practice of supervised classification is considerably more complex than unsupervised classification. The analyst must evaluate the situation at each of several steps, then return to an earlier point if it appears that refinements or corrections are necessary to ensure an accurate result.

1. *Prepare the menu of categories to be mapped.* Ideally these categories correspond to those of interest to the final users of the maps and data. But these requirements may not be clearly defined, or the user may require the assistance of the analyst to prepare a suitable plan for image classification.

2. *Select and define training data, as outlined above.* This step may be the most expensive and time-consuming phase of the project. It may be conducted in several stages as the analyst attempts to define a satisfactory set of training data. The analyst must outline the limits of each training site within the image, usually by using a mouse or trackball to define a polygon on a computer screen that displays a portion of the image to be classified.

This step requires careful comparison of the training fields, as marked on maps and photographs, with the digital image, as displayed on a video screen. It is for this reason that training fields must be easily located with respect to distinctive features visible both on maps and on the digital image. The image processing system then uses the image coordinates of the edges of the polygons to determine pixel values within the training field to be used in the classification process. After locating all training fields, the analyst can then display frequency distributions for training data as a means of evaluating the uniformity of each training field.

3. *Modify categories and training fields as necessary to define homogeneous training data.* As the analyst examines the training data, he or she may need to delete some categories from the menu, combine others, or subdivide still others into spectral subclasses. If training data for the modified categories meet the requirements of size and homogeneity, they can be used to enter the next step of the classification. Otherwise, the procedure (at least for certain categories) must start again at Step 1.

4. *Conduct classification.* Each image analysis system will require a different

series of commands to conduct a classification, but in essence the analyst must provide the program with access to the training data (often written to a specific computer file) and must identify the image that is to be examined. The results can be displayed on a video display terminal.

5. *Evaluate classification performance.* Finally, the analyst must conduct an evaluation using the procedures discussed in Chapter 14.

This sequence outlines the basic steps required for supervised classification; details may vary from individual to individual and with different image processing systems, but the basic principles and concerns remain the same: accurate definition of homogenous training fields.

12.5. Ancillary Data

Ancillary, or *collateral*, data can be defined as data acquired by means other than remote sensing that are used to assist in the classification or analysis of remotely sensed data. For manual interpretation, ancillary data have long been useful in the identification and delineation of features on aerial images. Such uses may be an informal, implicit application of an interpreter's knowledge and experience or may be more explicit references to maps, reports, and data. For digital remote sensing, uses of ancillary data are quite different. Ancillary data must be incorporated into the analysis in a structured and formalized manner that connects directly to the analysis of the remotely sensed data. Ancillary data may consist, for example, of topographic, pedologic, or geologic data. Primary requirements are (1) that the data be available in digital form, (2) that the data pertain to the problem at hand, and (3) that the data be compatible with the remotely sensed data.

In fact, a serious limitation to practically applying digital ancillary data is incompatibility with remotely sensed data. *Physical incompatibility* (matching digital formats, etc.) is, of course, one such incompatibility, but *logical incompatibility* is more subtle, though equally important. Seldom are ancillary data collected specifically for use with a specific remote sensing problem; usually they are data collected for another purpose, perhaps as a means of reducing costs in time and money. For example, digital terrain data gathered by the USGS and by the National Imagery and Mapping Agency (NIMA) are frequently used as ancillary data for remote sensing studies. A remote sensing project, however, is not likely to be budgeted to include the resources for digitizing, editing, and correcting these data, and therefore the data would seldom be compatible with the remotely sensed data with respect to scale, resolution, date, and accuracy. Some differences can be minimized by preprocessing the ancillary data to reduce effects of different measurement scales, resolutions, and so on, but without that, unresolved incompatibilities, possibly quite subtle, would detract from the effectiveness of the ancillary data.

Choice of ancillary variables is critical. In the mountainous regions of the western United States, elevation data have been very effective as ancillary data for mapping vegetation patterns with digital MSS data, due in part to the facts that there is dramatic variation in local elevation there and that change in vegetation is closely associated with change in elevation, slope, and aspect. In more level settings, elevation differences have subtle influences on vegetation distributions; therefore, elevation data would be less effective as ancillary variables. Although some scientists advocate the use of any and all available ancillary data, in the hope of deriving whatever advantage might be possible, common sense would seem to favor careful selection of variables with conceptual and practical significance to the mapped distributions.

The conceptual basis for the use of ancillary data is that the additional data, collected independently of the remotely sensed data, increase the information available for separating the classes and for performing other kinds of analysis. To continue the example from above, in some regions vegetation patterns are closely related to topographic elevation, slope, and aspect. Therefore, the combination of elevation data with remotely sensed data forms a powerful analytical tool because the two kinds of data provide separate, mutually supporting contributions to the subject being interpreted. Some techniques for working with ancillary data are outlined below.

Stratification refers to the subdivision of an image into regions that are easy to define using the ancillary data but might be difficult to define using only the remotely sensed data. For example, topographic data regarding elevation, slope, and aspect would be hard to extract from the spectral information in the remotely sensed image. But ancillary topographic data would provide a means of restricting the classes that would be necessary to do when using the spectral data.

The stratification is essentially an implementation of the layered classification strategy discussed in Section 12.7. Some analysts have treated each stratum as completely separate, selecting training data separately for each stratum and classifying each stratum independent of the others. The several classifications are then merged so as to present the final results as a single image. Hutchinson (1982) emphasized the dangers of ineffective stratification and warned of unpredictable, inconsistent effects of classification algorithms across different strata.

Ancillary data have also been used to form an additional channel. For example, digital elevation data could be incorporated with SPOT or Landsat MSS data as an additional band, in the hope that the additional data, independent of the remotely sensed data, would contribute to effective image classification. In general, it appears that this use of ancillary data has not been proven effective in producing substantial improvements in classification accuracy. Strahler (1980) implemented another approach, using ancillary data as a means of modifying prior probabilities in maximum likelihood classification. Digital elevation data, as an additional band of data, were used to refine the probabilities of observing specific classes of vegetation in Southern California mountains. For example, the probability of observing a specific class, such as "lodgepole pine," varies with topography and elevation. Therefore, a given set of spectral values can have a different meaning, depending on the elevation value in which a given pixel is situated. This approach appears to be effective, at least in how it has been used and studied thus far.

Finally, ancillary data can be used after the usual classifications have been completed, in a process known as *postclassification sorting*. Postclassification sorting examines the confusion matrix derived from a traditional classification. Confused classes are the candidates for postclassification sorting. Ancillary data are then used in an effort to improve discrimination between pairs of classes, thereby improving the overall accuracy by focusing on the weakest aspects of the classification. Hutchinson (1982) encountered several problems with his application of postclassification sorting but in general favors it for being simple, inexpensive, and convenient to implement.

On the whole, ancillary data, despite their benefits, present numerous practical and conceptual problems for the remote sensing analyst. Sometimes it may be difficult to select and acquire the data that may be most useful for a given problem. The most useful data might not be available in digital form; the analyst might be forced to invest considerable effort in identifying, acquiring, then digitizing the desired information. Therefore, many if not most uses of ancillary data rely on those data that can be acquired "off the shelf" from a library of data already acquired and digitized for another purpose. This presents a danger that the ancillary data may be used as much because of their availability as for their appropriateness for a specific situation.

Because ancillary data are generally acquired for a purpose other than the one at hand, the problem of compatibility can be important. Ideally, ancillary data should possess compatibility with respect to scale, level of detail, accuracy, geographic reference system, and, in instances in which time differences are important, date of acquisition. Because analysts are often forced to use off-the-shelf data, incompatibilities form one of the major issues in use of ancillary data. To be useful, ancillary data must be accurately registered to the remotely sensed data. Therefore, the analyst may be required to alter the geometry, or the level of detail, in the ancillary data to bring them into registration with the remotely sensed data. Efforts to alter image geometry or to resample data change the values and therefore are likely to influence the usefulness of the data.

Sometimes ancillary data may already exist in the form of discrete classes—that is, as specific geologic or pedologic classes, for example. Such data are typically represented on maps as discrete parcels, with abrupt changes at the boundaries between classes. Digital remote sensor data are continuous, and therefore differences between the discretely classed and continuous remote sensor data may influence the character of the results, possibly causing sharp changes where only gradual transitions should occur.

To summarize, ancillary data are nonimage information used to aid in classification of spectral data (Hutchinson, 1982). There is a long tradition of both implicit and explicit use of such information in the process of manual interpretation of images, including data from maps, photographs, field observations, reports, and personal experience. For digital analysis, ancillary data often consist of data available in formats consistent with the digital spectral data or in forms that can be conveniently transformed into usable formats. Examples include digital elevation data or digitized soil maps (Anuta, 1976).

Ancillary data can be used in either of two ways. They can be "added to" or "superimposed" over the spectral data to form a single multiband image; the ancillary data are treated simply as additional channels of data. Or the analysis can proceed in two steps using a layered classification strategy (as described below); the spectral data are classified in the first step, then the ancillary data form the basis for reclassification and refinement of the initial results.

Classification and Regression Tree Analysis

Classification and regression tree analysis (CART; Lawrence and Wright, 2001) is a method for incorporating ancillary data into image classification processes. Tools for applying CART are available in many statistical packages, which then can be employed in image processing packages. In this sense, it is more complicated to implement than many of the other classification techniques employed here. CART requires accurate training data, selected according to usual guidelines for training data delineation, but does not require a priori knowledge of the role of the variables. An advantage of CART, therefore, is that it identifies useful data and separates them from those that do not contribute to successful classification.

CART applies a recursive division of the data to achieve certain specified terminal results, set by the analyst according to the specifics of the particular algorithm employed. In a CART analysis, the dependent variable is the menu of classification classes, and the independent variables are the spectral channels and ancillary variables.

CART recursively examines binary divisions of the explanatory variables to derive an optimal assignment of the classes to the independent variables. Application of CART is sensitive to variations in numbers of pixels—it performs best when numbers of pixels in training data sets are approximately equal.

Often the use of a large number of spectral and ancillary variables leads to very accurate results, but results that are tailored to specific datasets—a condition known as *overfitting*. As a result, the application of CART often includes a pruning step, which reduces the solution to create a more concise, more robust, and generally applicable solution—division of the data into subsets, with results from some subsets cross-validated against others. Pal and Mather (2003) examined consequences of five alternative strategies for attribute selection and found that differences in accuracy were minimal.

12.6. Fuzzy Clustering

Fuzzy clustering addresses a problem implicit in much of the preceding material: that pixels must be assigned to a single discrete class. Although such classification attempts to maximize correct classifications, the logical framework allows only for direct one-to-one matches between pixels and classes. We know, however, that many processes contribute to prevent clear matches between pixels and classes, as noted in Chapter 10 and in the work of Robinove (1981) and Richards and Kelly (1984). Therefore, the focus on finding discrete matches between the pixels and informational classes ensures that many pixels will be incorrectly or illogically labeled. Fuzzy logic attempts to address this problem by applying a different classification logic.

Fuzzy logic (Kosko and Isaka, 1993) has applications in many fields, but has special significance for remote sensing. Fuzzy logic permits partial membership, a property that is especially significant in the field of remote sensing, as partial membership translates closely to the problem of mixed pixels (Chapter 10). So whereas traditional classifiers must label pixels as either "forest" or "water," for example, a fuzzy classifier is permitted to assign a pixel a membership grade of 0.3 for "water" and 0.7 for "forest," in recognition that the pixel may not be properly assigned to a single class. Membership grades typical vary from 0 (nonmembership) to 1.0 (full membership), with intermediate values signifying partial membership in one or more other classes (Table 12.5).

A fuzzy classifier assigns membership to pixels based on a membership function (Figure 12.19). Membership functions for classes are determined either by general relationships or by definitional rules describing the relationships between data and classes. Or, as is more likely in the instance of remote sensing classification, membership functions are derived from experimental (i.e., training) data for each specific scene to be examined. In the instance of remote sensing data, a membership function describes the relationship between class membership and brightness in several spectral bands.

Figure 12.19 provides contrived examples showing several pixels and their membership grades. (Actual output from a fuzzy classification is likely to form an image that shows varied levels of membership for specific classes.) Membership grades can be hardened (Table 12.6) by setting the highest class membership to 1.0 and all others to 0.0.

TABLE 12.5. Partial Membership in Fuzzy Classes

Class	Pixel						
	A	B	C	D	E	F	G
Water	0.00	0.00	0.00	0.00	0.00	0.00	0.00
Urban	0.00	0.01	0.00	0.00	0.00	0.00	0.85
Transportation	0.00	0.35	0.00	0.00	0.99	0.79	0.14
Forest	0.07	0.00	0.78	0.98	0.00	0.00	0.00
Pasture	0.00	0.33	0.21	0.02	0.00	0.05	0.00
Cropland	0.92	0.30	0.00	0.00	0.00	0.15	0.00

Hardened classes are equivalent to traditional classifications: Each pixel is labeled with a single label, and the output is a single image labeled with the identify of the hardened class. Programs designed for remote sensing applications (Bezdek et al., 1984) provide the ability to adjust the degree of fuzziness and thereby adjust the structures of classes and the degree of continuity in the classification pattern.

Fuzzy clustering has been judged to improve results, at least marginally, with respect to traditional classifiers, although the evaluation is difficult because the usual evaluation methods require the discrete logic of traditional classifiers. Thus the improvements noted for hardened classifications are probably conservative, as they do not reveal the full power of fuzzy logic.

12.7. Artificial Neural Networks

Artificial neural networks (ANNs) are computer programs that are designed to simulate human learning processes through establishment and reinforcement of linkages between input data and output data. It is these linkages, or pathways, that form the analogy with the human learning process in that repeated associations between input and output in the training process reinforce linkages, or pathways, that can then be employed to link input and output in the absence of training data.

FIGURE 12.19. Membership functions for fuzzy clustering. This example illustrates membership functions for the simple instance of three classes considered for a single spectral band, although the method is typically applied to multibands. The horizontal axis represents pixel brightness; the vertical axis represents degree of membership, from low, near the bottom, to high, at the top. The class "water" consists of pixels darker than brightness 20, although only pixels darker than 8 are likely to be completely occupied by water. The class "agriculture" can include pixels as dark as 22 and as bright as 33, although pure "agriculture" is found only in the range 27–29. A pixel of brightness 28, for example, can only be agriculture, although a pixel of brightness of 24 could be partially forested, partially agriculture. (Unlabeled areas on this diagram are not occupied by any of the designated classes.)

TABLE 12.6. "Hardened" Classes for Example Shown in Table 12.5

Class	Pixel						
	A	B	C	D	E	F	G
Water	0.00	0.00	0.00	0.00	0.00	0.00	0.00
Urban	0.00	0.00	0.00	0.00	0.00	0.00	1.00
Transportation	0.00	1.00	0.00	0.00	1.00	1.00	0.00
Forest	0.00	0.00	1.00	1.00	0.00	0.00	0.00
Pasture	0.00	0.00	0.00	0.00	0.00	0.00	0.00
Cropland	1.00	0.00	0.00	0.00	0.00	0.00	0.00

ANNs are often represented as composed of three elements. An *input layer* consists of the source data, which in the context of remote sensing are the multispectral observations, perhaps in several bands and from several dates. ANNs are designed to work with large volumes of data, including many bands and dates of multispectral observations, together with related ancillary data.

The *output layer* consists of the classes required by the analyst. There are few restrictions on the nature of the output layer, although the process will be most reliable when the number of output labels is small or modest with respect to the number of input channels. Included are training data in which the association between output labels and input data is clearly established. During the training phase, an ANN establishes an association between input and output data by establishing weights within one or more *hidden layers* (Figure 12.20). In the context of remote sensing, repeated associations between classes and digital values, as expressed in the training data, strengthen weights within hidden layers that permit the ANN to assign correct labels when given spectral values in the absence of training data.

Further, ANNs can also be trained by *back propagation* (BP). If establishment of the usual training data for conventional image classification can be thought of as "forward

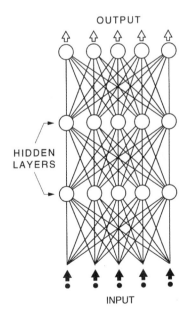

FIGURE 12.20. Artificial neural net.

propagation," then BP can be thought of as a retrospective examination of the links between input and output data in which differences between expected and actual results can be used to adjust weights. This process establishes *transfer functions*, quantitative relationships between input and output layers that assign weights to emphasize effective links between input and output layers. (For example, such weights might acknowledge that some band combinations may be very effective in defining certain classes and others effective for other classes.) In BP, hidden layers note errors in matching data to classes and adjust the weights to minimize errors.

ANNs are designed using less severe statistical assumptions than many of the usual classifiers (e.g., maximum likelihood), although in practice successful application requires careful application. ANNs have been found to be accurate in the classification of remotely sensed data, although improvements in accuracies have generally been small or modest. ANNs require considerable effort to train and are intensive in their use of computer time.

12.8. Contextual Classification

Contextual information is derived from spatial relationships among pixels within a given image. Whereas *texture* usually refers to spatial interrelationships among unclassified pixels within a window of specified size, *context* is determined by positional relationships between pixels, either classified or unclassified, anywhere within the scene (Gurney and Townshend, 1983; Swain et al., 1981).

Although contextual classifiers can operate on either classified or unclassified data, it is convenient to assume that some initial processing has assigned a set of preliminary classes on a pixel-by-pixel basis without using spatial information. The function of the contextual classifier is to operate on the preliminary classification to reassign pixels as appropriate in the light of contextual information.

Context can be defined in several ways, as illustrated in Figure 12.21a. In each instance the problem is to consider the classification of a pixel or a set of pixels (represented by the shaded pattern) using information concerning the classes of other, related pixels. Several kinds of links define the relationships between the two groups. The simplest link is that of distance. Perhaps the unclassified pixels are agricultural land, which is likely to be "irrigated cropland" if positioned within a certain distance of a body of open water; if the distance to water exceeds a certain threshold, the area might be more likely to be assigned to "rangeland" or "unirrigated cropland." Figure 12.21b illustrates the use of both distance and direction. Contiguity (Figure 12.21c) may be an important classification aid. For example, in urban regions specific land-use regions may be found primarily in locations adjacent to a specific category. Finally, specific categories may be characterized by their positions within other categories, as shown in Figure 12.21d.

Contextual classifiers are efforts to simulate some of the higher order interpretation processes used by human interpreters, in which the identity of an image region is derived in part from its location in relation to other regions of specified identity. For example, a human interpreter considers sizes and shapes of parcels in identifying land use, as well as the identities of neighboring parcels. The characteristic spatial arrangement of the central business district, the industrial district, and residential and agricultural land in an urban region permits the interpreter to identify parcels that might be indistinct if considered with conventional classifiers.

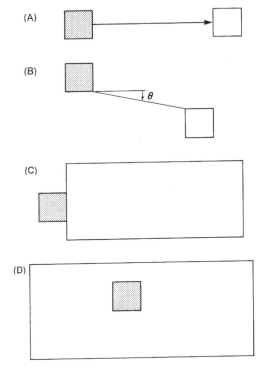

FIGURE 12.21. Contextual classification, from Gurney and Townshend (1983). The shaded regions depict pixels to be classified; the open regions represent other pixels considered in the classification decision. (a) Distance is considered; (b) Direction is considered; (c) Contiguity forms a part of the decision process; (d) Inclusion is considered. Copyright 1983 by the American Society of Photogrammetry and Remote Sensing. Reproduced by permission.

Contextual classifiers can also operate on classified data to reclassify erroneously classified pixels or to reclassify isolated pixels (perhaps correctly classified) that form regions so small and so isolated that they are of little interest to the user. Such uses may be essentially cosmetic operations, but they could be useful in editing the results for final presentation.

12.9. Object-Oriented Classification

Object-oriented classification uses a two-step process that is intended to mimic the higher order classification processes, each composed of many intermediate processes. Object-oriented classification applies a logic intended to mimic some of the higher order logic employed by human interpreters, who can use the sizes, shapes, and textures of regions, as well as the spectral characteristics used for conventional pixel-based classification.

The first of the two processes is *segmentation* of the image: identification of the edges of homogeneous patches that subdivide the image into interlocking regions. These regions are based on the spectral values of their pixels, as well as analyst-determined constraints. Segmentation is implemented hierarchically—regions are defined at several scales that nest within each other (Figure 12.22). These regions are the "objects" of objected-oriented classification. The analyst has the ability to set criteria that control the measures used to assess homogeneity and distinctness and the thresholds that apply to a specific classification problem.

These regions form the "objects" indicated by the term *object oriented*; they, rather than individual pixels, form the entities to be classified. Because the regions are composed

COARSE   FINE

FIGURE 12.22. Object-oriented classification. Through the segmentation process, pixels can be organized to form "objects" that correspond to features of interest in the image. These objects can then be treated as entities to be classified in the manner of the classification procedures discussed here. Here, the illustration shows two alternative segmentations—one at a coarse level, which creates large, diverse objects, and another at a fine level of detail, which creates more numerous, homogeneous objects.

of many pixels, they have a multitude of properties, such as standard deviations, maxima, and minima, that are possible for regions but not for individual pixels.

Further, object-oriented classification can consider distinctive topological properties of classes. For example, forest in an urban park and rural forest have similar, if not identical, spectral properties, even though they form very different classes. Object-oriented classification can use the forest's spectral properties to recognize it as forest but can also consider the effect of neighboring classes to recognize that regions of urban forest are surrounded by urban land use, whereas rural forest will border rural categories, such as open water, agricultural land, and other forest.

The second process is classification, using conventional classification procedures, usually nearest-neighbor or fuzzy classification. Each object or region is characterized by the numerous properties developed as a result of the segmentation process that can then be used in the classification process. The analyst can examine and select those properties that can be useful in classification.

Much of the conceptual and theoretical basis for object-oriented classification was developed in previous decades but was not widely used because of the difficulties of developing useful user interfaces that enable routine use for practical applications. One of the most successful of the application programs is Definiens Professional, developed by DEFINIENS Imaging AG, a Munich-based company that has specifically devoted its resources to the development of a practical object-oriented image classification system. eCognition's system provides a systematic approach and a user interface that permits implementation of concepts developed in previous decades.

Object-oriented classification can be used successfully with any multispectral imagery but has proven to be especially valuable for fine-resolution imagery, which is especially difficult to classify accurately using conventional classification techniques. eCognition has been used in defense, natural resources exploration, and facilities mapping.

Implementing object-oriented classification can be time-consuming, as the analyst often must devote considerable effort to trial and error, learning the most efficient approach to classification of a specific image for a specific purpose. Sometimes the segmentation itself is a product that can be used in other analytical processes.

12.10. Iterative Guided Spectral Class Rejection

Iterative guided spectral class rejection (IGSCR) (Wayman et al., 2001; Musy et al., 2006; Phillips et al., 2009) is a classification technique that minimizes user input and subjectivity characteristic of many other image classification approaches. IGSCR is based on training data, provided by the analyst, that represent the classes of interests and the application of unsupervised classification strategies. The unsupervised classification groups pixels together to define uniform spectral classes; IGSCR then attempts to match the spectral classes to classes as defined by the training data.

The IGSCR algorithm evaluates each of the spectral classes with respect to the training data and accepts or rejects each spectral class based on the closeness of each spectral class to the training data. IGSCR rejects spectral classes that are candidates for entering the classification if they do not meet set thresholds (often 90% homogeneity, with the minimum number of samples determined by a binomial probability distribution). Rejected pixels are regrouped into new spectral classes and again considered during the next iteration. The rejection process ensures that all pixels that enter the classification meet criteria set by the analyst and omits others at each iteration. This process ensures that all classes in the final classification will meet the analyst's criteria for uniformity. Classification stops when user-defined criteria are satisfied. Those remaining pixels that cannot meet criteria to enter the classification are left unclassified. IGSCR is a hybrid classifier—one that lends itself to accurate classification and ease of replication. The version described by Wayman et al. (2001) was designed to implement a binary (two-class) classification decision, such as forest–nonforest. The authors have since developed a multiclass version.

12.11. Summary

This chapter has described a few specific classifiers as a means of introducing the student to the variety of different classification strategies that are available today. Possibly the student may have the opportunity to use some of these procedures, so these descriptions may form the first step in a more detailed learning process. It is more likely, however, that the student who uses this text will never use many of the specific classifiers described here. Nonetheless, those procedures that are available for student use are likely to be based on the same principles outlined here using specific classifiers as examples. Therefore, this chapter should not be regarded as a complete catalog of image classification methods, but rather as an effort to illustrate some of the primary methods of image classification. Specific details and methods will vary greatly, but if the student has mastered the basic strategies and methods of image classification, he or she will recognize unfamiliar methods as variations on the fundamental approaches described here.

12.12. Some Teaching and Learning Resources

This list of resources is mainly focused on tutorials for object-oriented image classification. Others can be found on the Web, although many of these refer to outdated procedures

- What's Happening in Remote Sensing Today?
  www.youtube.com/watch?v=lYv6PhXVWVY&feature=related

- Object-Based Image Analysis Made Easy
  www.youtube.com/watch?v=PatMAWiV71M&NR=1

- eCognition Image Analysis: Urban Tree Canopy Assessment
  www.youtube.com/watch?v=pOi6hJW_vYY&feature=related

- eCognition Image Analysis: Extracting Tree Canopy from LIDAR
  www.youtube.com/watch?v=OR1Se18Zd4E&feature=related

- eCognition Image Analysis: Woolpert Webinar Part 2/3 (Analysis of Lidar Libraries)
  www.youtube.com/watch?v=P7_mkYNbnLs&feature=PlayList&p=A1DA9175 C127F443&playnext_from=PL&playnext=1&index=6

- microBRIAN (1987)
  www.youtube.com/watch?v=7pEFhusz_IQ

Review Questions

1. This chapter mentions only a few of the many strategies available for image classification. Why have so many different methods been developed? Why not use just one?

2. Why might the decision to use or not to use preprocessing be especially significant for image classification?

3. Image classification is not necessarily equally useful for all fields. For a subject of interest to you (geology, forestry, etc.), evaluate the significance of image classification by citing examples of how classification might be used. Also list some subjects for which image classification might be more or less useful.

4. Review this chapter and Chapter 6. Speculate on the course of further developments in image classification. Can you suggest relationships between sensor technology and the design of image classification strategies?

	Feb.	May		Feb.	May		Feb.	May		Feb.	May
(1)	F	W	(11)	P	C	(21)	F	C	(31)	F	C
(2)	F	F	(12)	F	C	(22)	P	P	(32)	W	F
(3)	C	W	(13)	W	P	(23)	W	P	(33)	P	F
(4)	P	C	(14)	F	P	(24)	W	W	(34)	F	W
(5)	P	F	(15)	F	C	(25)	P	W	(35)	W	C
(6)	C	W	(16)	F	W	(26)	P	F	(36)	F	C
(7)	F	F	(17)	P	F	(27)	F	F	(37)	W	F
(8)	F	P	(18)	C	W	(28)	C	F	(38)	P	C
(9)	C	F	(19)	F	W	(29)	C	P	(39)	F	P
(10)	W	C	(20)	F	F	(30)	P	W	(40)	C	F

F, forest; C, crop; P, pasture; W, water

5. The table above lists land-cover classes for the pixels given in Tables 12.2 and 12.3. (Note that individual pixels do not correspond for the two dates.) Select a few pixels

from Table 12.3; conduct a rudimentary feature selection "by inspection" (i.e., by selecting from the set of four bands). Can you convey most of the ability to distinguish between classes by choosing a subset of the four MSS bands?

6. Using the information given above and in Table 12.3, calculate normalized differences between water and forest, bands 5 and 7. (For this question and those that follow, the student should have access to a calculator.) Identify those classes easiest to separate and the bands likely to be most useful.

7. For both February and May, calculate normalized differences between forest and pasture, band 7. Give reasons to account for differences between the results for the two dates.

8. The results for questions 5 through 7 illustrate that normalized difference values vary greatly depending on data, bands, and classes considered. Discuss some of the reasons that this is true, both in general and for specific classes in this example.

9. Refer to Table 12.1. Calculate Euclidean distance between means of the four classes given in the table. Again, explain why there are differences from date to date and from band to band.

References

General

Anon. 2010. Remote Sensing Image Processing Software. *GIM International*, June, pp. 38–39.

Bazi, Y., and F. Melgani. 2006. Toward an Optimal SVM Classification System for Hyperspectral Remote Sensing Images. *IEEE Transactions on Geoscience and Remote Sensing*, Vol. 44, pp. 3374–3385.

Bryant, J. 1979. On the Clustering of Multidimensional Pictorial Data. *Pattern Recognition*, Vol. 11, pp. 115–125.

Duda, R. O., and P. E. Hart. 1973. *Pattern Classification and Scene Analysis*. New York: Wiley, 482 pp.

Foody, G. M., and A. Mathur. 2004. A Relative Evaluation of Multiclass Image Classification by Support Vector Machines. *IEEE Transactions on Geoscience and Remote Sensing*, Vol. 42, pp. 1335–1343.

Foody, G. M., and A. Mathur. 2004. Toward Intelligent Training of Supervised Image Classifications: Directing Training Data Acquisition for SVM Classification. *Remote Sensing of Environment*, Vol. 93, pp. 107–117.

Franklin, S. E., and M. A. Wulder. 2002. Remote Sensing Methods in Medium Spatial Resolution Satellite Data Land Cover Classification of Large Areas. *Progress in Physical Geography*, Vol. 26, pp. 173–205.

Gurney, C. M., and J. R. Townshend. 1983. The Use of Contextual Information in the Classification of Remotely Sensed Data. *Photogrammetric Engineering and Remote Sensing*, Vol. 49, pp. 55–64.

Jensen, J. R., and D. L. Toll. 1982. Detecting Residential Land Use Development at the Urban Fringe. *Photogrammetric Engineering and Remote Sensing*, Vol. 48, pp. 629–643.

Kettig, R. L., and D. A. Landgrebe. 1975. Classification of Multispectral Image Data by Extraction and Classification of Homogeneous Objects. In *Proceedings, Symposium on*

Machine Classification of Remotely Sensed Data. West Lafayette, IN: Laboratory for Applications in Remote Sensing, pp. 2A1–2A11.

Nikolov, H. S., D. I. Petkov, N. Jeliazkova, S. Ruseva, and K. Boyanov. 2009. Non-linear Methods in Remotely Sensed Multispectral Data Classification. *Advances in Space Research*, Vol. 43, pp. 859–868.

Pal, M., and P. M. Mather. 2005. Support Vector Machines for Classification in Remote Sensing. *International Journal of Remote Sensing*, Vol. 26, pp. 1007–1011.

Richards, J. A. 2005. Analysis of Remotely Sensed Data: The Formative Decades and the Future. *IEEE Transactions on Geoscience and Remote Sensing*, Vol. 43, pp. 422–432.

Richards, J. A., and D. J. Kelly. 1984. On the Concept of the Spectral Class. *International Journal of Remote Sensing*, Vol. 5, pp. 987–991.

Robinove, C. J. 1981. The Logic of Multispectral Classification and Mapping of Land. *Remote Sensing of Environment*, Vol. 11, pp. 231–244.

Strahler, A. H. 1980. The Use of Prior Probabilities in Maximum Likelihood Classification of Remotely Sensed Data. *Remote Sensing of Environment*, Vol. 10, pp. 135–163.

Swain, P. H., and S. M. Davis. (eds.). 1978. *Remote Sensing: The Quantitative Approach.* New York: McGraw-Hill, 396 pp.

Swain, P. H., S. B. Vardeman, and J. C. Tilton. 1981. Contextural Classification of Multispectral Image Data. *Pattern Recognition*, Vol. 13, pp. 429–441.

Ancillary Data

Anuta, P. E. 1976. Digital Registration of Topographic and Satellite MSS Data for Augmented Spectral Analysis. *Proceedings of the American Society of Photogrammetry.* 42nd Annual Meeting. Falls Church, VA: American Society of Photogrammetry.

Hutchinson, C. F. 1982. Techniques for Combining Landsat and Ancillary Data for Digital Classification Improvement. *Photogrammetric Engineering and Remote Sensing*, Vol. 48, pp. 123–130.

Training Data

Campbell, J. B. 1981. Spatial Correlation Effects upon Accuracy of Supervised Classification of Land Cover. *Photogrammetric Engineering and Remote Sensing*, Vol. 47, pp. 355–363.

Joyce, A. T. 1978. *Procedures for Gathering Ground Truth Information for a Supervised Approach to Computer-Implemented Land Cover Classification of Landsat-Acquired Multispectral Scanner Data* (NASA Reference Publication 1015). Houston, TX: National Aeronautics and Space Administration, 43 pp.

Todd, W. J., D. G. Gehring, and J. F. Haman. 1980. Landsat Wildland Mapping Accuracy. *Photogrammetric Engineering and Remote Sensing*, Vol. 46, pp. 509–520.

Image Texture

Arivazhagan, S., and L. Ganesan. 2003. Texture Segmentation using Wavelet Transform. *Pattern Recognition Letters*, Vol. 24, pp. 3197–3203.

Franklin, S. E., R. J. Hall, L. M. Moskal, A. J. Maudie, and M. B. Lavigne. 2000. Incorporating Texture into Classification of Forest Species Composition from Airborne Multispectral Images. *International Journal of Remote Sensing*, Vol. 21, pp. 61–79.

Haralick, R. M. 1979. Statistical and Structural Approaches to Texture. *Proceedings of the IEEE*, Vol. 67, pp. 786–804.

Haralick, R. M., et al. 1973. Textural Features for Image Classification. *IEEE Transactions on Systems, Man, and Cybernetics* (SMC-3), pp. 610–622.

Jensen, J. R. 1979. Spectral and Textural Features to Classify Elusive Land Cover at the Urban Fringe. *Professional Geographer*, Vol. 31, pp. 400–409.

Comparisons of Classification Techniques

Gong, P., and P. J. Howarth. 1992. Frequency-Based Contextual Classification and Gray-Level Vector Reduction for Land-Use Identification. *Photogrammetric Engineering and Remote Sensing*, Vol. 58, pp. 423–437.

Harvey, N. R., J. Theiler, S. P. Brumby, S. Perkins, J. J. Szymanski, J. J. Bloch, et al. 2002. Comparison of GENIE and Conventional Supervised Classifiers for Multispectral Image Feature Extraction. *IEEE Transactions on Geoscience and Remote Sensing*, Vol. 40, pp. 393–404.

Hixson, M., D. Scholz, and N. Fuhs. 1980. Evaluation of Several Schemes for Classification of Remotely Sensed Data. *Photogrammetric Engineering and Remote Sensing*, Vol. 46, pp. 1547–1553.

Lu, D., and Q. Weng. 2007. A Survey of Image Classification Methods and Techniques for Improving Classification Performance. *International Journal of Remote Sensing*, Vol. 28, pp. 823–870.

Scholz, D., N. Fuhs, and M. Hixson. 1979. An Evaluation of Several Different Classification Schemes, Their Parameters, and Performance. In *Proceedings, Thirteenth International Symposium on Remote Sensing of the Environment*. Ann Arbor: University of Michigan Press, pp. 1143–1149.

Story, M. H., J. B. Campbell, and G. Best. 1984. An Evaluation of the Accuracies of Five Algorithms for Machine Classification of Remotely Sensed Data. *Proceedings of the Ninth Annual William T. Pecora Remote Sensing Symposium*. Silver Spring, MD: IEEE, pp. 399–405.

Fuzzy Clustering

Bardossy, A., and L. Samaniego. 2002. Fuzzy Rule-Based Classification of Remotely Sensed Imagery. *IEEE Transactions on Geoscience and Remote Sensing*, Vol. 40, pp. 362–374.

Bezdek, J. C., R. Ehrilich, and W. Full. 1984. FCM: The Fuzzy c-Means Clustering Algorithm. *Computers and Geosciences*, Vol. 10, pp. 191–203.

Fisher, P. F., and S. Pathirana. 1990. The Evaluation of Fuzzy Membership of Land Cover Classes in the Suburban Zone. *Remote Sensing of Environment*, Vol. 34, pp. 121–132.

Kent, J. T., and K. V. Marida. 1988. Spatial Classification Using Fuzzy Membership Models. *IEEE Transactions on Pattern Analysis and Machine Intelligence*, Vol. 10, pp. 659–671.

Kosko, B., and S. Isaka. 1993. Fuzzy Logic. *Scientific American*, Vol. 271, pp. 76–81.

Melgani, F., B. A. R. Al Hashemy, and S. M. R. Taha. 2000. An Explicit Fuzzy Supervised Classification Method for Multispectral Remote Sensing Images. *IEEE Transactions on Geoscience and Remote Sensing*, Vol. 38, pp. 287–295.

Shackelford, A. K., and C. H. Davis. 2003. A Hierarchical Fuzzy Classification Approach for High-Resolution Multispectral Data over Urban Areas. *IEEE Transactions on Geoscience and Remote Sensing*, Vol. 41, pp. 1920–1932.

Wang, F. 1990a. Fuzzy Supervised Classification of Remote Sensing Images. *IEEE Transactions on Geoscience and Remote Sensing*, Vol. 28, pp. 194–201.

Wang, F. 1990b. Improving Remote Sensing Image Analysis through Fuzzy Information Representation. *Photogrammetric Engineering and Remote Sensing*, Vol. 56, pp. 1163–1169.

Artificial Neural Networks

Bischop, H., W. Schnider, and A. J. Pinz. 1992. Multispectral Classification of Landsat Images Using Neural Networks. *IEEE Transactions on Geoscience and Remote Sensing*, Vol. 30, pp. 482–490.

Chen, K. S., Y. C. Tzeng, C. F. Chen, and W. L. Kao. 1995. Land-Cover Classification of Multispectral Imagery Using a Dynamic Learning Neural Network. *Photogrammetric Engineering and Remote Sensing*, Vol. 61, pp. 403–408.

Miller, D. M., E. D. Kaminsky, and S. Rana. 1995. Neural Network Classification of Remote-Sensing Data. *Computers and Geosciences*, Vol. 21, pp. 377–386.

Combining Remotely Sensed Data with Ancillary Data

Davis, F. W., D. A. Quattrochi, M. K. Ridd, N. S.-N. Lam, S. J. Walsh, J. C. Michaelson, et al. 1991. Environmental Analysis Using Integrated GIS and Remotely Sensed Data: Some Research Needs and Priorities. *Photogrammetric Engineering and Remote Sensing*, Vol. 57, pp. 689–697.

Ehlers, M., G. Edwards, and Y. Bédard. 1989. Integration of Remote Sensing with Geographic Information Systems: A Necessary Evolution. *Photogrammetric Engineering and Remote Sensing*, Vol. 55, pp. 1619–1627.

Ehlers, M., D. Greenlee, T. Smith, and J. Star. 1992. Integration of Remote Sensing and GIS: Data and Data Access. *Photogrammetric Engineering and Remote Sensing*, Vol. 57, pp. 669–675.

Hutchinson, C. F. 1982. Techniques for Combining Landsat and Ancillary Data for Digital Classification Improvement. *Photogrammetric Engineering and Remote Sensing*, Vol. 48, pp. 123–130.

Lunnetta, R. S., R. G. Congalton, L. K. Fenstermaker, J. R. Jensen, K. C. McGwire, and L. R. Tinney. 1991. Remote Sensing and Geographic Information System Data Integration: Error Sources and Research Issues. *Photogrammetric Engineering and Remote Sensing*, Vol. 57, pp. 677–687.

Na, X., S. Zhang, X. Li, H. Yu, and C. Liu. 2010. Improved Land Cover Mapping Using Random Forests Combined with Landsat Thematic Mapper Imagery and Ancillary Geographic Data. *Photogrammetric Engineering and Remote Sensing*, Vol. 76, pp. 833–840.

Segmentation and Object-Cased Classification

Baatz, M., and A. Schape. 2000. Multi-Resolution Segmentation: An Optimization Approach of High-Quality Multi-Scale Image Segmentation (J. Strobul, T. Blaschke, and G. Griesebier, eds.). *Angewandte Geographische Informations verabietung*, Karlesruhe, pp. 12–23.

Barrile, V., and G. Bilotta. 2008. An Application of Remote Sensing: Object-Oriented Analysis of Satellite Data. International Archives of the Photogrammetry, Remote Sensing, and Spatial Information Sciences, Vol. 37, Part B8, Beijing, China.

Benz, U. C., P. Hoffman, G. Willhauk, I. Lingenfleder, and M. Heymen. 2004. Multi-Resolution, Object-Oriented Fuzzy Analysis of Remote Sensing Data for GIS-Ready Information. *ISPRS Journal of Photogrammetry and Remote Sensing*, Vol. 58, pp. 239–258.

Budreski, K. A., R. H. Wynne, J. O. Browder, and J. B. Campbell. 2007. Comparison of Segment and Pixel-Based Nonparametric Land Cover Classification in the Brazilian Amazon Using Multitemporal Landsat TM/ETM+ Imagery. *Photogrammetric Engineering and Remote Sensing*, Vol. 73, pp. 813–827.

Blaschke, T. 2010. Object-Based Image Analysis for Remote Sensing. *ISPRS Journal of Photogrammetry and Remote Sensing*, Vol. 65, pp. 2–16.

Burnett, C., and T. Blaschke. 2003. A Multi-Scale Segmentation/Object Relationship Modeling Methodology for Landscape Analysis. *Ecological Modelling*, Vol. 168, pp. 233–249.

Im, J., J. R. Jensen, and J. A. Tullis. 2008. Object-Based Change Detection using Correlation Image Analysis and Image Segmentation. *International Journal of Remote Sensing*, Vol. 29, pp. 399–423.

Forestry

Gislason, P. O., J. A. Benediktsson, and J. R. Sveinsson. 2006. Random Forests for Land Cover Classification. *Pattern Recognition Letters*, Vol. 27, pp. 294–300.

Makela, H., and A. Pekkarinen. 2004. Estimation of Forest Stand Volumes by Landsat TM Imagery and Stand-Level Field-Inventory Data. *Forest Ecology and Management*, Vol. 196, pp. 245–255.

Wynne, R. H., R. G. Oderwald, G. A. Reams, and J. A. Scrivani. 2000. Optical Remote Sensing for Forest Area Estimation. *Journal of Forestry*, Vol. 98, pp. 31–36.

Zimmermann, N. E., T. C. Edwards, G. G. Moisen, T. S. Frescino, and J. A. Blackard. 2007. Remote Sensing-Based Predictors Improve Distribution Models of Rare, Early Successional and Broadleaf Tree Species in Utah. *Journal of Applied Ecology*, Vol. 44, pp. 1057–1067.

Hybrid Classification

Bruzzone, L., M. M. Chi and M. Marconcini. 2006. A Novel Transductive SVM for Semisupervised Classification of Remote-Sensing Images. *IEEE Transactions on Geoscience and Remote Sensing*, Vol. 44, pp. 3363–3373.

Jacquin, A., L. Misakova, and M. Gay. 2008. A Hybrid Object-Based Classification Approach for Mapping Urban Sprawl in Periurban Environment. *Landscape and Urban Planning*, Vol. 84, pp. 152–165.

Kelly, M., D. Shaari, Q. H. Guo, and D. S. Liu. 2004. A Comparison of Standard and Hybrid Classifier Methods for Mapping Hardwood Mortality in Areas Affected by "Sudden Oak Death." *Photogrammetric Engineering and Remote Sensing*, Vol. 70, pp. 1229–1239.

Liu, W. G., S. Gopal, and C. E. Woodcock. 2004. Uncertainty and Confidence in Land Cover Classification Using a Hybrid Classifier Approach. *Photogrammetric Engineering and Remote Sensing*, Vol. 70, pp. 963–971.

Simpson, J. J., T. J. McIntire, and M. Sienko. 2000. An Improved Hybrid Clustering Algorithm for Natural Scenes. *IEEE Transactions on Geoscience and Remote Sensing*, Vol. 38, pp. 1016–1032.

KNN

Finley, A. O., R. E. McRoberts, and A. R. Ek. 2006. Applying an Efficient K-Nearest Neighbor Search to Forest Attribute Imputation. *Forest Science*, Vol. 52, pp. 130–135.

Holmstrom, H., and J. E. S. Fransson. 2003. Combining Remotely Sensed Optical and Radar Data in KNN-Estimation of Forest Variables. *Forest Science*, Vol. 49, pp. 409–418.

Liu, Y., S. T. Khu, and D. Savic. 2004. A Hybrid Optimization Method of Multi-Objective Genetic Algorithm (MOGA) and K-Nearest Neighbor (KNN) Classifier for Hydrological Model Calibration. *Intelligent Data Engineering and Automated Learning Ideal 2004. Proceedings*, Vol. 3177, pp. 546–551.

Thessler, S., S. Sesnie, Z. S. R. Bendana, K. Ruokolainen, E. Tomppo, and B. Finegan. 2008. Using K-NN and Discriminant Analyses to Classify Rain Forest Types in a Landsat TM Image over Northern Costa Rica. *Remote Sensing of Environment*, Vol. 112, pp. 2485–2494.

van Aardt, J. A. N., R. H. Wynne, and R. G. Oderwald. 2006. Forest Volume and Biomass Estimation Using Small-Footprint Lidar-Distributional Parameters on a Per-Segment Basis. *Forest Science*, Vol. 52, pp. 636–649.

Wynne, R. H., and D. B. Carter. 1997. Will Remote Sensing Live Up to its Promise for Forest Management? *Journal of Forestry*, Vol. 95, pp. 23–26.

CART

Brandtberg, T. 2002. Individual Tree-Based Species Classification in High Spatial Resolution Aerial Images of Forests Using Fuzzy Sets. *Fuzzy Sets and Systems*, Vol. 132, pp. 371–387.

Franklin, S. E., G. B. Stenhouse, M. J. Hansen, C. C. Popplewell, J. A. Dechka, and D. R. Peddle. 2001. An Integrated Decision Tree Approach (IDTA) to Mapping Landcover Using Satellite Remote Sensing in Support of Grizzly Bear Habitat Analysis in the Alberta Yellowhead Ecosystem. *Canadian Journal of Remote Sensing*, Vol. 27, pp. 579–592.

Lawrence, R., A. Bunn, S. Powell, and M. Zambon. 2004. Classification of Remotely Sensed Imagery Using Stochastic Gradient Boosting as a Refinement of Classification Tree Analysis. *Remote Sensing of Environment*, Vol. 90, pp. 331–336.

Lawrence, R. L., and A. Wright. 2001. Rule-Based Classification Systems Using Classification and Regression Tree Analysis. *Photogrammetric Engineering and Remote Sensing*, Vol. 67, pp. 1137–1142.

McIver, D. K., and M. A. Friedl. 2002. Using Prior Probabilities in Decision-Tree Classification of Remotely Sensed Data. *Remote Sensing of Environment*, Vol. 81, pp. 253–261.

Pal, M., and P. M. Mather. 2003. An Assessment of Decision Tree Methods for Land Cover Classification. *Remote Sensing of Environment*, Vol. 86, pp. 554–565.

Pantaleoni, E., R. H. Wynne, J. M. Galbraith, and J. B. Campbell. 2009. Mapping Wetlands Using ASTER Data: A Comparison between Classification Trees and Logistic Regression. *International Journal of Remote Sensing*, Vol. 30, pp. 3423–3440.

Xu, M., P. Watanachaturaporn, P. K. Varshney, and M. K. Arora. 2005. Decision Tree Regression for Soft Classification of Remote Sensing Data. *Remote Sensing of Environment*, Vol. 97, pp. 322–336.

Multitemporal

Townsend, P. A. 2001. Mapping Seasonal Flooding in Forested Wetlands Using Multi-Temporal Radarsat SAR. *Photogrammetric Engineering and Remote Sensing*, Vol. 67, pp. 857–864.

Iterative Guided Spectral Class Rejection

Musy, R. F., R. H. Wynne, C. E. Blinn, J. A. Scrivani, and R. E. McRoberts. 2006. Automated Forest Area Estimation Using Iterative Guided Spectral Class Rejection. *Photogrammetric Engineering and Remote Sensing*, Vol. 72, pp. 949–960.

Phillips, R. D., L. T. Watson, R. H. Wynne, and C. E. Blinn. 2009. Feature Reduction Using a Singular Value Decomposition for the Iterative Guided Spectral Class Rejection Hybrid Classifier. *ISPRS Journal of Photogrammetry and Remote Sensing*, Vol. 64, pp. 107–116.

Wayman, J. P., R. H. Wynne, J. A. Scrivani, and G. A. Burns. 2001. Landsat TM-Based Forest Area Estimation Using Iterative Guided Spectral Class Rejection. *Photogrammetric Engineering and Remote Sensing*, Vol. 67, pp. 1155–1166.

CHAPTER THIRTEEN

Field Data

13.1. Introduction

Every application of remote sensing must apply field observations in one form or another, if only implicitly. Each analyst must define relationships between image data and conditions at corresponding points on the ground. Although use of field data appears, at first glance, to be self-evident, there are numerous practical and conceptual difficulties to be encountered and solved in the course of even the most routine work (Steven, 1987).

Field data consist of observations collected at or near ground level in support of remote sensing analysis. Although it is a truism that characteristics of field data must be tailored to the specific study at hand, the inconvenience and expense of collecting accurate field data often lead analysts to apply field data collected for one purpose to other, perhaps different, purposes. Therefore, a key concern is not only to acquire field data but to acquire data suitable for the specific task at hand. Accurate field data permit the analyst to match points or areas on the imagery to corresponding regions on the ground surface and thereby to establish with confidence relationships between the image and conditions on the ground. In their simplest form, field data may simply permit an analyst to identify a specific region as *forest*. In more detailed circumstances, field data may identify a region as a specific form of forest, or specify height, volume, leaf area index, photosynthetic activity, or other properties according to specific purposes of a study.

13.2. Kinds of Field Data

Field data typically serve one of three purposes. First, they can be used to verify, to evaluate, or to assess the results of remote sensing investigations (Chapter 14). Second, they can provide reliable data to guide the analytical process, such as creating training fields to support supervised classification (Chapter 12). This distinction introduces the basic difference between the realms of qualitative (nominal labels) and quantitative analyses, discussed later in this chapter. Third, field data can provide information used to model the spectral behavior of specific landscape features (e.g., plants, soils, or water bodies). For example, analyses of the spectral properties of forest canopies can be based on quantitative models of the behavior of radiation within the canopy—models that require detailed and accurate field observations (Treitz et al., 1992).

Despite its significance, development of sound procedures for field data collection has not been discussed in a manner that permits others to easily grasp the principles that underlie them. Discussions of field data collection procedures typically include some combination of the following components:

- Data to record ground-level, or low-level, spectral data for specific features and estimates of atmospheric effects (e.g., Curtis and Goetz, 1994; Deering, 1989; Lintz et al., 1976).
- Broad guidelines for collection of data for use with specific kinds of imagery (e.g., Lintz et al., 1976).
- Discipline-specific guidelines for collection and organization of data (e.g., Congalton and Biging, 1992; Roy et al., 1991).
- Guidelines for collection of data for use with specific digital analyses (Foody, 1990; Joyce, 1978).
- Highlighting of special problems arising from methodologies developed in unfamiliar ecological settings (Campbell and Browder, 1995; Wikkramatileke, 1959).
- Evaluation of relative merits of quantitative biophysical data in relation to qualitative designations (Treitz et al., 1992).

The work by Joyce (1978) probably comes the closest to defining field data collection as a process to be integrated into a study's analytical plan. He outlined procedures for systematic collection of field data for digital classification of multispectral satellite data. Although his work focused on specific procedures for analysis of Landsat MSS data, he implicitly stressed the need for the data collection plan to be fully compatible with the resolution of the sensor, the spatial scale of the landscape, and the kinds of digital analyses to be employed. Further, his recognition of the need to prepare a specific document devoted to field data collection itself emphasized the significance of the topic.

Field data must include at least three kinds of information. The first consists of attributes or measurements that describe ground conditions at a specific place. Examples might include identification of a specific crop or land use or measurements of soil moisture content. Second, these observations must be linked to location and size (e.g., slope, aspect, and elevation) information so the attributes can be correctly matched to corresponding points in image data. And third, observations must also be described with respect to the time and date. These three elements—attributes, location, time—form the minimum information for useful field data. Complete field data also include other information, such as records of weather, illumination, identities of the persons who collected the data, calibration information for instruments, and other components as required for specific projects.

13.3. Nominal Data

Nominal labels consist of qualitative designations applied to regions delineated on imagery that convey basic differences with adjacent regions. Simple examples include *forest, urban land, turbid water*, and so on. Despite their apparent rudimentary character, accurate and precise nominal labels convey significant information. Those that are most precise—for example, *evergreen forest, single-family residential housing, maize, winter wheat*—convey more information than broader designations and therefore are more valuable.

Nominal labels originate from several alternative sources. One is an established classification system, such as that proposed by Anderson et al. (1976) (Chapter 20) and comparable systems established in other disciplines. These offer the advantage of acceptance by other workers, comparability with other studies, established definitions, and defined relationships between classes. In other instances, classes and labels have origins in local terminology or in circumstances that are specific to a given study. These may have the advantage of serving well the immediate purposes of the study but limit comparison with other studies. Collection of data in the field is expensive and inconvenient, so there are many incentives to use a set of field data for as many purposes as possible; this effect means that it is often difficult to anticipate the ultimate application of a specific set of data. Therefore, ad hoc classifications that are unique to a specific application often prove unsatisfactory if the scope of the study expands beyond its initial purpose. Nonetheless, analysts must work to address the immediate issues at hand without attempting to anticipate every potential use for their data. The objective should be to maintain flexibility rather than to cover every possibility.

In the field, nominal data are usually easy to collect at points or for small areas; difficulties arise as one attempts to apply the labeling system to larger areas. For these reasons, it is usually convenient to annotate maps or aerial photographs in the field as a means of relating isolated point observations to areal units. As the system is applied to broader regions, requiring the work of several observers, it becomes more important to train observers in the consistent application of the system and to evaluate the results to detect inconsistencies and errors at the earliest opportunity. Because physical characteristics of some classes vary seasonally, it is important to ensure that the timing of field observations match those of the imagery to be examined.

13.4. Documentation of Nominal Data

The purpose of field data is to permit reconstruction, in as much detail as possible, of ground and atmospheric conditions at the place and time that imagery was acquired. Therefore, field data require careful and complete documentation because they may be required later to answer questions that had not been anticipated at the time they were collected. Nominal data can be recorded as field sketches or as annotated aerial photographs or maps (Figure 13.1). The inherent variability of landscapes within nominal labels means that careful documentation of field data is an important component of the field data collection plan. Field observations should be collected using standardized procedures designed specifically for each project to ensure that uniform data are collected at each site (Figure 13.2). Reliable documentation includes careful notes, sketches, ground photographs, and even videotapes. Workers must keep a log to identify photographs in relation to field sites and to record dates, times, weather, orientations, and related information (Figure 13.3). Photographs must be cross-indexed to field sketches and notes. In some instances, small-format aerial photographs might be useful if feasible within the constraints of the project (see Chapter 3, Section 3.11).

13.5. Biophysical Data

Biophysical data consist of measurements of physical characteristics collected in the field describing, for example, the type, size, form, and spacing of plants that make up the

FIGURE 13.1. Record of field sites. The rectangular units represent land holdings within the study area in Brazil studied by Campbell and Browder (1995). Each colonist occupied an 0.5 km × 2.0 km lot (strip-numbered at the left and right). Within each colonist's lot, the irregular parcels outlined and numbered here show areas visited by field teams who prepared notes and photographs. See Figure 13.2. Copyright 1995 by the International Society for Remote Sensing. Reproduced by permission.

vegetative cover. Or, as another example, biophysical data might record the texture and mineralogy of the soil surface. The exact character of biophysical data collected for a specific study depends on the purposes of the study, but typical data might include such characteristics as leaf area index (LAI), biomass, soil texture, and soil moisture. Many such measurements vary over time, so they often must be coordinated with image acquisition; in any event, careful records must be made of time, date, location, and weather.

Biophysical data typically apply to points, so often they must be linked to areas by averaging of values from several observations within an area. Further, biophysical data must often be associated with nominal labels, so they do not replace nominal data but rather document the precise meaning of nominal labels. For example, biophysical data often document the biomass or structure of vegetation within a nominal class rather than completely replacing a nominal label. In their assessment of field data recording forest

FIGURE 13.2. Field sketch illustrating collection of data for nominal labels. The V-shaped symbols represent the fields of view of photographs documenting land cover within cadastral units such as those outlined in Figure 13.1, keyed to notes and maps not shown here. From Campbell and Browder (1995). Copyright 1995 by the International Society for Remote Sensing. Reproduced by permission.

GROUND DATA INVENTORY FORM

Kansas, Morton County

Data record 7 / 12 / and 7 / 15 / 96

Field #	Acreage	Land use Crop Code			Irrigated	Fertilized	Planting date Month/Day	
1	160.4	7	0	0	YES	NO		WELLSITE SOUTH SIDE
2	163.3	7	0	0		NO		
3	158.8	4	0	4	NO	NO	9-20-95	HARVESTED
4	77.7	4	0	2	NO	NO	9-7-95	GRAZED - STUBBLED PLOWED
5	81.9	7	0	0	YES	NO		WELLSITE EAST SIDE
6	19.6	7	0	0				
7	72.4	4	0	2	NO	NO	SEPT 95	HARVESTED
8	65.2	7	0	0				
9	153.7	4	0	2	NO	NO	9-16-95	HARVESTED - YIELD 14 BL? - SPRAYED
10	156.6	7	0	0				
11	154.6	7	0	0				WELLSITE IN SW 40 AC
12	152.8	4	0	0	NO	NO	SEPT 95	STUBBLE BEING PLOWED
13	156.8	7	0	0	NO	NO	SEPT 95	STUBBLE BEING PLOWED -WELLSITE E SIDE
14	73.0	4	0	2				
15	89.4	4	0	0	NO	YES	9-15-95	STUBBLE PLOWED
16	162.8	4	0	0	NO	NO	SEPT 95	GRAZED JAN-APRIL 96; RESIDUE PLOWED
17	120.0	5	0	0			SEPT 95	DESTROYED BY GREEN BUGS; RESIDUE PLOWED
18	35.3	7	0	0				WELLSITE NW CORNER
19	163.2	6	1	6			SEPT 95	HARVESTED - YIELD 16
20	94.8	7	0	0	YES	YES	5-28-96	SPRAYED IN JULY

FIGURE 13.3. Log recording field observations documenting agricultural land use.

sites, Congalton and Biging (1992) found that purely visual assessment provided accurate records of species present, but that plot and field measurements were required for accurate assessment of dominant-size classes.

13.6. Field Radiometry

Radiometric data permit the analyst to relate brightnesses recorded by an aerial sensor to corresponding brightnesses near the ground surface. A field spectroradiometer consists of a measuring unit with a handheld probe connected to the measuring unit by a fiber optic cable (Figure 13.4). The measuring unit consists of an array of photosensitive detectors, with filters or diffraction gratings to separate radiation into several spectral regions. Radiation received by the probe can therefore be separated into separate spectral regions and then projected onto detectors similar to those discussed in Chapter 4. Brightnesses can usually be presented either as radiances (absolute physical values) or reflectances (proportions).

Direct measurements can be recorded in physical units, although spectra are often reported as reflectances to facilitate comparisons with other data. Reflectance (relative brightness) can be determined with the aid of a reference target of known reflectance/ transmittance over the range of brightnesses examined. Such a target is essentially a pure white surface. Everyday materials, despite their visual appearance, do not provide uniform brightness over a range of nonvisible wavelengths, so reference panels must be prepared from materials with well-known reflectance properties, such as barium sulphate (BaSO₄), or proprietary materials, such as Spectralon®, that have been specifically manufactured to present uniform brightness over a wide range of wavelengths. If readings are made from such targets just before and just after readings are acquired in the field, then the field measurements can be linked to measurements of the reference target under the same illumination conditions. Usually, a reported field reading is derived from the average of a set of 10 or more spectroradiometer readings collected under specified conditions, to minimize the inevitable variations in illumination and field practice. Readings from a reflectance standard form the denominator in calculating the *relative reflectance* ratio:

REFLECTANCE PROBE

BATTERY
PACK

SPECTRORADIOMETER
INSTRUMENTATION
WITH NOTEBOOK
COMPUTER

FIGURE 13.4. Field radiometry.

$$\text{Relative reflectance ratio} = \frac{\text{Measured radiation, sample } (\lambda)}{\text{Measured brightness, reference source } (\lambda)} \qquad \text{(Eq. 13.1)}$$

where (λ) signifies a specific wavelength interval. Because reflectance is an inherent characteristic property of a feature (unlike radiance, which varies according to illumination and weather conditions), it permits comparisons of samples collected under different field conditions.

Some instruments have the ability to match their spectral sensitivities to specific sensors, such as SPOT HRV, MSS, or TM. The results can be displayed as spectral plots or as an array of data. Many units interface to notebook computers, which permit the analyst to record spectra in the field and write data to minidisks, which can then be transferred to office or laboratory computers for analysis or incorporation into databases. Typically, a field spectroradiometer fits within a portable case designed to withstand inclement weather and other rigors of use in the field.

Analysts can select from several designs for probes with differing fields of view or for immersion in liquids (e.g., to record radiation transmitted by water bodies). Sometimes measuring units are situated on truck-mounted booms that can be used to acquire overhead views that simulate the viewing perspective of airborne or satellite sensors (Figure 13.5).

Field spectroradiometers must be used carefully, with specific attention devoted to the field of view in relation to the features to be observed, background surfaces, direction of solar illumination, and the character of diffuse light from nearby objects (Figure 13.6). The field of view for the field unit is determined by the *foreoptic*, a removable attachment that permits the operator to carefully aim the field of view for the instrument. Foreoptics might have fields of view between 1° and 10°; some have pistol grips and sighting scopes. Diffuse light (Chapter 2) is influenced by surrounding features, so measurements should be acquired at sites well clear of features of contrasting radiometric properties. Likewise, the instrument may record indirect radiation reflected from the operator's clothing, so

FIGURE 13.5. Boom truck equipped for radiometric measurements in the field.

FIGURE 13.6. Careful use of the field radiometer with respect to nearby objects. Redrawn from Curtis and Goetz (1994).

special attention must be devoted to orientation of the probe with respect to surrounding objects.

The most useful measurements are those coordinated with simultaneous acquisition of aircraft or satellite data. Operators must carefully record the circumstances of each set of measurements to include time, date, weather, conditions of illumination, and location. Some instruments permit the analyst to enter such information through the laptop's keyboard as annotations to the radiometric data. As data are collected, the operator should supplement spectral data with photographs or videos to permit clear identification of the region represented by the radiometric observations.

In the field it is often difficult to visualize the relationship between ground features visible at a given location and their representations in multispectral data. Not only is it sometimes difficult to relate specific points in the field to the areas visible in overhead views, but often it is equally difficult to visualize differences between representations in visible and nonvisible regions of the spectrum and at the coarse resolutions of some sensors. Therefore, analysts must devote effort to matching point observations to the spectral classes that will be of interest during image analysis. Robinove (1981) and Richards and Kelley (1984) emphasize difficulties in defining uniform spectral classes.

13.7. Unmanned Airborne Vehicles

Unmanned airborne vehicles (UAVs) are remotely piloted light aircraft that can carry cameras or other sensors in support of remote sensing applications. Current UAVs extend design concepts originally developed by hobbyists for recreation and by military services for reconnaissance under difficult or dangerous conditions. Although the basic concept of designing small, remotely piloted aircraft has been pursued for decades, recent advances in miniaturization, communications, strengths of lightweight materials, and power supplies have recently permitted significant advances in UAV design. This discussion outlines applications of UAVs in support of civil remote sensing operations. There is a much larger body of knowledge and experience that pertains specifically to military UAVs. UAVs are powered either by small gasoline engines or by electric motors. Smaller UAVs typically have wingspans of perhaps of a meter or so, whereas wingspans of larger vehicles might reach several meters (Figure 13.7). Navigation is controlled by remote radio signals, usually given by an operator who can directly observe the UAV in flight or use remote television images to view the terrain observed by the UAV.

Recently, a variation of the UAV, known as the *micro aerial vehicle* (MAV), has been developed and equipped with GPS receivers and imaging systems (Winkler and Vörsmann, 2004). MAVs have the approximate dimensions of a small bird, with wingspans of about 15 or 16 in. (40 cm) and a weight of about 18 oz. (500 g). The GPS receivers are used to navigate the MAV remotely, while a miniaturized sensor relays a signal to a ground display. MAVs have been used for aerial reconnaissance, to monitor traffic flows on highways, and to conduct aerial assessments of crop health and vigor. Relative to UAVs, MAVs are easier to launch and control but are sensitive to weather conditions, especially wind.

Although it may seem incongruous to consider aerial imagery as constituting a form of field data, such imagery can provide detailed data that permits interpretation of other imagery collected at coarser levels of detail. UAVs can carry cameras, digital cameras, or video cameras, as required. Images can be recorded on film or disk for retrieval after the UAV has landed or can be transmitted by telemetry to a ground receiver.

Because UAVs can fly close to the surfaces they observe, they can collect ancillary data, such as temperature, CO_2 concentrations, humidity level, or other information related to the subject under investigation. Furthermore, because they can be deployed in the field at short notice, they have the ability to monitor sites more often than would be possible with other systems and to respond quickly to unexpected events. Their value resides in the opportunity they provide for investigators to closely coordinate UAV operations with those of broader scale imagery acquired by aircraft or satellite systems.

Despite their potential, there are still obstacles to the use of UAVs for routine remote sensing applications. For example, in rugged terrain or densely vegetated regions, suitable landing surfaces may not be within range of the areas to be observed. Piloting may be difficult, and experienced UAV pilots may not always be at hand. Accidents can destroy expensive equipment, perhaps in inaccessible terrain. Local weather conditions at or near the ground surface may limit flight operations.

UAVs are unlikely to replace more conventional remote sensing platforms, but they

FIGURE 13.7. Unmanned airborne vehicle (UAV).

do offer advantages as means of supplementing conventional field data, especially by providing data that can be coordinated with broad-scale imagery from aircraft and satellite platforms. Civilian applications of UAVs for remote sensing are still under development, principally for applications in agriculture, forestry, and related areas. A classic difficulty encountered in the collection of field data for such applications is ensuring that variable data can be collected simultaneously with image data. For example, measurements such as temperature, humidity, or illumination can vary over time and space so rapidly that the values vary as data are collected throughout a large region. Therefore, an investigator who desires to capture a snapshot of temperature conditions, for example, simultaneously with the acquisition of an aircraft or satellite image may find that the temperature values may be changing even as they are being collected. UAVs can assist with this kind of data collection task by facilitating the reliable acquisition of such data rapidly over rather large regions.

Aerial Videography

Aerial videography is the use of video images to record aerial imagery, typically using the usual commercial models of video cameras, supported by global positioning receivers and equipment that can record time codes on the imagery as it is collected. Aerial videography is usually acquired using a light aircraft at altitudes of 1800–2000 ft. above the terrain. Imagery can be recorded as videotape or in digital form.

Advantages include the ability to acquire aerial imagery quickly and inexpensively. It provides relatively fine spatial resolution and the opportunity to collect imagery at frequent intervals. An especially valuable capability is flexibility in scheduling—the ability to acquire imagery on short notice without the constraints and long lead times typical for other forms of aerial imagery. Aerial videography is often employed as part of a broader survey using other forms of imagery to provide additional detail, training data, or ground truth data for other systems (Wulder et al., 2005). Because it usually uses commercial models of common video cameras, equipment costs are inexpensive relative to alternatives.

The principal difficulty in using aerial videography lies in preserving the ability to identify precise locations and to match video images to other maps and imagery of the same region. The use of the light aircraft and commercial video equipment means that it is difficult to closely control the orientation of the camera, leading to difficulties in registering it to other imagery (Figure 13.8).

There is a range in sophistication, but equipment for aerial videography is basically simple and inexpensive. Often a system might include two digital video cameras, one used for wide angle and the other zoomed to provide fine detail within the wide-angle view. Or, alternatively, one camera can be oriented for a vertical view, the other for forward-looking oblique views. Some cameras can provide color infrared images. Depending on specifics, it may be necessary to use special mounts and other modifications to provide desired orientation for the cameras.

Because of the difficulty in matching the video imagery to maps and images, special efforts are devoted to recording that permit the analyst to match images to locations. This task can be accomplished using a device that generates a time code that permits the imagery to be matched to GPS signals and a record of the orientation of the camera. The real-time GPS receiver and a time-code generator send data to a computer and also send the time to the digital video audio track. A gyroscope can record the orientation of the

start

end

19:42:13.05	19:42:10.28	19:42:08.19

mixed deciduous forest   **residential land use**   **harvested agricultural land**

FIGURE 13.8. Aerial videography. A sample sequence of aerial videography. Top: strip of aerial video. The irregular edges reflect the difficulty of controlling the orientation of the video camera in a light aircraft. Bottom: enlargement of sections illustrating varied landscape features. From Conservation Management Institute.

aircraft and write it to the computer. The GPS time code is used to synchronize imagery from the two cameras and to match imagery with the locational information from the GPS. These data permit extrapolation of GPS positions to the area imaged on the ground, offering the ability to create mosaics.

Aerial videography missions are usually planned to achieve either of two different objectives. If complete coverage of a region is desired, imagery can be collected through parallel transects. This strategy may be used in situations in which it is not possible to acquire higher quality data or in situations in which high-quality imagery is available but successive images are required to record patterns that are changing rapidly, such as crops within a growing season or flooding, fires, and other natural disasters. Or, alternatively, videography can be planned as a series of transects that capture representative regions of interest to the analyst.

For analysis, images and GPS data are synchronized and imported into a GIS, where the video data can be superimposed over maps or images representing the same region but in coarser detail. The video data then provide a detailed view of selected regions within the lower resolution images.

13.8. Locational Information

Locational data permit attributes or measurements gathered in the field to be matched to imagery of the same region. Traditionally, many of the principal problems in collection of field data for remote sensing analysis have focused on the difficulty of acquiring locational information that permits field observations to be matched to corresponding points on imagery. Accurate observations have little use unless they can be confidently placed in their correct locations. Because of the difficulty and inaccuracies of deriving absolute locational information (e.g., latitude and longitude or comparable coordinates) in the field, most field data have been positioned with reference to distinctive landmarks (e.g., water bodies, road intersections, topographic features) visible on imagery. Such methods, of course, can introduce positional errors that negate the effort invested in accurate measurement of other qualities. In well-mapped regions of developed nations, it is often possible to acquire accurate maps to use in positioning of field data, but in unmapped or poorly mapped regions the absence of reliable positional information is a major difficulty for the use of field observations.

In some situations, locational information can be acquired by positioning easily recognized targets within the area to be imaged. If positions of these targets are known with precision, and if they are easily recognized on the imagery, then it is possible to derive an accurate network of GCPs within an image. For photogrammetric surveys, high-contrast (e.g., black figures against white backgrounds) cloth targets are often employed to mark positions of known locations. For radar imagery, portable corner reflectors can be positioned in the field to form a network of easily recognizable points within an image.

Global Positioning System

The availability of GPS technology has permitted convenient, inexpensive, and accurate measurement of absolute location. GPS has greatly enhanced the usefulness of remote sensing data, especially when it is necessary to integrate image data with field data. These instruments now are inexpensive and easy to use, and they can be employed in almost any area on the Earth's surface.

A *GPS receiver* consists of a portable receiving unit sensitive to signals transmitted by a network of Earth-orbiting satellites. These satellites are positioned in orbits such that each point on the Earth's surface will be in view of at least four, and perhaps as many as twelve, satellites at any given time. A system of 24 satellites is positioned at an altitude of about 13,500 mi., to circle the Earth at intervals of 12 hrs., spaced to provide complete coverage of the Earth's surface (Figure 13.9).

These satellites continuously broadcast signals at two carrier frequencies within the L-band region of the microwave spectrum (Chapter 7). Although at ground level these signals are very weak, they are designed so that they can be detected even under adverse

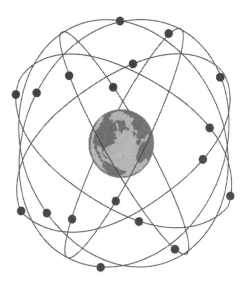

FIGURE 13.9. GPS satellites.

conditions (e.g., severe weather or interference from other signals). The frequency of each of these carrier signals is modulated in a manner that both (1) identifies the satellite that broadcasts the signal and (2) gives the exact time that the signal was broadcast. A receiver, therefore, can calculate the time delay for the signal to travel from a specific satellite and then accurately estimate the distance from the receiver to a specific satellite.

One reason that it is possible to employ such a weak signal is that the time and identification information each satellite transmits is very simple, and the receiver can listen for long periods to acquire it accurately. Because a receiver is always within range of multiple satellites, it is possible to combine positional information from three or more satellites to accurately estimate geographic position on the Earth's surface. A network of ground stations periodically recomputes and uploads new positional data to the GPS satellites.

GPS Receivers

A *GPS receiver* consists of an antenna, power supply, electronic clock, and circuitry that can translate signals into positional information. Handheld receivers, intended for recreational or leisure use, typically are about the size of a wireless phone. Survey-grade GPS units, designed for scientific or commercial use, are larger, use large antennae, and include connections to permit use of a keypad and a computer (e.g., the usual laptop or notebook models). Survey-grade units are usually designed to be carried in a small backpack (Figure 13.10). Smaller units achieve portability by conserving power or using less accurate clocks. More sophisticated GPS receivers use very accurate clocks and employ more substantial power supplies. Both handheld and survey-grade units can permit the analyst to enter coding information in the field and to download information to laboratory or office computers for use with GIS or image processing systems. Other GPS units have been specifically designed as miniaturized circuits (about the size of a credit card) to be used as components in digital cameras or related equipment. Images or data acquired with such instruments can be referenced directly to maps or images.

Pseudoranging

Pseudoranging is one of two ways that GPS signals can be used to estimate location. Pseudoranging is based on the use of time differences to estimate distances. GPS satellites

FIGURE 13.10. Field-portable survey-grade GPS receiver.

are equipped with accurate clocks. The system maintains its own time standard, with the capability to update times as signals are received from five ground stations. GPS receivers simultaneously generate codes that match codes generated by GPS satellites. As GPS receivers receive coded signals from satellites, they estimate the temporal displacement required to synchronize the two codes. The amount of displacement forms the basis for an estimate of distances between the receiver and each of the satellites within range. Pseudoranging is less accurate than other methods, but its simplicity, convenience, and cost effectiveness mean that it is widely used for determining positions of fixed points—the kind of information usually required for applications in remote sensing and GIS.

Carrier Phase Measurement

An alternative strategy for using GPS signals to estimate distance is known as *carrier phase measurement*, which is based on a detailed examination of signals broadcast by the satellites. GPS receivers detect either or both of the two L-band signals and add them to a signal generated by the receiver. This process is, in essence, the application of the Doppler principle (Chapter 7), using observed frequency shifts to derive positional information. If a receiver can acquire data from as many as four satellites, it is possible to estimate both vertical and horizontal positions. Survey-grade GPS units often are based on the use of carrier-phase measurement.

Precise Positioning Service

Because of the implications for national security of universal availability of quick, accurate locational data, some aspects of the satellite signal were encrypted. Exploitation of the system's full accuracy required access to a cryptographic key, provided only to authorized users. The system's most accurate data, known as *precise positioning service* (PPS), is provided to military services and to other governmental agencies with requirements for high levels of positional accuracy.

GPS Accuracy

GPS errors arise from a combination of several sources:

- Orbital position (ephemeris)
- Clock error
- Ionospheric effects (atmospheric effects encountered in the high atmosphere)
- Tropospheric effects (atmospheric effects in the lower atmosphere)
- Noise contributed by the GPS receiver itself
- Multipath error (reflection of GPS signals from nearby surfaces)

In addition, designers of the GPS system deliberately added errors to create *selective availability* (SA) to preserve the advantage of full precision of the GPS system for authorized users while allowing the civil community access to the degraded signal. SA introduced deliberate errors in both the ephemeris and the clock to significantly increase positional errors. Access to the signal's full accuracy was provided by a cryptographic key that allowed friendly military services and other authorized users to receive the undegraded signal.

In May 2000 the U.S. government announced the decision to turn off SA, allowing the civil GPS community to receive the undegraded GPS signal. This decision leaves iono-spheric effects as the largest source of positional error. This policy recognized the growing significance of accurate GPS signals for civil applications and the potential impact if foreign governments were to build competing systems using different technical specifications. The removal of SA opens the possibility that foreign systems would be compatible with the U.S. system, thereby increasing the accuracy of both. Conservative estimates of GPS accuracy suggest that horizontal errors should be within about 22 m 95% of the time, and vertical errors should be about 33 m 95% of the time (Shaw et al., 2000). However, observed accuracies suggest that performance is usually superior to these estimates, perhaps 10 m or less in some instances.

Differential GPS

When a GPS receiver can be stationed at a fixed position of known location, it becomes possible to derive estimates of some of the errors that influence GPS accuracy and then to apply these estimates to improve the accuracy of GPS positions at unknown locations. This process is known as *local differential GPS*. Differential GPS requires that a GPS receiver be established at a fixed point for which the geographic location is known with confidence. This location forms the *base station*. Information from the base station can be applied to locational information from GPS receivers to derive more accurate estimates of location. A carefully established differential network can permit estimates of location to within centimeters of the true position.

The basic concept of differential correction undergirds the wide area augmentation system (WAAS). Originally developed for civil aviation, WAAS-enabled GPS receivers are now common even among recreational users of the technology. Precisely surveyed wide area reference stations (WRS) enable detection of errors in the received GPS signals. These errors are forwarded to one of two WAAS master stations (WMS) where augmentation messages are generated and then uplinked to geostationary communications satellites. The augmentation messages are broadcast from space at 1575.42 MHz. The augmentation messages contain corrections for errors caused by clock drift and atmospherically induced signal delay. Horizontal and vertical positional accuracy must be 25 ft. or less at least 95% of the time to meet the WAAS specification. However, most users can expect horizontal accuracy to be less than 10 ft. 95% of the time—and horizontal accuracies of under 3 ft. are now increasingly common even with low-cost WAAS-enabled receivers.

Careful use of GPS receivers can provide locational information adequate for use in many if not most remote sensing studies, especially if it is possible to employ differential location. Atmospheric effects (Chapter 2) constitute one of the major sources of error in GPS measurements. Atmospheric interference can delay transmission of signals from the satellites, and these delays can be interpreted by the system as a difference in distance. Electrically charged particles in the ionosphere (30–300 mi., 48–483 km, above

FIGURE 13.11. Panoramic ground photograph. This photograph was prepared using specialized software to knit together a sequence of photographs that provide a 360° view. From Peter Sforza.

the Earth's surface) and unusually severe weather in the troposphere (ground level to about 7.5 mi, or 12 km) can combine to cause errors of as much as 1–5 m. Some of these errors cancel each other out or can be removed by filtering, whereas others remain to create errors in the estimated positions.

13.9. Using Locational Information

Locational information can serve several purposes. One of the most important is to assist in identification of GCPs (Chapter 11) that allow analysts to resample image data to provide accurate planimetric location and correctly match image detail to maps and other images. Accurate locational information also permits correct positioning of field data with image data so that field data match to the correct regions within the image.

GPS permits analysts to specify coordinates for distinctive landmarks clearly visible both on imagery and on the ground. Examples include highway intersections, water bodies, distinctive terrain features, and the like. A network of such points then provides a locational framework that permits specification of locations of other points that might not be so distinctive. In other instances, field teams may acquire data in remote regions, far from distinctive landmarks. Prior to the availability of GPS, these data could be positioned within imagery only approximately, using relative locations from distant landmarks as the basis for rather rough estimates. GPS provides the capability to specify positions of such observations precisely within the imagery, provided that the image itself can be positioned within a coordinate system. That is, even with the accurate information provided by the GPS, it is still important to devote care to establishing the link between the image data and the ground observations.

13.10. Ground Photography

Systematic ground photographs form one of the best means of recording field conditions. Usefulness of photographs depends on detailed notes that record the location, time, and conditions depicted by the photographs. Stereo photographs acquired using handheld cameras (Figure 3.29) can be especially effective. Figures 13.2 and 13.3 illustrate how careful records can key photographs to specific locations and record the circumstances depicted by the photographs. The starlike symbols in Figure 13.2 show sets of photographs designed to record 360° fields of view from a specific location of significance. If digital photographs are available, specialized software can match the edges to create a 360° panorama (Figure 13.11).

13.11. Geographic Sampling

Although ideally it would be useful to acquire complete information concerning a study area, practical constraints require that analysts sample the patterns they study. Even modest-sized images can show such large areas that complete inventories are not feasible,

FIGURE 13.11. *(cont.)*

and large areas of the study area may be inaccessible. As a result, field data must usually be collected for a sample of pixels from the image. Here the word *observation* signifies the selection of a specific cell or pixel; the word *sample* is used here to designate a set of observations that will be used to construct an error matrix.

Any effort to sample a map or any other spatial pattern must consider three separate decisions: (1) determination of the number of observations to be used, (2) choice of the sampling pattern to position observations within an image, and (3) the selection for a given sampling pattern of the spacing of observations.

Numbers of Observations

Clearly, determination of the number of observations to be collected is fundamental. The issue is much more complex than it might seem to be at first consideration, and scientists have employed different strategies to estimate the numbers of observations. However, it is possible to outline the main considerations:

1. The *number of observations* determines the confidence interval of an estimate of the accuracy of a classification. Thus, from elementary statistics, we know that a large sample size decreases the width of the confidence interval of our estimate of a statistic—that is, we can have greater confidence that our estimate is close to the actual value when we employ larger numbers of samples.

2. For most purposes, we must consider not only the observations for the entire image but also those that are allocated to each class, as it is necessary to have some minimum number of observations assigned to each class.

Sampling Pattern

Sampling pattern specifies the arrangement of observations. The most commonly used sampling patterns include the *simple random sampling* pattern, the *stratified random* pattern, the *systematic* pattern, and the *systematic stratified unaligned* pattern. A simple random sample (Figure 13.12) positions observations such that the choice of any specific location as the site for an observation is independent of the selection of any other location as an observation. Randomness ensures that all portions of a region are equally subject to selection for the sample, thereby yielding data that accurately represent the area examined and satisfying one of the fundamental requirements of inferential statistics.

Typically the *simple random* sample of a geographic region is defined by first dividing the region to be studied into a network of cells. Each row and column in the network is numbered, then a random number table is used to select values that, taken two at a time,

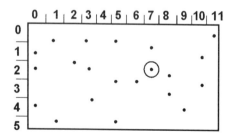

FIGURE 13.12. Simple random sampling pattern. Samples (represented by dots) are positioned at random within a study area by selecting pairs of values from a random sequence of values. These pairs then form the coordinates to [position the sample within the study area. The circled sample is positioned at location (row 2, column 7).

form coordinate pairs for defining the locations of observations. Because the coordinates are selected at random, the locations they define should be positioned at random.

The random sample is probably the most powerful sampling strategy available as it yields data that can be subjected to analysis using inferential statistics. However, in a geographic context, it has some shortcomings. Often a random sample is not uniform, but instead positions observations in patterns that tend to cluster together. Also, observations are unlikely to be proportionately distributed among the categories on the image, and classes of small areal extent may not be represented at all in the sample. In addition, observations may fall at or near the boundaries of parcels; in the context of accuracy assessment, such observations should probably be avoided, as they may record registration errors rather than classification errors. Thus the pure random sample, despite its advantages in other situations, may not be the best choice for comparing two images. It is, however, important to include an element of randomness in any sampling strategy, as described above.

A *stratified sampling pattern* assigns observations to subregions of the image to ensure that the sampling effort is distributed in a rational manner (Figure 13.13). For example, a stratified sampling effort plan might assign specific numbers of observations to each category on the map to be evaluated. This procedure would ensure that every category would be sampled. If a purely random pattern is used, on the other hand, many of the smaller categories might escape the sampling effort. Often the positions of observations within strata are selected at random—a process that yields the stratified random sampling pattern. Thus stratification allocates observations to categories (the strata, in this instance) in proportion to their size or significance; the random element ensures that observations are located within categories by chance, thereby ensuring a representative sample within each class.

Systematic sampling positions observations at equal intervals according to a specific strategy (Figure 13.14). Because selection of the starting point predetermines the positions of all subsequent observations, data derived from systematic samples will not meet the requirements of inferential statistics for randomly selected observations.

Another potential problem in the application of systematic samples is the presence of systematic variation in the underlying pattern that is to be sampled. For example, in the central United States, the Township–Range Survey System introduces a systematic component to the sizes, shapes, and positions of agricultural fields. Therefore, a purely systematic sample of errors in the classification of such landscape could easily place all

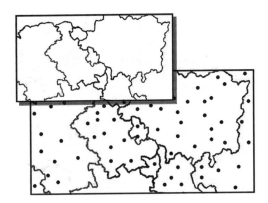

FIGURE 13.13. Stratified random sampling pattern. Here a region has been subdivided into subregions (*strata*) shown at the top. Samples (dots) are then allocated to each stratum in proportion to its expected significance to the study, then positioned randomly within each subarea.

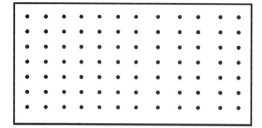

FIGURE 13.14. Systematic sampling pattern.

observations at the centers of fields, thereby missing errors that might occur systematically near the edges of fields and giving an inflated estimate of accuracy. Or systematic samples could oversample the edges and overestimate errors. There may be other, more subtle periodicities in landscapes that could interact with a systematic pattern to introduce other errors. Systematic sampling may nevertheless be useful if it is necessary to ensure that all regions within a study area are represented; a purely random sample, for example, may place observations in clustered patterns and leave some areas only sparsely sampled.

The *stratified systematic nonaligned sampling pattern* combines features of both systematic and stratified samples while simultaneously preserving an element of randomness (Figure 13.15). These features combine to form a sampling strategy that can be especially versatile for sampling geographic distributions.

The entire study area is divided into uniform cells, usually by means of a square grid. The grid cells, of course, introduce a systematic component to the sample and form the basis for stratification; one observation is placed in each cell. An element of randomness is contributed by the method of placement of observations within each cell. A number is randomly selected as the east–west coordinate for placement of the observation within the first cell; this value forms the east–west coordinate not only within the first cell but also for all cells in that column (Figure 13.15). A second number is randomly selected to

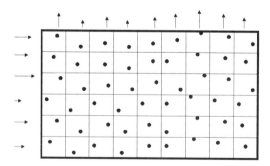

FIGURE 13.15. Stratified systematic nonaligned sampling pattern. Each stratum (in this application, a cell within the grid) includes a signal sample, represented by the dots. The position of the dot varies according to a value for each row that sets the distance of the sample from the left edge of each cell (represented by the length of each arrow on the left) and another independent value for each column that specifies the distance above the lower edge of each cell (represented by the length of each arrow at the upper edge of the grid). The values for each row and column are determined randomly.

define the north–south coordinate within the first cell; this value also specifies the north–south placement of observations within all cells in the first row of cells. Each row and each column is assigned a new random number that is applied to locating observations within each cell.

The resulting placement of observations within the study area has many favorable characteristics. The use of grid cells as the basis for stratification means that observations are distributed evenly within an image; no segment of the area will be unrepresented. The random element destroys the rigid alignment caused by a purely systematic design and introduces an element of chance that increases the probability that the observations will accurately represent the categories within the image. The primary pitfall is that the analyst must specify an appropriate size for the grid, which determines the number of observations and the spacing between them. These factors determine the effectiveness of a given sample in representing a specific pattern. If the analyst knows enough about the region to make a good choice of grid size, the stratified systematic nonaligned sample is likely to be among the most effective.

In some situations, the sampling effort must be confined to certain sites, perhaps because of difficult access to ground observation of remote areas or because financial restrictions confine the analyst's ability to apply the other sampling patterns mentioned above. *Cluster sampling* selects points within a study area, then uses each point as a center to determine locations of additional "satellite" observations placed nearby, so that the overall distribution of observations forms the clustered pattern illustrated in Figure 13.16. Centers can be located randomly or be assigned specifically to certain areas or classes on a reference map. Satellite observations can be located randomly, within constraints of distance. Cluster sampling may be efficient with respect to time and resources devoted to the sampling effort. If the pattern to be sampled is known beforehand, it may provide reasonably accurate results. However, if the distribution to be sampled is not known beforehand, it is probably best not to use a clustered pattern if it is at all possible to employ alternatives. Because observations are constrained to be close to one another, there is a distinct danger that the observations will replicate information rather than provide new information. If so, the analyst receives an inflated confidence in the assessment because the number of observations will suggest that more information is available than is actually the case.

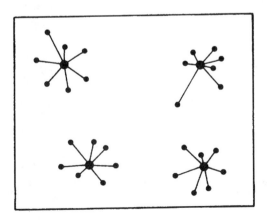

FIGURE 13.16. Clustered sampling pattern. Larger dots represent focal points for clusters selected randomly or possibly according to access or availability. Smaller dots represent satellite samples positioned at randomly selected distances and directions from the central point.

Spacing of Observations

All efforts to define geographic sampling patterns must consider the effect of spacing of observations. If observations are spaced too close together, they tend to replicate information already provided by their neighbors and thereby provide the analyst with an unjustified confidence in the reliability of his or her conclusions. Use of 100 observations, if closely spaced, may in fact convey the information of only 50 evenly spaced observations. Closely spaced samples may underestimate variability and provide a poor indication of the number of categories present.

Examination of only a few images should be sufficient to reveal the significance of sample spacing. Some landscapes have a coarse fabric of landscape parcels—for example, the large agricultural fields of the North American plains can be represented by observations spaced at rather large distances. Other landscapes, typically those of complex topography or densely occupied regions, display a finer, more intricate fabric and require more intense sampling patterns.

Spatial autocorrelation describes the tendency of measurements at one location to resemble those at nearby locations. Thus a measurement of soil moisture at one place is likely to resemble another measurement made 6 in. distant and less likely to resemble one made 6 ft. away. Spatial autocorrelation provides quantitative estimates of the rate at which similarities between paired observations change over distance. Spatial autocorrelation assesses the degree to which samples provide independent information.

Without prior knowledge of the pattern to be studied, it is impossible to decide whether a given sampling pattern positions observations too close to one another. However, in some instances we can apply the results of experience acquired in similar situations and avoid using clustered samples unless they are dictated by other considerations. Even a purely random sample will often place observations in clusters, so many investigators prefer the systematic stratified nonaligned pattern, which retains a random component but forces a degree of uniformity in spacing samples. Even when the most effective sampling patterns are used for accuracy assessment, it is usually necessary to augment the sample with observations selected at random from the less significant categories (in terms of area) as a means of ensuring that all categories are represented by an adequate number of independent observations.

Sampling Summary

It is difficult to recommend sampling strategies that might prove best for assessing accuracy because there has been little study of the comparative advantages of sampling procedures in the context of accuracy assessment. Congalton (1984) evaluated five sampling patterns to assess their performance in representing agricultural, range, and forested landscapes (as is shown in Chapter 14). For the agricultural pattern, he concluded that the random sample performed the best. He found that the stratified random sample was best for the rangeland scene and that the systematic random sample provided the most accurate information regarding the forest image. In general, he concluded that the systematic and systematic nonaligned samples performed well on the forest scene but poorly on other scenes.

His results highlight the difficulty in making broad recommendations for application of sampling strategies: The optimum sampling pattern is likely to vary according to the nature of the pattern to be sampled. In some applications, we can examine the pattern to be sampled, then select the sampling strategy likely to provide the best results. However, in many instances we are likely to face situations in which it is difficult or impossible to know beforehand the best sampling strategy, and we will be forced to develop experience on a case-by-case basis.

13.12. Summary

Full exploitation of field data can be considered one of the unexplored topics in the field of remote sensing. In recent years the availability of GPS, field spectroradiometers, and laptop computers (among other technological advances) has greatly expanded the nature and accuracy of field data. These developments have provided data that are vastly superior to those of previous eras with respect to quantity of accuracy by:

- Effective integration of field data within an image processing system
- Compatibility of varied forms of field data
- Coordination of field acquisition with image data
- Evaluation and assessment
- Questions of efficient and accurate sampling
- Access to remote regions

13.13. Some Teaching and Learning Resources

- Space: Global Positioning System (GSP) Constellation
  /www.youtube.com/watch?v=v_6yeGcpoyE&NR=1&feature=fvwp

- Launchpad: How GPS Works
  www.youtube.com/watch?v=DsmvTzw3GP4

- Galileo Satellites 3
  www.youtube.com/watch?v=JRLJOJnwl7o&feature=related

- Teaching Video 1
  www.youtube.com/watch?v=RUzlYJ58bxI&feature=related

- Jaz Spectrometers—The Ultimate in Field Analysis
  www.youtube.com/watch?v=pGwtAXfQfF8

- Geographic Information Specialists
  www.youtube.com/watch?v=cHBfQJ-Y3n0

- New German UAV—Microdrone
  www.youtube.com/watch?v=Y4jtguSF0n4&feature=related

Review Questions

1. Typically, one might expect that a thorough field data collection effort might take much longer than the time required to collect image data. During the interval between

the two, elements within the scene are likely to change, and certainly weather and illumination will change. How can field data compensate for these changes?

2. Many kinds of field observation necessarily record characteristics of *points*. Yet on remotely sensed data we normally examine *areas*. How can point data be related to areas? How can field observations be planned to accurately record characteristics of areas?

3. Sometimes field data are referred to as "ground truth." Why is this term inappropriate?

4. Prepare a rough estimate of the time and the number of people required to collect field data for a region near your school. Assume they will collect data to support a specific study in your discipline.

5. Prepare an itinerary and schedule for the field crew. Use local road maps or topographical maps to plan a route to observe field sites.

6. Refine your estimates for Question 4 and 5 to include costs of (a) equipment, (b) transportation, (c) wages, and (d) supplies and materials.

7. Further consider your analyses of the questions posed in 4, 5, and 6 to include tabulation and analysis of field data. Develop a plan that can permit you to match field data to image data. How can you be sure that the two match accurately? How long will it take to prepare the field observations in a form in which they can be used?

8. Devise also a sound sampling plan for field data collection in your discipline and your local area.

9. Some areas are clearly inaccessible, as a practical matter, for field data collection. How should these areas be treated in a field data collection plan?

10. Some features on the ground are highly variable over time and over distance. How can they be accurately observed for field data collection?

References

Field Data: General

Anderson, J. R., E. E. Hardy, J. T. Roach, and R. E. Witmer. 1976. *A Land Use and Land Cover Classification for Use with Remote Sensor Data* (U.S. Geological Survey Professional Paper 964). Washington, DC: U.S. Government Printing Office, 28 pp.

Bechtel, R., A. Sanchez-Azofeifa, and B. Rivard. 2004. Associations between Woodland Caribou Telemetry Data and Landsat TM Spectral Reflectance. *International Journal of Remote Sensing*, Vol. 25, pp. 4813–4827.

Campbell, J. B., and J. O. Browder. 1995. Field Data for Remote Sensing Analysis: SPOT Data, Rondonia, Brazil. *International Journal of Remote Sensing*, Vol. 16, pp. 333–350.

Congalton, R. G., and G. S. Biging. 1992. A Pilot Study Evaluating Ground Reference Data for Use in Forest Inventory. *Photogrammetric Engineering and Remote Sensing*. Vol. 58, pp. 1701–1703.

Foody, G. M. 1990. Directed Ground Survey for Improved Maximum Likelihood Classification of Remotely Sensed Data. *International Journal of Remote Sensing*, Vol. 11, pp. 1935–1940.

Heinsch, F. A., M. S. Zhao, S. W. Running, J. S. Kimball, R. R. Nemani, K. J. Davis, et al.

2006. Evaluation of Remote Sensing Based Terrestrial Productivity from MODIS using Regional Tower Eddy Flux Network Observations. *IEEE Transactions on Geoscience and Remote Sensing*, Vol. 44, pp. 1908–1925.

Herlyn, M. 2005. Quantitative Assessment of Intertidal Blue Mussel (*Mytilus edulis* L.) Stocks: Combined Methods of Remote Sensing, Field Investigation and Sampling. *Journal of Sea Research*, Vol. 53, pp. 243–253.

Joyce, A. T. 1978. *Procedures for Gathering Ground Truth Information for a Supervised Approach to Computer-Implemented Land Cover Classification of Landsat-Acquired Multispectral Scanner Data* (NASA Reference Publication 1015). Houston, TX: NASA, 43 pp.

Lintz, J., P. A. Brennan, and P. E. Chapman. 1976. Ground Truth and Mission Operations. Chapter 12 in *Remote Sensing of Environment* (J. Lintz and D. S. Simonett, eds.). Reading, MA: Addison-Wesley, pp. 412–437.

McCoy, R., M. 2005. *Field Methods in Remote Sensing*, New York: Guilford Press. 159 pp.

Peterson, D. L., M. A. Spencer, S. W. Running, and K. B. Teuber. 1987. Relationship of Thematic Mapper Simulator Data to Leaf Area of Temperate Coniferous Forest. *Remote Sensing of Environment*, Vol. 22, pp. 323–341.

Roy, P. S., B. K. Ranganath, P. G. Diwaker, T. P. S. Vophra, S. K. Bhan, I. S. Singh, et al. 1991. Tropical Forest Type Mapping and Monitoring Using Remote Sensing. *International Journal of Remote Sensing*, Vol. 11, pp. 2205–2225.

Sader, S. A., R. B. Waide, W. T. Lawrence, and A. T. Joyce. 1989. Tropical Forest Biomass and Successional Age Class Relationships to a Vegetation Index Derived from Landsat TM Data. *Remote Sensing of Environment*, Vol. 28, pp. 143–156.

Steven, M. D. 1987. Ground Truth: An Underview. *International Journal of Remote Sensing*, Vol. 8, pp. 1033–1038.

Treitz, P. M., P. J. Howarth, R. C. Suffling, and P. Smith. 1992. Application of Detailed Ground Information to Vegetation Mapping with High Resolution Digital Imagery. *Remote Sensing of Environment*, Vol. 42, pp. 65–82.

Wikkramatileke, R. 1959. Problems of Land-Use Mapping in the Tropics—An Example from Ceylon. *Geography*, Vol. 44, pp. 79–95.

Winkler, S., and P. Vörsmann, 2004. Bird's-Eye View: GPS and Micro Aerial Vehicles. *GPS World*, Vol. 15, No. 10, pp. 14–22.

Field Radiometry

Asner, G. P., J. M. O. Scurlock, and J. A. Hicke. 2003. Global Synthesis of Leaf Area Index Observations: Implications for Ecological and Remote Sensing Studies. *Global Ecology and Biogeography*, Vol. 12, pp. 191–205.

Curtis, B., and A. Goetz. 1994. Field Spectroscopy: Techniques and Instrumentation. In *Proceedings, International Symposium on Spectral Sensing Research (ISSSR '94)*. Alexandria, VA: U.S. Army Corps of Engineers Topographic Engineering Center, pp. 195–203.

Daughtry, C. S. T., C. L. Walthall, M. S. Kim, E. B. de Colstoun, and J. E. McMurtrey. 2000. Estimating Corn Leaf Chlorophyll Concentration from Leaf and Canopy Reflectance. *Remote Sensing of Environment*, Vol. 74, pp. 229–239.

Deering, D. W. 1989. Field Measurement of Bidirectional Reflectance. Chapter 2 in *Theory and Applications of Optical Remote Sensing* (G. Asor, ed.). New York: Wiley, pp. 14–65.

Dehaan, R. L., and G. R. Taylor. 2002. Field-Derived Spectra of Salinized soils and Vegetation as Indicators of Irrigation-Induced Soil Salinization. *Remote Sensing of Environment*, Vol. 80, pp. 406–417.

Gallego, F. J. 2004. Remote Sensing and Land Cover Area Estimation. *International Journal of Remote Sensing*, Vol. 25, pp. 3019–3047.

He, Y. H., X. L. Guo, and J. F. Wilmshurst. 2007. Comparison of Different Methods for Measuring Leaf Area Index in a Mixed Grassland. *Canadian Journal of Plant Science*, Vol. 87, pp. 803–813.

Liu, Y., Y. Zha, J. Gao, and S. Ni. 2004. Assessment of Grassland Degradation near Lake Qinghai, West China, using Landsat TM and in Situ Reflectance Spectra Data. *International Journal of Remote Sensing*, Vol. 25, pp. 4177–4189.

Pedros, R., J. A. Martinez-Lozano, M. P. Utrillas, J. L. Gomez-Amo, and F. Tena. 2003. Column-Integrated Aerosol Optical Properties from Ground-Based Spectroradiometer Measurements at Barrax (Spain) during the Digital Airborne Imaging Spectrometer Experiment (DAISEX) Campaigns. *Journal of Geophysical Research—Atmospheres*, Vol. 108, p. 4571.

van Aardt, J. A. N., and R. H. Wynne. 2001. Spectral Separability among Six Southern Tree Species. *Photogrammetric Engineering and Remote Sensing*, Vol. 67, pp. 1367–1375.

GPS

Berry, J. K. 2000. Capture "Where and When" on Video-Based GIS. *GEOWorld*, Vol. 14, No. 9, p. 26.

Divis, D. 2000. SA: Going the Way of the Dinosaur. *GPS World*, Vol. 11, No. 6, pp. 16–19.

Harrington, A. 2000. How Will Selective Availability Affect GPS/GIS Integration? *GEOWorld*, Vol. 13, No. 8, p. 28.

Hurn, J. 1989. *GPS: A Guide to the Next Utility*, Sunnyvale, CA: Trimble Navigation, 76 pp.

Rizos, C. 2010. Introducing the Global Positioning System. Chapter 7 in *Manual of Geospatial Sciences and Technology*, 2nd ed. (J. D. Bossler, ed.). London: Taylor & Francis, pp. 95–114.

Shaw, M. K. Sandhoo, and T. Turner. 2000. Modernization of the Global Positioning System. *GPS World*, Vol. 11, No. 9, pp. 36–40.

Spectral Classes

Richards, J. A., and D. J. Kelley. 1984. On the Concept of the Spectral Class. *International Journal of Remote Sensing*, Vol. 5, pp. 987–991.

Robinove, C. J. 1981. The Logic of Multispectral Classification and Mapping of Land. *Remote Sensing of Environment*, Vol. 11, pp. 231–244.

Sampling

Ayeni, O. O. 1982. Optimum Sampling for Digital Terrain Models: A Trend Towards Automation. *Photogrammetric Engineering and Remote Sensing*, Vol. 40, pp. 1687–1694.

Bellhouse, D. R. 1977. Some Optimal Designs for Sampling in Two Dimensions. *Biometrika*, Vol. 64, pp. 605–611.

Campbell, J. B. 1981. Spatial Correlation Effects upon Accuracy of Supervised Classification of Land Cover. *Photogrammetric Engineering and Remote Sensing*, Vol. 47, pp. 355–363.

Cochran, W. G. 1961. Comparison of Methods for Determining Stratum Boundaries. *Bulletin of the International Statistical Institute*, Vol. 58, pp. 345–358.

Congalton, R. G. 1984. *A Comparison of Five Sampling Schemes Used in Assessing the Accuracy of Land Cover/Land Use Maps Derived from Remotely Sensed Data.* PhD dissertation, Virginia Polytechnic Institute and State University, Blacksburg, 147 pp.

Congalton, R. G. 1988. A Comparison of Sampling Schemes Used in Generating Error Matrices for Assessing the Accuracy Maps Generated from Remotely Sensed Data. *Photogrammetric Engineering and Remote Sensing*, Vol. 54, pp. 593–600.

Franklin, S. E., D. P. Peddle, B. A. Wilson, and C. Blodget. 1991. Pixel Sampling of Remotely Sensed Digital Imagery. *Computers and Geosciences*, Vol. 17, pp. 759–775.

Ginevan, M. E. 1979. Testing Land-Use Map Accuracy: Another Look. *Photogrammetric Engineering and Remote Sensing*, Vol. 45, pp. 1371–1377.

Halme, M., and E. Tomppo. 2001. Improving the Accuracy of Multisource Forest Inventory Estimates by Reducing Plot Location Error: A Multicriteria Approach. *Remote Sensing of Environment*, Vol. 78, pp. 321–327.

Hay, A. 1979. Sampling Designs to Test Land Use Map Accuracy. *Photogrammetric Engineering and Remote Sensing*, Vol. 45, pp. 529–533.

Marignani, M., E. Del Vico, and S. Maccherini. 2007. Spatial Scale and Sampling Size Affect the Concordance between Remotely Sensed Information and Plant Community Discrimination in Restoration Monitoring. *Biodiversity and Conservation*, Vol. 16, pp. 3851–3861.

Oderwald, R. G., and R. H. Wynne. 2000. Field Applications for Statistical Data and Techniques. *Journal of Forestry*, Vol. 98, pp. 58–60.

Rosenfield, G. H., and K. Fitzpatrick-Lins. 1986. A Coefficient of Agreement as a Measure of Thematic Classification Accuracy. *Photogrammetric Engineering and Remote Sensing*, Vol. 52, pp. 223–227.

Rosenfield, G. H., K. Fitzpatrick-Lins, and H. S. Ling. 1982. Sampling for Thematic Map Accuracy Testing. *Photogrammetric Engineering and Remote Sensing*, Vol. 48, pp. 131–137.

Sawaya, K. E., L. G. Olmanson, N. J. Heinert, P. L. Brezonik, and M. E. Bauer. 2003. Extending Satellite Remote Sensing to Local Scales: Land and Water Resource Monitoring Using High-Resolution Imagery. *Remote Sensing of Environment*, Vol. 88, pp. 144–156.

Skidmore, A. K., and B. J. Turner. 1992. Map Accuracy Assessment Using Line Intersect Sampling. *Photogrammetric Engineering and Remote Sensing*, Vol. 58, pp. 1453–1457.

Stehman, S. V. 1992. Comparison of Systematic and Random Sampling for Estimating the Accuracy of Maps Generated from Remotely Sensed Data. *Photogrammetric Engineering and Remote Sensing*, Vol. 58, pp. 1343–1350.

Van der Wel, F. J. M., and L. L. F. Janssen. 1994. A Short Note on Pixel Sampling of Remotely Sensed Digital Imagery. *Computers and Geosciences*, Vol. 20, pp. 1263–1264.

Wulder, M., J. White, and S. McDonald. 2005, November. Truth in the Air. *GeoWorld*, pp. 28–31.

CHAPTER FOURTEEN

Accuracy Assessment

14.1. Definition and Significance

Prospective users of maps and data derived from remotely sensed images quite naturally ask about the accuracy of the information they will use. Yet questions concerning accuracy are surprisingly difficult to address in a convincing manner. This chapter describes how the accuracy of a thematic map can be evaluated and how two maps can be compared to determine whether they are statistically different from one another.

Accuracy and Precision

Accuracy defines "correctness"; it measures the agreement between a standard assumed to be correct and a classified image of unknown quality. If the image classification corresponds closely with the standard, it is said to be "accurate." There are several methods for measuring the degree of correspondence, all described in later sections.

Precision defines a measure of the sharpness (or certainty) of a measurement (Figure 14.1). In remote sensing, precision has two connotations: (1) categorical specificity of a thematic map and (2) the confidence interval within which estimates of map accuracy or area are likely to be contained. For continuous variables, such as aboveground biomass or Secchi depth (a measure of clarity of a waterbody, described in Chapter 19, Section 19.2) this latter connotation can also be expressed by the root mean square error or standard error of the estimator. In either case, however, calculation of explicit confidence intervals is beyond the scope of this text, but established techniques (e.g., Card, 1982) are in wide use. With respect to categorical specificity, one may be able to increase accuracy by decreasing precision—that is, by being vague in the classification. For example, as the analyst classifies a stand of trees as *forest, coniferous forest, pine forest, shortleaf pine forest*, or *mature shortleaf pine forest*, detail increases, and so too does the opportunity for error. (It is clearly more difficult to assign detailed classes correctly than to assign general classes.) Evaluation of accuracy seldom explicitly considers precision, yet we must always ask whether the precision is appropriate for the purpose at hand. Accuracy of 95% in separating *water* and *forest* is unlikely to be useful if we need to know the distributions of *evergreen* and *deciduous* categories.

In a statistical context, high accuracy means that *bias* is low (that estimated values are consistently close to an accepted reference value). High precision means that the variability of estimates (independent of their bias) is low (Figure 14.1). The usefulness of a map is related not only to its correctness but also to the precision with which the user

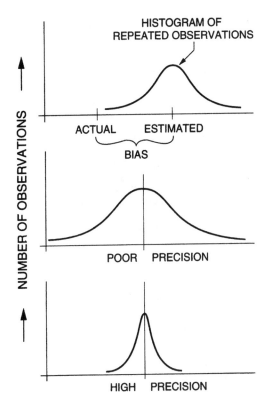

FIGURE 14.1. Bias and precision. Accuracy consists of bias and precision. Consistent differences between estimated values and true values create bias (top diagram). The lower diagram illustrates the concept of precision. High variability of estimates leads to poor precision; low variability of estimates creates high precision.

can make statements about specific points depicted on the map. A map that offers only general classes (even if correct) enables users to make only vague statements about any given point represented on the map; one that uses detailed classes permits the user to make more precise statements (Webster and Beckett, 1968).

Significance

Accuracy has many practical implications; for example, it affects the legal standing of maps and reports derived from remotely sensed data, the operational usefulness of such data for land management, and their validity as a basis for scientific research. Analyses of accuracies of alternative classification strategies have significance for everyday uses of remotely sensed data. There have, however, been few systematic investigations of relative accuracies of manual and machine classifications, of different individuals, of the same interpreter at different times, of alternative preprocessing and classification algorithms, or of accuracies associated with different images of the same area. As a result, accuracy studies would be valuable research in both practical and theoretical aspects of remote sensing.

Often people assess accuracy from the appearance of a map, from past experience, or from personal knowledge of the areas represented. These can all be misleading, as overall accuracy may be unrelated to the map's cosmetic qualities, and often personal experience may be unavoidably confined to a few unrepresentative sites. Instead, accuracy should be evaluated through a well-defined effort to assess the map in a manner that permits quantitative measure of accuracy and comparisons with alternative images of the same area.

Evaluation of accuracies of information derived from remotely sensed images has long been of interest, but a new concern regarding the accuracies of digital classifications has stimulated research on accuracy assessment. When digital classifications were first offered in the 1970s as replacements for more traditional products, many found the methods of machine classification to be abstract and removed from the direct control of the analyst; their validity could not be accepted without evidence. This concern prompted much of the research outlined in this chapter.

Users should not be expected to accept at face value the validity of any map, regardless of its origin or appearance, without supporting evidence. We shall see in this chapter how difficult it can be to compile the data necessary to support credible statements concerning map accuracy.

14.2. Sources of Classification Error

Errors are present in any classification. In manual interpretations, errors are caused by misidentification of parcels, excessive generalization, errors in registration, variations in detail of interpretation, and other factors. Perhaps the simplest causes of error are related to the misassignment of informational categories to spectral categories (Chapter 12). Bare granite in mountainous areas, for example, can be easily confused with the spectral response of concrete in urban areas. However, most errors are probably more complex. Mixed pixels occur as resolution elements fall on the boundaries between landscape parcels; these pixels may well have digital values unlike either of the two categories represented and may be misclassified even by the most robust and accurate classification procedures. Such errors may appear in digital classification products as chains of misclassified pixels that parallel the borders of large, homogeneous parcels (Figure 14.2).

In this manner the character of the landscape contributes to the potential for error through the complex patterns of parcels that form the scene. A very simple landscape composed of large, uniform, distinct categories is likely to be easier to classify accurately than one with small, heterogeneous, indistinct parcels arranged in a complex pattern. Key landscape variables are likely to include:

- Parcel size
- Variation in parcel size
- Parcel identities
- Number of categories
- Arrangement of categories
- Number of parcels per category
- Shapes of parcel
- Radiometric and spectral contrast with surrounding parcels

These variables change from one region to another (Podwysocki, 1976; Simonett and Coiner, 1971) and within a given region from season to season. As a result, errors present in a given image are not necessarily predictable from previous experience in other regions or on other dates.

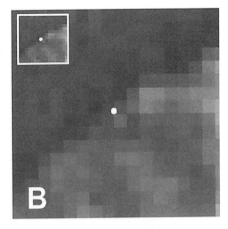

FIGURE 14.2. Incorrectly classified border pixels at the edges of parcels. Both examples, based on Musy et al. (2006), show validation points that were classified as nonforest but were labeled as forest in the reference data. The images are black-and-white infrared composites (Landsat TM bands 4, 3, and 2). In both cases, the forest lies at the edge of a water body.

14.3. Error Characteristics

Classification error is the assignment of a pixel belonging to one category (as determined by ground observation) to another category during the classification process. There are few if any systematic studies of geographic characteristics of these errors, but experience and logic suggest that errors are likely to possess at least some of the characteristics listed below:

- Errors are not distributed over the image at random but display a degree of systematic, ordered occurrence in space. Likewise, errors are not assigned at random to the various categories on the image but are likely to be preferentially associated with certain classes.
- Often erroneously assigned pixels are not spatially isolated but occur grouped in areas of varied size and shape (Campbell, 1981).
- Errors may have specific spatial relationships to the parcels to which they pertain; for example, they may tend to occur at the edges or in the interiors of parcels.

Figure 14.3 shows three error patterns from Landsat classifications reported by Congalton (1984). Each image shows an area corresponding to a USGS 7.5-minute topographic quadrangle; the distributions show errors in land-cover classifications derived from Landsat images. These three areas were specifically selected to represent contrasting landscapes of predominantly forested, agricultural, and range land use in the rural United States. Because accurate ground observations were available for each region, it was possible to compare the classification based on the Landsat data with the actual landscapes and to produce a complete inventory of errors for each region on a pixel-by-pixel basis. Dark areas represent incorrectly classified pixels; white areas show correct classification. (Here the scale of the images is so small that we can see only the broad outlines of the pattern without resolving individual pixels.)

(a)   (b)   (c)

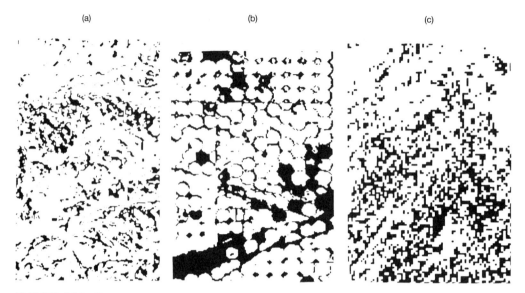

FIGURE 14.3. Error patterns. The dark areas are erroneously classified pixels for three areas as given by Congalton (1984): (a) Forested region. (b) Agricultural scene. (c) Rangeland. These images are shown at very small scale; at larger scale, it is possible to recognize the dark areas as individual pixels.

Although we cannot identify sources of these errors, we can clearly see the contrasting patterns. None of the images show random patterns—the misclassified pixels tend to form distinctive patterns. In the forest image, errors form crescent-shaped strips formed by shadowed terrain; in the agricultural scene, center-pivot irrigation systems create ring-shaped patches left by fallow land or land planted with contrasting crops that have been erroneously classified. On the rangeland image, we clearly see the effect of the systematic Township and Range Survey System, probably through the influence of mixed pixels at the edges of rectangular parcels.

In our own image classifications we have no knowledge of the corresponding error patterns because seldom do we have the opportunity to observe the total inventory of classification errors for areas as large as those shown in Congalton's study. Yet these patterns of error do exist, even though we can seldom know their abundance or distribution.

14.4. Measurement of Map Accuracy

The accuracy assessment task can be defined as one of comparing two sources of information, one based on analysis of remotely sensed data (the *map*) and another based on a different source of information, assumed to be accurate, the *reference data*. The reference data are of obvious significance; if they are in error, then the attempt to measure accuracy will be in error. For information that varies seasonally or over time, it is important that the reference data match with respect to time—otherwise, the comparison may detect differences that are not caused solely by inaccuracies in the classification. For example, some of the differences may not really be errors but simply changes that have occurred during the interval that elapsed between image acquisition and reference data acquisition. In some instances, we may examine two maps simply to decide whether there is a difference, without concluding that one is more accurate than the other. For example, we may compare maps of the same area made from data acquired by different sensors or using different classification protocols. More commonly in such cases, however, we will use reference data applicable to both classifications to determine whether one map is more accurate than the other.

To assess the accuracy of a map, it is necessary that the map and reference data be coregistered, that they both use the same classification system and minimum mapping unit, and that they have been classified at comparable levels of detail. The strategies described here are not appropriate if the two data sources differ with respect to detail, number of categories, or meanings of the categories.

In the rare case of suitable reference data formed by another map of the same area, the simplest method of evaluation is to compare the two maps with respect to the areas assigned to each category. The result of such a comparison is to report the areal proportions of categories (Figure 14.4). These values report the extent of the agreement between the two maps in respect to total areas in each category but do not take into account compensating errors in misclassification that cause this kind of accuracy measure to be itself inaccurate. For example, underestimation of *forest* in one part of the image can compensate for overestimation of *forest* in another part of the image; serious errors in classification have been made but are not revealed in the simple report of total areas in each category.

This form of error assessment is sometimes called *non-site-specific accuracy* because it does not consider agreement between the two maps at specific locations but only the overall figures for the two maps. Figure 14.4 illustrates this point. The two patterns are clearly different, but the difference is not revealed by the overall percentages for each category. Gersmehl and Napton (1982) refer to this kind of error as *inventory error*, as the process considers only the aggregate areas for classes rather than the placement of classes on the map.

Site-Specific Accuracy

The second form of accuracy, *site-specific accuracy*, is based on the detailed assessment of agreement between the map and reference data at specific locations (Figure 14.5).

 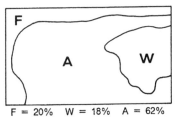

F = 20%   W = 18%   A = 62%   F = 20%   W = 18%   A = 62%

FIGURE 14.4. Non-site-specific accuracy. Here two images are compared only on the basis of total areas in each category. Because total areas may be similar even though placement of the boundaries differs greatly, this approach can give misleading results, as shown here.

FIGURE 14.5. Site-specific accuracy. Reference data (shown here as black or white crosses) with known locations are used to assess map accuracy.

Gersmehl and Napton (1982) refer to this kind of error as *classification error*. In many accuracy assessments of maps derived from moderate resolution imagery, the samples are pixels, but this strategy is impractical for very large (and usually mixed) or very small pixels (for which, among other issues, accurate registration between the mapped pixel and the reference data becomes more difficult). Therefore, instead of pixel-by-pixel comparisons, accuracy comparison often uses pixel groups (commonly a square 3 × 3 window) or polygons.

Once the sample unit has been chosen, the next two questions that immediately arise are (1) What sampling scheme should be used? and (2) How many samples should be collected? With regard to sampling scheme, the most critical issue is that it needs to have a random element. Remember that the goal of accuracy assessment is to estimate the accuracy of the image classification (the *population*, in this case) using a *sample* of reference data. With a large enough sample size, a *simple random sample* is best. However, as sample size decreases, categories that are less prevalent on the landscape can be missed using a simple random sample. For this reason stratification based on map categories (resulting in a *stratified random sample*) is often used, but this makes it difficult to sample for both training and validation concomitantly. Further, the stratification will be most appropriate for the map from which strata were developed, making the validation set less appropriate for other maps of the same area (such as when one is comparing maps made by different analysis and/or different classification decision rules). Combinations of systematic and random sampling are often employed, such as stratified systematic non-aligned sampling, to address both of these concerns (see additional discussion in Chapter 13). However, a simple random sample of sufficient size clearly affords the most robust inference of map accuracy.

So what is a "sufficient" sample size? Congalton and Green (2009) suggest acquiring 50 samples per category as a general guideline, with increases to 75–100 samples per category as either the area or the number of categories increases beyond certain thresholds (1 million acres and/or 12 categories). However, they also note that to be certain the sample size should be calculated based on the multinomial distribution and a desired precision for the classification (usually 5%). As shown by the solid line in Figure 14.6, this results in a sample that is larger than what would result by taking 50 samples per class (the dotted line) with smaller numbers of classes. However, as the number of classes continues to increase, this "worst case" scenario actually results in fewer necessary samples.

Foody (2009) notes that one of the principal objectives of accuracy assessment is typically comparison, whether it be among classification results or against a reference standard. Statistically valid comparisons must have a sample size appropriate to the type of test, effect size (which Foody defines as the "minimum meaningful difference in accuracy"), power (see discussion later in the chapter), significance level, and confidence limits of the comparison. In many studies there are too few samples to permit statistically robust comparisons.

The Error Matrix

The standard form for reporting site-specific error is the *error matrix*, sometimes referred to as the *confusion matrix* because it identifies not only overall errors for each category but also misclassifications (due to confusion between categories) by category. Compilation of an error matrix is required for any serious study of accuracy. It consists of an $n \times n$ array, where n represents the number of categories (Table 14.1).

The left-hand side (y axis) is labeled with the categories on the map; the upper edge (x axis) is labeled with the same n categories from the reference data. (Note that the meanings of the two axes can be reversed in some applications, as the convention is not universal.) Here values in the matrix represent numbers of pixels for which the analyst has been able to compare the evaluated and reference images. Sometimes the matrix is constructed using percentages rather than absolute values, but here we show absolute values counted directly from the image. Even when percentages are used, it is helpful to also show the sample numbers, especially as they tend to highlight matrix elements with few samples.

Inspection of the matrix shows how the classification represents actual areas on the

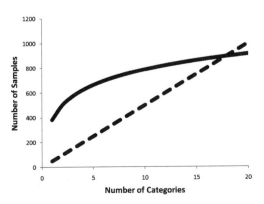

FIGURE 14.6. Accuracy assessment sample size. The solid line shows a conservative "worst case" sample size (Congalton and Green, 2009) that results in a sample that is larger than what would result by taking 50 samples per class (the dotted line) with smaller numbers of informational classes (categories). The "worst case" sample size was calculated using the chi-square distribution with one degree of freedom, a significance level (α) of 5%, an assumed class proportion of 50%, and a 5% desired classification precision.

TABLE 14.1. Example of an Error Matrix

	Reference data						
Classification data	Row crops	Early succession	Pasture or hay	Coniferous forest	Deciduous forest	Open water	Row total
Row crops	46	3	11	0	0	0	60
Early succession	5	58	4	5	4	0	76
Pasture or say	7	1	55	0	0	0	63
Coniferous forest	0	1	0	64	3	0	68
Deciduous forest	0	3	0	9	1	0	13
Open water	0	0	0	1	1	10	12
Column total	58	66	70	79	9	10	292

landscape. For example, in Table 14.1, 58 reference samples were in *row crops*. Of these 58 points, 46 were classified as such (row 1, column 1); these, of course, are the pixels of *row crops* correctly classified. Reading succeeding values down the first column, we next find incorrectly classified *row crop* pixels and the categories to which they were assigned: *early succession*, 5; *pasture or hay*, 7; *coniferous forest*, 0; *deciduous forest*, 0, *open water*, 0. Reading across each row, we see what pixels mapped as a given category actually were when field-verified. The diagonal from upper left to lower right gives numbers of correctly classified pixels for each class. Note that some off-diagonal elements of the matrix have no samples; this is normal, particularly for maps with larger number of classes.

Further inspection of the matrix reveals a summary of other information. Row totals give the total number of pixels in each class as recorded on the reference image. Column totals show the numbers of pixels assigned to each class on the thematic map.

Compiling the Error Matrix

To construct the error matrix, the analyst must compare two sources of data—the reference samples and the classified image—on a point-by-point basis to determine exactly how each of the validation samples is represented in the classification. It is very important that the validation samples and the map be well coregistered to one another. Errors in registration will appear as errors in classification, so registration problems will create errors in the assessment of accuracy. This is also true with respect to timing; the validation data and the image should be acquired as close in time as possible.

The categories for the map and reference data are compared systematically for each sample. The analyst or software maintains a count of the numbers in each reference category as they are assigned to classes on the map. A summation of this tabulation becomes the error matrix.

Let's walk through the whole process using a very simple dataset. A Landsat image has been classified, using two different techniques (labeled as *Map 1* and *Map 2*), into *forest* and *nonforest*. Twenty samples, representing Landsat pixels at ground resolution, have been randomly chosen (less than the 100 that our rule of thumb would tell us are required) and categorized using the same classification scheme and minimum mapping unit using a combination of aerial photography and field visits at or near the time of image acquisition. The resulting table is Table 14.2, where 1 = *nonforest* and 2 = *forest*.

The resulting error matrix (for *Map 1*) is shown as Table 14.3.

TABLE 14.2. Accuracy Assessment Data

Point number	Map 1	Map 2	Reference
1	1	1	1
2	1	2	1
3	1	1	1
4	1	1	2
5	1	1	1
6	1	2	1
7	1	1	1
8	2	2	2
9	2	1	1
10	2	2	2
11	2	2	1
12	2	2	1
13	2	2	2
14	2	1	2
15	2	1	1
16	2	2	2
17	2	2	2
18	2	2	2
19	2	2	2
20	2	2	2

Errors of Omission and Errors of Commission

Examination of the error matrix reveals, for each category, *errors of omission* and *errors of commission*. Errors of omission are, for example, the assignment of pixels of *forest* on the ground to *nonforest* on the map (in other words, an area of "real" forest on the ground has been omitted from the map). Using the same example, an error of commission would be to assign an area of *nonforest* on the ground to the *forest* category on the map. The analyst in this instance has actively committed an error by assigning a region of forest to a wrong category. (This error of commission for one category will, of course, also be tabulated as an error of omission for another category.) The distinction is essential, because if every pixel in the map were labeled *forest*, then no forest on the ground would be omitted from the map. However, many forest pixels would be improperly committed to *nonforest*.

By examining relationships between the two kinds of errors, the map user gains insight about the varied reliabilities of classes on the map, and the analyst learns about the performance of the process that generated the maps. Examined from the user's perspective, the matrix reveals *user's accuracy* (or *consumer's accuracy*); examined from the analyst's point of view, the matrix reveals *producer's accuracy*. (The omission error

TABLE 14.3. Error Matrix Derived from Map 1 in Table 14.2

Map	Reference		Row total
	NF	F	
NF	6	1	7
F	4	9	13
Column total	10	10	20

is 1 – producer's accuracy; the commission error is 1 – user's accuracy.) The difference between the two lies in the base from which the error is assessed. For producer's accuracy, the base is the area in each class on the final map. Thus, for the example in Table 14.3, producer's accuracy for *forest* is 9/10, or 90%. For the same class, user's accuracy is 9/13, or 69%. User's accuracy forms a guide to the reliability of the map as a predictive device—it tells the user of the map that in this example, of the area labeled *forest* on the map, 69% actually corresponds to *forest* on the ground. Producer's accuracy informs the analyst who prepared the classification that, of the actual forested landscape, 90% was correctly classified. In both instances the error matrix permits identification of the classes erroneously labeled *forest* and forested areas mislabeled as *nonforest*.

14.5. Interpretation of the Error Matrix

Tables 14.1 and 14.3 show examples of error matrices. The column of sums on the right-hand edge (*row totals*) of the matrix gives total numbers of pixels in each category on the map; the row of sums at the bottom shows total pixels in each category in the classified scene. The sequence of values that extends from the upper left corner to the lower right corner is referred to as the *main diagonal* (here we are not interested in the opposite, or "anti" diagonal). These main diagonal entries show the number of correctly classified pixels: *row crops* classified as *row crops*, *open water* as *open water*, and so on. Nondiagonal values in each column give errors of omission. For example, as we read down the third column (Table 14.1), we see that 11 of 70 reference samples in *pasture or hay* were mapped as *row crops*. This is an "error of omission" because the image analyst has erred by omitting some *pasture or hay* from the interpreted image. More specifically, we know that *pasture or hay* is most likely to be misclassified as *row crops* on the classified image.

In contrast, nondiagonal values along the rows give errors of commission. To continue the example with the *pasture or hay* category, errors of commission are caused by active misassignment of other categories to the *pasture or hay* class. These errors are found by reading across the third row (Table 14.1): *row crops*, 7; *early succession*, 1; *coniferous forest*, 0; *deciduous forest*, 0; and *open water*, 0. These values reveal that the classification of *pasture or hay* erred most often by assigning true *pasture or hay* to the *row crops* class. Here errors of omission and commission both reveal that *pasture or hay* is most often confused with *row crops* and that the map is likely to confuse these two categories. Table 14.4 summarizes the producer's and user's accuracies, and resulting errors of omission and commission, for the matrix in Table 14.3. Note that even though *forest* was well classified from the producer's standpoint, this was at the expense of incorrectly labeling many *nonforest* pixels as *forest*.

TABLE 14.4. User's and Producer's Accuracies from Table 14.3

	User's accuracy	Commission error	Producer's accuracy	Omission error
Nonforest	6 ÷ 7 = 85.7%	1 –85.7% = 14.3%	6 ÷ 10 = 60.0%	1 – 60% = 40.0%
Forest	9 ÷ 13 = 69.2%	1 –69.2% = 30.8%	9/10 90.0%	1 – 90% = 10.0%

Overall Accuracy

One of the most widely used measures of accuracy is the *overall accuracy*, a report of the overall proportion of correctly classified pixels in the image or in the sample used to construct the matrix. The overall accuracy, of course, can be easily found; the number correct is the sum of the diagonal entries (also called the *trace*). Dividing this value by the total number of pixels examined gives the proportion of samples that have been correctly classified. For Table 14.1, this would be

$$\frac{46 + 58 + 55 + 64 + 1 + 10}{292} = \frac{234}{292} = 80.1\% \qquad \text{(Eq. 14.1)}$$

For Table 14.2 the overall accuracy is

$$\frac{6 + 9}{20} = \frac{15}{20} = 75.0\% \qquad \text{(Eq. 14.2)}$$

This value estimates the unknowable "true" value; the closeness of this value to the true value depends, in part, on the sampling strategy, as explained below. The overall accuracy can be reported with a confidence interval (Hord and Brooner, 1976).

Often the overall accuracy is used alone, without the error matrix, as a simple measure of accuracy. By itself, the overall accuracy may suggest the relative effectiveness of a classification, but in the absence of an opportunity to examine the full error matrix, it cannot provide convincing evidence of the classification's accuracy. A full evaluation must consider the categories used in the classification. For example, it would be easy to achieve high values of overall accuracy by classifying a scene composed chiefly of open water—a class that is easy to classify correctly. Furthermore, variations in the accuracies of specific classes should be noted, as should the precision of the classes. A classification that uses only broadly defined classes can achieve high accuracies but will not be useful for someone who requires more detail. Finally, later sections of this chapter will show that raw values for the overall accuracy can be inflated by chance assignment of pixels to classes—yet another reason for careful inspection of the assessment task.

Hay (1979) stated that it is necessary to consider five questions to thoroughly understand the accuracy of a classification:

- What proportion of the classification decision is correct?
- What proportion of assignments to a given category is correct?
- What proportion of a given category is correctly classified?
- Is a given category overestimated or underestimated?
- Are errors randomly distributed?

"Overall accuracy" can answer only the first of these questions; the others can be answered only by examination of the full error matrix.

Quantitative Assessment of the Error Matrix

After an initial inspection of the error matrix reveals the overall nature of the errors present, there is often a need for a more objective assessment of the classification. For example, we may ask if the two maps are in agreement—a question that is very difficult to answer because the notion of "agreement" may be difficult to define and implement. The error matrix is an example of a more general class of matrices, known as *contingency tables*, which summarize classifications analogous to those considered here. Some of the procedures that have been developed for analyzing contingency tables can be applied to examination of the error matrix.

Chrisman (1980), Congalton and Mead (1983), and Congalton et al. (1983) propose application of techniques described by Bishop et al. (1975) and Cohen (1960) as a means of improving interpretation of the error matrix. A shortcoming of usual interpretations of the error matrix is that even chance assignments of pixels to classes can result in surprisingly good results, as measured by overall accuracy. Hord and Brooner (1976) and others have noted that the use of such measures is highly dependent on the samples, and therefore on the sampling strategy used to derive the observations used in the analysis.

κ (kappa) is a measure of the difference between the observed agreement between two maps (as reported by the diagonal entries in the error matrix) and the agreement that might be attained solely by chance matching of the two maps. Not all agreement can be attributed to the success of the classification. κ attempts to provide a measure of agreement that is adjusted for chance agreement. κ is estimated by $\hat{\kappa}$ ("k hat"):

$$\hat{\kappa} = \frac{\text{Observed} - \text{expected}}{1 - \text{expected}} \qquad \text{(Eq. 14.3)}$$

This form of the equation, given by Chrisman (1980) and others, is a simplified version of the more complete form given by Bishop et al. (1975). Here *observed* designates the accuracy reported in the error matrix, and *expected* designates the correct classification that can be anticipated by change agreement between the two images.

Observed is the value for "overall accuracy" defined previously: the sum of the diagonal entries divided by the total number of samples. *Expected* is an estimate of the contribution of chance agreement to the observed overall accuracy. Expected values are calculated using the row and column totals. Products of row and column totals (Table 14.5) estimate the numbers of pixels assigned to each cell in the confusion matrix, given that pixels are assigned by chance to each category. This is perhaps clearer if we consider the situation in its spatial context (Figure 14.7). The row and column totals can be seen

TABLE 14.5. Calculating Products of Row and Column Totals Using Table 14.3

	Nonforest column total	Forest column total
Nonforest row total	10 × 7 = 70	10 × 7 = 70
Forest row total	10 × 13 = 130	10 × 13 = 130

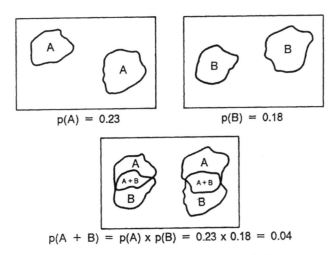

$$p(A + B) = p(A) \times p(B) = 0.23 \times 0.18 = 0.04$$

FIGURE 14.7. Contrived example illustrating computation of chance agreement of two categories when two images are superimposed. Here $p(A)$ and $p(B)$ would correspond to row and column marginals; their product estimates the agreement expected by chance matching of the two areas.

to represent areas occupied by each of the categories on the reference map and the image, and the chance mapping of any two categories is the product of their proportional extents on the two maps.

The *expected* correct, based on chance mapping, is found by summing the diagonals in the (row total × column total) product matrix, then dividing by the grand total of products of row and column totals, for example,

$$\frac{\text{Trace of product matrix}}{\text{Cumulative sum of product matrix}} = \frac{70 + 130}{70 + 70 + 130 + 130} = \frac{200}{400} = 50.0\% \quad \text{(Eq. 14.4)}$$

This estimate for *expected* correct can be compared with the overall accuracy or observed value (75%) calculated in Equation 14.2. $\hat{\kappa}$ for the error matrix shown in Table 14.2 is then calculated as follows:

$$\hat{\kappa} = \frac{\text{Observed} - \text{expected}}{1 - \text{expected}} = \frac{75\% - 50\%}{100\% - 50\%} = \frac{25\%}{50\%} = 50.0\% \quad \text{(Eq. 14.5)}$$

$\hat{\kappa}$ in effect adjusts the *overall accuracy* measure by subtracting the estimated contribution of chance agreement. Thus $\hat{\kappa} = 0.50$ can be interpreted to mean that the classification achieved an accuracy that is 50% better than would be expected from chance assignment of pixels to categories. As the overall accuracy approaches 100, and as the contribution of chance agreement approaches 0, the value of $\hat{\kappa}$ approaches positive 1.0 (100%), indicating the perfect effectiveness of the classification (Table 14.6). On the other hand, as the effect of chance agreement increases and the overall accuracy decreases, $\hat{\kappa}$ can assume negative values.

TABLE 14.6. Contrived Matrices Illustrating $\hat{\kappa}$

10				
	10			
		10		
			10	
				10

$\hat{\kappa} = +1.00$

10				
	10			
			10	
		10		
				10

$\hat{\kappa} = 0.50$

				25
25				

$\hat{\kappa} = -1.00$

				10
			10	
		10		
	10			
10				

$\hat{\kappa} = 0.00$

2	2	2	2	2
2	2	2	2	2
2	2	2	2	2
2	2	2	2	2
2	2	2	2	2

$\hat{\kappa} = 0.00$

				10
			10	10
10				
10				

$\hat{\kappa} = -0.25$

$\hat{\kappa}$ and related coefficients have been discussed by Rosenfield and Fitzpatrick-Lins (1986), Foody (1992), and Ma and Redmond (1995). According to Cohen (1960), any negative value indicates a poor classification, but the possible range of negative values depends on the specific matrix evaluated. Thus the magnitude of a negative value should not be closely interpreted as an indication of the performance of the classification. Values near zero suggest that the contribution of chance is equal to the effect of the classification and that the classification process yields no better results than would a chance assignment of pixels to classes. $\hat{\kappa}$ is asymptotically normally distributed.

One of the substantial benefits of calculating kappa is that the large sample variance can be estimated, then used in a z test to determine whether individual $\hat{\kappa}$ scores differ significantly from one another. Students should read the article by Hudson and Ramm (1987) as they consider using these methods, especially as the variance equation was incorrect in the original (Cohen, 1960) paper (Foody, 2009). Unfortunately, even if one uses the appropriate method for sample variance calculation, comparing two classifications using $\hat{\kappa}$ and its variance $\hat{\sigma}_\kappa^2$ for each requires sample independence. The McNemar test (Foody, 2004; Foody, 2009) is extensively used as a nonparametric alternative for comparing thematic maps using the same validation dataset. The test requires a particular "cross-tabulation" of the thematic maps, as shown in Table 14.7.

The McNemar calculation of z (the *standard score*) is then given as Equation 14.6:

$$z = \frac{b-c}{\sqrt{b+c}} \qquad \text{(Eq. 14.6)}$$

Let's look at an example using the data in Table 14.2, repeated as Table 14.8 with an additional column added to denote the appropriate matrix element (denoted by letter a, b, c, or d) in Table 14.7 (again with the caveat that the sample size is far too small). Let's look at the first and second rows as examples. In the first row, both maps are correct (both they and the validation point were placed in category 1, *nonforest*). Looking at Table 14.7, if both maps are correct, then the appropriate matrix element is a, so we place an a in the *Letter* column. In the second row, the first map is correct, and the second map is not. In Table 14.7, this would make it matrix element c.

The McNemar statistic is now easily calculated using equation 14.6 and the number of b's (2) and c's (3) we found in constructing Table 14.8, to wit:

$$z = \frac{b-c}{\sqrt{b+c}} = \frac{2-3}{\sqrt{2-3}} = -0.45 \qquad \text{(Eq. 14.7)}$$

Let's assume, per Foody (2009), that the maps will be stated to be statistically different from one another at α (the predetermined probability of making a Type I error, defined in this instance as incorrectly stating that the maps are different when in fact they are not) = 0.05 if $|z| > 1.96$ (two-tailed test). As such, since $|-0.45|$ is not greater than 1.96, the maps are not different from one another. It should be noted that although this is a common α level, there is no one level that is appropriate for all studies. Besides $\alpha = 0.05$, another common value for α is 0.01, in which case (for the two-tailed test) $|z|$ must be greater than 2.58.

Although it is often ignored by researchers and practitioners, Type II error (its probability denoted by β) occurs when there really is a noteworthy difference between maps that is not detected. The *power* of the test is then $1 - \beta$. As one increases the desired power and decreases the desired α, a larger sample size is required. Many, if not most, accuracy assessments contain too few samples, though it is possible (when, say, using another

TABLE 14.7. McNemar's Test Cross Tabulation

	Map 1	
Map 2	Correct	Incorrect
Correct	a	b
Incorrect	c	d

TABLE 14.8. Accuracy Assessment Data Labeled for McNemar Test

Point number	Map 1	Map 2	Reference	Letter
1	1	1	1	a
2	1	2	1	c
3	1	1	1	a
4	1	1	2	d
5	1	1	1	a
6	1	2	1	c
7	1	1	1	a
8	2	2	2	a
9	2	1	1	b
10	2	2	2	a
11	2	2	1	d
12	2	2	1	d
13	2	2	2	a
14	2	1	2	c
15	2	1	1	b
16	2	2	2	a
17	2	2	2	a
18	2	2	2	a
19	2	2	2	a
20	2	2	2	a

map as reference data) to have so many samples that very small differences between and among maps are declared to be statistically significant (Foody, 2009). The McNemar test presented here often requires sample sizes larger than the "rule of thumb" earlier presented to minimize the possibility of Type II error given typical α levels. Readers are encouraged to consult references such as Foody (2009) to determine the appropriate sample sizes for their particular applications.

14.6. Summary

Accuracy assessment is a very complex process. This chapter cannot address all the relevant topics in detail, because even the most complete discussion leaves many issues unresolved. Research continues, and despite agreement on many important aspects of accuracy evaluation, other issues are likely to be debated for a long time before they are resolved. For example, there is no clearly superior method of sampling an image for accuracy assessment, and there is disagreement concerning the best way to compare two thematic maps. For many problems, there may be no single "correct" way to conduct the analysis, but we may be able to exclude some alternatives and to speak of the relative merits and shortcomings of others.

This chapter should provide the background for assessing the accuracies of classifications using procedures that, if not perfect, are at least comparable in quality to those in common use today. Furthermore, the student should now be prepared to read some of the current research on this topic and possibly to contribute to improvements in the study of accuracy assessment. Many of the problems in this field are difficult but are not beyond the reach of interested and informed students.

Review Questions

Matrix A

Actual land use	Interpreted land use					
	Urban	Agriculture	Range	Forest	Water	Total
Urban	510	110	85	23	10	738
Agriculture	54	1,155	235	253	35	1,732
Range	15	217	930	173	8	1,343
Forest	37	173	238	864	27	1,339
Water	5	17	23	11	265	321
Total	621	1,672	1,511	1,324	345	5,473

Matrix B

Actual land use	Interpreted land use					
	Urban	Agriculture	Range	Forest	Water	Total
Urban	320	98	320	64	26	828
Agriculture	36	1,451	112	85	48	1,732
Range	98	382	514	296	53	1,343
Forest	115	208	391	539	86	1,339
Water	28	32	68	23	170	321
Total	597	2,171	1,405	1,007	383	5,563

1. Refer to Matrix A. Which class shows the highest error of commission?

2. Based on your examination of Matrix A, what class was most often confused with agricultural land?

3. For Matrix A, which class was most accurately classified? Which class has the lowest accuracy?

4. For Matrix A, which class shows the highest number of errors of omission?

5. Compare Matrices A and B, which are error matrices for alternative interpretations of the same area, perhaps derived from different images or compiled by different analysts. Which image is the most accurate? Can you see problems in responding with a simple answer to this question?

6. Refer again to Matrices A and B. If you were interested in accurate delineation of agricultural land, which image would you prefer? If accurate classification of forested land is important, which image should be used?

7. What might be the consequence if we had no effective means of assessing the accuracy of a classification of a remotely sensed image?

8. Make a list of applications of the accuracy assessment technique, including applications within remote sensing and photointerpretation, as well as others that might fall in other disciplines. That is, think of several different circumstances or purposes for

which it might be useful to determine which of two alternative classifications is the more accurate.

9. Discuss key considerations for deciding the sample size appropriate for accuracy assessment of a specific remote sensing project.

10. Calculate producer's and consumer's accuracies for agriculture and forested lands as reported in Matrix A.

11. Inspect the contrived matrices shown in Table 14.6, and make a general statement concerning the overall appearance of the matrix and its relationship to the corresponding value of kappa for that matrix.

12. Why might the usual statements concerning desired sample sizes for an experiment not be satisfactory in the context of accuracy assessment?

13. If producer's accuracy and user's accuracy differ (as they usually do), how can makers and users ever have a common interest in the preparation and use of a specific map?

14. Can you make an argument that, for accuracy assessment, there might be practical value in favoring Type I error over Type II error?

References

Anderson, J. R., E. E. Hardy, J. T. Roach, and R. E. Witmer. 1976. *A Land Use and Land Cover Classification for Use with Remote Sensor Data* (U.S. Geological Survey Professional Paper 964). Washington, DC: U.S. Government Printing Office, 28 pp.

Bishop, Y. M. M., S. E. Fienber, and P. W. Holland. 1975. *Discrete Multivariate Analysis: Theory and Practice*. Cambridge, MA: MIT Press, 557 pp.

Campbell, J. B. 1981. Spatial Correlation Effects upon Accuracy of Supervised Classification of Land Cover. *Photogrammetric Engineering and Remote Sensing*, Vol. 47, pp. 355–363.

Card, D. H. 1982. Using Known Map Category Marginal Frequencies to Improve Estimates of Thematic Map Accuracy. *Photogrammetric Engineering and Remote Sensing*, Vol. 48. pp. 431–439.

Chrisman, N. R. 1980. Assessing Landsat Accuracy: A Geographic Application of Misclassification Analysis. In *Second Colloquium on Quantitative and Theoretical Geography*. Cambridge, UK.

Cohen, J. 1960. A Coefficient of Agreement for Nominal Scales. *Educational and Psychological Measurement*, Vol. 20, No. 1, pp. 37–40.

Congalton, R. G. 1984. *A Comparison of Five Sampling Schemes Used in Assessing the Accuracy of Land Cover/Land Use Maps Derived from Remotely Sensed Data*. PhD. Dissertation, Virginia Polytechnic Institute and State University, Blacksburg, VA. 147 pp.

Congalton, R. G. 1988. A Comparison of Sampling Schemes Used in Generating Error Matrices for Assessing the Accuracy Maps Generated from Remotely Sensed Data. *Photogrammetric Engineering and Remote Sensing*, Vol. 54, pp. 593–600.

Congalton, R. G. 1988. Using Spatial Autocorrelation Analysis to Explore the Errors in Maps Generated from Remotely Sensed Data. *Photogrammetric Engineering and Remote Sensing*, Vol. 54, pp. 587–592.

Congalton, R. G., and K. Green. 2009. *Assessing the Accuracy of Remotely Sensed Data: Principles and Practices* (2nd ed.). Boca Raton, FL: CRC Press, 183 pp.

Congalton, R. G., and R. A. Mead. 1983. A Quantitative Method to Test for Consistency and Correctness in Photointerpretation. *Photogrammetric Engineering and Remote Sensing*, Vol. 49, pp. 69–74.

Congalton, R. G., R. G. Oderwald, and R. A. Mead. 1983. Assessing Landsat Classification Accuracy Using Discrete Multivariate Analysis Statistical Techniques. *Photogrammetric Engineering and Remote Sensing*, Vol. 49, pp. 1671–1687.

Fitzpatrick-Lins, K. 1978. Accuracy and Consistency Comparisons of Land Use and Land Cover Maps Made from High-Altitude Photographs and Landsat Multispectral Imagery. *Journal of Research, U.S. Geological Survey*, Vol. 6, pp. 23–40.

Foody, G. M. 1992. On the Compensation for Chance Agreement in Image Classification Accuracy Assessment. *Photogrammetric Engineering and Remote Sensing*, Vol. 58, pp. 1459–1460.

Foody, G. M. 2004. Thematic Map Comparison: Evaluating the Statistical Significance of Differences in Classification Accuracy. *Photogrammetric Engineering and Remote Sensing*, Vol. 70, pp. 627–633.

Foody, G. M. 2009. Sample size determination for image classification accuracy assessment and comparison. *International Journal of Remote Sensing*, Vol. 30, pp. 5273–5291.

Ginevan, M. E. 1979. Testing Land-Use Map Accuracy: Another Look. *Photogrammetric Engineering and Remote Sensing*, Vol. 45, pp. 1371–1377.

Hay, A. 1979. Sampling Designs to Test Land Use Map Accuracy. *Photogrammetric Engineering and Remote Sensing*, Vol. 45, pp. 529–533.

Hord, R. M., and W. Brooner. 1976. Land Use Map Accuracy Criteria. *Photogrammetric Engineering and Remote Sensing*, Vol. 46, pp. 671–677.

Hudson, W. D., and C. W. Ramm. 1987. Correct Formulation of the Kappa Coefficient of Agreement. *Photogrammetric Engineering and Remote Sensing*, Vol. 53, pp. 421–422.

Janssen, L. L. F., and F. J. M. van der Wel. 1994. Accuracy Assessment of Satellite Derived Land Cover Data: A Review. *Photogrammetric Engineering and Remote Sensing*, Vol. 60, pp. 419–426.

Lhermitte, S., J. Verbesselt, I. Jonckheere, K. Nachaerts, J. A. N. Van Aardt, W. W. Verstraeten, and P. Coppin. 2008. Hierarchical Image Segmentation Based upon Similarity of NDVI Time Series. *Remote Sensing of Environment*, Vol. 112, pp. 506–521.

Ma, Z., and R. L. Redmond. 1995. Tau Coefficients for Accuracy Assessment of Classification of Remotely Sensed Data. *Photogrammetric Engineering and Remote Sensing*, Vol. 61, pp. 435–439.

McNemar, Q. 1947. Note on the Sampling Error of the Difference between Correlated proportions or Percentages. *Psychometrika*, Vol. 12, pp. 153–157.

Musy, R. F., R. H. Wynne, C. E. Blinn, and J. A. Scrivani. 2006. Automated Forest Area Estimation via Iterative Guided Spectral Class Rejection. *Photogrammetric Engineering and Remote Sensing*, Vol. 72 pp. 949–960.

Podwysocki, M. H. 1976. *An Estimate of Field Size Distribution for Selected Sites in Major Grain Producing Countries* (Publication No. X-923-76-93). Greenbelt, MD: Goddard Space Flight Center, 34 pp.

Quirk, B. K., and F. L. Scarpace. 1980. A Method of Assessing Accuracy of a Digital Classification. *Photogrammetric Engineering and Remote Sensing*, Vol. 46, pp. 1427–1431.

Rosenfield, G. H., and K. Fitzpatrick-Lins. 1986. A Coefficient of Agreement as a Measure of Thematic Map Classification Accuracy. *Photogrammetric Engineering and Remote Sensing*, Vol. 52, pp. 223–227.

Rosenfield, G. H., K. Fitzpatrick-Lins, and H. S. Ling. 1982. Sampling for Thematic Map Accuracy Testing. *Photogrammetric Engineering and Remote Sensing*, Vol. 48, pp. 131–137.

Simonett, D. S., and J. C. Coiner. 1971. Susceptibility of Environments to Low Resolution Imaging for Land Use Mapping. In *Proceedings of the Seventh International Symposium on Remote Sensing of Environment*. Ann Arbor: University of Michigan Press, pp. 373–394.

Skidmore, A. K., and B. J. Turner. 1992. Map Accuracy Assessment Using Line Intersect Sampling. *Photogrammetric Engineering and Remote Sensing*, Vol. 58, pp. 1453–1457.

Stehman, S. V. 1992. Comparison of Systematic and Random Sampling for Estimating the Accuracy of Maps Generated from Remotely Sensed Data. *Photogrammetric Engineering and Remote Sensing*, Vol. 58, pp. 1343–1350.

Stehman, S. 1997. Selecting and Interpreting Measures of Thematic Map Accuracy. *Remote Sensing of Environment*, Vol. 62, pp. 77–89.

Stehman, S. V. 2004. A Critical Evaluation of the Normalized Error Matrix in Map Accuracy Assessment. *Photogrammetric Engineering and Remote Sensing*, Vol. 70, pp. 743–751.

Stehman, S. V., and R. L. Czaplewski. 1998. Design and Analysis for Thematic Map Accuracy Assessment: Fundamental Principles. *Remote Sensing of Environment*, Vol. 64, pp. 331–344.

Story, M., and R. G. Congalton. 1986. Accuracy Assessment: A User's Perspective. *Photogrammetric Engineering and Remote Sensing*, Vol. 52, pp. 397–399.

Todd, W. J., D. G. Gehring, and J. F. Haman. 1980. Landsat Wildland Mapping Accuracy. *Photogrammetric Engineering and Remote Sensing*, Vol. 46, pp. 509–520.

Tom, C., L. D. Miller, and J. W. Christenson. 1978. *Spatial Land-Use Inventory, Modeling, and Projection: Denver Metropolitan Area, with Inputs from Existing Maps, Airphotos, and Landsat Imagery* (NASA Technical Memorandum 79710). Greenbelt, MD: Goddard Space Flight Center, 225 pp.

Turk, G. 1979. GT Index: A Measure of the Success of Prediction. *Remote Sensing of Environment*, Vol. 8, pp. 65–75.

Van der Wel, F. J. M., and L. L. F. Janssen. 1994. A Short Note on Pixel Sampling of Remotely Sensed Digital Imagery. *Computers and Geosciences*, Vol. 20, pp. 1263–1264.

Van Genderen, J., and B. Lock. 1977. Testing Map Use Accuracy. *Photogrammetric Engineering and Remote Sensing*, Vol. 43, pp. 1135–1137.

Webster, R., and P. H. T. Beckett. 1968. Quality and Usefulness of Soil Maps. *Nature*, Vol. 219, pp. 680–682.

CHAPTER FIFTEEN

Hyperspectral Remote Sensing

15.1. Introduction

Remote sensing involves examination of features observed in several regions of the electromagnetic spectrum. Conventional remote sensing, as outlined in previous chapters, is based on use of several rather broadly defined spectral regions. *Hyperspectral remote sensing* is based on examination of many narrowly defined spectral channels. Sensor systems such as SPOT 1 HRV, Landsat MSS, and Landsat TM provide 3, 4, and 7 spectral channels, respectively. The hyperspectral sensors described below can provide 200 or more channels, each only 10 nm wide. In the context of the discussion of image resolution presented in Chapter 10, hyperspectral sensors implement the concept of "spectral resolution" to the extreme. Although hyperspectral remote sensing applies the same principles and methods discussed previously, it requires such specialized data sets, instruments, field data, and software that it forms a specialized field of inquiry.

15.2. Spectroscopy

Hyperspectral data have detail and accuracy that permit investigation of phenomena and concepts that greatly extend the scope of traditional remote sensing. For example, analysts can begin to match observed spectra to those recorded in spectral libraries and to closely examine relationships between brightnesses in several spectral channels to estimate atmospheric effects using data within the image itself. Such capabilities present opportunities for much more precise identification of Earth surface phenomena than is possible with broadband sensors, for investigation of phenomena such as the blue shift (Chapter 18) and the red shift (Chapter 17), and for correction of data in some bands using other bands that convey information about atmospheric transmission.

These capabilities extend the reach of remote sensing into the field of *spectroscopy*, the science devoted to the detailed examination of very accurate spectral data. Classical spectroscopy has its origins in experiments conducted by Isaac Newton (1642–1727), who used glass prisms to separate visible light into the spectrum of colors. Later, another English physicist, William Wollaston (1766–1828), noted that spectra displayed dark lines when light is projected through a narrow slit. The meaning of these lines was discovered through the work of Joseph Fraunhofer, a German glassmaker who discovered distinctive lines in spectra of light from the Sun and from stars. Dark lines (*absorption spectra*) are observed as radiation passes through gases at low pressure; bright lines (*emission spectra*) form as heated gases (e.g., in the sun's atmosphere) emit radiation. These lines have origins in the chemical elements present in the gases, a discovery that has permitted astronomers to investigate differences in chemical compositions of stars and planets. The Danish physicist Neils Bohr (1885–1962) found that the character of Fraunhofer lines is ultimately determined by the atomic structure of gases.

Instruments used in spectroscopy—*spectroscopes, spectrometers, spectrographs*—are designed to collect radiation with a lens and to divide it into spectral regions (using a prism or diffraction grating) that are then recorded on film or measured electronically. This form of spectroscopy is now a standard method not only in astronomy but also for laboratory analyses to characterize unidentified materials.

15.3. Hyperspectral Remote Sensing

Hyperspectral remote sensing uses the practice of spectroscopy to examine images of the Earth's surface. Although hyperspectral remote sensing sometimes applies the techniques of classical spectroscopy to the study of atmospheric gases and pollutants, for example, more commonly it applies these techniques to the making of precise, accurate, detailed spectral measurements of the Earth's surface (*imaging spectrometry*). Such data have accuracy and detail sufficient to begin to match observed spectra to those stored in databases known as *spectral libraries*. Instruments for hyperspectral remote sensing differ from those of conventional spectroscopy in that they gather spectra not only for point targets but also for areas—not for stars or laboratory samples but for regions of the Earth's surface. Instruments for hyperspectral remote sensing differ from other remote sensing instruments in terms of their extraordinarily fine spectral, spatial, and radiometric resolutions and their careful calibration. Some hyperspectral instruments collect data in 200 or more channels at 10–12 bits. Because of their calibration and ability to collect data having fine detail, such instruments greatly extend the reach of remote sensing not only by extending the range of applications but also by defining new concepts and analytical techniques.

Although the techniques of classical spectroscopy can be used in hyperspectral remote sensing to examine, for example, atmospheric gases, hyperspectral remote sensing typically examines very detailed spectra for images of the Earth's surface, applies corrections for atmospheric effects, and matches them to spectra of known features.

15.4. The Airborne Visible/Infrared Imaging Spectrometer

One of the first airborne hyperspectral sensors was designed in the early 1980s by the Jet Propulsion Laboratory (JPL; Pasadena, California). The *airborne imaging spectrometer* (AIS) greatly extended the scope of remote sensing by virtue of its number of spectral bands; their fine spatial, spectral, and radiometric detail; and the accuracy of its calibration. AIS collected 128 spectral channels, each about 10 nm wide in the interval 1.2–2.4 µm. The term *hyperspectral remote sensing* recognizes the fundamental difference between these data and those of the usual broadband remote sensing instruments. (Sensor systems with even finer spectral resolution, designed primarily to study atmospheric gases, are known as *ultraspectral* sensors.)

Although several hyperspectral instruments are now in operational use, an important pioneer in the field of hyperspectral remote sensing is the *airborne visible/infrared imaging spectrometer* (AVIRIS; *aviris.jpl.nasa.gov/*). AVIRIS was developed by NASA and JPL from the foundations established by AIS. AVIRIS was first tested in 1987, was placed in service in 1989, and has since been modified at intervals to upgrade its reliability and performance. It has now acquired hundreds of images of test sites in North America and Europe.

Hyperspectral sensors necessarily employ designs that differ from those of the usual sensor systems. An objective lens collects radiation reflected or emitted from the scene; a colluminating lens then projects the radiation as a beam of parallel rays through a diffraction grating that separates the radiation into discrete spectral bands (Figure 15.1).

Energy in each spectral band is detected by linear arrays of silicon and indium antimonide (Chapter 9). Because of the wide spectral range of AVIRIS, detectors are configured in four separate panels (0.4–0.7 µm, 0.7–1.3 µm, 1.3–1.9 µm, and 1.8–2.8 µm), each calibrated independently (Figure 15.2). AVIRIS operates over the spectral range of 400–2500 nm (0.4–2.45 µm), producing 224 spectral channels, each 10 nm wide (Figure 15.3). After processing, data consist of 210 spectral channels. At its usual operating altitude, each image records a strip 11 km wide, processed to form scenes recording areas about 11 km × 11 km each. Each line of data conveys about 614 pixels. If operated at low altitude, pixels might each represent ground areas about 4 m on a side; if operated at higher altitudes, as is common practice, pixels each represent ground areas as large as about 20 m on a side.

15.5. The Image Cube

The *image cube* refers to the representation of hyperspectral data as a three-dimensional figure, with two dimensions formed by the *x* and *y* axes of the usual map or image display and the third (*z*) formed by the accumulation of spectral data as additional bands are superimposed on each other. In Figure 15.4 and Plate 17, the top of the cube is an image composed of data collected at the shortest wavelength (collected in the ultraviolet), and the bottom of the cube is an image composed of data collected at the longest wavelength (2.5 µm). Intermediate wavelengths are found as horizontal slices through the cube at intermediate positions. Values for a single pixel observed along the edge of the cube form a spectral trace describing the spectra of the surface represented by the pixel.

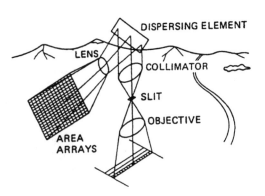

FIGURE 15.1. Imaging spectrometer. NASA diagram.

FIGURE 15.2. AVIRIS sensor design. NASA diagram.

15.6. Spectral Libraries

The development of hyperspectral remote sensing has been accompanied by the accumulation of detailed spectral data acquired in the laboratory and in the field. These data are organized in *spectral libraries*, databases maintained primarily by governmental agencies, but also by other organizations. These libraries assemble spectra that have been acquired at test sites representative of varied terrain and climate zones, observed in the field under natural conditions. Included also are other data describing, for example, construction materials, minerals, vegetation, and fabrics as observed in laboratories under standardized conditions.

Such data are publicly available to the remote sensing community and have been

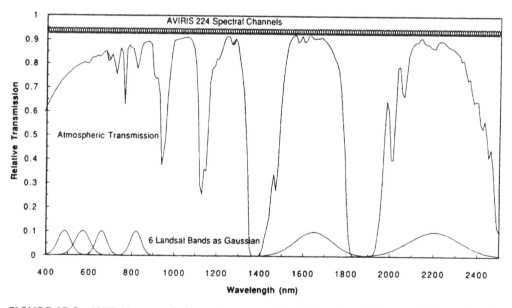

FIGURE 15.3. AVIRIS spectral channels compared with Landsat TM spectral channels. The 224 narrow bands at the top represent AVIRIS spectral channels. For comparison, Landsat TM channels are shown at the bottom. From Green and Simmonds (1993).

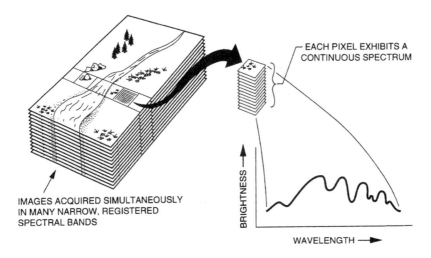

IMAGES ACQUIRED SIMULTANEOUSLY
IN MANY NARROW, REGISTERED
SPECTRAL BANDS

EACH PIXEL EXHIBITS A
CONTINUOUS SPECTRUM

BRIGHTNESS →

WAVELENGTH →

FIGURE 15.4. Image cube. Based on a NASA diagram.

incorporated into software designed for use in hyperspectral remote sensing. Maintenance of a spectral library requires specialized effort to bring data into a common format that can be used by a diverse community. Spectral data are typically collected by diverse instruments under varied conditions of illumination. These many differences must be resolved to prepare data in a format that permits use by a diverse community.

Because of the fine spectral, spatial, and radiometric detail of hyperspectral analysis, identification and cataloging are special problems for design of spectral libraries. Therefore, each spectral record must be linked to detailed information specifying the instruments used, meteorological conditions, nature of the surface, and the circumstances of the measurement. These kinds of ancillary data are required for successful interpretation and analysis of the image data.

15.7. Spectral Matching

Figure 15.5, derived from Barr (1994), illustrates a sequence of analysis for hyperspectral data. Analysis begins with acquisition and preprocessing (1) to remove known system errors and ensure accurate calibration. Any of several methods can be used to correct for atmospheric effects (2), as outlined in Chapter 11. The lower portion of Figure 15.5 represents the analytical process. Data are organized to permit convenient display and manipulation of specified bands or combinations of bands. Selected bands can be used to prepare color composites for display. *Alarm* (5) means that the analyst can place the cursor on the image to mark a specific pixel or region of pixels, then instruct the computer to highlight (alarm) other regions on the image that are characterized by the same spectral response. Then (6) the analyst can attempt to match the alarmed areas to spectra from a spectral library accessible to the image processing system. If a match can be made, then the analyst can examine another region and attempt to match it to entries from the spectral library.

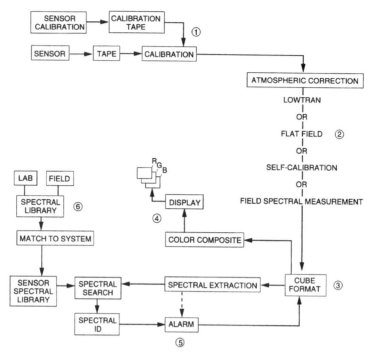

FIGURE 15.5. Spectral matching. Based on a diagram by Barr (1994).

15.8. Spectral Mixing Analysis

Fine spectral resolution, like fine spatial resolution, does not overcome the enduring obstacles to the practice of remote sensing. Surface materials recorded by the sensor are not always characteristic of subsurface conditions. Atmospheric effects, shadowing, and topographic variations contribute to observed spectra to confuse interpretations. Even when observed at fine spatial and spectral detail, surfaces are often composed of varied materials. Therefore, the sensor observes composite spectra that may not clearly match to the pure spectra of spectral libraries. *Linear mixing* refers to additive combinations of several diverse materials that occur in patterns too fine to be resolved at the resolution of the sensors. This is the effect of mixed pixels outlined in Chapter 10. As long as the radiation from component patches remains separate until it reaches the sensor, it is possible to estimate proportions of component surfaces from the observed pixel brightness using methods such as those illustrated by Equation 10.2. Linear mixing might occur when components of a composite surface are found in a few compact areas (Figure 15.6b). *Nonlinear mixing* occurs when radiation from several surfaces combines before it reaches the sensor. Nonlinear mixing occurs when component surfaces arise in highly dispersed patterns (Figure 15.6c). Nonlinear mixing cannot be described by the techniques addressed here.

 Spectral mixing analysis (also known as *spectral unmixing*) is devoted to extracting pure spectra from the complex composites of spectra that by necessity form each image. It assumes that pixels are formed by linear mixing and further assumes that it is possible to identify the components contributing to the mixture. It permits analysts to define key components of a specific scene and forms an essential component in the process of spectral matching, discussed below. Analysts desire to match data from hyperspectral images to corresponding laboratory data to identify surfaces from their spectral data much more precisely than previously was possible. Whereas conventional image analysis (Chapter 12) matches pixels to broad classes of features, hyperspectral image matching attempts to make more precise identifications—to specific mineralogies of soils or rocks, for example.

Therefore, spectral matching requires the application of techniques that enable analysts to separate pure pixels from impure pixels. This problem is well matched to the capabilities of *convex geometry*, which examines multidimensional data envisioned in *n* dimensions. Individual points (pixels) within this data space can be examined as linear combinations of an unknown number of pure components. Convex geometry can solve such problems provided that the components are linearly weighted, sum to unity, and are positive. We assume also that the data have greater dimensionality (more spectral bands) than the number of pure components.

The illustrations here, for convenience and legibility, show only two dimensions, although the power of the technique is evident only with much higher dimensionality. In Figure 15.7, the three points *A*, *B*, and *C* represent three spectral observations at the extreme limits of the swarm of data points, represented by the shaded pattern. That is, the shaded pattern represents values of all the pixels within a specific image or subimage, which is simplified by the triangle shape. (Other shapes can be defined as appropriate to approximate the shape of a data swarm, although an objective is to define the simplest shape [*simplex*] that can reasonably approximate the pattern of the data swarm.)

These three points (for the example in Figure 15.7a) form *endmembers*, defined as the pure pixels that contribute to the varied mixtures of pixels in the interior of the data swarm. Once the simple form is defined, the interior pixels can be defined as linear combinations of the pure endmembers. (The discussion of mixed pixels in Chapter 10 anticipates this concept.) In general, interior points can be interpreted as positive unit-sum combinations of the pure variables represented by endmembers at the vertices. In general, a shape defined by *n* + 1 vertices is the simplest shape that encompasses interior

FIGURE 15.6. Linear and nonlinear spectral mixing. (a) If a pixel represents a uniform ground area at the resolution of the sensor, the pixel represents a pure spectrum. (b) If a pixel represents two or more surfaces that occur in patches that are large relative to the sensor's resolution, mixing occurs at the sensor. The pattern of the composite surfaces can never be resolved, but because mixing occurs in a linear manner, proportions of the components can be estimated. (c) If the composite occurs at a scale that is fine relative to the resolution of the sensor, mixing occurs before the radiation reaches the sensor, and components of the composite cannot be estimated using the methods described here.

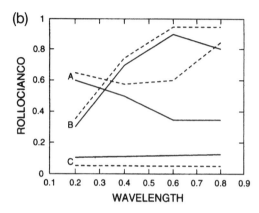

FIGURE 15.7. Spectral mixing analysis (Tompkins et al., 1993). (a) Simplex; (b) End-members.

points (i.e., for two dimensions [$n = 2$], the simplex is a triangle [i.e., $3 = n + 1$]). The faces of the shape are *facets*, and the exterior surface is a *convex hull*. In Figure 15.7, A, B, and C are the observed approximations of the idealized spectra A, B, and C that may not be observed on any specific image.

In the application of convex geometry to hyperspectral data, it is first necessary to define the dimensionality of the data. An image is typically subset to define a region of relative homogeneity and to remove bands with high variance (assumed to include high levels of error). The original hyperspectral data are converted to ground-level reflectances (from radiances) using atmospheric models. The data are condensed by applying a version of principal components analysis (PCA) (Chapter 11). Although hyperspectral data may include many bands (224 for AVIRIS), duplication from one channel to another means that inherent dimensionality is much less (perhaps as few as 3 to 10), depending on the specifics of each scene.

The analyst then examines the transformed data in a data space to define the smallest simplex that fits the data. This process defines the $n + 1$ facets that permit identification of the n endmembers (Figure 15.7a). These vertices, when projected back into the original spectral domain, estimate the spectra of the endmembers. These spectra are represented by Figure 15.7b. The objective is to match these endmembers to spectra from spectral libraries, then to prepare maps and images that reveal the varied mixtures of surfaces that contribute to the observed spectra in each image.

Typical endmembers in arid regions have included bare soil, water, partially vegetated surfaces, fully vegetated surfaces, and shadows. Endmembers can be investigated in the field to confirm or revise identifications made by computer. Software for hyperspectral analysis often includes provisions for accessing spectral libraries (and for importing additional spectra as acquired in the field or laboratory) and the ability to search for matches with endmembers identified. Although it may not always be possible to uniquely identify matches in spectral libraries, such analyses can narrow the range of alternatives. In some instances mathematical models can assist in defining poorly established endmembers.

15.9. Spectral Angle Mapping

Spectral angle mapping (SAM) is a classification approach that examines multispectral or hyperspectral data by evaluating the relationships between pixel values projected in multispectral data space. Envision a pixel projected into multispectral data space—its position can be described by a vector with an angle in relation to the measurement axes (Figure 15.8). Its position relative to another pixel (or, perhaps, a set of reference or training data) can be evaluated by assessing the difference between the angles of the two vectors. Small angles indicate a close similarity; large angles indicate lower similarity.

The effectiveness of this technique is achieved when it is applied to many more pixels and dimensions than can be represented in a single diagram. SAM is most useful when extended to tens or hundreds of dimensions (Kruse et al., 1993). SAMs differ from the usual classification approaches because they compare each pixel in the image with each spectral class, then assign a value between 0 (low resemblance) and 1 (high resemblance) to each pair. Although SAMs are valuable for hyperspectral analyses in many dimensions, their effectiveness is based on the implicit assumption that each pixel is being compared with a pure spectra—in fact, of course, most images include many mixed pixels (Chapter 10), so the results are influenced by the presence of mixed pixels.

15.10. Analyses

Other investigations have attempted to understand relationships between spectral data and specific physical or biological processes. For example, Curran (1994) reviews efforts

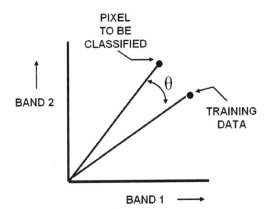

FIGURE 15.8. Spectral angle mapping.

to use hyperspectral data to monitor botanical variables, such as the chlorophyll, lignin, cellulose, water, and nitrogen content of plant tissues. Others have examined, at high levels of spectral, radiometric, and spatial resolution, observed spectra of laboratory samples of plant tissues influenced by atomic and molecular structures of water and by specific concentrations of organic compounds (such as chlorophyll, lignin, and cellulose). These relationships have formed the basis for research devoted to the examination of hyperspectral data of vegetation canopies to derive estimates of foliar chemistry of plant tissues in situ. Such estimates would support agricultural, forestry, and ecological studies by providing indications of nutrient availability, rates of productivity, and rates of decomposition. Green (1993) and Rivard and Arvidson (1992) report some of the applications of hyperspectral data to lithologic and mineralogic analyses.

15.11. Wavelet Analysis for Hyperspectral Imagery

A *wavelet* can be defined as a wave-like oscillation of a fixed duration, characterized by an amplitude that has a value of zero, then increases, then declines again to zero—analogous perhaps to oscillations such as those that we encounter on seismographs or heart monitors. (See Figure 15.9 for an example of a Haar wavelet.) *Wavelet analysis* decomposes images, sounds, or spectra into wavelets of varying duration. Given the focus of this chapter, we discuss wavelet analysis of spectra, but the basic techniques are the same for images and sounds. Wavelet analysis is inherently a *multiscale* (or *multiresolution*) analysis because the spectrum is decomposed into subspectra of ever-finer scales. In practice, we first determine the finest scale wavelets (from the first *detail*) present in a spectrum, then remove them, resulting in a less detailed *approximation* of the spectrum. The next finest scale wavelets (second detail) are then removed, and so on, until we end up with a spectrum in which only the broadest absorption and reflection features are present. The details are illustrated by Figure 15.10, in which on the right side, d1–d6 represent ever-coarser scales. The resulting approximations of the spectrum, once each level of detail has been removed, are shown on the left. To use a musical analogy, for which we are indebted to Walker (2008), think of the process as successively stripping off "ornaments" to an essential melody. A thorough discussion of wavelets is well beyond the reach of our discussion here, so readers are referred to sources such as Walker (2008) for a proper introduction to this topic.

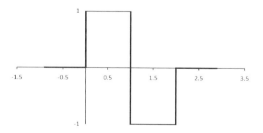

FIGURE 15.9. Illustration of a Haar wavelet.

FIGURE 15.10. Wavelet analysis. Original hyperspectral radiance (in $W \cdot m^{-2} \cdot sr^{-1}$) curve of a randomly chosen shortleaf pine spectrum (acquired using the AVIRIS sensor) and its approximation ($a_1, a_2, \ldots a_6$) and details ($d_1, d_2 \ldots d_6$), up to six levels of decomposition. From Banskota et al. (in press).

15.12. Summary

The vast amounts of data collected by hyperspectral systems and the problems they present for both collection and analysis prevent routine use of hyperspectral data in the same way that we might collect TM or SPOT data on a regular basis. More likely, hyperspectral data will provide a means for discovering and refining the knowledge needed to develop improved sensors and analytical techniques that can be applied on a more routine basis. A second important role for hyperspectral data lies in the monitoring of long-term research sites, especially those devoted to study of biophysical processes and other phenomena that change over time. The fine detail of hyperspectral data will provide enhanced data for ecological monitoring and for understanding patterns recorded on lower resolution data for the same sites.

Several plans to deploy hyperspectral sensors for satellite observation are under way. One is NASA's EO–1 satellite, launched in November 2000, and designed as a vehicle to test the feasibility of advanced imaging systems. One of EO–1's sensors is the Hyperion, a high-resolution hyperspectral sensor capable of resolving 220 spectral bands from 0.4 to 2.5 m at a 30-m resolution (Figure 15.11). The instrument can image a 7.5 km × 100 km land area to provide detailed spectral mapping across all 220 channels with high radiometric accuracy. EO–1 has been placed in an orbit that permits both Landsat 7 and

FIGURE 15.11. Hyperion image, Mt. Fuji, Japan. Right: A single band of the 220-channel Hyperion image collected in February 2003. Left: A single band of the ALI sensor described in Chapter 6 representing the same region. From USGS EROS Data Center.

EO–1 to image the same ground area at least once a day, permitting both systems to collect images under identical viewing conditions. The paired images are then available for evaluation of the imaging technology.

Review Questions

1. Discuss the *advantages* of hyperspectral remote sensing in relation to more conventional remote sensing instruments.

2. List some *disadvantages* of hyperspectral remote sensing relative to use of systems such as SPOT or Landsat.

3. Prepare a plan to monitor an agricultural landscape in the Midwest using *both* hyperspectral data *and* SPOT or Landsat data.

4. How would use of hyperspectral data influence collection of field data, compared with similar studies using SPOT or Landsat imagery?

5. Discuss how *preprocessing* and *image classification* differ with hyperspectral data, compared with more conventional multispectral data.

6. How would *equipment* needs differ for image processing of hyperspectral data, compared with more conventional multispectral data?

7. The question of choosing between broad-scale coverage at coarse detail and focused coverage at fine detail recurs frequently in many fields of study. How does availability of hyperspectral data influence this discussion?

8. Can you think of ways that availability of hyperspectral data will influence the concepts and theories of remote sensing?

9. Hyperspectral data have so much volume that it is not feasible to accumulate geographic coverage comparable to that of SPOT or Landsat, for example. What value, then, can hyperspectral data have?

10. Discuss the problems that arise in attempting to design and maintain a spectral library. Consider, for example, the multitude of different materials and surfaces that must be considered, each under conditions of varied illumination.

References

Spectral Unmixing

Asner, G. P., and K. B. Heidebrecht. 2002. Spectral Unmixing of Vegetation, Soil and Dry Carbon Cover in Arid Regions: Comparing Multispectral and Hyperspectral Observations. *International Journal of Remote Sensing*, Vol. 23, pp. 3939–3958.

Landgrebe, D. 2002. Hyperspectral Image Data Analysis. *IEEE Signal Processing Magazine*, Vol. 19, pp. 17–28.

Manolakis, D., C. Siracusa, and G. Shaw. 2001. Hyperspectral Subpixel Target Detection Using the Linear Mixing Model. *IEEE Transactions on Geoscience and Remote Sensing*, Vol. 39, pp. 1392–1409.

Plaza, A., P. Martinez, R. Perez, and J. Plaza. 2004. A Quantitative and Comparative Analysis of Endmember Extraction Algorithms from Hyperspectral Data. *IEEE Transactions on Geoscience and Remote Sensing*, Vol. 42, pp. 650–663.

General

Barr, S. 1994. Hyperspectral Image Processing. In *Proceedings of the International Symposium on Spectral Sensing Research*. Ft. Belvior, VA: U.S. Army Corps of Engineers, pp. 447–490.

Crippen, R. R., and R. G. Blom. 2001. Unveiling the Lithology of Vegetated Terrains in Remotely Sensed Imagery. *Photogrammetric Engineering and Remote Sensing*, Vol. 67, pp. 935–942.

Curran, P. J. 1989. Remote Sensing of Foliar Chemistry. *Remote Sensing of Environment*, Vol. 29, pp. 271–278.

Curran, P. J. 1994. Imaging Spectrometry. *Progress in Physical Geography*, Vol. 18, pp. 247–266.

Curran, P. J., and J. A. Kupiec. 1995. Imaging Spectrometry: A New Tool for Ecology. Chapter 5 in *Advances in Environmental Remote Sensing* (F. M. Danson and S. E. Plummer, eds.). New York: Wiley, pp. 71–88.

Goa, B. C., K. B. Heidebrecht, and A. F. H. Goetz. 1993. Derivation of Scaled Surface Reflectances from AVIRIS Data. *Remote Sensing of Environment*, Vol. 44, pp. 165–178.

Goetz, A. F. H., G. Vane, J. E. Solomon, and B. N. Rock. 1985. Imaging spectrometry for earth remote sensing. *Science*, Vol. 228, pp. 1147–1153.

Green, R. O. (ed.). 1993. *Summaries of the Fourth Annual JPL Airborne Geoscience Workshop, October 25–29, 1993*. Pasadena, CA: NASA and the Jet Propulsion Laboratory, 209 pp.

Green, R. O., and J. J. Simmonds. 1993. A Role for AVIRIS in Landsat and Advanced Land Remote Sensing System Program. In *Summaries of the Fourth Annual JPL Airborne Geoscience Workshop, October 25–29, 1993* (R. O. Green, ed.). Pasadena, CA: NASA and the Jet Propulsion Laboratory, pp. 85–88.

Kruse, F. A., A. B. Lefkoff, J. B. Boardman, K. B. Heidebrecht, A. T. Shapiro, P. J. Barloon, et al. 1993. The Spectral Image Processing System (SIPS): Interactive Visualization and Analysis of Imaging Spectrometer Data. *Remote Sensing of Environment*, Vol. 44, pp. 145–163.

Lee, C., and D. A. Landgrebe. 1992. Analyzing High Dimensional Data. In *Proceedings, International Geoscience and Remote Sensing Symposium (IGARS `93)*, Vol. 1. New York: Institute of Electrical and Electronics Engineers, pp. 561–563.

Rivard, B., and D. E. Arvidson. 1992. Utility of Imaging Spectrometry for Lithologic Mapping in Greenland. *Photogrammetric Engineering and Remote Sensing*, Vol. 58, pp. 945–949.

Roberts, D. A., M. O. Smith, and J. B. Adams. 1993. Green Vegetation, Nonphotosynthetic Vegetation, and Soils in AVIRIS Data. *Remote Sensing of Environment*, Vol. 44, pp. 255–269.

Tompkins, S., J. F. Mustard, C. M. Peters, and D. W. Forsyth. 1993. Objective Determination of Image End-Members in Spectral Mixture Analysis of AVIRIS Data. In *Summaries of the Fourth Annual JPL Airborne Geoscience Workshop, October 25–29, 1993* (R. O. Green, ed.). Pasadena, CA: NASA and the Jet Propulsion Laboratory, pp. 177–180.

Tsai, F., and W. D. Philpot. 1998. Derivative Analysis of Hyperspectral Data. *Remote Sensing of Environment*, Vol. 66, pp. 41–51.

Vegetation Analysis

Broge, N. H., and E. Leblanc. 2001. Comparing Prediction Power and Stability of Broadband and Hyperspectral Vegetation Indices for Estimation of Green Leaf Area Index and Canopy Chlorophyll Density. *Remote Sensing of Environment*, Vol. 76, pp. 156–172.

Chaudhari, A. J., F. Darvas, J. R. Bading, R. A. Moats, P. S. Conti, D. J. Smith, et al. 2005. Hyperspectral and Multispectral Bioluminescence Optical Tomography for Small Animal Imaging. *Physics in Medicine and Biology*, Vol. 50, pp. 5421–5441.

Haboudane, D., J. R. Miller, E. Pattey, P. J. Zarco-Tejada, and I. B. Strachan. 2004. Hyperspectral Vegetation Indices and Novel Algorithms for Predicting Green LAI of Crop Canopies: Modeling and Validation in the Context of Precision Agriculture. *Remote Sensing of Environment*, Vol. 90, pp. 337–352.

Haboudane, D., J. R. Miller, N. Tremblay, P. J. Zarco-Tejada, and L. Dextraze. 2002. Integrated Narrow-Band Vegetation Indices for Prediction of Crop Chlorophyll Content for Application to Precision Agriculture. *Remote Sensing of Environment*, Vol. 81, pp. 416–426.

Lefsky, M. A., W. B. Cohen, and T. A. Spies. 2001. An Evaluation of Alternate Remote Sensing Products for Forest Inventory, Monitoring, and Mapping of Douglas-Fir Forests in Western Oregon. *Canadian Journal of Forest Research (Revue Canadienne De Recherche Forestiere)*, Vol. 31, pp. 78–87.

Melgani, F., and L. Bruzzone. 2004. Classification of Hyperspectral Remote Sensing Images with Support Vector Machines. *IEEE Transactions on Geoscience and Remote Sensing*, Vol. 42, pp. 1778–1790.

Okin, G. S., D. A. Roberts, B. Murray, and W. J. Okin. 2001. Practical Limits on Hyperspectral Vegetation Discrimination in Arid and Semiarid Environments. *Remote Sensing of Environment*, Vol. 77, pp. 212–225.

Pearlman, J. S., P. S. Barry, C. C. Segal, J. Shepanski, D. Beiso, and S. L. Carman. 2003. Hyperion: A Space-based Imaging Spectrometer. *IEEE Transactions on Geoscience and Remote Sensing*, Vol. 41, pp. 1160–1173.

Phillips, R. D., C. E. Blinn, L. T. Watson, and R. H. Wynne. 2009. An Adaptive Noise-Filtering Algorithm for AVIRIS Data with Implications for Classification Accuracy. *IEEE Transactions on Geoscience and Remote Sensing*, Vol. 47, pp. 3168–3179.

Serrano, L., S. L. Ustin, D. A. Roberts, J. A. Gamon, and J. Penuelas. 2000. Deriving Water Content of Chaparral Vegetation from AVIRIS data. *Remote Sensing of Environment*, Vol. 74, pp. 570–581.

Strachan, I. B., E. Pattey, and J. B. Boisvert. 2002. Impact of Nitrogen and Environmental Conditions on Corn as Detected by Hyperspectral Reflectance. *Remote Sensing of Environment*, Vol. 80, pp. 213–224.

Thenkabail, P. S., R. B. Smith, and E. De Pauw. 2000. Hyperspectral Vegetation Indices and Their Relationships with Agricultural Crop Characteristics. *Remote Sensing of Environment*, Vol. 71, pp. 158–182.

Underwood, E., S. Ustin, and D. DiPietro. 2003. Mapping Non-Native Plants using Hyperspectral Imagery. *Remote Sensing of Environment*, Vol. 86, pp. 150–161.

Ustin, S. L., D. A. Roberts, J. A. Gamon, G. P. Asner, and R. O. Green. 2004. Using Imaging Spectroscopy to Study Ecosystem Processes and Properties. *Bioscience*, Vol. 54, pp. 523–534.

Van Aardt, J. A. N., and R. H. Wynne. 2007. Examining Pine Spectral Separability Using Hyperspectral Data from an Airborne Sensor: An Extension of Field-Based Results. *International Journal of Remote Sensing*, Vol. 28, pp. 431–436.

Zarco-Tejada, P. J., J. R. Miller, T. L. Noland, G. H. Mohammed, and P. H. Sampson. 2001. Scaling-Up and Model Inversion Methods with Narrowband Optical Indices for Chlorophyll Content Estimation in Closed Forest Canopies with Hyperspectral Data. *IEEE Transactions on Geoscience and Remote Sensing*, Vol. 39, pp. 1491–1507.

Wavelets

Banskota, A., R. H. Wynne, and N. Kayastha. in press. Improving Within-Genus Tree Species Discrimination Using the Discrete Wavelet Transform Applied to Hyperspectral Data. *International Journal of Remote Sensing*.

Bradshaw, G. A., and T. A. Spies. 1992. Characterizing Canopy Gap Structure in Forests Using Wavelet Analysis. *Journal of Ecology*, Vol. 80, pp. 205–215.

Bruce, L. M., C. H. Koger, and L. Jiang. 2002. Dimensionality Reduction of Hyperspectral Data Using Discrete Wavelet Transform Feature Extraction. *IEEE Transactions on Geoscience and Remote Sensing*, Vol. 40, pp. 2331–2338.

Bruce, L. M., C. Morgan, and S. Larsen. 2001. Automated Detection of Subpixel Hyperspectral Targets with Continuous and Discrete Wavelet Transforms. *IEEE Transactions on Geoscience and Remote Sensing*, Vol. 39, pp. 2217–2226.

Green, A. A., M. Berman, P. Switzer, and M. D. Craig. 1988. A Transformation for Ordering Multispectral Data in Terms of Image Quality with Implications for Noise Removal. *IEEE Transactions on Geoscience and Remote Sensing*, Vol. 26, pp. 65–74.

Hsu, P. H. 2007. Feature Extraction of Hyperspectral Images Using Wavelet and Matching Pursuit. *ISPRS Journal of Photogrammetry and Remote Sensing*, Vol. 62, pp. 78–92.

Kaewpijit, S., Le Moigne, J., and T. El-Ghazawi. 2003. Automatic Reduction of Hyperspectral Imagery Using Wavelet Spectral Analysis. *IEEE Transactions on Geoscience and Remote Sensing*, Vol. 41, pp. 863–871.

Lee, J. B., A. S. Woodyatt, and M. Berman. 1990. Enhancement of High Spectral Resolution Remote-Sensing Data by a Noise-Adjusted Principal Components Transform. *IEEE Transactions on Geoscience and Remote Sensing*, Vol. 28, pp. 295–304.

Lisowski, J. J., C. A. Cook, and S. S. Shen. 1996. A SVD Method for Spectral Decomposition and Classification of ARES Data. *Proceedings of the Society of Photo-Optical Instrumentation Engineers (SPIE)*, Vol. 2281, pp. 14–29.

Moon, T., and E. Merenyi. 1995. Classification of Hyperspectral Images Using Wavelet Transforms and Neural Networks. *Proceedings of the Society of Photo-Optical Instrumentation Engineers (SPIE)*, Vol. 2569, pp. 725–735.

Peng, Z. K., M. R. Jackson, J. A. Rongong, F. L. Chu., and R. M. Parkin. 2009. On the Energy Leakage of Discrete Wavelet Transform. *Mechanical Systems and Signal Processing*, Vol. 23, pp. 330–343.

Phillips, R. D., C. E. Blinn, L. T. Watson, and R. H. Wynne. 2009. An Adaptive Noise Filtering Algorithm for AVIRIS Data with Implications for Classification Accuracy. *IEEE Transactions on Geoscience and Remote Sensing*, Vol. 47, pp. 3168–3179.

Pu, R., and P. Gong. 2004. Wavelet Transform Applied to EO-1 Hyperspectral Data for Forest LAI and Crown Closure Mapping. *Remote Sensing of Environment*, Vol. 91, pp. 212–224.

Pu, R. L. .2009. Broadleaf Species Recognition with In Situ Hyperspectral Data. *International Journal of Remote Sensing*, Vol. 30, pp. 2759–2779.

Walker, J. S. 2008. *A Primer on Wavelets and Their Scientific Applications.* 2nd ed. Boca Raton FL: Chapman and Hall/CRC, 155 pp.

Zhang, J., B. Rivard, A. Sanchez-Azofeiza, and K. Casto-Esau. 2006. Intra- and Interclass Spectral Variability of Tropical Tree Species at La Selva, Costa Rica: Implications for Species Identification Using HYDICE Imagery. *Remote Sensing of Environment*, Vol. 105, pp. 129–141.

CHAPTER SIXTEEN

Change Detection

16.1. Introduction

Repeated imaging enables assessment of changes in the type or condition of surface features. This is one of the most important of all analyses in remote sensing, typically called *change detection*. Many of these analyses use images acquired at two points in time, known as *bitemporal change detection*, the primary focus of this chapter. Increasingly, however, more widely available image time series have afforded the possibility of *multitemporal change detection*, addressed in the last section.

Comparing images subsequent to classifying each is called *postclassification change detection*. Postclassification change detection consists only of comparing the "from" class and "to" class for each pixel or segment. Even though this operation is a simple raster GIS analysis, it is often not the first choice of most analysts. The reason is that the resulting accuracy is typically low, as it incorporates any errors present in the original classifications, approximated as the product of the overall accuracies of the individual classifications (Coppin et al. 2004).

In part because of the inherently low accuracy of postclassification change detection, analysis of spectral changes between two (or more) dates, called *spectral change detection* (Plate 18), is often preferred for analysis of traditional optical imagery (rather than lidar or radar data). Whether the unit of analysis is a pixel, a neighborhood, a multitemporal segment, or even (rarely) a spectral class, it is very important to follow the image preprocessing steps that minimize signal from variation that could be confused with the change detection signal of interest. Ideally, this means that the images being compared be:

1. Acquired from the same or well intercalibrated sensors and acquired at the same time of day using the same IFOV and look angle.

2. For interannual analyses, acquired during the same season to minimize differences due to phenological changes.

3. Well coregistered, preferably to within two-tenths of a pixel or less (Dai and Khorram 1999).

4. Free of clouds in the area of analysis.

5. Corrected to top-of-atmosphere or (preferably) surface reflectance (see Chapter 11).

6. Free of other conditions not deemed part of the signal of interest. Examples of this last point could include the presence of soil moisture differences when assess-

ing changes in forest wetland canopy cover or differences in harvest dates when assessing changes in vegetative cover using anniversary date images acquired in autumn.

It is often difficult in practice to meet all of these conditions, but maximizing signal from the change of interest requires minimizing signal from anything that could be confused with it.

16.2. Bitemporal Spectral Change Detection Techniques

Spectral change detection techniques can be classified (not mutually exclusively) as follows: (1) visual interpretation, (2) image algebra, (3) transformation and data reduction, (4) classification, and (5) statistical. The categories and techniques described here are common in both remote sensing research and applications but are by no means exhaustive. The reader should refer to edited compilations such as Lunetta and Elvidge (2000) or reviews such as Coppin et al. (2004) and Lu et al. (2004) for additional information.

Visual Interpretation

Comparison of images from two dates is always the first place to start and is easily accomplished using one of three basic techniques. In the first, known as a multidate color image composite, two suitable radiometrically calibrated and coregistered images are *composited* (in which all bands from both dates are stacked to form one image), then bands from different dates are chosen for display to highlight the change of interest. In the second, the images are not composited but are displayed with one on top of the other. Tools commonly available in most image processing packages can then be used to visualize change, including swiping one image over the other and flickering between images. The third technique is simply to view coregistered and geolinked images side by side, using the typical pan, zoom, and cursor inquiry tools for targeted analysis.

It should be noted that one of the primary goals of visual interpretation is often to help select the best change detection techniques based on the goals of the analysis. In some instances, however, it might be the mechanism by which the analysis is conducted. If the latter, heads-up digitizing is often used to both identify and classify change between the images. In some instances a sample of points is interpreted. In either instance, however, the analyst is able to use elements of image interpretation besides hue (e.g., texture, shadows, pattern, shape, size, and association; Chapter 5), combined with knowledge of the area of application, with the potential for very accurate results (albeit through a time-consuming and subjective process).

Image Algebra

In image algebra, arithmetic operations are applied to corresponding pixels in each image, with the change image formed from the resulting values. Although there are many possible algebraic operations, the most common are image differencing, image ratios, Euclidean distance, and change vector analysis.

Image differencing, particularly using vegetation indices, is one of the most widely used (and most effective) spectral change detection techniques. It consists of simply subtracting one or more bands from the same band(s) of the same area acquired at a different date (Figure 16.1). Values at or near zero identify pixels that have similar spectral values and therefore presumably have experienced no change between the two dates. The analyst must specify some threshold value, often heuristically, beyond which change is considered to have occurred. The results are very sensitive to the threshold chosen, and, as such, it is often desirable to use an objective process for its selection. A "salt and pepper" appearance also often results because of small spectral changes caused by other factors previously discussed, including slight misregistration.

Taking the ratio of two bands, one from one image date and the other from the same band in the second image date, is another popular change detection technique. A simple numerical example using two Landsat images can illustrate the two approaches. On Date 1 we have a pixel in class *forest* with digital numbers of 100 and 24 in the near infrared and red, respectively. On Date 2, the digital numbers for the same pixel (now in class *urban or bare soil*) are 61 and 36 in the near infrared and red. Let's take a look at the image difference and image ratio for the near infrared band. The image difference would be 100 – 61 = 39. The image ratio would be 100 ÷ 61 = 1.64.

Now, what happens if we want to assess the magnitude of change using more than one band? One common means of doing this is the Euclidean distance introduced in Section 12.3, Equation 12.2. Stated in words, this calculation is nothing more than the square root of the sum of the squares of the distances. Returning to the previous example, the difference in the red band for this pixel is 24 –36 = –12. The square root of the sum of the squares of the distances between these two brightness value vectors is then just

$$D = \sqrt{(39)^2 + (-12)^2} = \sqrt{1665} = 40.8 \qquad \text{(Eq. 16.1)}$$

FIGURE 16.1. Example of an NDVI difference image in which the 1992 Landsat TM-derived NDVI was subtracted from the 2002 Landsat-TM derived NDVI in Rondônia, Brazil. Areas that are brighter showed an increase, those that are darker, a decrease, in NDVI.

Stated in the more easily extendable matrix form, the Euclidean distance is

$$D = \sqrt{(x_1 - x_2)'(x_1 - x_2)} \qquad \text{(Eq. 16.2)}$$

where x is the brightness value vector for the given pixel and the prime represents a matrix transposition. Working this through for our example gives us

$$D = \sqrt{(x_1 - x_2)'(x_1 - x_2)} = \sqrt{[39 - 12]\begin{bmatrix} 39 \\ -12 \end{bmatrix}} = \sqrt{39^2 + (-12)^2} = \sqrt{1665} = 40.8 \qquad \text{(Eq. 16.3)}$$

The Euclidean distance, however, gives only the *magnitude* of the multiband change. What if we also want to know the direction of the change? In this instance, *change vector analysis* (of which Euclidean distance can be considered a subset) is needed. This operation is best examined graphically. Consider Figure 16.2, in which the two pixels we have been examining are shown as vectors, with the near infrared digital number as the y axis and the red digital number as the x axis . The angle between them is a (shown in Figure 16.2), defined as

$$\alpha = \cos^{-1}\left(\frac{x_1 \cdot x_2}{\|x_1\| + \|x_2\|}\right) \qquad \text{(Eq. 16.4)}$$

where

$$x_1 \cdot x_2 = \sum_{i=1}^{n} x_{1,n} x_{2,n} = x_{1,1}x_{2,1} + x_{1,2}x_{2,2} + \ldots + x_{1,n}x_{2,n}$$

which is the *dot product* of the two vectors. For our two vectors, the dot product would be as follows:

$$x_1 \cdot x_2 = \begin{bmatrix} 100 \\ 24 \end{bmatrix}\begin{bmatrix} 61 \\ 36 \end{bmatrix} = (100)(61) + (24)(36) = 6100 + 864 = 6,964$$

The other term in Equation 16.4 that needs explanation is the vector *length* (also called the *norm*), defined as the sum of the squares of the vector elements; that is,

$$\|x\| = x_1^2 + x_2^2 + \ldots + x_n^2 \qquad \text{(Eq. 16.5)}$$

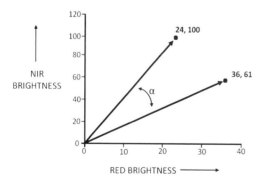

FIGURE 16.2. Change vectors. Two pixels are shown as vectors. The near infrared digital number forms the y axis, and the red digital number is the x axis. The angle between them is α.

For our two vectors, the lengths are then as follows:

$$\|x_1\| = \begin{Vmatrix} 100 \\ 24 \end{Vmatrix} = 100^2 + 24^2 = 10{,}000 + 576 = 10{,}576$$

$$\|x_2\| = \begin{Vmatrix} 61 \\ 36 \end{Vmatrix} = 61^2 + 36^2 = 3{,}721 + 1{,}296 = 5{,}017$$

Note that the vector length in two dimensions is just the hypotenuse calculated using the familiar Pythagorean theorem.

Substituting the dot product and vector lengths into Equation 16.4, we are able to compute the angle between the two vectors as follows:

$$\alpha = \cos^{-1}\left(\frac{x_1 \cdot x_2}{\|x_1\| + \|x_2\|}\right) = \cos^{-1}\left(\frac{6{,}964}{10{,}576 + 5{,}017}\right) = \cos^{-1}(0.45) = 63.5°$$

Let's look at this now from the perspective of the land use change we are trying to detect. The Euclidean distance between the *forest* and *urban or bare soil* was 40.8, giving the magnitude of the change vector. The angle between the two vectors is 63.5°. Note that, for this example, only two bands and two pixels were used for simplicity, but there is nothing to preclude use of any number of spectral or derived bands or alternative units of analysis (e.g., image segments). Further analysis of the change between the vectors can proceed many ways, such as thresholding, discussed earlier in the context of image differencing, or classification, discussed later. Figure 16.3 shows a visual comparison of the spectral angle strategy in relation to a distance measure derived from the same pair of images.

Transformation/Data Reduction

There are a variety of techniques by which the data in the original image can be transformed to new axes composed of linear combinations of the existing bands. The most widely used of these are principal components analysis (PCA) and the tasseled cap transformation, both discussed in Chapters 11 and 17. Other possibilities, more widely utilized in hyperspectral data analysis, include the singular value decomposition and maximum noise fraction.

Recall that PCA reorients the axes of multidimensional data space such that there is no longer any remaining covariance among the PC bands, which are linear combinations of the original bands. This operation results in a de facto ability to provide the same information content using fewer bands. The result has value for change detection in two very particular ways. First, a PCA can be implemented on the composite image. If the change the analyst wishes to detect represents a large portion of the variance between the two dates, then one or more of the resulting PC bands may be oriented such that it highlights that change. When this effect occurs, it is very powerful. Unfortunately, there is no guarantee that PCA will produce this in every application. There are alternative transformations (such as canonical discriminant analysis) that can reorient the axes to maximize discrimination among classes, but these are beyond the scope of this discussion.

The second way in which PCA can be used in bitemporal spectral change detection is to perform the transformation on each image first, then use one of the other techniques discussed in this section (e.g., image differencing, change vector analysis, classification, image cross-correlation) to assess whether there are important changes in the PC bands between dates. Again, however, this technique is not foolproof, as the axis orientations will be different for each date, producing results difficult to interpret. That being said, this can still be a valuable technique, particularly when the images are very similar (thus ensuring a similar axis rotation) except for the change being assessed.

The tasseled cap transformation, because it uses the same coefficients to reorient the axes every time, circumvents many of the issues raised above with respect to the image-dependent nature of PCA. Most commonly, particularly when a mid infrared band is available, the first three axes (*brightness*, *greenness*, and *wetness*) or some derivation therefrom (such as the greenness–brightness ratio) are used. Analysis can then proceed using one or more of the techniques discussed in this section.

Classification

Classification is often underutilized in spectral change detection, but it is extremely powerful, alone or in combination with other techniques discussed in this section. One of the reasons for its use is that we often want not only to *detect* change but also to *attribute* it. As an example, we may want to know not only if there is a loss of forest cover but also, if so, whether the lost forest became a pasture, strip mine, or suburban development. Clas-

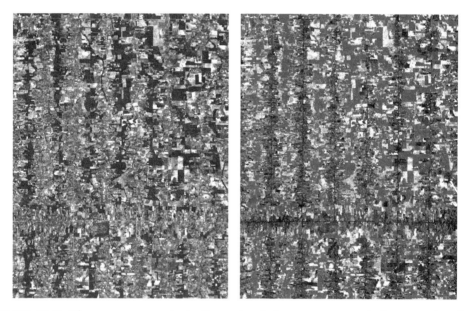

FIGURE 16.3. Change vector analysis. Example of change vector analysis output in which the distance (using all six nonthermal bands) between a 1992 Landsat TM image and a 2002 Landsat TM image acquired over Rondônia, Brazil, is shown at left. The right-hand image is the spectral angle image of the same area, in which the gray tones indicate the value of the spectral angle, with brighter tones indicating higher spectral angles and, therefore, higher spectral differences between the two dates. In this instance, at least, the spectral angle map seems to provide a more nuanced representation of the changes, with fewer areas represented as the dark gray or black tones indicating that no changes had occurred.

sification can occur using the whole area of interest, or it can be used only to attribute change once the areas in which change have occurred have been identified using some other technique.

If classification is used alone on the whole area of interest, a common technique is to use a composite image of either the spectral or transformed bands (e.g., tasseled cap) as the source of data. The brightness value vectors from the composite, whether they be from pixels or multitemporal segments or are made up of bands from two dates. When the change of interest results in separable bitemporal spectral signatures, the classification protocol and decision rule (be it supervised or unsupervised, parametric or nonparametric) can then identify the change of interest. However, there may be other precursors to the classification other than the spectral or transformed bands. These might include, for example, the magnitude and direction from change vector analysis or the Pearson product–moment correlation coefficient from image cross-correlation (see the next subsection). As such, there are a variety of approaches or combinations of approaches that can be used depending on the specific needs of the analysis.

Statistical Techniques

There are numerous ways to compare images statistically, but one of the most widely used techniques is image cross-correlation using pixel neighborhoods or multitemporal segments. The mathematics is the same regardless, but this discussion will presume use of the latter. The equation forms used are those of Im and Jensen (2005, 2008) and use only a single band for simplicity. Multiband correlations are feasible but, absent some method of combining the bands a priori (such as a vegetation index), will result in a correlation matrix, thus increasing the complexity of interpretation.

The Pearson product–moment correlation, r, between brightness values (or reflectances) from a single band across two dates of imagery is defined as

$$r = \frac{\text{cov}_{12}}{s_1 s_2} \tag{Eq. 16.6}$$

where s_1 and s_2 are the standard deviations of the brightness values found in each segment and cov_{12} is the covariance between brightness values of the two dates for the segment.

The standard deviation for a given band, date, and segment is as follows:

$$SD = \sqrt{\sum \frac{(y - \bar{y})^2}{n - 1}} \tag{Eq. 16.7}$$

where n is the number of pixels in the segment, y is the brightness value of each pixel, and $\bar{y}$ is the mean of all brightness values in the segment.

Standard tests for the significance of the resulting correlation(s) can be used as an additional variable, as can the slope and intercept derived from a least-squares analysis of the brightness value pairs within the segment (Im and Jensen 2005, 2008).

Choosing a Technique

The first and most important choice is usually whether or not to first classify the data from each date and then proceed with a postclassification change detection. If the classification accuracy will be high and the classification process objective with both images, this is a clear choice. However, if, as is often the case, the classification accuracy will neither be high nor the process objective, then spectral change detection presents a very viable alternative as long as the change of interest is spectrally distinguishable. Visual interpretation can often help the analyst ascertain (1) whether the change of interest is distinguishable and (2) if so, which of the several techniques presented here, alone or in combination, might be most useful. Knowledge of the spectral properties of the material type or conditions of interest—and how they change—is essential.

16.3. Multitemporal Spectral Change Detection

No-Cost Data Enabling New Class of Analyses

Data policies affording access to no-cost satellite data at a coarse resolution have been common within the meteorological community for decades. However, the land remote sensing community, typically desiring to address the scale at which land is used and managed (and thus changes) has required moderate- to high-resolution data. As discussed in other chapters, even satellites built, launched, and operated using public funds have had cost recovery mechanisms in place that effectively limited data availability (and thus algorithm development). However, terrain-corrected Landsat data are now available for preview and download at no cost through the USGS Global Visualization Viewer or USGS EarthExplorer (for more details, see the more detailed discussion elsewhere in this volume). The availability of these data is opening up a rich set of applications never before possible, as analysts can now use *multitemporal* stacks of images, be they of original or derived bands, to conduct their analyses. Differences in the seasonal expression of vegetation or other features used within a single year—or *phenology*—have been addressed elsewhere in the book. This section addresses comparable analyses using interannual image chronosequences, referred to as *interannual multitemporal change detection*.

Tradeoffs

Before getting into some of the details on how these analyses can be conducted, it is worth thinking about some of the advantages and disadvantages of interannual multi-temporal change detection. Advantages include the following (Robert Kennedy, personal communication, 2010):

- Low-magnitude change can be better separated from scene differences not arising from changes in land use or management (e.g., sun or sensor angle, phenology, and atmospheric variability).
- Slow rates of change (e.g., vegetation growth, ecological succession, and drought- or pathogen-induced mortality) can often be detected just as well as abrupt phenomena (e.g., vegetation clearing or fire).
- Annual time steps are often better suited to many socioeconomic and natural processes than longer time steps.

There are also some disadvantages associated with multitemporal spectral change detection, including the following:

- Persistent cloud cover over a given area that spans multiple years of images must be explicitly addressed for most algorithms to function (this is often the largest constraint to analysis).
- Images must be geometrically and radiometrically the same as much as possible.
- Phenology, drought, uncorrected atmospheric or sensor effects, and other factors can affect reflectance in a manner unrelated to the signal of interest (e.g., vegetation recovery).
- Temporal signatures, although useful, require calibration to the timing of an event of interest (e.g., vegetation clearing) that can occur across a range of years.
- Data volume can be very high.
- Algorithm development is still in its nascent stages, with few commercial remote sensing image processing packages fully supporting this type of analysis.

Preprocessing

Although approaches to analyzing interannual multitemporal chronosequences of (primarily Landsat) images are rapidly evolving, a close look at some of the seminal early studies characterizing forest disturbance and recovery can help inform the preprocessing and algorithm selection for any given problem and dataset. Although steps can and do differ from application to application, the following preprocessing steps are widely utilized:

1. Acquire images at or near the same time of year and phenological condition.
2. Ensure that the images are well coregistered.
3. Convert to top-of-atmosphere or surface reflectance (Chapter 11).
4. Remove clouds (usually a combination of semiautomated and manual techniques).
5. (Optionally) choose or derive a single band with which to work.
6. Normalize the image to remove remnant atmospheric, phenological, or sensor effects.
7. (Optionally) segment the image into objects in which spatially juxtaposed pixels with similar spectral trajectories are grouped (*sensu* Lhermitte et al. 2008)

As noted earlier, cloud cover often forms a major constraint. The analyst will, of course, choose the best image for a given year that minimizes cloud cover, undesired phenological effects, and the like. However, for most areas the requirement of having one image per year inevitably leads to having to use some images that are not ideal, particularly with respect to cloud cover. Once the clouds are removed, the first question that arises is how to address the resulting gaps in the time series for a given area. This is an active area of research, but in general there are only three principal choices, as follows: (1) (for small areas) interpolate using adjacent pixel values in the same image, (2) interpolate using images acquired near the same date as the preferred image, or (3) use techniques by which missing values are explicitly handled in the subsequent analysis.

A second constraint to multitemporal analysis is often the remnant variability in the surface reflectance not due to targeted changes in surface material type or condition. Often these effects are the result of subtle changes in phenology, but there are a host of other possible sources of this variability. The image normalization mentioned in Step 6 is designed to address this remnant variability in reflectance at the preprocessing stage. There are a variety of tools by which this can be achieved, but the analyst should be very careful in their selection and use, as all change the data from their original values. Although the hope is that these changes help separate the signal of interest from the noise, there are many instances in which use of these tools reduces the quality of the resulting analysis. Discussion of all the possibilities is beyond the scope of this text, but area-based image matching techniques include histogram equalization, image cross-correlation, image-to-image regression, phase correlation, and maximization of mutual information (Zitová and Flusser 2003). The reader should consult Zitová and Flusser (2003) plus standard remote sensing image processing texts (e.g., Richards and Jia 2006) for additional information.

The particular demands of multitemporal image analysis have also given rise to their own class of image matching techniques. These are based on the idea of normalizing to a reference material type or condition. For example, for deforestation studies it might be helpful to first identify forest that is undisturbed throughout the period of inquiry. In the next step, the difference between the mean brightness value or reflectance vector for the reference forest and each pixel's or object's brightness value or reflectance vector is calculated, as shown in the following equation (Browder et al. 2005):

$$d = \sqrt{(\mathbf{f} - \mathbf{m}_k)(\mathbf{f} - \mathbf{m}_k)'} \qquad \text{(Eq. 16.8)}$$

where d is the Euclidean distance, $\mathbf{f}$ is the mean brightness value vector for primary forest, and $\mathbf{m}$ is the mean brightness value vector for each pixel, k. Browder et al. (2005) used this approach as a preprocessing step to assess whether the species planted in agroforestry plots in Rondônia made any difference to forest recovery. As shown in Figure 16.4, the onset of active management in 1993 resulted in a notable increase in the distance from forest in all groups due to the clearing of vegetation in the plot establishment phase. The growth in the year following plot establishment resulted in a decrease in distance to forest as vegetation cover returned to the plot, but the degree to which the groups were temporally coherent after the establishment phase varied. The plot composition (and resulting management) differences in the three groups appear to be associated with different land use strategies, with the *timber* group being most proximate to the spectral characteristics of primary forest toward the end of the project period.

Healey et al. (2005) took a different (but related) approach to both the image normalization and the creation of a single derived band, the *disturbance index* (DI), based on the first three tasseled cap (Crist and Cicone, 1984; Kauth and Thomas, 1976) bands: brightness, B; greenness, G; and wetness, W (Chapter 17). Each tasseled cap band is rescaled to its standard deviation above or below the scene's mean undisturbed forest value, as follows:

$$B_r = \frac{(B - \mu_B)}{\sigma_B}$$

$$G_r = \frac{(G - \mu_G)}{\sigma_G}$$

$$W_r = \frac{(W - \mu_W)}{\sigma_W}$$

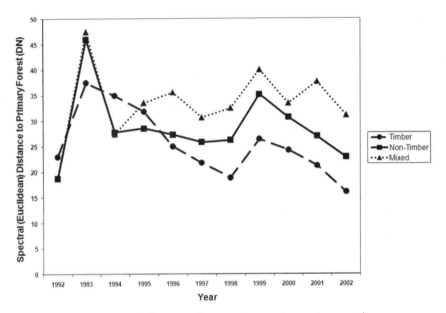

FIGURE 16.4. Mean spectral distance from primary forest by agroforestry group type. Euclidean distance between reference forest condition and agroforestry practices using Landsat TM/ETM+ imagery from 1992 to 2002. After Browder et al. (2005).

where B_r, G_r, and W_r are the rescaled tasseled cap indices; μ_B, μ_G, and μ_W are the means of the tasseled cap indices in undisturbed forest; and σ_B, σ_G, and σ_W are the standard deviations of the tasseled cap indices in undisturbed forest. Note that this method explicitly addresses interband differences in variance. Healey et al. (2005) point out that recently cleared forest exhibits high brightness and low greenness and wetness in relation to undisturbed forest. *DI* is defined to take advantage of these disturbance-related changes in the tasseled cap indices, as follows:

$$DI = B_r^{10} - (G_r + W_r)$$

Regardless of the technique chosen, the analyst should realize that in many cases some sort of image normalization or matching procedure may be necessary to maximize signal and minimize noise across multiple scenes.

Time-Series Analysis

A *time series* is a sequence of data points evenly sampled through time. Classical analysis of time series can be used with image chronosequences. Although specificity on these techniques is beyond the scope of this text, there is merit in a brief discussion (especially of key constraints to implementation) given the clear relevance to this chapter. For readers who wish to follow up, we recommend any of several good introductory texts on time-series analysis (e.g., Montgomery et al. 2008).

Assuming an evenly sampled data set, classical time-series analysis can separate trends in the data from periodic fluctuations (often due to seasonality) and measurement error (noise). However, two major constraints to time-series analysis using imagery are (1) the presence of step changes in reflectance caused by land cover change, such as forest clearing; and (2) the need to have the data points evenly sampled. The latter is typically quite unlikely using moderate-resolution Earth resource satellite data, as the analyst typically has to choose the best possible image each year under other constraints (such as being within a certain part of the growing season). This has two major deleterious effects on time-series analysis. First, the time series is then inherently unevenly sampled. Second, and perhaps more perniciously, subtle differences in reflectance due to phenology are often introduced. New methods to predict daily surface reflectance at moderate (e.g., Landsat) resolution have been introduced (Gao et al. 2006), and these show promise for addressing both these issues when appropriate imagery is available and landscape fragmentation is not overly high.

Analysis of Vegetation Disturbance and Recovery

Time-series analysis will form an ever-larger portion of the suite of tools used for multitemporal change analysis. However, in the last few years simpler tools for analysis of vegetation *disturbance* and *recovery* have been developed and have merit by themselves or with simple extensions to other multitemporal change analysis problems.

There are many ways in which the analysis of the resulting time series can take place, most of which are far beyond the scope of this text. However, work carried out by faculty and students at Virginia Tech on identifying surface coal mines in the southern Appalachians and analyzing their postreclamation vegetation recovery using interannual, multitemporal Landsat data provides a good illustration of typical steps. It should be noted that this is but an illustrative example; a wide range of both applications and analytical approaches are feasible. Figure 16.5 shows a stereotyped representation of some of the key features of such an analysis through the life cycle of a surface coal mine in the southern Appalachians.

Figure 16.6 shows an actual trajectory for the interval 1984 to 2008. Most important are the features that are diagnostic of the disturbance and its type. Notice the substantial drop (in 1997) as the clearing of the native hardwoods precedes mining. The 3 years of the mining period have only minor fluctuations in the vegetation index value. However,

FIGURE 16.5. Interannual multitemporal trajectory of a vegetation index for a surface coal mine in the southern Appalachians. Notice the significant drop in vegetation as the area to be mined is cleared of trees, then its gradual recovery after reclamation. From C. E. Zipper, R. H. Wynne, and S. Sen, Virginia Tech.

FIGURE 16.6. Vegetation index trajectory (1984 to 2008) for a mining disturbance that is reclaimed promptly compared with other trajectories common in the area. (Vegetation index images are shown at key points in the trajectory.) Vegetation index trajectories contrasted to the background undisturbed forest of surrounding areas (left) and (right) typical recovery trajectories of other common disturbances. The horizontal time dimension constitutes several decades. From C. E. Zipper, R. H. Wynne, and S. Sen, Virginia Tech.

the area is reclaimed after the coal has been extracted, leading to a rapid recovery (starting in 2000) of the vegetation. Note, however, that even after 8 years the vegetation index has not yet recovered its value prior to the onset of mining.

Both the magnitude of the drop and the slope of the vegetation recovery are diagnostic of the type of disturbance (Figure 16.6). The mined areas are more completely cleared of vegetation and thus have the most precipitous drop; the drop for mined areas is greater than those caused by either forest fires or clearing for urban development (in this area). Further, recovery does not begin immediately in mined areas, but only after the cessation of mining and onset of reclamation. The recovery is steep for mines but more gradual (with a lower asymptote) for urban development. As such, the following three components of a typical mining trajectory can be used to discriminate mining from other common disturbances:

1. The minimum value that equals or is less than a threshold (the mining delineation *threshold*): The sharp drop of the vegetation index, caused by the rapid clearing of vegetation and surface soil to expose underlying geologic material, can discriminate mining from less drastic landscape disturbances such as forest harvests and forest fires.

2. The *recovery trajectory*: Following the drop, the trajectory takes an upward trend due to vegetation development. The recovery is steepest during its first 7–9 years, and the slope of the trajectory in these years can be used to discriminate mining

from construction disturbances such as urban development and road construction.

3. The *recovery maximum*: The maximum value in the recovery phase can be used to help assess the quality of the reclamation process or to separate mining from other disturbances.

The same type of disturbance and recovery reflectance or vegetation index trajectory can lend itself to many other parameters describing its shape. For example, Kennedy et al. (2007) defined four disturbance classes (*disturbance, disturbance and recovery, recovery*, and *recovery to stable state*). For hypothesized trajectories describing each of these classes, they defined three shape parameters, as follows: (1) the end year of the disturbance, (2) the predisturbance mean, and (3) the postdisturbance mean (Figure 16.7). Each pixel was classified based on the best nonlinear least-squares fit between the observed spectral trajectory (from Landsat TM/ETM+ band 5) and each of the four hypothesized trajectories.

Focusing only on recovery trajectories, Schroeder et al. (2007) first mapped clearcuts, then used ISODATA clustering to develop and map four forest-regrowth classes (*little to no, slow, moderate*, and *fast*) developed using Landsat TM/ETM+ data. Third-order polynomial curves fit to each of three periods' forest-regrowth classes are shown as Figure 16.8.

The preceding examples are helpful in understanding the important role that multi-

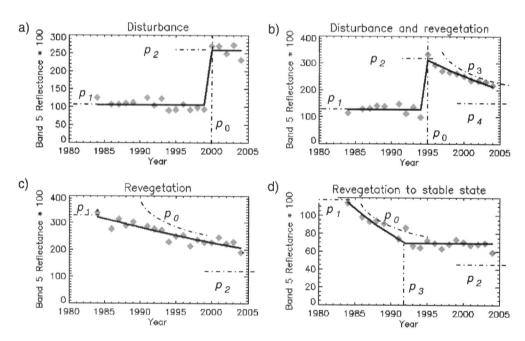

FIGURE 16.7. A detailed view of fitted values from four pixels representing four hypothesized models of disturbance or recovery. (a) Simple disturbance. (b) Disturbance followed by exponential revegetation. (c) Revegetation from disturbance prior to observation record. (d) Revegetation from prior disturbance, reaching a stable point during the observation period. From Kennedy et al. (2005).

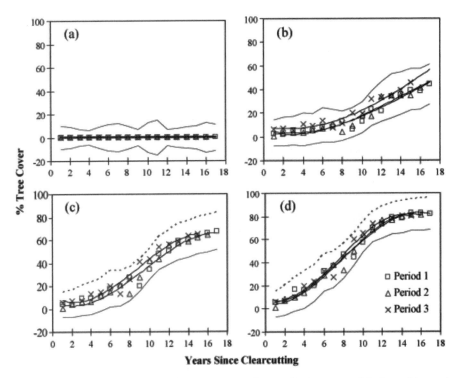

FIGURE 16.8. Mean forest regrowth trajectories for (a) little to no, (b) slow, (c) moderate, and (d) fast regrowth classes. Solid lines are fitted third-order polynomial curves; dashed lines are the average across period standard deviations. From Schroeder et al. (2007).

temporal change detection can play in analyzing disturbance and recovery. The important points to remember are as follows:

- Local knowledge of the land use and land cover transitions that are typical in the area is essential.
- Analysis of the resulting time series requires a close understanding of the spectral dynamics associated with land use and land cover change.
- Spectral trajectories require a new way of thinking about and analyzing remotely sensed data.

It is important to recognize also that although these three examples focus primary on vegetation disturbance and recovery, there is nothing to preclude using similar approaches for more traditional classification problems, including vegetation mapping. In Figure 16.9, for example, invasive shrub species are separable from native hardwoods and pines because of their growth trajectories.

For the analyst or student interested in analyzing spectral trajectories, we must reiterate the importance of appropriate preprocessing. After that, it is extremely important to know your area and problem well, as is true with all remote sensing. This type of analysis will burgeon with both innovation and increasingly diverse applications in the coming years, as both data and appropriate analytical tools become increasingly available.

16.4. Summary

Change detection, whether it be bitemporal or multitemporal, is a mainstay of remote sensing analyses. Its importance is magnified by the extensive changes in land use occurring with swelling—and increasingly well off—human populations. No-cost, well-preprocessed data, increasingly available from programs such as Landsat and Sentinel (European Space Agency), are essential to ensuring the data quantity and quality necessary for these analyses. Remotely sensed data are essential to our ability to monitor and model the changes we are making to the Earth's ecosystems—and to ensure their sustainability.

Review Questions

1. Prepare a list of reasons why development of effective change detection capability is an important capability to support public policy at international, national, and local scales.

2. In your next job, you are called on to testify in court to defend the rigor of a change detection analysis that has been challenged as unscientific. Prepare notes that you would use to defend the validity of the analysis.

3. You do such a good job in preparing your responses for Question 2 that the opposition has now hired you to attack the validity of a change analysis in dispute in their next case. Prepare notes that you will use to question the validity of this new change analysis.

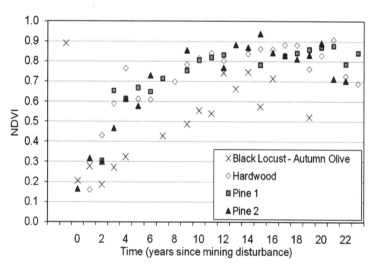

FIGURE 16.9. Invasive species detected using interannual, multitemporal image chronosequences. NDVI trajectories from disturbed sites in Virginia and West Virginia that have developed full canopies of differing woody species types. The trajectories were not coincident in time but have been plotted with "Year 0" as the estimated date of the original disturbance. Missing annual values, which are evident in the Black Locust—Autumn Olive data series, occur because of occasional obscuration by clouds. From C. E. Zipper, R. H. Wynne, and S. Sen, Virginia Tech.

4. Can you identify landscapes that might be inherently difficult to analyze using spectral change detection?

5. Outline ideal circumstances for conducting change analysis using remotely sensed data.

6. Review the text for this chapter that describes preprocessing necessary for execution of change detection. Prepare a list of the most important considerations.

7. Even if images have been selected carefully and processed, change detection can be subject to errors. Review the text for this chapter to identify some of the sources of error that are inherent to the process of change detection.

8. Why are established techniques for analyzing time series seldom used in analysis of remotely sensed data?

9. For a given spatial resolution, there is often a level of landscape fragmentation that precludes accurate spectral change detection. Why might this be the case?

10. List three advantages and three disadvantages of multitemporal change detection as compared with bitemporal change detection.

References

Browder, J., R. H. Wynne, and M. Pedlowski. 2005. Agroforestry Diffusion and Secondary Forest Regeneration in the Brazilian Amazon: Further Findings from the Rondônia Agroforestry Pilot Project (1992–2002). *Agroforestry Systems*, Vol. 65, pp. 99–111.

Coppin, P., I. Jonckheere, K. Nackaerts, B. Muys, and E. Lambin. 2004. Digital Change Detection Methods in Ecosystem Monitoring: A Review. *International Journal of Remote Sensing*, Vol. 25, pp. 1565–1596.

Crist, E. P., and R. C. Cicone. 1984. A Physically-Based Transformation of Thematic Mapper Data: The TM Tasseled Cap. *IEEE Transactions on Geoscience and Remote Sensing*, Vol. GE-22, pp. 256–263.

Crist, E. P., and R. C. Cicone. 1984. Application of the Tasseled Cap Concept to Simulated Thematic Mapper Data. *Photogrammetric Engineering and Remote Sensing*, Vol. 50, pp. 343–352.

Dai, X., and S. Khorram. 1999. Data Fusion Using Artificial Neural Networks: A Case Study on Multitemporal Image Analysis. *Journal of Computers, Environment, and Urban Systems*, Vol. 23, pp. 19–31.

Donoghue, D. N. M. 2002. Remote Sensing: Environmental Change. *Progress in Physical Geography*, Vol. 26, pp. 144–151.

Franklin, S. E., M. B. Lavigne, M. A. Wulder, and G. B. Stenhouse. 2002. Change Detection and Landscape Structure Mapping Using Remote Sensing. *Forestry Chronicle*, Vol. 78, pp. 618–625.

Gao, F., J. Masek, M. Schwaller, and F. Hall. 2006. On the Blending of the Landsat and MODIS Surface Reflectance: Predicting Daily Landsat Surface Reflectance. *IEEE Transactions on Geoscience and Remote Sensing*, Vol. 44, pp. 2207–2218.

Healey, S. P., W. B. Cohen, Y. Zhiqiang, and O. N. Krankina. 2005. Comparison of Tasseled Cap-Based Landsat Data Structures for Use in Forest Disturbance Detection. *Remote Sensing of Environment*, Vol. 97, pp. 301–310.

Hilker, T., M. A. Wulder, N. C. Coops, N. Seitz, J. C. White, F. Gao, et al. 2009. Generation of Dense Time Series Synthetic Landsat Data through Data Blending with MODIS Using

462   III. ANALYSIS

a Spatial and Temporal Adaptive Reflectance Fusion Model. *Remote Sensing of Environment*, Vol. 113, pp. 1988–1999.

Im, J., and J. R. Jensen. 2005. A Change Detection Model Based on Neighborhood Correlation Image Analysis and Decision Tree Classification. *Remote Sensing of Environment*, Vol. 99, pp. 326–340.

Im, J., J. R. Jensen, and J. A. Tullis. 2008. Object-Based Change Detection Using Segmentation Correlation Image Analysis and Image Segmentation. *International Journal of Remote Sensing*, Vol. 29, pp. 399–423.

Kauth, R. J., and G. S. Thomas. 1976. The Tasseled Cap: A Graphic Description of the Spectral–Temporal Development of Agricultural Crops as Seen by Landsat. In *LARS: Proceedings of the Symposium on Machine Processing of Remotely Sensed Data*. West Lafayette, IN: Perdue University Press, pp. 4B-41–4B-51.

Kennedy, R. E, W. B. Cohen, and T. A. Schroeder. 2007. Trajectory-Based Change Detection for Automated Characterization of Forest Disturbance Dynamics. *Remote Sensing of Environment*, Vol. 10, pp. 370–386.

Lu, D. P., Mausel, E. Brondizio, and E. Moran. 2004. Change Detection Techniques. *International Journal of Remote Sensing*, Vol. 26, pp. 2365–2401.

Lunetta, R., and C. Elvidge. 2000. *Remote Sensing Change Detection: Environmental Monitoring Methods and Applications*. London: Taylor and Francis, 320 pp.

Montgomery, D. C., C. L. Jennings, and M. Kulahci. 2008. *Introduction to Time Series Analysis and Forecasting*. Hoboken, N.J.: Wiley-Interscience, 392 pp.

Richards, J. A., and X. Jia. 2006. *Remote Sensing Digital Image Analysis: An Introduction*. Berlin: Springer, 439 pp.

Schroeder, T. A., W. B. Cohen, and Z. Q. Yang. 2007. Patterns of Forest Regrowth Following Clearcutting in Western Oregon as Determined by a Landsat Time-Series. *Forest Ecology and Management*, Vol. 243, pp. 259–273.

Todd A., T. A. Schroeder, W. B. Cohen, and Z. Yang. 2007. Patterns of Forest Regrowth Following Clearcutting in Western Oregon as Determined from a Landsat Time Series. *Forest Ecology and Management*, Vol. 243, pp. 259–273.

Vogelmann, J. E. 1988. Detection of Forest Change in the Green Mountains of Vermont Using Multispectral Scanner Data. *International Journal of Remote Sensing*, Vol. 9, pp. 1187–1200.

Zitová, B., and J. Flusser. 2003. Image Registration Methods: A Survey. *Image and Vision Computing*, Vol. 21, pp. 977–1000.

PART FOUR

APPLICATIONS

CHAPTER SEVENTEEN

Plant Sciences

17.1. Introduction

The Earth's vegetative cover is often the first surface encountered by the energy we use for remote sensing. So, for much of the Earth's land area, remote sensing imagery records chiefly the character of the vegetation at the surface. Therefore, our ability to interpret the Earth's vegetation canopy forms the key to knowledge of other distributions, such as geologic and pedologic patterns that are not directly visible but do manifest themselves indirectly through variations in the character and distribution of the vegetation cover.

In other situations, we have a direct interest in the vegetation itself. Remote sensing can be useful for monitoring areas planted to specific crops, for detecting plant diseases and insect infestations, and for contributing to accurate crop production forecasts. In addition, remote sensing has been used to map forests, including assessments of timber volume, insect infestation, and site quality.

Further, remote sensing provides the only practical means of mapping and monitoring changes in major ecological regions that, although not directly used for production of food or fiber, have great long-term significance for mankind. For example, the tropical forests that cover significant areas of the Earth's surface have never been mapped or studied except in local regions that are unlikely to be representative of the unstudied regions. Yet these regions are of critical importance to mankind due to their role in maintaining the Earth's climate (Rouse et al., 1979) and as a source of genetic diversity. Humans are rapidly destroying large areas of tropical forests: it is only by means of remote sensing that we are ever likely to understand the nature and locations of these changes. Similar issues exist with respect to other ecological zones; remote sensing provides a means to observe such regions at global scales and to better understand the interrelationships between the many factors that influence such patterns.

Vegetation Classification and Mapping

Vegetation classification can proceed along any of several alternative avenues. The most fundamental is simply to separate vegetated from nonvegetated regions or forested from open lands. Such distinctions, although ostensibly very simple, can have great significance in some contexts, especially when data are aggregated over large areas or are observed over long intervals of time. Thus national or state governments, for example, may have an interest in knowing how much of their territories are covered by forest or may want to make changes in forested land from one 10-year period to the next, even though there may be no data available regarding the different kinds of forest.

However, it is usually important to acquire information at finer levels of detail. Although individual plants can be identified on aerial photographs, seldom if ever is it practical to use the individual plant as the unit for mapping vegetation. Instead, it is more useful to define mapping units that represent groups of plants. A *plant community* is an aggregation of plants with mutual interrelationships among each other and with the environment. Thus, an *oak–hickory forest* is a useful designation because we know that communities are not formed by random collections of plants but by consistent associations of the same groups of plants—plants that tend to prefer the same environmental conditions and to create the environments that permit certain other plants to exist nearby. A community consists of many *stands*—specific, individual occurrences of a given community (Figure 17.1).

Plants within communities do not occur in equal proportions. Certain species tend to dominate; these species are often used to name communities (e.g., *hickory* forest), although others may be present. Dominant species may also dominate physically, forming the largest plants in a sequence of layers, or *strata*, that are present in virtually all communities. *Stratification* is the tendency of communities to be organized vertically, with some species forming an upper canopy, others a middle stratum, with shrubs, mosses, lichens, and so on forming other layers nearer to the ground. Even ostensibly simple vegetation communities in grasslands or arctic tundra, for example, can be shown to consist

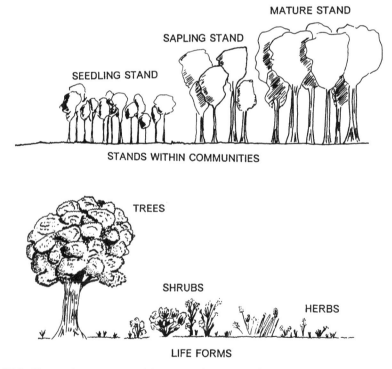

FIGURE 17.1. Vegetation communities, stands, vertical stratification, and different life forms.

of distinct strata. Specific plants within each community tend to favor distinctive positions within each strata because they have evolved to grow best under the conditions of light, temperature, wind, and humidity that prevail in their strata.

Floristics refers to the botanical classification of plants, usually based on the character of the reproductive organs, using the system founded by Carolus Linnaeus (1707–1778). Linnaeus, a Swedish botanist, created the basis for the binomial system of designating plants by Latin or Latinized names that specify a hierarchical nomenclature, of which the *genus* and *species* (Table 17.1) are most frequently used. The *Linnaean system* provides a distinctive name that places each plant in relationship with others in the taxonomy. Floristic classification reveals the genetic character and evolutionary origin of individual plants.

In contrast, the *life form*, or *physiognomy*, describes the physical form of plants (Table 17.2). For example, common physiognomies might include "tree," "shrub," "herbaceous vegetation," and so on. Physiognomy is important because it reveals the *ecological role* of the plant, the nature of its relationship with the environment and other plants. Floristics and physiognomy often have little direct relationship to one another. Plants that are quite close floristically may have little similarity in their growth form, and plants that are quite similar in their ecological roles may be very different floristically. For example, the rose family (*Rosaceae*) includes a wide variety of trees, shrubs, and herbaceous plants that occupy diverse environments and ecological settings. Conversely, a single environment can be home to hundreds of different species; for example, the alpine meadows of New England are composed of some 250 species, including such diverse families as *Primulaceae* (primrose), *Labiatae* (mint), *Araliaceae* (ginseng), and *Umbelliferae* (parsley). Thus a purely floristic description of plants (e.g., "the rose family") seldom conveys the kind of information we would like to know about plants, their relationships with one another, and their environment.

Although it is often possible to identify specific plants and to assign taxonomic designations from large-scale imagery, vegetation studies founded on remotely sensed images typically employ the structure and physiognomy of vegetation for classification purposes. That is, usually we wish to separate forest from grassland, for example, or to distinguish between various classes of forest. Although it is important to identify the dominant species for each class, our focus is usually on separation of vegetation communities based on their overall form and structure rather than on floristics alone.

No single approach to vegetation classification can be said to be universally superior to others. At given levels of detail, and for specific purposes, each of the approaches mentioned above serves important functions. Floristic classification is useful when scale is large and mapping is possible in fine detail that permits identification of specific plants. For example, analyses for forest management often require a large scale, both for measurement

TABLE 17.1. Floristic Classification

Level	Example[a]
Class	Angiospermae (broad/general)
Order	Sapindales
Family	Aceraccae
Genus	*Acer*
Species	*saccharum* (narrow/specific)

[a]Example = sugar maple (*Acer saccharum*).

TABLE 17.2. Classification by Physiognomy and Structure

Woody plants	Broadleaf evergreen
	Broadleaf deciduous
	Needleleaf evergreen
	Leaves absent
	Mixed
	Semideciduous
Herbaceous plants	Graminoids
	Forbs
	Lichens and mosses
Special life forms	Climbers
	Stem succulents
	Tuft plants
	Bamboos
	Epiphytes

Note. Data from Küchler (1967). Küchler's complete classification specifies plant height, leaf characteristics, and plant coverage.

of timber volume and for identification of individual trees. Physiognomy or structure is important whenever image scale is smaller, detail is coarser, and the analyst focuses more clearly on the relationships of plants to environment. Ecological classification may be used at several scales for analyses that require consideration of broader aspects of planning for resource policy, wildlife management, or inventory of biological resources.

Another approach to classification of vegetation considers vegetation as the most easily observed component of an environmental complex, including vegetation, soil, climate, and topography. This approach classifies regions as *ecological zones*, usually in a hierarchical system comparable to that shown in Table 17.3. Bailey (1998b, p. 145) defines his units as *ecoregions*: "major ecosystem[s], resulting from large-scale predictable patterns of solar irradiation and moisture, which in turn affect the kinds of local ecosystems and animals and plants found there." At the very broadest scales, ecological classification is based on long-term climate and very broad-scale vegetation patterns, traditionally derived from information other than remotely sensed data. However, later sections of this chapter show how it is now possible to use remotely sensed data to derive these classifications with much more precision and accuracy than was previously possible.

At finer levels of detail, remotely sensed imagery is essential for delineating ecore-

TABLE 17.3. Bailey's Ecosystem Classification

Level	Example
Domain	Humid temperate
Division	Hot continental
Province	Eastern deciduous forest
Section	Mixed mesophytic forest
District	
Land-type association	
Land type	
Land-type phase	
Site	

Note. Data from Bailey (1978).

gions. The interpreter considers not only vegetation cover but also elevation, slope, aspect, and other topographic factors in defining units on the map.

Biophysical Measures of Vegetation Cover

A contrasting approach applies multispectral remote sensing to estimate basic biophysical properties of the Earth's vegetation cover. That is, instead of labeling a region to a particular class or category as explained above, this approach attempts to provide specific measures of the biophysical function of that region. Three common biophysical measures follow.

Vegetation fraction (VF) is defined as the percentage of vegetation occupying a pixel as viewed in vertical projection. It's a comprehensive quantitative index in forest management and vegetation community cover conditions, and it's also an important parameter in many remote sensing ecological models.

Leaf area index (LAI) is the ratio between the total area of the upper leaf surface of vegetation and the surface area of the pixel in question. LAI is a dimensionless value, typically ranging between 0 (for a pixel composed of bare soil) and values as high as 6 (for a dense forest). See Plate 19 for an application that applies remote sensing imagery to estimate LAI for the continental United States.

Net primary production (NPP) is a measure of the inherent productivity of a region or ecological system—mainly the Earth's production of organic matter, principally through the process of photosynthesis in plants. The organisms responsible for primary production, *primary producers*, form the base of the food chain. Whereas *gross primary productivity* estimates the total productivity of a region, NPP is the value that remains after allowing for the amount of organic material used to sustain the producers. Because of the basic significance of NPP to understanding agricultural systems and ecosystems in general, the use of remote sensing is of interest as a tool to measure place-to-place and seasonal variations in NPP (Ruimy et al., 1994).

Kinds of Aerial Imagery for Vegetation Studies

Aerial imagery enables the analyst to conduct quick and accurate delineation of major vegetation units, providing at least preliminary identification of their nature and composition. Note, however, that interpretation from aerial images cannot replace ground observations; an accurate interpretation assumes that the analyst has field experience and knowledge of the area to be examined and will be able to evaluate his or her interpretation in the field.

Stereoscopic coverage provides advantages for interpretation of vegetation coverage. Optimum choice of image scale, if the analyst has control over such variables, depends on the nature of the map and the complexity of the vegetation pattern. Detailed studies have been conducted using photography at scales as large as 1:5,000, but scales from about 1:15,000 to 1:24,000 are probably more typical for general-purpose vegetation studies. Of course, if photographs at several different scales, dates, or seasons are available, multiple coverage can be used to good advantage to study changes. Small-scale images can be used as the basis for delineation of extents of major regions; the greater detail of large-scale images can be used to identify specific plants and plant associations.

Panoramic photographs have been successfully used for a variety of purposes pertaining to vegetation mapping and forest management. Modern panoramic cameras permit acquisition of high-resolution imagery over very wide regions, such that large areas can be surveyed quickly. Such images have extreme variations in scale and perspective, so they cannot be used for measurements, but their wide areal coverage permits a rapid, inexpensive inventory of wide areas and identification of areas that might require examination using more detailed imagery.

Timing of flights, which may not always be under the control of the analyst, can be a critical factor in some projects. For example, mapping the understory in forested areas can be attempted only in the early spring when shrubs and herbaceous vegetation have already bloomed but the forest canopy has not fully emerged to obscure the smaller plants from overhead views. Because not all plants bloom at the same time, a succession of carefully timed photographic missions in the spring can sometimes record the sequence in which specific species bloom, thereby permitting mapping of detail that could not be reliably determined by a single image showing all trees in full bloom.

Infrared and color infrared imagery have obvious advantages for vegetation studies of all kinds, as emphasized previously. The CIR photography acquired at high altitudes for the United States (usually during leaf-off conditions) is routinely available at modest cost through the NAPP (Section 3.13). Custom-flown CIR imagery is likely to be considered expensive; many users will find black-and-white imagery satisfactory for many purposes.

17.2. Structure of the Leaf

Many applications of remote sensing to vegetation patterns depend on a knowledge of the spectral properties of individual leaves and plants. These properties are best understood by examining leaf structure at a rather fine level of detail.

The cross-section of a typical leaf reveals its essential elements (Figure 17.2). The uppermost layer, the *upper epidermis*, consists of specialized cells that fit closely together without openings or gaps between cells. This upper epidermis is covered by the *cuticle*, a translucent waxy layer that prevents moisture loss from the interior of the leaf. The underside of the leaf is protected by the *lower epidermis*, similar to the upper epidermis except that it includes openings called *stomates* (or *stomata*), which permit movement of air into the interior of the leaf. Each stomate is protected by a pair of *guard cells* that can open and close as necessary to facilitate or prevent movement of air to the interior

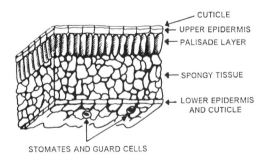

FIGURE 17.2. Diagram of a cross-section of a typical leaf.

of the leaf. The primary function of stomata is apparently to allow atmospheric CO_2 to enter the leaf for photosynthesis. Although the guard cells and the epidermis appear to be small and inefficient, they are in fact very effective in transmitting gases from one side of the epidermis to the other. Their role in permitting atmospheric carbon dioxide (CO_2) to enter the leaf is obviously essential for the growth of the plant, but they also play a critical role in maintaining the thermal balance of the leaf by permitting movement of moisture to and from the interior of the leaf. The guard cells can close to prevent excessive movement of moisture and thereby economize moisture use by the plant. Apparently the positions of stomata on the lower side of the leaf favor maximum transmission of light through the upper epidermis and minimize moisture loss when stomata are open.

On the upper side of the leaf, just below the epidermis, is the *palisade tissue* consisting of vertically elongated cells arranged in parallel, at right angles to the epidermis. Palisade cells include *chloroplasts*, cells composed of chlorophyll and other ("accessory") pigments active in photosynthesis, as described below. Below the palisade tissue is the spongy *mesophyll tissue*, which consists of irregularly shaped cells separated by interconnected openings. The surface of the mesophyll has a very large surface area; it is the site for the oxygen (O_2) and CO_2 exchange necessary for photosynthesis and respiration. Although leaf structure is not identical for all plants, this description provides a general outline of the major elements common to most plants, especially those that are likely to be important in agricultural and forestry studies.

In the visible portion of the spectrum, chlorophyll controls much of the spectral response of the living leaf (Figure 17.3). *Chlorophyll* is the green pigment that is chiefly responsible for the green color of living vegetation. Chlorophyll enables the plant to absorb sunlight, thereby making photosynthesis possible; it is located in specialized lens-shaped structures, known as *chloroplasts*, found in the palisade layer. Light that passes through the upper tissues of the leaf is received by chlorophyll molecules in the palisade layer, which is apparently specialized for photosynthesis, as it contains the largest chloroplasts in greater abundance than other portions of the plant. CO_2 (as a component of the natural atmosphere) enters the leaf through stomata on the underside of the leaf and then diffuses throughout cavities within the leaf. Thus photosynthesis creates carbohydrates from CO_2 and water (H_2O), using the ability of chloroplasts to absorb sunlight as a source of energy. Chloroplasts include a variety of pigments, some known as *accessory pigments*, that can absorb light, then pass its energy to chlorophyll. Chlorophyll occurs in two forms. The most common is chlorophyll *a*, the most important photosynthetic agent in most green plants. A second form, known as chlorophyll

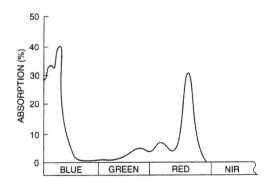

FIGURE 17.3. Absorption spectrum of a typical leaf. The leaf absorbs blue and red radiation and reflects green and near infrared radiation.

b, has a slightly different molecular structure; it is found in most green leaves but also in some algae and bacteria.

17.3. Spectral Behavior of the Living Leaf

Chlorophyll does not absorb all sunlight equally. Chlorophyll molecules preferentially absorb blue and red light for use in photosynthesis (Figure 17.4). They may absorb as much as 70–90% of incident light in these regions. Much less of the green light is absorbed and more is reflected, so the human observer, who can see only the visible spectrum, sees the dominant reflection of green light as the color of living vegetation (Figures 17.3 and 17.5).

In the near infrared spectrum, reflection of the leaf is controlled not by plant pigments but by the structure of the spongy mesophyll tissue. The cuticle and epidermis are almost completely transparent to infrared radiation, so very little infrared radiation is reflected from the outer portion of the leaf. Radiation passing through the upper epidermis is strongly scattered by mesophyll tissue and cavities within the leaf. Very little of this infrared energy is absorbed internally; most (up to 60%) is scattered upward (which we call "reflected energy") or downward ("transmitted energy"). Some studies suggest that palisade tissue may also be important in infrared reflectance. Thus the internal structure of the leaf is responsible for the bright infrared reflectance of living vegetation.

At the edge of the visible spectrum, as the absorption of red light by chlorophyll pigments begins to decline, reflectance rises sharply (Figure 17.5). Thus, if reflectance is considered not only in the visible but also across the visible and the near infrared, peak reflectance of living vegetation is not in the green but in the near infrared. This behavior explains the great utility of the near infrared spectrum for vegetation studies and facilitates separation of vegetated from nonvegetated surfaces, which are usually much darker in the near infrared. Furthermore, differences in reflectivities of plant species often are more pronounced in the near infrared than they are in the visible, meaning that discrimination of vegetation classes is sometimes made possible by use of the near infrared reflectance (Figure 17.6).

As a plant matures or is subjected to stress by disease, insect infestation, or moisture shortage, the spectral characteristics of the leaf may change. In general, these changes apparently occur more or less simultaneously in both the visible and the near infrared regions, but changes in near infrared reflectance are often more noticeable. Reflectance

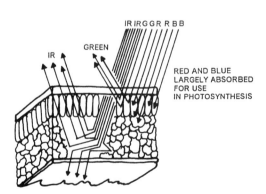

FIGURE 17.4. Interaction of leaf structure with visible and infrared radiation.

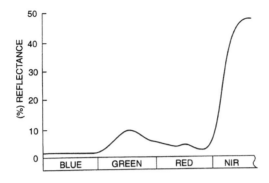

FIGURE 17.5. Typical spectral reflectance from a living leaf. The leaf is brightest in the near infrared region, although in the visible region the maximum is in the green region.

in the near infrared region is apparently controlled by the nature of the complex cavities within the leaf and internal reflection of infrared radiation within these cavities. Although some scientists suggest that moisture stress or natural maturity of a leaf causes these cavities to "collapse" as a plant wilts, others maintain that it is more likely that decreases in near infrared reflection are caused by deterioration of cell walls rather than physical changes in the cavities themselves. Thus changes in infrared reflectance can reveal changes in vegetative vigor; infrared images have been valuable in detecting and mapping the presence, distribution, and spread of crop diseases and insect infestations. Furthermore, changes in leaf structure that accompany natural maturing of crops are subject to detection with infrared imagery, so that it is often possible to monitor the ripening of crops as harvest time approaches. CIR imagery is valuable for observing such changes because of its ability to show spectral changes in both the visible and near infrared regions and to provide clear images that show subtle tonal differences.

In the longer infrared wavelengths (beyond 1.3 μm) leaf water content appears to control the spectral properties of the leaf (Figure 17.7). The term *equivalent water thick-*

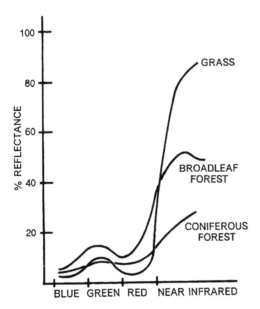

FIGURE 17.6. Differences between vegetation classes are often more distinct in the near infrared than in the visible.

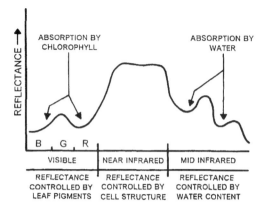

FIGURE 17.7. Changes in leaf water content may be pronounced in the mid infrared region. This diagram illustrates *differences* in reflectances between equivalent water thicknesses of 0.018 cm and 0.014 cm. Changes in reflectance are most evident in the mid infrared region. Diagram based on simulated data reported by Tucker (1979, p. 10).

ness (EWT) has been proposed to designate the thickness of a film of water that can account for the absorption spectrum of a leaf within the interval 1.4–2.5 μm.

Reflection from Canopies

Knowledge of the spectral behavior of individual leaves is important for understanding the spectral characteristics of vegetation canopies, but it cannot itself completely explain reflectance from areas of complete vegetative cover. Vegetation canopies are composed of many separate leaves that may vary in their size, orientation, shape, and coverage of the ground surface. In the field, a vegetation canopy (e.g., in a forest or a cornfield) is composed of many layers of leaves; the upper leaves form shadows that mask the lower leaves, creating an overall reflectance created by a combination of leaf reflectance and shadow.

Shadowing tends to decrease canopy reflectance below the values normally observed in the laboratory for individual leaves. Knipling (1970) cited the following percentages reported in several previously published studies:

	Percent reflected	
	Visible	Near infrared
Single leaf	10%	50%
Canopy	3–5%	35%

Thus the reflectance of a canopy is considerably lower than reflectances measured for individual leaves. But the relative decrease in the near infrared region is much lower than that in the visible. The brighter canopy reflection for the near infrared is attributed to the fact that plant leaves transmit near infrared radiation, perhaps as much as 50–60%. Therefore, infrared radiation is transmitted through the upper layers of the canopy, reflected in part from lower leaves, and retransmitted back through the upper leaves, resulting in bright infrared reflectance (Figure 17.8). Physicists and botanists have been able to develop mathematical models for canopy reflectances by estimating the optical properties of the leaves and the canopy as a whole. Because some of the transmitted infrared radiation may in fact be reflected from the soil surface below the canopy, such models have formed an important part of the research in agricultural remote sensing.

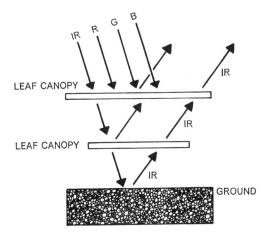

FIGURE 17.8. Simplified cross-sectional view of behavior of energy interacting with a vegetation canopy. (See Figures 17.3 and 17.4.) In some portions of the spectrum, energy transmitted through the upper layer is available for reflection from lower layers (or from the soil surface).

The Red Shift

Collins (1978) reports the results of studies that show changes in the spectral responses of crops as they approach maturity. His research used high-resolution multispectral scanner data of numerous crops at various stages of the growth cycle. Collins's research focused on examination of the far red region of the spectrum, where chlorophyll absorption decreases and infrared reflection increases (Figure 17.9). In this zone, the spectral response of living vegetation increases sharply as wavelength increases—in the region from just below 0.7 µm to just above 0.7 µm brightness increases by about 10 times (Figure 17.9).

Collins observed that as crops approach maturity, the position of the chlorophyll absorption edge shifts toward longer wavelengths, a change he refers to as the "red shift" (Figure 17.9). The red shift is observed not only in crops but also in other plants. The magnitude of the red shift varies with crop type (it is a pronounced and persistent feature in wheat).

FIGURE 17.9. Red shift. The absorption edge of chlorophyll shifts toward longer wavelengths as plants mature. The shaded area represents the magnitude of this shift as the difference between the spectral response of headed wheat and mature alfalfa. Similar but smaller shifts have been observed in other plants. The red shift is important for distinguishing the headed stage of a crop from earlier stages of the same crop and for distinguishing headed grain crops from other green, nongrain crops. From Collins (1978, p. 47). Copyright 1978 by the American Society of Photogrammetry and Remote Sensing. Reproduced by permission.

Collins observed the red shift along the entire length of the chlorophyll absorption edge, although it was most pronounced near 0.74 m, in the infrared region, near the shoulder of the absorption edge. He suggests that very narrow bands at about 0.745 μm and 0.780 μm would permit observation of the red shift over time and thereby provide a means of assessing differences between crops and the onset of maturity of a specific crop.

Causes of the red shift appear to be very complex and are not understood in detail. Chlorophyll a appears to increase in abundance as the plant matures; increased concentrations change the molecular form in a manner that adds absorption bands to the edge of the chlorophyll a absorption region, thereby producing the red shift. (In Chapter 18 we shall see that certain factors can alter the spectral effect of chlorophyll, thereby shifting the edge of the absorption band toward shorter wavelengths; this is the "blue shift" observed in geobotanical studies.)

17.4. Forestry

Identification of individual plants can be accomplished by close examination of crown size and shape (Figure 17.10). At the edges of forested areas, the shadows of trees can form clues to their identification. At smaller scales, individual plants are not recognizable, and the interpreter must examine patterns formed by the aggregate appearance of individual stands, in which individual crowns form a distinctive tone and texture. Identification may be relatively straightforward if stands are pure or are composed of only a few species that occur in consistent proportions. If many species are present and their proportions vary, then specific identification may not be possible, and the use of such broad designations as *mixed deciduous forest* may be necessary.

At these smaller scales, an interpreter identifies separate cover types from differences in image tone, image texture, site, and shadow. Identification of cover types can be considered the interpreter's attempt to implement one of the classification strategies mentioned above in the context of a specific geographic region, using a specific image.

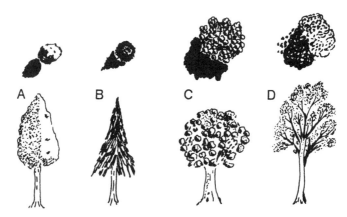

FIGURE 17.10. Identification of individual plants by crown size and shape. Sizes and shapes of individual tree canopies form distinctive shadows, as do the structures and densities of the individual branches.

Cover types may not match exactly to the categories in a classification system, but they form the interpreter's best approximation of these categories for a specific set of imagery. Cover classes can be considered as rather broad vegetation classes, based perhaps on the predominant species present, their age, and their degree of crown closure. Thus forest-cover-type classes might include *aspen/mixed conifer* and *Douglas fir*, indicating the predominant species without precluding the residence of others. Subclasses or secondary descriptors could indicate sizes of the trees, crown closure, and presence of undergrowth.

Such classifications can be prepared by manual interpretation of aerial photographs if the interpreter has field experience and knowledge of the area to be mapped. Each region has only a finite number of naturally occurring cover types, which usually prefer specific topographic sites, so the interpreter can bring several kinds of knowledge to bear on the problem of recognition and classification. Image tone, texture, shadow, and other photographic features permit separation of major classes. Rather broad classes can be separated using small-scale photography, but finer distinctions require high-quality, large-scale, stereo photography. For example, the classes given in Table 17.4a might be appropriate for information interpreted from aerial photographs at 1:60,000, whereas classes in Table 17.4b could be interpreted only from much larger scale photographs, perhaps at 1:15,840 to 1:24,000. At larger scales, stereoscopic coverage is important for identifying and distinguishing between classes.

Forest Photogrammetry

Foresters desire to identify specific *stands* of timber, areas of forest with uniform species composition, age, and density that can be treated as homogeneous units. Stands are the basic unit of forest management, so forest managers wish to monitor their growth time to detect the effects of disease, insect damage, and fire or drought. Even when stands have been planted from seedlings by commercial foresters, aerial photography, by virtue of its map-like perspective and its wealth of environmental information, provides accurate and economical information concerning the condition of the stand at a specific time.

If large-scale, high-quality images are available (preferably in stereo), it is possible

TABLE 17.4. Examples of Cover Types

a. At small scale (1:1,000,000 to 1:50,000), cover types must be defined at coarse levels of detail because it will not be possible to distinguish or to delineate classes that are uniform at the species level:

"evergreen forest"
"deciduous forest"
"chaparral"
"mixed broadleaf forest"
"mangrove forest"

b. At large scale (larger than about 1:20,000), cover types may be defined at fine levels of detail. Sometimes, when stands are very uniform, cover types can be mapped at the detail of individual species:

"balsam fir"
"shortleaf pine"
"aspen and white birch"
"oak–hickory forest"

to apply the principles of photogrammetry to measurement of factors of significance in forestry. Tree height can be measured using lengths of shadows projected onto level, open ground, although there are numerous practical problems that make this method inconvenient for routine use (level, open ground is not often easy to find; it is necessary to determine sun elevations for the latitude, day of the year, and time of day; and often shadows are too short for reliable measurements).

Crown Closure

One of the most important variables that contributes to estimates of stand density is assessment of *crown closure*—the proportion of the area of a stand that is covered by crowns of trees. Crown closure measures the density of trees in a stand and indicates the degree of competition between trees as a stand matures. Because crown closure is also related to stand volume, it assists predictions of economic dimensions of a stand at harvest. Because the size of the crown of a tree is closely related to many of its physiological characteristics (such as its ability to conduct photosynthesis), crown closure, monitored over time, permits foresters to assess the growth of the forest.

Although crown closure can be measured at ground level using any of several procedures, aerial photography forms a valuable tool for assessing crown closure. It assesses the percentage of a stand that is covered by the crowns of the dominant and codominant species. (Crowns of understory trees are not included in the estimate.) Crown closure is reported as a percentage (e.g., 60%), or as a decimal in the form 0.6 or 0.7.

With experience, crown closure can be accurately estimated by eye. Use of stereo is especially useful when understory crowns are visible. The quality of the photography can be important, including scale, shadows, and haze. So also can variables related to the stand, including the presence of irregular topography, nature of spacing of trees, and backgrounds of other vegetation. Some examples illustrate varied degrees of crown closure.

Timber Volume Estimation

Foresters are routinely interested in estimation of timber volume for a specific stand as a means of monitoring its growth over time, assessing management practices, and determining the amount of timber to be obtained at harvest. Volume measurement consists of estimates of board-foot or cubic-foot volumes to be obtained from a specific tree or, more often, from a specific stand. In the field, the forester measures the diameter of the tree at breast height (dbh) and the height of the straight section of the trunk (bole) as the two basic components of volume estimation. Measurements made for each tree are summed to give volume estimates for an entire stand.

There are several approaches to estimation of volume from measurements derived from aerial photographs; there can be no universally applicable relationship between photo measurements and timber volume, as species composition, size, age, soil, and climate vary so greatly from place to place. Typically, the interpreter must apply a *volume table*, a tabular summary of the relationship between timber volume and photo-derived measurements such as crown diameter and tree height (Table 17.5). The interpreter can make measurements directly from the photograph, then use the table to find volumes for each tree or for entire stands. New volume tables must be compiled as stand composition and the environmental setting change from place to place. Many variations on this

TABLE 17.5. Example of Aerial Stand Volume Table

Average stand height (ft.)	Crown cover (%)				
	15	35	55	75	95
8- to 12-ft. crown diameter					
30	15	50	70	95	135
40	35	70	95	125	175
50	50	95	125	170	215
60	90	150	195	235	265
70	160	230	270	300	330
80	250	310	345	380	420
13- to 17-ft. crown diameter					
30	20	55	80	105	135
40	40	80	100	145	195
50	60	100	135	185	235
60	100	160	205	245	285
70	170	240	275	310	350
80	260	320	360	390	430
90	360	415	450	485	525
100	450	515	555	595	635
18- to 22-ft. crown diameter					
30	30	70	90	120	160
40	50	90	120	160	200
50	75	120	160	205	245
60	115	180	225	265	305
70	190	255	290	330	370
80	275	330	370	410	450
90	370	430	470	505	550
100	470	530	570	610	650
110	600	660	700	740	780

Note. Volume (tens of cubic feet) per acre, given average stand height, average crown diameter, and crown cover, for Rocky Mountain conifer species. Derived from field measurements in test plots. Abbreviated from a more extensive table given in Wilson (1960, pp. 480–481). Copyright 1960 by the American Society for Photogrammetry and Remote Sensing. Reproduced by permission.

basic strategy have been developed, but most rely on the same kinds of estimates from the photograph.

17.5. Agriculture

Aerial imagery provides a powerful tool for understanding dynamics of agricultural systems, especially if it is employed with knowledge of the regional agricultural context.

Crop Calendar

The *crop calendar* is the cycle of crops grown during the year in harmony with the regional climate, local practices, and economic incentives. Each farmer adopts a specific sequence of preparing the field, planting, and harvesting to minimize labor and risk and to maximize production. Although each farmer decides specifics for a given farm, over broad regions farmers as a group tend to follow a specific sequence of agricultural activities for their region.

As an example, the crop calendar for principal crops of western Kansas can be broadly sketched as in Table 17.6. Analysts who examine images of agricultural landscapes must apply knowledge of a region's agricultural calendar to understand the meaning of the patterns they see on the imagery. At any given time, the landscape shows a variety of crops, each at different growth stages, due to varied planting dates, as determined by the local agricultural calendar.

The opportunities for regular coverage provided by Earth observation satellites mean that agricultural areas can be observed over time and that knowledge of the crop calendar can be applied in a manner that is not always practical if there is imagery for only a single date. For example, the winter wheat crop is characterized by bare fields in September and October, then by a mixture of soil and newly emerged vegetation in mid- to late autumn. During spring and early summer plant cover increases, completely masking the soil by late May or early June. Then, as the crop matures, spectral evidence of senescence records the approach of harvest in late June or early July. In contrast, the corn crop (Table 17.7) is not planted until spring, does not attain complete coverage until June, and is not mature until August. Thus each crop displays a characteristic behavior over time, which permits tracking of the progress of the crop. Knowledge of the local crop calendar permits selection of a single date that will provide maximum contrast between two crops. In the instance of discrimination between corn and wheat, for example, selection of a date in late spring should show complete vegetative cover in wheat fields, whereas corn fields show only newly emerged plants against a background of bare soil.

Crop Identification

The *growth stage* of any crop influences its appearance on aerial imagery. Specific crops do not display distinct spectral signatures because their appearance changes throughout the agricultural calendar. Agricultural scientists often define many stages for specific crops. These vary from crop to crop, and not all stages can be determined from aerial photography. However, it is often possible to interpret principal growth stages using large-scale aerial photography. Typically, when the crop is planted, the field is mainly bare, tilled soil. Initially, there is little evidence of the presence of plants. *Emergence*

TABLE 17.6. Crop Calendar for Winter Wheat, Western Kansas

September	Plowing and planting
October	Emergence and early growth
November	
December	
January	Dormancy
February	
March	
April	Growth resumes
May	
June	Heading
July	Harvest
August	

Note. Variations in timing are caused by local differences in weather and climate.

TABLE 17.7. Crop Calendar for Corn, Central Midwestern United States

April ⎫	Plowing and planting
May ⎭⎫	Emergence and early growth
June ⎭	
July	Tasseling
August ⎫	Maturity and harvest
September ⎭	

indicates that seedlings have sprouted, providing visible evidence of the plants. Although plants at this stage are usually too small to identify as individual crops, their identities may be indicated by planting pattern, the presence or absence of irrigation, and so on. Further growth progressively presents larger areas of leaf, shielding the soil from view and indicating the increasing maturity of the crop.

Many grains will exhibit *senescence* (aging of the plant, usually indicated by drying of the leaves) as the crop reaches *maturity*. *Harvest* occurs, weather permitting, at or near maturity. After harvest, the field is usually partially covered with residue from the harvest process; this residue typically remains on the field until the field is tilled just prior to planting of the next crop.

For many purposes, photointerpreters need only to distinguish between cereals (such as wheat, rye, and oats) with small seeds and coarser grains (such as barley, maize, sorghum) and as they are represented on aerial photography (Figure 17.11). Often, it may

FIGURE 17.11. Crop identification. S = sorghum; W = wheat; C = corn; F = summer fallow.

not be possible to distinguish between specific crops within each class, so remotely sensed imagey is seldom the principal source of information regarding the specific identity of crops within an area. However, some broad distinctions can be made using aerial imagery, provided the scale is large enough and plants are approaching maturity.

Once the crop approaches maturity, coarse grains usually can be recognized by their rougher textures as they appear in the field, due to their larger stalks, coarser leaves, complex structure, and increased microshadowing. Small grains, once mature, have an even, smooth, texture that tends to masks row and irregularities in terrain. Row crops, such as soybeans, are usually recognizable by the latent linear appearance that remains even after the crops are mature enough to mask the rows themselves.

Although remotely sensed imagery can provide valuable information about the agricultural landscape, it generally is not used as the principal means of identifying specific crops. Thus, aside from the ability of aerial imagery to provide a broad sketch of the kinds of crops grown in a region, its value is usually to assess acreages planted to croplands or left in fallow, the progress of crops as they mature through the growing season, detection of disease and insect infestation, effects of droughts, and other data relating to crop status.

Crop Status

Throughout the growing season, agronomists monitor the progression of the crop regionally through the agricultural calendar to assess disruptions from the usual timetable. Aerial imagery can form a valuable aid to documenting damage for insurance, guiding response to infestations, and assisting commodity markets.

Growth Stage

As crops emerge after planting and germination, they present leafy growth above ground. Initially, this growth is usually barely detectable on aerial photography (on CIR photography, possibly by a light pink tint). At this stage the field, as seen from above, is primarily bare soil with only slight evidence of the emerging crop. As the crop matures, its leaves begin the shade the soil formerly visible between plants, presenting an increasingly complete cover as seen from above. Depending on the specific crop and its method of cultivation, the linear pattern of rows from initial planting may be visible. Assuming that the crop matures without drought or attack by disease or insects, usually the ground surface will be completely covered as the crop matures. At this stage, on CIR photography, such field will appear as bright red.

At maturity, the crop experiences *senescence*—plant growth stops or slows, plant tissues begin to lose moisture, and the mesophyll tissue starts to deteriorate. As seen on CIR aerial photography, mature crops eventually attain a lighter color and often a grayish, or greenish, cast as the NIR reflectance decreases. After harvest, stubble and vegetative debris left on the soil give fields the same greenish cast, although the bare soil usually brightens the overall tone of harvested fields.

Aerial imagery can be a significant capability for observing effects of disease, weather damage, and insect infestation. A severe form of storm damage occurs as *lodging*. Lodging damages the stems of mature wheat (for example), as they are beaten down by heavy rain, hail, or winds. Mature crops are especially vulnerable to lodging; as crops approach maturity, the plants are larger and the stems weaker. Even if the lodged plants survive to continue growth, the broken stems prevent harvesting. *Hail damage* to some crops completely shreds the leaves, physically destroying the plant beyond hope of recovery. Yet some plants can survive hail damage, especially if they are young enough, because smaller, more compact leaves and stems are strong in comparison with those of the mature plant.

Aerial photography and photointerpretation are especially important tools for assessing crop status. Although farmers and agricultural analysts usually know the crops planted in a particular region, it is more difficult to know the status of the crops at a given time because of the difficulty of observing variations within such large regions. The spatial perspective and the use of the nonvisible spectrum offered by aerial photography provide the agricultural community with a valuable means of understanding the crop status.

17.6. Vegetation Indices

Vegetation indices (VIs), based on digital brightness values, attempt to measure biomass or vegetative vigor. A VI is formed from combinations of several spectral values that are added, divided, or multiplied in a manner designed to yield a single value that indicates the amount or vigor of vegetation within a pixel. High values of the VI identify pixels covered by substantial proportions of healthy vegetation. The simplest form of VI is a ratio between two digital values from separate spectral bands. Some band ratios have been defined by applying knowledge of the spectral behavior of living vegetation.

Band ratios are quotients between measurements of reflectance in separate portions of the spectrum. Ratios are effective in enhancing or revealing latent information when there is an inverse relationship between two spectral responses to the same biophysical phenomenon . If two features have the same spectral behavior, ratios provide little additional information; but if they have quite different spectral responses, the ratio between the two values provides a single value that concisely expresses the contrast between the two reflectances.

For living vegetation, the ratioing strategy can be especially effective because of the inverse relationship between vegetation brightness in the red and infrared regions. That is, absorption of red light (R) by chlorophyll and strong reflection of infrared (IR) radiation by mesophyll tissue ensures that the red and near infrared values will be quite different and that the IR/R ratio of actively growing plants will be high. Nonvegetated surfaces, including open water, man-made features, bare soil, and dead or stressed vegetation, will not display this specific spectral response, and the ratios will decrease in magnitude. Thus the IR/R ratio can provide a measure of photosynthetic activity and biomass within a pixel.

The IR/R ratio is only one of many measures of vegetation vigor and abundance. The green/red (G/R) ratio is based on the same concepts used for the IR/R ratio, although it is considered less effective. Although ratios can be applied with digital values from any remote sensing system, much of the research on this topic has been conducted using Landsat MSS and TM data. For TM data this ratio would be implemented as (TM4/TM3). One of the most widely used VIs is known as the *normalized difference vegetation index* (NDVI):

$$\mathrm{NDVI} = \frac{\mathrm{NIR} - \mathrm{R}}{\mathrm{NIR} + \mathrm{R}} = \frac{\mathrm{TM4} - \mathrm{TM3}}{\mathrm{TM4} + \mathrm{TM3}} \qquad \text{(Eq. 17.1)}$$

This index in principle conveys the same kind of information as the IR/R and G/R ratios but is constrained to vary within limits that preserve desirable statistical properties in the resulting distributions. Tucker (1979a, 1976b) and Perry and Lautenschlager (1984) suggest that in practice there are few differences between the many VIs that have been proposed.

17.7. Applications of Vegetation Indices

Many of the first studies examining applications of vegetation indices attempted to "validate" their usefulness by establishing that values of the VIs are closely related to biological characteristics of plants. For example, such studies might examine test plots during an entire growing season, then compare values of the VIs, measured throughout the growing season, with samples of vegetation taken at the same times. The objective of such studies is ultimately to establish use of VIs as a means of remote monitoring of the growth and productivity of specific crops or of seasonal and yearly fluctuations in productivity.

Often values of the VIs have been compared with in situ measurements of LAI or with above-ground biomass (the weight of vegetative tissue, excluding roots.). A number of VIs appear to be closely related to LAI (at least for specific crops), but no single VI seems to be equally effective for all plants and all agricultural conditions. Results of such studies have in general confirmed the utility of the quantitative uses of VIs, but details vary with the specific crop considered, with atmospheric conditions, and with local agricultural practices.

A second category of applications uses VIs as a mapping device—that is, much more as a qualitative than a quantitative tool. Such applications use VIs to assist in image classification, to separate vegetated from nonvegetated areas, to distinguish between different types and densities of vegetation, and to monitor seasonal variations in vegetative vigor, abundance, and distribution (e.g., Plates 20 and 21).

Although such studies have established the utility of VIs for practical applications, there are many precautions for uses of VIs that are to be interpreted quantitatively rather than qualitatively. Values of ratios and VIs can be influenced by many factors external to the plant leaf, including viewing angle, soil background, and differences in row direction and spacing in the case of agricultural crops. Ratios may be sensitive to atmospheric degradation: Because atmospheric effects have greater impact on bands at shorter wavelengths than on those at longer wavelengths, atmospheric effects can alter the ratios from their true values (Figure 17.12). Because atmospheric path length varies with viewing angle, values calculated using off-nadir satellite data (Chapter 6) vary according to position within the image. Clevers and Verhoef (1993) found that the relationship between LAI and the VI they studied was very sensitive to differences in leaf orientation. Although preprocessing can sometimes address such problems, it may still be difficult to compare values of VIs over a period of time because of variation in external factors. Price (1987) and others have noted that efforts to compare ratios or indices over time or from one sensor to another should reduce digital values to radiances before calculating ratios, to account for differences in calibration of sensors.

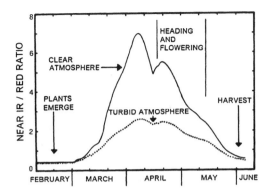

FIGURE 17.12. Influence of atmospheric turbidity on IR/R ratio, as estimated from simulated data. From Jackson et al., 1983, p. 195. Copyright 1983 by Elsevier Science Publishing Co., Inc. Reproduced with permission.

Further, although NDVI has been validated as a measure that is effective in separating pixels characterized as active vegetation growth from those that are unvegetated or characterized by senescent vegetation, it may not be appropriate for applications in which the NDVI values are to be interpreted as continuous measures of biophysical measures. Myneni, Hall, Sellecs, and Marshak (1995), Myneni et al. (2002), and Huete et al. (2002), for example, have shown that NDVI exhibits nonlinear relationships with biophysical measures such as VF or LAI, for example. Values of NDVI approach saturation (their maximum) before the biomass of the observed target reaches a maximum— that is, NDVI values reach a maximum and do not increase despite continued increases in LAI, biomass, or similar measures. Such effects do not negate the qualitative utility of NDVI and related measures but signify danger in application as quantitative measures of biomass. Researchers have proposed modified indices to compensate for such effects.

17.8. Phenology

Phenology is the study of the relationship between vegetative growth and environment; often phenology refers specifically to seasonal changes in vegetative growth and decline. Many phenological changes can be monitored by means of remote sensing because plants change in appearance and structure during their growth cycle. Of special significance are spectral and physiological changes that occur as a plant matures. Each season, plants experience chemical, physical, and biological changes, known as *senescence*, that result in progressive deterioration of leaves, stems, fruit, and flowers. In the midlatitudes, annual plants typically lose most or all of their roots, leaves, and stems each year. Woody perennial plants typically retain some or all of their roots, woody stems, and branches but shed leaves. Evergreens, including tropical plants, experience much more elaborate phenological cycles; individual leaves may experience senescence separately (i.e., trees do not necessarily shed all leaves simultaneously), and individual trees or branches of trees may shed leaves on cycles quite distinct from others in the same forest (Koriba, 1958).

During the onset of senescence, deterioration of cell walls in the mesophyll tissue produces a distinctive decline in infrared reflectance; an accompanying increase in visible brightness may be the result of decline in the abundance and effectiveness of chlorophyll as an absorber of visible radiation. Changes in chlorophyll produce the red shift mentioned earlier. Such changes can be observed spectrally, so remote sensing imagery can be an effective means of monitoring seasonal changes in vegetation.

The phenology of a specific plant defines its seasonal pattern of growth, flowering, senescence, and dormancy. Remotely sensed images can expand the scope of study to include overviews of vegetation communities or even of entire biomes. (*Biomes* are broad-scale vegetation regions that correspond roughly to the Earth's major climate regions.) Dethier et al. (1973), for example, used several forms of imagery and data to observe the geographic spread of the emergence of new growth in spring, as it progressed from south to north in North America. This phenomenon has been referred to as the *green wave*. Then, in late summer, the *brown wave* sweeps across the continent, as plant tissues mature, dry, and are harvested.

Figure 1.1, first discussed at the beginning of Chapter 1, illustrates local phenological differences in the spread of the green wave in the early spring. The image shows a Landsat MSS band 4 (near infrared) radiation, which is sensitive to variations in the density, type, and vigor of living vegetation. The region represented by Figure 1.1, and shown again in somewhat different form in Figure 17.13, has uneven topography covered by a mixture of forested and open land, including large regions of cropland and pasture. When this image was acquired in mid-April, vegetation in the open land was just starting to emerge. Grasses and shrubs in these regions have bright green leaves, but leaves on the larger shrubs and trees have not yet emerged. Therefore, the pattern of white depicts the regions occupied primarily by early blooming grasses and shrubs in lower elevations. Within a week or so after this image was acquired, leaves on trees began to

FIGURE 17.13. Landsat MSS band 4 image of southwestern Virginia. From EROS Data Center.

emerge, first at the lower elevations, then later at higher elevations. Thus a second image acquired only 2 weeks or so after the first would appear almost completely white, due to the infrared brightness of the almost complete vegetation cover. If it were possible to observe this region on a daily basis, under cloud-free conditions, we could monitor the movement of the green wave upward from lower to higher elevations and from south to north, as springtime temperatures prevail over more and more of the region. In reality, of course, we can see only occasional snapshots of the movement of the green wave, as the infrequent passes of Landsat and cloud cover prevent close observation during the short period when we must watch its progress. In an agricultural context, phenology manifests itself through the local crop calendar, which will vary from region to region in response to interactions between the genetic character of the crop, the local climate, local agricultural technology, and local practices.

17.9. Advanced Very-High-Resolution Radiometer

VIs and knowledge of phenology can also be applied in a much different context, using data acquired by a meteorological satellite first introduced in Chapter 6. The advanced very-high-resolution radiometer (AVHRR) is a multispectral radiometer carried by a series of meteorological satellites operated by NOAA in near-polar, sun-synchronous orbits. They can acquire imagery over a swath width of approximately 2,800 km, providing global coverage on a daily basis. Although AVHRR was designed primarily for meteorological studies, it has been successfully used to monitor vegetation patterns over very broad geographic regions. Areal coverage is extensive enough that entire biomes, or major ecological zones, can be directly observed and monitored in ways that were not previously feasible.

The satellite makes about 14 passes in a 24-hour period, collecting data for each 2,800-km swath twice daily, at 12-hour intervals. Resolution at the nadir is about 1.1 km, but an onboard computer can generalize data to 4-km resolution before data are transmitted to ground stations to permit broader geographic coverage for a given volume of data. Data are recorded at 10 bits. The wide angular view means that areas recorded near the edges of images suffer from severe geometric and angular effects. As a result, AVHRR data selected from the regions near the nadir provide the most accurate information. Although designed initially for much narrower purposes, Tucker et al. (1985) report that 800 to 900 km of the 2,800-km swath are usable. (The unusable data are from the edges of the image, where angular effects of perspective and atmospheric path length present problems.)

Details of spectral coverage vary with the specific mission, but in general AVHRR sensors have been designed to attain meteorological objectives, including discrimination of clouds, snow, ice, land, and open water. Nonetheless, AVHRR has been employed on an ad hoc basis for land resource studies. AVHRR collects five channels of data at a spatial resolution of 1.1 km in the visible, near infrared, and thermal infrared (Table 17.8). One channel is in the visible, one in the near infrared, and three in the thermal infrared. Channel 1 was originally proposed as 0.55–0.90 m, but after use on the prototype it was redefined to 0.58–0.68 m to improve separation of snow-free and snow-covered land areas. For agricultural scientists, this change provided the benefit of offering a channel positioned in the red region that permits calculation of vegetation indices.

TABLE 17.8. Spectral Channels for AVHRR

Channel	Spectral limits	Region
1	0.58–0.68 μm	Visible
2	0.72–1.10 μm	Near infrared
3	3.55–3.93 μm	Thermal infrared
4	10.30–11.30 μm	Thermal infrared
5	11.5–12.5 μm	Thermal infrared

Note. Spectral definitions differ between different AVHRR missions.

The visible and near infrared data can be ratioed to form a *greenness index* or VI similar to those discussed above:

$$\text{NDVI} = \frac{\text{Channel 2} - \text{Channel 1}}{\text{Channel 2} + \text{Channel 1}} \qquad \text{(Eq. 17.2)}$$

Although the spectral bands differ from those mentioned earlier, the meaning of the ratio is the same: High values reveal pixels dominated by high proportions of green biomass. The resolution is much coarser than that of Landsat MSS and TM data, but areal coverage is much broader, and the opportunity for repeat coverage is much greater. The frequent repeat coverage permits gathering of data for areas that may be obscured by clouds on one date but not the next. Therefore, over time, cloud-free coverage of continental or subcontinental areas can be acquired. Such images do not show the detail of MSS or TM data, but they do give a broad geographic perspective that is not portrayed by other images.

Compiled AVHRR Data

Full-resolution data, at about 1.1 km at nadir, can be acquired during any given day from a restricted region of the Earth, usually only when the satellite is within line of sight of a receiving station. This process produces local area coverage (LAC). LAC archives have been compiled for selected regions of the Earth.

Eidenshink (1992) describes the AVHRR dataset for the United States derived from 1990 AVHRR data, since repeated for other years. These composites are designed to record phenological variations throughout a growing season. Full-resolution AVHRR scenes were selected to form a set of 19 biweekly composites from March to December. Compositing over the 2-week period permits selection of a least one cloud-free date for each pixel. A special processing program was designed to select data based on viewing geometry, solar illumination, sensor calibration, and cloud cover and then to prepare geometrically registered NDVI composites. The process produces composite images composed of the original AVHRR bands, NDVI data, data describing the compositing process, and a statistical summary.

Global area coverage (GAC) is compiled by onboard sampling and processing, transmitted to ground stations on a daily basis. A GAC pixel is formed by averaging the first four pixels a row, skipping the fifth pixel, averaging the next four pixels, skipping the next, continuing to the end of the line. The algorithm skips the next two lines, then processes the fourth line in the same manner as it did the first line, and so on, until the entire scene is processed. These data provide spatial resolution of about 4 km.

Global vegetation index (GVI) data are created from the daily AVHRR data by examining the differences between the red and near infrared channels; the highest values for a 7-day interval are then used to selected the dates for which to calculate an NDVI value that represents a pixel for the entire 7-day period. The preliminary selection using the red–near infrared difference is designed both to select cloud-free pixels (if possible) for the 7-day interval and to minimize atmospheric effects. NDVI data are processed in such a manner that they represent about 15-km resolution. These data have been compiled since 1982, although algorithms and formats have varied considerably during this interval (Townshend, 1994).

Some AVHRR Examples

These broad-scale images provide an opportunity to observe major ecological zones and seasonal changes in a manner that was not previously possible. Figures 17.14 and 17.15 show such images for two dates for North America. Darker areas show high values for the VI (a reversal of the usual convention of representing high values as bright tones). The October scene shows the continent blanketed with a broad cover of vegetation. The lighter tones indicate sparsely vegetated arid regions in the Great Basin and southwestern United States and the tundra zone north of the arctic tree line. The March scene shows (in addition to snow, ice, and cloud cover in northern areas) much lower biomass, except for localized patches in tropical central America, southern California, and the Gulf Coast of the United States, where winter temperature and moisture permit year-round vegetative growth.

Such products are produced on a weekly basis by NOAA. Although they are not used to study specific sites, they provide an excellent overview of ecological conditions and zones and permit identification of regions worthy of more detailed study with more detailed data.

AVHRR data have been archived to form a global database and have been processed to provide a variety of products depicting seasonal changes in vegetative cover over areas of continental size. Some of these data are described in Chapter 20; Internet access to some of them has been outlined previously in Chapter 4.

17.10. Conservation Tillage

Conservation tillage consists of practices that minimize cultivation, retain stubble from the previous season's crop at the soil surface, reduce uses of herbicides, retain soil moisture, protect against erosion, and retain organic matter within the soil. Although conservation tillage can have disadvantages, its use has been encouraged to improve sustainability and soil management relative to more conventional practices. Further, the practice of conservation tillage can influence the retention of soil carbon within surface soil. Within the United States, the Conservation Technology Information Center (CTIC; *www.ctic. purdue.edu/CTIC%20HOME/*) conducts annual country-level inventories of conservation tillage through field surveys and questionnaires. The ability to analyze remote sensed imagery to extract information concerning conservation tillage practices supplements

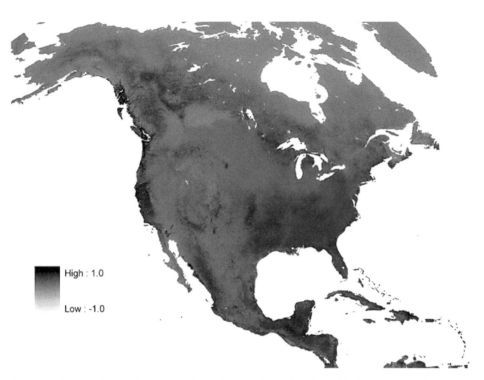

FIGURE 17.14. AVHRR image of North America, March 2001. This image shows the values of the NDVI calculated from red and near infrared bands of the AVHRR (Table 17.8). The key shows that the highest values of the NDVI are shown as dark tones; the lowest values are bare earth, snow, ice, or sparse vegetation. This image shows much of the continent to be covered by dormant vegetation, and the far north by snow and ice. Areas along the Gulf Coast, the central valley of California, and the tropical forests of Mexico, Central America, and the Caribbean Islands are covered by green or evergreen vegetation. The western deserts and midcontinental grasslands are shown to be sparsely vegetated. From National Space Science Data Center.

and improves such surveys. Practitioners of remote sensing have proposed a variety of indices designed to identify the presence of crop debris at the soil surface.

As an example, the normalized tillage index (van Deventer et al., 1997) is defined as:

$$NDVI = (TM5 - TM7)/(TM5 + TM7) \qquad \text{(Eq. 17.3)}$$

TM spectral bands are not well positioned to separate crop residues from soil background, so there have been many studies to evaluate the effectiveness of such indices. Whereas the NDVI and related vegetation indices can exploit the sharp contrast in brightness of vegetation in the red and near infrared regions, there is no such dramatic contrast between crop debris and the soil background. Moreover, the contrast that does exist decreases as it weathers and decomposes, and effectiveness of indices varies in effectiveness regionally in part because of variations in the brightnesses of the soil background (sandy soils are brighter than darker upland soils). A variety of other measures have been proposed, some

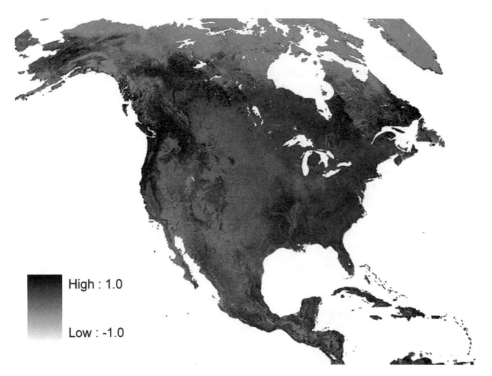

High : 1.0

Low : -1.0

FIGURE 17.15. AVHRR image of North America, October 2001. Green vegetation has emerged through most of the humid climates of the continent. At this date, the central grasslands, western deserts, and northern barren land are represented at low NDVI values. From National Space Science Data Center.

using other instruments; Serbin et al. (2009), Sullivan et al. (2008), Arsenault and Bonn (2005), and Daughtry et al. (2005) review many of these indices.

17.11. Land Surface Phenology

Traditionally, the study of phenology focused largely on individual plants. However, as the field of remote sensing developed the tools to assess seasonal changes over rather large areas, it became possible to evaluate the phenological behavior at regional, or even continental, scales using metrics to be described below. This broad-scale focus is known as *land surface phenology* (de Beurs & Henebry, 2004; Morisette et al., 2009).

NDVI, as observed by satellite sensors, permits tracking the seasonal rise and decline of photosynthetic activity within large regions over several growing seasons. Reed et al. (1994) were among the first to systematically apply broad-scale remote sensing instruments to derive phenology for very large regions. They accumulated cloud-free AVHRR of the U.S. composites over a 4-year period, calculating NDVI for each pixel. Examination of NDVI over such long periods and such large areas permitted inspection of phenological patterns of varied land cover classes and responses to climatic and meteorological events, including drought, floods, and freezes. They used AVHRR 14-day composites, basically collected twice a month during the March–October growing season and once a month during the winter season. Such products are useful for monitoring agricultural regions, assessing climatic impact on ecosystems—for example, production of food and cover for wildlife. Agricultural systems can be observed at a broad scale to assess crops' response to meteorological and climatic variations throughout the growing season and to forecast yields. Subsequent applications of the same approach, some using MODIS data (Chapters 7 and 21), provide additional spatial detail within the framework of broad coverage (Kang et al., 2003; Stockli and Vidale, 2004).

Some precautions are necessary in making detailed interpretations of such data. NDVI is, of course, subject to atmospheric effects, and the coarse resolution of AVHRR data encompasses varied land cover classes (Chapter 10) with varied phenological responses, especially in regions in which landscapes are locally complex and diverse. Atmospheric effects can create noise that contaminates the brightness values, so it is necessary to smooth the observations using a moving average to filter effects of the atmosphere, snow cover, and the like on NDVI values (Figure 17.16).

Observed over time, values of the NDVI depict a seasonal pattern of increase and decrease (Figure 17.17.). During the winter season, NDVI reach minimum values as vegetation enters dormancy, although in some regions, residual islands of green vegetation may remain. As springtime approaches, NDVI begins to rise. The onset of longer days, warmer temperatures, and spring rains promotes budding of vegetation and emergence of leaves, first with grasses and shrubs, then later with trees. The seasonal response in NDVI is known as *green-up*. The rise in NDVI marks *the start of season* (SOS). Eventually green-up peaks at *maximum NDVI*. Mature plants invest energy in growing the seeds and fruits necessary to prepare the species for success in the next year's growing season. As the season progresses, green-up slows, stops, and then declines, marking the *end of season* (EOS). The decline from mid- to late summer, known as *senescence*, records the decline in vigor as senescence and harvest time approaches. The regional decline in NDVI in the fall is sometime known as *brown-down*. The interval between the SOS and EOS

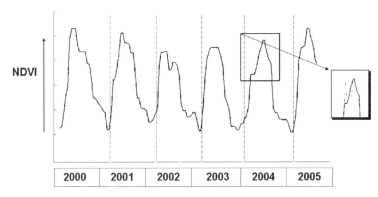

FIGURE 17.16. Seasonal phenological variation, derived from NDVI for a single MODIS pixel tracked over the interval 2000–2005. The pattern shows the seasonal variation in NDVI in response to phonological behavior of the plant cover within the pixel. Because of variations in weather and atmosphere, the NDVI values can vary, so the values are statistically filtered and smoothed to improve the continuity of the record. The inset shows a portion of the original data, before filtering, superimposed over the filtered curve. From Aaron Dalton and Suming Jin, VirginiaView.

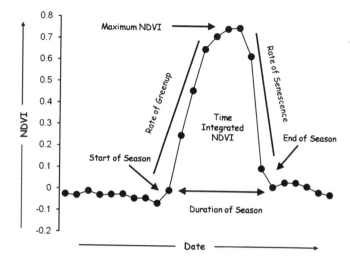

FIGURE 17.17. Idealized phenological diagram for a single season. The vertical axis represents NDVI; the horizontal axis represents time, beginning and ending in the winter season. Annotations identify key metrics, including start of season (SOS) and end of season (EOS), defined by quantitative assessment of the sequence of individual NDVI measurements throughout the year. From Kirsten de Beurs.

forms one definition of a region's growing season, sometimes measured as the *duration of season*, or by integrating the area under the phonological curve as the *time-integrated NDVI*. Thus these data can provide a means of assessing place-to-place variation in the length of growing seasons. The result produces a typical seasonal pattern as depicted in Figures 17.17 and 17.18.

Some useful resources of interest for both teachers and students can be found at

- NASA: *nasadaacs.eos.nasa.gov/articles/2009/2009_plants.html*
- National Phenology Network: *www.usanpn.org/*
- USGS—Remote Sensing Phenology: *phenology.cr.usgs.gov/*
- Project BudBurst: *www.windows.ucar.edu/citizen_science/budburst/*
- Multimedia resources (NASA and NOAA): *svs.gsfc.nasa.gov/search/DataSet/id_0426.html*

17.12. Separating Soil Reflectance from Vegetation Reflectance

Digital data collected by satellites, especially at the coarse resolutions of the Landsat MSS and AVHRR, mix vegetation reflectance with that of the soil surface. Some mixing of soil and vegetation reflectances occurs because of mixed pixels (Chapter 10). In agricultural scenes, however, reflections from individual plants or individual rows of plants are closely intermingled with the bare soil between plants and between rows of plants, so that reflectances are mixed even at the finest resolutions. Of course, this mixing is especially important after plants have emerged, because large proportions of soil are exposed to the sensor. But even after leaves have fully emerged, soil can still contribute to reflec-

FIGURE 17.18. Phenological map of Virginia and neighboring areas compiled from MODIS data. From Peter Sfroza, VirginiaView. Darker pixels represent areas of increasing greenness, here principally near the lower coastal plain, Piedmont, and the Central Valley.

tance because of penetration of some wavelengths through the vegetation canopy. Mixing of soil and vegetation reflectances can therefore be a serious barrier to implementation of the concepts presented earlier in this chapter. Hutchinson (1982) found that vegetation information is difficult to extract when vegetation cover is less than 30%. In arid and semiarid regions, vegetative cover may always be this low, but even in agricultural settings plant cover is low when crops are emerging early in the growing season.

If many spectral observations are made of surfaces of bare soil, the values of red and near infrared brightnesses resemble those along axis *ABC* in Figure 17.19. The soil brightnesses tend to fall on a straight line—as a soil becomes brighter in the near infrared, so does it tend to get brighter in the red. Dry soils tend to be bright in both spectral

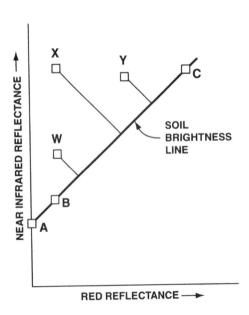

FIGURE 17.19. Perpendicular vegetation index. Based on Richardson and Weigand (1977, p. 1547). Copyright 1977 by the American Society for Photogrammetry and Remote Sensing. Reproduced with permission.

regions and appear at the high end of the line (C); wet soils tend to be dark and are positioned at the low end (B). Richardson and Wiegand (1977) defined this relationship, known as the *soil brightness line*, and recognized that the spectral response of living vegetation will always have a consistent relationship to the line. Soils typically have high or modest response in the red and infrared regions, whereas living vegetation must display low values in the red (due to the absorption spectra of chlorophyll) and high values in the near infrared (due to IR brightness of mesophyll tissue). Thus points representing "pure" vegetation response will be positioned in the upper left of Figure 17.19, where values on the red axis are low and those on the IR axis are high. Furthermore, Richardson and Wiegand defined an index to portray the relative magnitudes of soil background and vegetative cover to a given spectral response. Thus point X typifies a "pure" vegetation pixel, with a spectral response determined by vegetation alone, with no spectral contribution from soil. In contrast, point Y typifies a response from a partially vegetated pixel—it is brighter in the red and darker in the near infrared than is X. Richardson and Wiegand quantified this difference by defining the *perpendicular vegetation index* (PVI) as a measure of the distance of a pixel (in spectral data space) from the soil brightness line. The PVI is simply a Euclidean distance measure similar to those discussed in Chapter 12:

$$\mathrm{PVI} = \sqrt{(S_R - V_R)_{IR}^2 + (S_{ir} - V_{IR})^2} \qquad \text{(Eq. 17.4)}$$

where S is the soil reflectance, V is the vegetation reflectance, subscript R represents red radiation, and subscript IR represents near infrared radiation.

In practice, the analyst must identify pixels known to be composed of bare soil to identify the local soil brightness line and pixels known to be fully covered by vegetation to identify the local value for full vegetative cover (point X in Figure 17.19). Then intermediate values of PVI indicate the contributions of soil and vegetation to the spectral response. The soil line has been used to indicate soils that are free of vegetation at the time that the image was acquired and to separate partially vegetated pixels from those that have complete vegetative cover. The position of any specific pixel on the soil line is likely determined by the nature of the soil surface, to include soil properties (physical and chimerical), roughness, and moisture status (Baret et al., 1993). Because such properties can be transient, it is not likely that a single soil line can precisely represent all soils.

Baret et al. (1993) investigated the soil brightness line from both experimental and theoretical perspectives. They concluded that the theoretical basis for the soil brightness line is sound, although effects of variations in soil moisture and surface roughness are still not well understood. Their experimental data revealed that it is not feasible to define a single, universally applicable soil brightness line, as local variations in soil types lead to spectral variations. However, these variations were found to be minor in the red and infrared regions. They concluded that the use of a single soil brightness line is a reasonable approximation, especially in the context of analysis of coarse-resolution satellite data. Nanni and Dematte (2006) and Ladoni et al. (2009) provide more recent reviews of research on these topics.

17.13. Tasseled Cap Transformation

The "tasseled cap" transformation (Kauth and Thomas, 1976) (sometimes known as the K–T transform) is a linear transformation of Landsat MSS data that projects soil and vegetation information into a single plane in multispectral data space—a plane in which the major spectral components of an agricultural scene are displayed in two dimensions. Although defined initially for MSS data, subsequent research has extended the concept to the six nonthermal bands of the TM (Crist, 1983). The transformation can be visualized as a rotation of a solid multidimensional figure (representing all spectral bands) in a manner that permits the analyst to view the major spectral components of an agricultural scene as a two-dimensional figure.

The transformation consists of linear combinations of the original spectral channels to produce a set of four new variables, each describing a specific dimension of the agricultural scene. The following equations describe the transformation for the Landsat 2 MSS (Kauth and Thomas, 1976), presented here as a concise example.

$$TC1 = +0.433 \text{ MSS } 4 + 0.632 \text{ MSS } 5 + 0.586 \text{ MSS } 6 + 0.264 \text{ MSS } 7 \qquad \text{(Eq. 17.5)}$$

$$TC2 = -0.290 \text{ MSS } 4 - 0.562 \text{ MSS } 5 + 0.600 \text{ MSS } 6 + 0.491 \text{ MSS } 7 \qquad \text{(Eq. 17.6)}$$

$$TC3 = -0.829 \text{ MSS } 4 + 0.522 \text{ MSS } 5 - 0.039 \text{ MSS } 6 + 0.194 \text{ MSS } 7 \qquad \text{(Eq. 17.7)}$$

$$TC4 = +0.223 \text{ MSS } 4 + 0.012 \text{ MSS } 5 - 0.543 \text{ MSS } 6 + 0.810 \text{ MSS } 7 \qquad \text{(Eq. 17.8)}$$

(The equations use the Landsat 2 designations for MSS channels, MSS 4 = MMS 1, MMS 5 = MSS 2, MSS 6 = MSS 3, MSS 7 = MSS 4.) Here the new bands are designated as "TC1" for "tasseled cap band 1," and so on. The coefficients are calculated by means of an iterative procedure (Jackson, 1983). The transformation is based on the calibration information for each specific sensor, so it requires a dedicated effort to derive the values for a specific sensor. Once the values are available, they can be applied to imagery from that sensor—some are reported in the scientific literature, and some are incorporated into imaging processing packages.

Although these four new tasseled cap bands do not match directly to observable spectral bands, they do carry specific information concerning agricultural scenes. Kauth and Thomas interpret TC1 as *brightness*, a weighted sum of all four bands. TC2 is designated as *greenness*, a band that conveys information concerning abundance and vigor of living vegetation. It can be considered analogous to the PVI mentioned above. TC3 has been designated as *yellowness*, derived from the contrast between the red and green bands; it is an axis oriented at right angles to the other two. Finally, TC4 is referred to as *nonesuch* because it cannot be clearly matched to observable landscape features and is likely to carry system noise and atmospheric information.

The first two bands (TC1, brightness, and TC2, greenness) usually convey almost all the information in an agricultural scene—often 95% or more. Therefore, the essential components of an agricultural landscape are conveyed by a two-dimensional diagram, using TC1 and TC2 (Figure 17.20), which is in many respects similar to that defined by Richardson and Wiegand (Figure 17.19).

Over the interval of an entire growing season, TC1 and TC2 values for a specific field follow a stereotyped trajectory (Figure 17.20). Initially, the spectral response of a field is dominated by soil, as the field is plowed, disked, and planted (point 1 in Figure 17.20). The field has a position near the soil brightness line. As the crop emerges (2) and grows, it simultaneously increases in greenness and decreases in soil brightness as the green can-

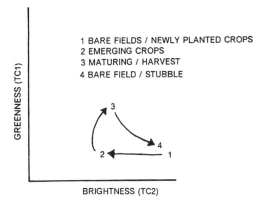

FIGURE 17.20. Seasonal variation of a field in data space defined by the greenness and brightness axes. From Crist (1983). Copyright 1983 by Purdue Research Foundation, West Lafayette, IN 47907. Reproduced with permission.

opy covers more and more of the soil surface (2 to 3). Then, as senescence, maturity, and harvest occur, the field decreases in greenness and increases in soil brightness to return the field back near its original position on the diagram (3 to 4).

A plot of data for an entire growing season, or for an image that shows many fields at various points in the crop cycle, has a distinctive shape similar to that shown in Figure 17.21; the resemblance of this shape to a tasseled ("Santa Claus-type") cap provides the basis for the name that Kauth and Thomas gave to their technique. The analogy holds as the data are examined in other TC dimensions (Figure 17.21). Crist and Cicone (1984) show that the intermediate values are chiefly vegetated pixels in which plants are extending their canopy cover as they mature. Pixels representing senescent plant cover retain high greenness values until the canopy cover is so low that bare soil is again exposed.

Crist and Cicone (1984) extended the transformation from the MSS to the six nonthermal bands of the TM. Because the TM employs spectral definitions that differ from

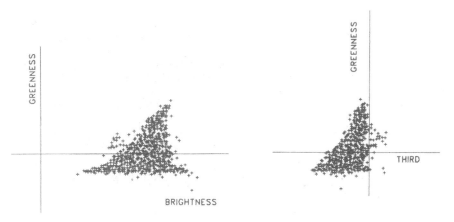

FIGURE 17.21. Tasseled cap viewed in multidimensional data space. Spectral values from an agricultural scene have been plotted on TC1 and TC2 axes (left) and TC2 and TC3 (right). These data are simulated TM data for agricultural regions in the midwestern United States. From Crist and Cicone (1984, p. 347). Copyright 1984 by the American Society for Photogrammetry and Remote Sensing. Reproduced with permission.

those of the MSS, they require a different set of coefficients. The resulting transformation is analogous to that reported above for the MSS but differs in detail. For the TM, as for the MSS, TC1 corresponds to *brightness* and TC2 to *greenness*. However, TC3, weighted by the TM's mid infrared bands, portrays *wetness*, based on soil moisture and moisture in plant tissues.

The first two components of the TM tasseled cap transformation, brightness and greenness, usually account for majority of variation in any vegetated scene, but, despite their names, they are not always precise matches to equivalents from MSS. The third TM component, wetness, forms a new landscape component. Yellowness and nonesuch, the remaining features of less significance, are characterized by lack of a clear match with physical features of the landscape; they form basically statistical artifacts of variation not accounted for by the primary features (possibly effects of microshadowing, atmospheric haze, or atmospheric water vapor).

Figure 17.22 illustrates the application of the TM tasseled cap model to TM imagery of an irrigated agricultural region in California, showing the original image, TC1 (brightness), TC2 (greenness), and TC3 (wetness). TC4 and TC6, although they do not have assigned interpretations, are shown here to illustrate the declining information content of the higher TC bands. Unlike the PC transformation described in Chapter 12, which must be calculated individually for each new image to be examined and reinterpreted for each individual image, the tasseled cap, once calculated for a specific sensor, applies to all images acquired by that sensor, and the meanings of the TC components are consistent across such images. As a result, the tasseled cap transformation has been used for understanding images of not only agricultural scenes but also a variety of other ecological settings.

17.14. Foliar Chemistry

The development of operational hyperspectral sensors (Chapter 15) has opened possibilities for application of remote sensing to observation of characteristics of vegetated surfaces not feasible using instruments previously at hand. Hyperspectral data have created the capability to investigate *foliar chemistry*, the chemical composition of living leaves. If such a capability could be applied in an operational context, it could assist in the identification of plants at levels of detail not previously feasible and in the monitoring of the growth and health of crops and forests.

Although a considerable body of research is devoted to remote sensing of foliar chemistry, much of this work is specific to particular sensors, individual test sites, or specific processing algorithms and has been acquired by varied means (laboratory analyses, field observations, and airborne data). Therefore, much work remains to be completed to develop the suggestions offered by existing studies.

The following paragraphs outline some of the efforts to investigate hyperspectral sensing of foliar chemistry.

• *Lignin and starch content.* Kokaly and Clark (1999) applied a normalization procedure that permitted detection of lignin and cellulose in dried ground leaves in a laboratory setting. Peterson et al. (1988) found that the spectral region in the range 1,500 nm–1,750 nm in AIS data was linked to levels of lignin and starch content. Martin and

FIGURE 17.22. The tasseled cap transformation applied to a TM image of an irrigated agricultural landscape in California. Top left shows visible bands from the original image. Upper right, tasseled cap band 1, brightness; then in sequence, TC2, greenness, TC3, wetness. TC4 and TC6 are not named but show decreasing levels of information relative to TC1, TC2, and TC3.

Aber (1997) reported positive results for predicting lignin using four AVIRIS bands in the range bands 1,660 nm–2,280 nm at test sites in the northeastern United States.

  • *Nitrogen.* Card et al. (1988) found promising relationships between spectroradiometer data and nitrogen content of ground dried laboratory samples. Wessman et al. (1989) investigated forest canopy chemistry using AIS imagery; they found strong relationships between foliar nitrogen and reflectance in bands between 1,265 nm and

1,555 nm. Martin and Aber (1997) examined AVIRIS data (20-m resolution) of a mixed broadleaf forest of oak (*Quercus rubra*), maple (*Acer rubrum*), and needle-leaved species; they found a relationship between nitrogen content and brightness in the intervals 750 nm and 2,140 nm and in the intervals 950 and 2,290 nm.

• *Chlorophyll concentration.* Kupiec and Curran (1993) examined AVIRIS data to find that brightness at 723 nm was strongly correlated with chlorophyll content in the needles of slash pine (*Pinus elliotti*).

17.15. Lidar Data for Forest Inventory and Structure

Chapter 8 introduced the application of laser altimetry to acquisition of data characterizing vegetation structure. An example is the SLICER (Scanning Lidar Imager of Canopies by Echo Recovery) program (Blair et al., 1994; Harding et al., 1994) that applies lidar altimetry to provide detailed information about the vertical structure of vegetation canopies—information that is essential for understanding the function of ecosystems but very difficult to obtain by the usual field measurements. Information about SLICER is available at:

> *ltpwww.gsfc.nasa.gov/eib/projects/airborne_lidar/slicer.html*
>
> *www.fsl.orst.edu/~lefsky/slicerpage.html*

The SLICER instrument uses radiation in the near infrared (1.06 μm) to illuminate regions 10–15 m in diameter. Some of the radiation is reflected from the canopy, and some reaches the ground through gaps in the canopy. The complete returned laser signal is examined to reveal the vertical distribution of the backscatter of laser illumination from all canopy elements (foliar and woody) and the ground reflection. The relatively large footprints (5–15m) are designed to simultaneously recover returns from the top of the canopy and from the ground surface yet record details of the structure of individual crowns. The time difference between initial and last returns provides the basis for estimation of overall canopy height (Nelson et al., 1988a; Nelson et al., 1988b). Figure 17.23 presents the results of the SLICER lidar system flown over a 1-km section of deciduous forest (Duke Forest, October 1994). The base of each line represents the varying topography at ground level—the light fringe depicts the height of the tallest trees.

Because lidar data directly record data characterizing the physical structure of a forest (heights, crown closure, crown size, etc.), they provide an opportunity to directly assess the three-dimensional structure of vegetation formations in ways that are not possible with other sensors. For example, Means et al. (2000) examined lidar data of Douglas fir (*Pseuotsuga taxifolia*) stands in the western Cascades of Oregon using airborne lidar with a footprint of 0.2 m–0.9 m. They found that lidar data were effective in estimating basal area, height, volume, and canopy closure for a range of growth states, and they concluded that lidar survey should be an effective inventory tool for forest management. Figure 17.24 illustrates the use of lidar imagery to derive estimates of crown diameter by applying an algorithm that calculates the elevation profiles along two perpendicular axes, thereby revealing the shape and dimensions of the crown.

FIGURE 17.23. Vegetation canopy data captured by the SLICER sensor. From NASA—Goddard Space Flight Center.

17.16. Precision Agriculture

Traditional agricultural practices treat entire fields as single management units. A farmer applies seed, fertilizer, and pesticides at the same rate for an entire field. This practice prevails despite inherent variation in natural agronomic components, including soil, solar radiation, moisture, and drainage, which are known to vary markedly within fields. Although farmers understand well that such characteristics vary considerably even within

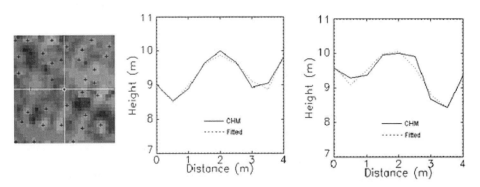

FIGURE 17.24. Calculating crown diameter from a CHM. The crown shape is estimated as the average of two values measured along two perpendicular directions fitting a fourth-degree polynomial on both profiles. Left: Lidar data, as extrapolated from the lidar returns (indicated by the +'s). Center: Profile along the east–west axis, representing the canopy height as derived from lidar returns. Right: Transect along the north–south axis, also revealing the top and edges of the crown. From S. C. Popescu.

short distances, traditionally they have lacked both the information they need to under-stand the details and the means to implement any understanding.

Precision agriculture is based on the premise that farmers can control agricultural management practices (e.g., seeding rate, fertilization, tillage, pesticides, irrigation) at varying rates within fields to optimize production. Thus fertilizer, for example, would be applied at heavier rates where soil conditions require intense fertilization but at lower rates elsewhere within the same field, thereby economizing on fuel, materials, and labor and minimizing unintended environmental effects.

Such measures are possible because of the increased precision and detail of informa-tion that can be acquired in part from remotely sensed data and the ability to monitor changes throughout the growing season, as moisture conditions, insect infestations, and the like vary within fields and over time as the growing season progresses. The practice of precision agriculture is based on use of agricultural equipment with GPS to determine accurate position, multispectral imagery to provide current information on crop condi-tions, and agricultural machinery designed to employ varying rates of application and even to sense field conditions on the spot.

Remote sensing forms an important component of precision agriculture, first because it provides a map-like representation of agricultural landscapes and second because it can provide information pertaining to growth and health of crops throughout the growing sea-son. Remote sensing can reveal variations in surface soil texture (Post et al., 1988), guiding collection of soil samples required to support the precision agriculture effort. Remotely sensed images can detect anomalies, revealing, for example, localized effects of drought or insect infestations. In some instances, data from calibrated instruments can provide direct measurement of physical values, such as brightness in specific portions of the spectrum.

17.17. Remote Sensing for Plant Pathology

The Disease Triangle

Plant pathology is the study of plant diseases and their occurrence. *Plant pathologists* study the biochemistry and biology of pathogens and the plants they infect to understand the life cycle of the pathogens and the means by which they infect hosts. Plant patholo-gists counter the effects of pathogens by proposing management techniques, designing pesticides, and recommending countermeasures to minimize effects of the disease and to prevent or slow its spread to other regions.

A few well-known examples can remind us of the significance of plant diseases. Dur-ing 1846–1850, the Irish potato famine killed at least one million Irish citizens, caused immense social disruption, and created emigration that itself influenced social systems in the United States and elsewhere. In the United States, Dutch elm disease and the chestnut blight have destroyed trees that once constituted North America's most distinctive and beautiful trees. Specifically, the decline of the American chestnut tree changed the ecol-ogy of large regions of Appalachian forests. The tree once constituted perhaps one-fourth of the trees of Appalachian forests and provided food for both humans and wildlife. Both exist now only as isolated remnants. Often such catastrophes are connected closely to human behavior, especially through the introduction of plants and diseases from other regions, the tendency to depend on single crops, usually densely planted, and the uses of genetically uniform hybrids that may increase susceptibility to attacks by pathogens.

Infection depends on the co-occurrence of three interrelated factors (known as the *disease triangle*):

1. A pathogen (bacteria, viruses, fungi, mycoplasmas, nematodes)
2. A susceptible host plant (a plant's resistance to disease)
3. Favorable environmental conditions (including temperature, wind, solar illumination, soil)

Pathogens spread through the environment via wind, water, and soil and by biologic transmission, including through insects, animals, and humans.

Remote sensing offers many ways to supplement and reinforce the efforts of the detailed, ground-based expertise of the plant pathologist. Applications of remote sensing employ the plant pathologist's understanding of the dynamics of an infestation in its spatial context at the landscape level and the use of ground-based surveys to understand the behavior of the pathogen at the level of the plant and the cell.

Remote sensing offers the ability to conduct systematic surveys to detect mortality associated with diseases or declines in vegetation health and to document spread from one region to another and the effectiveness of countermeasures. Specifically, remote sensing has been useful in:

- Detection of infestation- symptoms from aerial imagery
- Detection of host plants, identification of crop residue and volunteer growth that can harbor pathogens
- Phenology—understanding the onset of seasonal growth in hosts or vectors
- Assessment of environmental conditions favorable for the pathogen
- Spatial perspective on the occurrence and spread of the disease and monitoring of its spread.

A few case studies can highlight some of the issues pertaining to the role of remote sensing in assessing and monitoring the occurrence of plant diseases.

Corn Blight Watch Experiment, 1971

One of the earliest applications of remote sensing to a broad-scale infestation occurred during the 1971 Corn Blight Watch Experiment. In previous years, corn breeders had introduced varieties of maize that had been optimized for hybridization but that were vulnerable to specific varieties of the fungus that causes southern corn leaf blight (SCLB; a fungus, *Bipolaris maydis*).

SCLB exists in nature, mainly as a minor disease of corn. SCLB produces elongated brown spots on leaves and stalks during the interval mid-June to mid-October. If the disease attacks before the ears have formed, crop losses can be severe. The fungus resides in soil and in crop debris left after harvest and survives the winter season to infest the crops of the succeeding year. SCLB thrives in warm, moist weather, so it spreads easily in the summer weather of central North America. Its occurrence is very difficult to control.

Once the vulnerable hybrid became widely used, SCLB rose from a minor nuisance to a major threat to U.S. agriculture. The 1970 corn crop had been reduced by 15% of the previous crop, creating havoc in commodities markets and political pressures for agricultural officials. By the early months of 1971, seed carrying the genetic weakness had already been widely distributed to farmers throughout the continent. As the 1971 growing season began, agricultural scientists became increasingly concerned about the vulnerability of the North American corn crop. Because of the widespread use of the vulnerable seed and the dense occurrence of maize cultivation throughout central North America, the right weather conditions could lead to rapid spread of the blight and destruction of the nation's corn production, with ominous implications for agricultural economy and a major portion of the nation's food production.

The Corn Blight Watch Experiment was based on uses of remotely sensed data to monitor the occurrence and spread of SCLB during the 1971 growing season. A series of flightlines was designed to assess the spread of the infection throughout the growing season. Biweekly flights collected CIR photography, as well as multispectral scanner data, for principal corn-growing regions within the United States and Canada (Figure 17.25). Ground data were collected in coordination with the collection of aerial imagery. Sampling design allowed ground data to be extended to broad-scale estimates for the principal corn-producing regions of the United States and Canada.

Photointerpretation estimates varied in accuracy when compared against the data collected at ground level. Photointerpretation was effective in distinguishing between healthy or slight levels of infection for more severely infected fields but was not effective in assessing slight or mild levels of infection. Early in the growing season, there was a large discrepancy between the estimates of crop damage from ground-based observations and those derived from photointerpretation (Figure 17.26). Some difficulties were experienced in assessing the effects of blight when other causes influenced the health of the corn crop.

The Corn Blight Watch Experiment is significant today because it is the only example of a broad-scale attempt to use remote sensing as a tool to contribute to the understanding of the spatial behavior of a plant pathogen. As globalization and the concerns about bioterrorism raise awareness about the possible impacts of introduced pathogens, this experiment from several decades past may deserve examination.

FIGURE 17.25. Flight lines for the 1971 Corn Blight Watch Experiment. From Sharples (1973). Reproduced by permission of Laboratory for Application of Remote Sensing, Purdue University.

FIGURE 17.26. Photointerpretation accuracy for the Corn Blight Watch Experiment. From Johannsen et al. (1972). Reproduced by permission of Laboratory for Application of Remote Sensing, Purdue University.

Sudden Death Oak Syndrome

A newly defined pathogen (*Phytophthora ramorum*) has been killing a variety of oak species in coastal California and Oregon since the mid-1990s. The disease complex associated with *P. ramorum* is called sudden oak death (SOD). Previously, this fungus had been known principally as an infection of European rhododendron.

In eastern North America, forest species known or suspected to be hosts include red oak (*Quercus rubra*) and mountain laurel (*Kalmia latifolia*). Studies completed by the USDA Forest Service estimate that over half of the forests in the eastern United States contain some susceptible tree hosts. The greatest impact would be on forests in areas that contain large percentages of red or live oak species, including the Ozark–Ouachita Highlands, the pin oak sand flats in the Lake states, and live oak areas in Florida. Several other species, including some ornamental nursery plants, are known to be reservoirs of infection, although it is not now clear whether they are harmed by the disease. *P. ramorum* can be spread through shipment of infected nursery plants in the wild through rain and wind.

The most obvious evidence of infection on trees is the presence of cankers that ooze sap through the bark on trunks or large limbs of infected trees. The pathogen can girdle trees, causing tree crowns to suddenly turn brown. Trees can die within a few months to several years after infection is evident. Infected trees are often subject to attack by insects and other pathogens, which may themselves be the immediate cause of death.

Kelly et al. (2004) report their use of remotely sensed data to assess and monitor the occurrence of mortality associated with SOD through time. They used these data to model locations to which *P. ramorum* might spread over time. They also established a convenient system to encourage county-level workers and the general public who live and recreate in the coastal forests of California to use GPS receivers to record the locations of potentially infected trees and to report them to a website. Reporting is coordinated with a training program for field samplers and the submission of samples to a laboratory for analysis. The Web-based system permits workers in local governmental offices, as well as individual citizens, to submit timely reports of the occurrence and spread of SOD. These data are used to guide planning for remotely sensed data collections that contribute additional information concerning SOD.

www.suddenoakdeath.org

kellylab.berkeley.edu/SODmonitoring/sitemap.htm

Their system forms a model of how remotely sensed data, geospatial data in general, and the Internet can contribute to a framework that permits integration of a landscape-level perspective with plant-level data.

17.18. Summary

For a large proportion of the Earth, vegetation cover forms the surface observed by remote sensing instruments. In some instances, we have a direct interest in examining this vegetative blanket to map the patterns of different forests, rangelands, and agricultural production. In other instances, we must use vegetation as a means of understanding those patterns that may lie hidden beneath the plant cover. In either instance, it is essential that we are able to observe and understand the information conveyed by the vegetated surface.

Given the obvious significance of agriculture and forestry and the immense difficulties of monitoring use of these resources, the techniques outlined in this chapter are likely to form one of the most significant contributions of remote sensing to the well-being of mankind. The ability to examine vegetation patterns using the VIs described here, combined with the synoptic view and repetitive coverage of satellite sensors, provides an opportunity to survey agricultural patterns in a manner that was not possible even a few years ago. If the information gathered by remote sensing systems can be integrated into the decision making and the transportation and distribution systems so important for agricultural production, there is a prospect for improvements in the effectiveness of food production.

Chapter 18 demonstrates how some of the information introduced here has significance in geological remote sensing. Full pursuit of extraction of geological information from an image requires intimate knowledge of how the spectral behavior of living plants responds to variations in the geologic substratum—a good example of how the fabric of remote sensing is so closely woven that it is not possible to isolate individual components as separate units.

17.19. Some Teaching and Learning Resources

- NASA: Science for a Hungry World: Part 1
  www.youtube.com/watch?v=1RJ6AqWAOEg

- NASA: Science for a Hungry World: Part 2
  www.youtube.com/watch?v=3rucWGuj0Lk

- NASA: Science for a Hungry World: Part 3
  www.youtube.com/watch?v=GXuS9gX7CRM

- NASA: Science for a Hungry World: Part 4
  www.youtube.com/watch?v=HDQ7XIOIAfY

- Landsat—Growing Concerns (1977)
  www.youtube.com/watch?v=CeqDUvI6CjI&feature=SeriesPlayList&p=7C8C9 C30AE62EBD5

This resource is a composite of the three films—*Growing Concerns* (agriculture), *Pollution Solution* (environmental quality), and *Fractured Look* (geosciences)—produced by NASA to publicize applications of Landsat data. Each topic is about 15 minutes in length; all date from 1977. This URL links to copies provided by the U.S. National Archives (through Public.Resoures.Org).

- NASA Earth Observatory: Vegetation Index
  www.youtube.com/watch?v=FO6ztYUAvhY

- SeaWiFS Biosphere Data over the North Pacific
  www.youtube.com/watch?v=sQvYvlonu7Y&NR=1

- The Earth Breathing
  www.youtube.com/watch?v=ZwzPhRvOgiA&feature=related

- SeaWiFS Biosphere Data over the North Atlantic
  www.youtube.com/watch?v=sGB-sqrLZ9U&feature=related

- Phenology Movie @ the Santa Rita Experimental Range
  www.youtube.com/watch?v=GBmvSjL78gM

- Hopkins Memorial Forest Phenology
  www.youtube.com/watch?v=bGDuxffCTeg&feature=related

Review Questions

1. Summarize differences between classification of vegetation from ground observations and classification from aerial images. Consider such factors as (a) basis of the classification and (b) the units classified. Identify distinctions for which remote sensing is especially well suited and those for which it is not likely to be useful.

2. Remote sensing to monitor crop development is much more difficult than it might appear initially. List some of the practical problems you might encounter as you plan an experiment to use satellite data to study the development of the winter wheat crop in western Kansas (or another crop region specified by your instructor).

3. Return to Tables 12.2 and 12.3 and to the category labels tabulated in the review questions to Chapter 12. Use at least two of the vegetation indices discussed in this chapter (as specified by your instructor) to assess the forested, pasture, and crop pixels. Examine differences between classes and between dates. Do some indices seem more effective than others?

4. To apply some of the knowledge presented in this chapter (e.g., the red shift), it is necessary to have data with very fine spectral, radiometric, and spatial detail. From your knowledge of remote sensing, discuss how this requirement presents difficulties for operational applications.

5. List some of the reasons why an understanding of the spectral behavior of an individual plant leaf is not itself sufficient to conduct remote sensing of vegetation patterns.

6. List some reasons why multispectral satellite images might be especially well suited for observation of vegetation patterns. Or, if you prefer, list some reasons why it is not quite so useful for vegetation studies. Briefly compare the MSS, TM, and SPOT with respect to utility for vegetation studies.

7. Write a short description of a design for a multispectral sensor tailored specifically for recording information about living vegetation and vegetation patterns, disregarding all other applications. Suggest optimum timing for a satellite to carry the sensor in a sun-synchronous orbit.

8. Describe some of the ways in which image classification (Chapter 12) might be useful in the study of vegetation patterns. Also identify some of the limitations of such methods in the study of vegetation.

9. Study of vegetative patterns by remote sensing is greatly complicated by the effects of soil background. Summarize contributions of soil background to spectral responses of a variety of natural and man-made landscapes, including cultivated crops, dense forest, orchards, and pasture. How might effects differ as season changes?

10. How do changes in sun angle and sun azimuth (due to differences in season and time of day) influence the way in which vegetative patterns are recorded on remotely sensed images?

11. List some of the ways in which monthly phenologic information might be useful for study of (a) cropped agricultural land, and (b) rangeland or forested land.

References

General

Atherton, B. C., M. T. Morgan, S. A. Sheraer, T. S. Stombaugh, and A. D. Ward. 1999. Site-Specific Farming: A Perspective on Information Needs, Benefits, and Limitations. *Journal of Soil and Water Conservation*, Second Quarter, pp. 339–355.

Badhwar, G. D. 1980. Crop Emergence Date Determination from Spectral Data. *Photogrammetric Engineering and Remote Sensing*, Vol. 46, pp. 369–377.

Badhwar, G. D., J. G. Carnes, and W. W. Austin. 1982. Use of Landsat-Derived Temporal Profiles for Corn–Soybean Feature Extraction and Classification. *Remote Sensing of Environment*, Vol. 12, pp. 57–79.

Bailey, R. G. 1976. *Ecoregions of the United States* (Map at 1:7,500,000). Ogden, UT: U.S. Forest Service.

Bailey, R. G. 1978. *Description of Ecoregions of the United States*. Ogden, UT: U.S. Forest Service, 77 pp.

Bailey, R. G. 1996. *Ecosystem Geography*. New York: Springer-Verlag, 204 pp.

Bailey, R. G. 1998a. *Ecoregions Map of North America* (Misc. Publication 1548). Washington, DC: U.S. Forest Service, 10 pp.

Bailey, R. G. 1998b. *Ecoregions: The Ecosystem Geography of the Oceans and the Continents.* New York: Springer-Verlag, 176 pp.

Collins, W. 1978. Remote Sensing of Crop Type and Maturity. *Photogrammetric Engineering and Remote Sensing*, Vol. 44, pp. 43–55.

Condit, H. R. 1970. The Spectral Reflectance of American Soils. *Photogrammetric Engineering*, Vol. 36, pp. 955–966.

Eidenshink, J. C. 1992. The 1990 Conterminous U.S. AVHRR Data Set. *Photogrammetric Engineering and Remote Sensing*, Vol. 58, pp. 809–813.

Gausman, H. W. 1974. Leaf Reflectance of Near-Infrared. *Photogrammetric Engineering*, Vol. 40, pp. 183–191.

Howard, J. A. 1970. *Aerial Photo-Ecology.* New York: Elsevier, 325 pp.

Hutchinson, C. F. 1982. Techniques for Combining Landsat and Ancillary Data for Digital Classification Improvement. *Photogrammetric Engineering and Remote Sensing*, Vol. 48, pp. 123–130.

Hyyppa, J., H. Hyyppa, M. Inkinen, M. Engdahl, S. Linko, and Y. H. Zhu. 2000. Accuracy Comparison of Various Remote Sensing Data Sources in the Retrieval of Forest Stand Attributes. *Forest Ecology and Management*, Vol. 128, pp. 109–120.

Jackson, R. D. 1983. Spectral Indices in N-Space. *Remote Sensing of Environment*, Vol. 13, pp. 409–421.

Jackson, R. D., P. J. Pinter, S. B. Idso, and R. J. Reginato. 1979. Wheat Spectral Reflectance: Interactions between Crop Configuration, Sun Elevation, and Azimuth Angle. *Applied Optics*, Vol. 18, pp. 3730–3732.

Jasinski, M. F., and P. S. Eagleson. 1990. Estimation of Subpixel Vegetation Cover Using Red-Infrared Scattergrams. *IEEE Transactions on Geoscience and Remote Sensing*, Vol. 28, pp. 253–267.

Jensen, J. R. 1978. Digital Land Cover Mapping Using Layered Classification Logic and Physical Composition Attributes. *American Cartographer*, Vol. 5, pp. 121–132.

Jensen, J. R. 1983. Biophysical Remote Sensing. *Annals of the Association of American Geographers*, Vol. 73, pp. 111–132.

Kanemasu, E. T. 1974. Seasonal Canopy Reflectance Patterns of Wheat Sorghum and Soybean. *Remote Sensing of Environment*, Vol. 3, pp. 43–57.

Kelly, M., D. Shaari, Q. Guo, and D. Liu. 2004. A comparison of standard and hybrid classifier methods for mapping hardwood mortality in areas affected by sudden oak death. *Photogrammetric Engineering and Remote Sensing*, Vol. 70, pp. 1229–1239.

Kerr, J. T., and M. Ostrovsky. 2003. From Space to Species: Ecological Applications for Remote Sensing. *Trends in Ecology and Evolution*, Vol. 18, pp. 299–305.

Kimes, D. S., W. W. Newcomb, R. F. Nelson, and J. B. Schutt. 1986. Directional Reflectance Distributions of a Hardwood and Pine Forest Canopy. *IEEE Transactions on Geoscience and Remote Sensing*, Vol. GE–24, pp. 281–293.

Kimes, D. S., J. A. Smith, and K. J. Ranson. 1980. Vegetation Reflectance Measurements as a Function of Solar Zenith Angle. *Photogrammetric Engineering and Remote Sensing*, Vol. 46, pp. 1563–1573.

Knipling, E. B. 1970. Physical and Physiological Basis for the Reflectance of Visible and Near Infrared Radiation from Vegetation. *Remote Sensing of Environment*, Vol. 1, pp. 155–159.

Küchler, A. W. 1967. *Vegetation Mapping.* New York: Ronald Press. 472 pp.

Means, J. E., S. E. Acker, B. J. Fitt, M. Renslow, L. Emerson, and C. J. Hendrix. 2000. Predicting Forest Stand Characteristics with Airborne Scanning Lidar. *Photogrammetric Engineering and Remote Sensing*, Vol. 66, pp. 1367–1371.

Meyers, V. I. 1975. Crops and Soils. Chapter 22 in *Manual of Remote Sensing* (R. Reeves, ed.). Falls Church, VA: American Society of Photogrammetry, pp. 1715–1816.

Myneni, R. B., J. Dong, C. J. Tucker, R. K. Kaufmann, P. E. Kauppi, J. Liski, et al. 2001. A Large Carbon Sink in the Woody Biomass of Northern Forests. *Proceedings of the National Academy of Sciences of the United States of America*, Vol. 98, pp. 14784–14789.

National Research Council. 1997. *Precision Agriculture in the 21st Century.* Washington, DC: National Academy Press, 149 pp.

Phillips, T. L., and staff. 1972. *1971 Corn Blight Watch Experiment Data Processing, Analysis, and Interpretation* (LARS Information Note 012272). West Lafayette, IN: Purdue University, Laboratory for Applications of Remote Sensing, 21 pp. *www.lars.purdue.edu/home/references/LTR_012272.pdf*

Post, D. F., C. Mack, P. D. Camp, and A. S. Suliman. 1988. Mapping and Characterization of the Soils on the University of Arizona Maricopa Agricultural Center. *Proceedings, Hydrology and Water Resources in Arizona and the Southwest*, Vol. 18. Tucson: Arizona–Nevada Academy of Science, pp. 49–60.

Rouse, J. W., R. H. Haas, J. A. Schell, D. W. Deering, J. C. Sagan, O. B. Toon, et al. 1979. Anthropogenic Changes and the Earth's Climate. *Science*, Vol. 206, pp. 1363–1368.

Ruimy, A., B. Saugier, and G. Dedieu. 1994. Methodology for the Estimation of Terrestrial Net Primary Production from Remotely Sensed Data. *Journal of Geophysical Research*, Vol. 99, No. D3, pp. 5263–5283.

Running, S. W., R. R. Nemani, F. A. Heinsch, M. S. Zhao, M. Reeves, and H. Hashimoto. 2004. A Continuous Satellite-Derived Measure of Global Terrestrial Primary Production. *Bioscience*, Vol. 54, pp. 547–560.

Sims, D. A., and J. A. Gamon. 2002. Relationships between Leaf Pigment Content and Spectral Reflectance across a Wide Range of Species, Leaf Structures and Developmental Stages. *Remote Sensing of Environment*, Vol. 81, pp. 337–354.

Townshend, J. R. G. 1994. Global Data Sets for Land Applications from the Advanced Very High Resolution Radiometric: An Introduction. *International Journal of Remote Sensing*, Vol. 15(17), pp. 3319–3332.

Tucker, C. J., H. H. Elgin, and J. E. McMurtrey. 1979. *Relationship of Red and Photographic Infrared Spectral Radiances to Alfalfa Biomass, Forage Water Content, Percentage Canopy Cover, and Severity of Drought Stress* (NASA Technical Memorandum 80272). Greenbelt, MD: Goddard Space Flight Center, 13 pp.

Tucker, C. J., J. R. G. Townshend, and T. E. Goff. 1985. African Land-Cover Classification Using Satellite Data. *Science*, Vol. 227, pp. 369–374.

Turner, W., S. Spector, N. Gardiner, M. Fladeland, E. Sterling, and M. Steininger. 2003. Remote Sensing for Biodiversity Science and Conservation. *Trends in Ecology and Evolution*, Vol. 18, pp. 306–314.

Vanden Heuvel, R. M. 1996. The Promise of Precision Agriculture. *Journal of Soil and Water Conservation*, Vol. 51, pp. 38–40.

Wilson, R. C. 1960. Photo Interpretation in Forestry. Chapter 7 in *Manual of Photographic Interpretation* (R. N. Colwell, ed.). Falls Church, VA: American Society of Photogrammetry, pp. 457–520.

Yool, S. R., D. W. Eckhardt, J. L. Star, T. L. Becking, and J. E. Estes. 1985. Image Processing for Surveying Natural Vegetation: Possible Effects on Classification Accuracy. In *Techni-

cal Papers, American Society of Photogrammetry. Falls Church, VA: American Society of Photogrammetry, pp. 595–603.

Land Surface Phenology

de Beurs, K. M., and G. M. Henebry. 2004. Land Surface Phenology Climatic Variation and Institutional Change: Analyzing Agricultural Land Cover Change in Kazakhstan. *Remote Sensing of Environment*, Vol. 89, pp. 497–509.

Dethier, B. E., M. D. Ashley, B. Blair, and R. J. Hopp. 1973. Phenology Satellite Experiment. *Symposium on Significant Results from the Earth Resources Technology Satellite-1*, Vol. 1. National Aeronautics and Space Administration. SP-327, pp. 157–165.

Harlan, C. 1974. *Monitoring the Vernal Advancement and Retrogradation (Greenwave Effect) of Natural Vegetation* (Type III Final Report). Greenbelt, MD: Goddard Space Flight Center, 371 pp.

Hopkins, A. D. 1938. *Bioclimatics: A Science of Life and Climate Relations* (U.S. Department of Agriculture Miscellaneous Publication 36). Washington, DC: U.S. Department of Agriculture, 188 pp.

Kang, S., S. W. Running, J. Lim, M. Zhao, and C. Park. 2003. A Regional Phenology Model for Detecting Onset of Greenness in Temperate Mixed Forests, Korea: An Application of MODIS Leaf Area Index (LAI). *Remote Sensing of Environment*, Vol. 86, pp. 232–242.

Koriba, K. 1958. On the Periodicity of Tree-Growth in the Tropics, with Special Reference to the Mode of Branching, the Leaf Fall, and the Formation of the Resting Bud. *Gardens' Bulletin* (Singapore), Series 3, Vol. 17. pp. 11–81.

Loveland, T. R., J. W. Merchant, J. F. Brown, D. O. Ohlen, B. Read, and P. Olsen. 1995. Seasonal Land Cover Regions of the United States. *Annals of the Association of American Geographers*, Vol. 85, pp. 339–355.

Morisette, J. T., A. D. Richardson, A. K. Knapp, J. I. Fisher, E. A. Graham, J. Abatzoglou, et al. 2009. Tracking the Rhythm of the Seasons in the Face of Global Change: Phenological Research in the 21st Century. *Frontiers in Ecology and the Environment*, Vol. 7, pp. 253–259.

Reed, B. C., J. Brown, D. Vander Zee, T. Loveland, J. Merchant, and D. Ohlen. 1994. Measuring Phenological Variability from Satellite Imagery. *Journal of Vegetation Science*, Vol. 3, pp. 703–714.

Reed, B. C., J. F. Brown, D. VanderZee, T. R. Loveland, J. W. Merchant, and D. O Ohlen. 1994. Variability of Land Cover Phenology in the United States. *Journal of Vegetation Science*, Vol. 5, pp. 703–714.

Reed, B. C., and L. Yang. 1997. Seasonal Vegetation Characteristics of the United States. *GeoCarto International*, Vol. 12, No. 2, pp. 65–71.

Rouse, J. W., R. H. Haas, J. A. Schell, and D. W. Deering. 1973. Monitoring Vegetation Systems in the Great Plains with ERTS. In *Third ERTS Symposium* (NASA Special Publication SP-351I), pp. 309–317.

Schwartz, M. D., and B. C. Reed. 1999. Surface Phenology and Satellite Sensor-Derived Onset of Greenness: An Initial Comparison. *International Journal of Remote Sensing*, Vol. 17, pp. 3451–3457.

Stockli, R., and P. L. Vidale. 2004. European Plant Phenology and Climate as Seen in a 20-Year AVHRR Land-Surface Parameter Dataset. *International Journal of Remote Sensing*, Vol. 17, pp. 3303–3330.

Weber, K. T. 2001. A Method to Incorporate Phenology into Land Cover Change Analysis. *Journal of Range Management*, Vol. 54, pp. A1–A7.

Wilson, J. S., T. S. Brothers, and E. J. Marcano. 2001. Remote Sensing of Spatial and Temporal Vegetation Dynamics in Hispaniola: A Comparison of Haiti and the Dominican Republic. *GeoCarto International*, Vol. 16, No. 2, pp. 5–16.

Zhang, X., M. A. Friedl, C. B. Schaaf, A. H. Strahler, J. C. F. Hodges, F. Gao, et al. 2003. Monitoring Vegetation Phenology Using MODIS. *Remote Sensing of Environment*, Vol. 84, pp. 471–475.

Vegetation Indices

Clevers, J. G. P. W., and W. Verhoef. 1993. LAI Estimation by Means of the WDVI: A Sensitivity Analysis with a Combined PROSPECT-SAIL Model. *Remote Sensing Reviews*, Vol. 7, pp. 43–64.

Cohen, W. B. 1991. Response of Vegetation Indices to Changes in Three Measures of Leaf Water Stress. *Photogrammetric Engineering and Remote Sensing*, Vol. 57, pp. 195–202.

Curran, P. 1980. Multispectral Remote Sensing of Vegetation Amount. *Progress in Physical Geography*, Vol. 4, pp. 315–341.

Haboudane, D., J. R. Miller, N. Tremblay, P. J. Zarco-Tejada, and L. Dextraze. 2002. Integrated Narrow-Band Vegetation Indices for Prediction of Crop Chlorophyll Content for Application to Precision Agriculture. *Remote Sensing of Environment*, Vol. 81, pp. 416–426.

Morain, S. A. 1978. *A Primer on Image Processing Techniques* (TAC TR #78-009). Albuquerque: University of New Mexico Technology Applications Center, 54 pp.

Perry, C. R., and L. F. Lautenschlager. 1984. Functional Equivalence of Spectral Vegetation Indices. *Remote Sensing of Environment*, Vol. 14, pp. 169–182.

Price, J. C. 1987. Calibration of Satellite Radiometers and Comparison of Vegetation Indices. *Remote Sensing of Environment*, Vol. 18, pp. 35–48.

Tucker, C. J. 1979a. Red and Photographic Infrared Linear Combinations for Monitoring Vegetation. *Remote Sensing of Environment*, Vol. 8, pp. 127–150.

Tucker, C. J. 1979b. *Remote Sensing of Leaf Water Content in the Near Infrared* (NASA Technical Memorandum 80291). Greenbelt, MD: Goddard Space Flight Center, 17 pp.

Saturation of Vegetation Indices

Cicone, R. C., and J. A. Olsenholler. 1997. A Summary of Asian Vegetation Using Annual Vegetation Dynamic Indicators. *Geocarto International*, Vol. 12, pp. 13–25.

Gitelson, A. A. 2004. Wide Dynamic Range Vegetation Index for Remote Quantification of Biophysical Characteristics of Vegetation. *Journal of Plant Physiology*, Vol. 161, pp. 165–173.

Huete, A. R., K. Didan, T. Miura, E. P. Rodriguez, X. Gao, and L. G. Ferreira. 2002. Overview of Radiometric and Biophysical Performance of the MODIS Vegetation Indices. *Remote Sensing of Environment*, Vol. 83, pp. 195–213.

Myneni, R. B., F. G. Hall, P. J. Sellers, and A. L. Marshak. 1995. The Interpretation of Spectral Vegetation Indexes. *IEEE Trans Geoscience and Remote Sensing*, Vol. 33, pp. 481–486.

Myneni, R. B., S. Hoffman, Y. Knyazikhin, J. L. Privette, J. Glassy, Y. Tian, et al. 2002. Global Products of Vegetation Leaf Area and Fraction Absorbed PAR from Year One of MODIS Data. *Remote Sensing Environment*, Vol. 83, pp. 214–231.

Hyperspectral Sensing of Vegetation

Card, D. H., D. L. Peterson, and P. A. Matson. 1988. Prediction of Leaf Chemistry by the Use of Visible and Near Infrared Reflectance Spectroscopy. *Remote Sensing of Environment*, Vol. 26, pp. 123–147.

Curran, P. J. 1989. Remote Sensing of Foliar Chemistry. *Remote Sensing of Environment*, Vol. 30, pp. 271–278.

Kokaly, R. F., and R. N. Clarke. 1999. Spectroscopic Determination of Leaf Biochemistry Using Band-Depth Analysis of Absorption Features and Stepwise Multiple Linear Regression. *Remote Sensing of Environment*, Vol. 67, pp. 267–287.

Kupiec, J., and P. J. Curran. 1993. AVIRIS Spectra Correlated with the Chlorophyll Concentration of a Forest Canopy. In *Summaries of the Fourth Annual JPL Airborne Geoscience Workshop* (JPL Publication 93-26). Pasadena, CA: Jet Propulsion Laboratory, Vol. 1, pp. 105–108.

Martin, M. E., and J. D. Aber. 1997. High Spectral Resolution Remote Sensing of Forest Canopy Lignin, Nitrogen, and Ecosystem Processes. *Ecological Applications*, Vol. 7, pp. 431–443.

Peterson, D. L., J. D. Aber, P. A. Matson, D. H. Card, N. Swanberg, C. Wessman, et al. 1988. Remote Sensing of Forest Canopy and Leaf Biochemistry Contents. *Remote Sensing of Environment*, Vol. 24, pp. 85–108.

Wessman, C. A., J. D. Aber, and D. L. Peterson. 1989. An Evaluation of Imaging Spectrometry for Estimating Forest Canopy Chemistry. *Remote Sensing of Environment*, Vol. 10, pp. 1293–1316.

Plant Pathology

Johannsen, C. J., M. E. Bauer, and Staff. 1972. *Corn Blight Watch Experiment Results* (LARS Information Note 012372). West Lafayette, IN: Purdue University, Laboratory for Applications of Remote Sensing, 24 pp. *www.lars.purdue.edu/home/references/LTR_012372.pdf*.

Kelly, M., D. Shaari, Q. Guo, and D. Liu. 2004. A Comparison of Standard and Hybrid Classifier Methods for Mapping Hardwood Mortality in Areas Affected by Sudden Oak Death. *Photogrammetric Engineering and Remote Sensing*, Vol. 70, pp. 1229–1239.

Kelly, M., K. Tuxin, and F. Kearns. 2004. Geospatial Informatics for Management of a New Forest Disease: Sudden Oak Death. *Photogrammetric Engineering and Remote Sensing*, Vol. 70, pp. 1001–1004.

Lefsky, M. A., D. Harding, W. B. Cohen, G. Parker, and H. H. Shugart. 1999. Surface Lidar Remote Sensing of Basal Area and Biomass in Deciduous Forests of Eastern Maryland, USA. *Remote Sensing of Environment*, Vol. 67, pp. 83–98.

Marshall, G. F., ed. 2005. *Handbook of Optical and Laser Scanning*. New York: Dekker, 792 pp.

Nelson, R., W. Krabill, and J. Tonelli. 1988a. Estimating Forest Biomass and Volume using Airborne Laser Data. *Remote Sensing of Environment*, Vol. 24, pp. 247–267.

Nelson, R., R. Swift, and W. Krabill. 1988b. Using Airborne Lasers to Estimate Forest Canopy and Stand Characteristics. *Journal of Forestry*, Vol. 86, pp. 31–38.

Nilsson, M. 1996. Estimation of Tree Heights and Stand Volume Using an Airborne Lidar System. *Remote Sensing of Environment*, Vol. 56, pp. 1–7.

Sharples, J. A. 1973. *The Corn Blight Watch Experiment: Economic Implications for Use of Remote Sensing Data Collecting Data on Major Crops* (LARS Information Note

110173). West Lafayette, IN: Purdue University, Laboratory for Applications of Remote Sensing, 12 pp. *www.lars.purdue.edu/home/references/LTR_110173.pdf*.

Ullstrup, A. J. 1972. The Impacts of the Southern Corn Leaf Blight Epidemics of 1970–1971. *Annual Review of Phytopathology*, Vol. 10, pp. 37–50.

Tasseled Cap

Crist, E. P. 1983. The Thematic Mapper Tasseled Cap: A Preliminary Formulation. In *Proceedings, Symposium on Machine Processing of Remotely Sensed Data 1983*. West Lafayette, IN: Purdue University, Laboratory for Applications of Remote Sensing, pp. 357–364.

Crist, E. P. 1985. A TM Tasseled Cap Equivalent Transformation for Reflectance Factor Data. *Remote Sensing of Environment*, Vol. 17, pp. 301–306.

Crist, E. P, and R. C. Cicone. 1984. A Physically Based Transformation of Thematic Mapper Data: The TM Tasseled Cap. *IEEE Transactions on Geoscience and Remote Sensing*, Vol. GE-22, pp. 256–263.

Crist, E. P., and R. C. Cicone. 1984. Application of the Tasseled Cap Concept to Simulated Thematic Mapper Data. *Photogrammetric Engineering and Remote Sensing*, Vol. 50, pp. 343–352.

Healey, S. P., W. B. Cohen, Y. Zhiqiang, and O. N. Krankina. 2005. Comparison of Tasseled Cap-Based Landsat Data Structures for Use in Forest Disturbance Detection. *Remote Sensing of Environment*, Vol. 97, pp. 301–310.

Kauth, R. J., and G. S. Thomas. 1976. The Tasseled Cap: A Graphic description of the Spectral–Temporal Development of Agricultural Crops as Seen by Landsat. In *LARS: Proceedings of the Symposium on Machine Processing of Remotely Sensed Data*. West Lafayette, IN: Purdue University Press, pp. 4B-41–4B-51.

Lidar Applications for Forestry

Blair, J. B., D. Coyle, J. Bufton, and D. Harding. 1994. Optimization of an Airborne Laser Altimeter for Remote Sensing of Vegetation and Tree Canopies. *Proceedings of IGARSS'94*, Vol. 2, pp. 939–941.

Bortolot, Z. J., and R. H. Wynne. 2005. Estimating Forest Biomass Using Small Footprint LiDAR Data: An Individual Tree-Based Approach that Incorporates Training Data. *ISPRS Journal of Photogrammetry and Remote Sensing*, Vol. 59, pp. 342–360.

Gong, P., G. S. Biging, S. M. Lee, X. Mei, Y. Sheng, R. Pu, et al. 1998. Photo Ecometrics for Forest Inventory. *Geographic Information Sciences*, Vol. 4, pp. 16.

Hyyppä, J., and M. Inkinen. 2002. Detecting and Estimating Attributes for Single Trees Using Laser Scanner. *Photogrammetric Journal of Finland*, Vol. 18, pp. 43–53.

Lefsky, M. A., W. B. Cohen, G. G. Parker, and D. J. Harding. 2002. Lidar Remote Sensing for Ecosystem Studies. *Bioscience*, Vol. 52, pp. 19–30.

Lefsky, M. A., D. Harding, W. B. Cohen, G. Parker, and H. H. Shugart. 1999. Surface Lidar Remote Sensing of Basal Area and Biomass in Deciduous Forests of Eastern Maryland, USA. *Remote Sensing of Environment*, Vol. 67, pp. 83–98.

Lim, K., P. Treitz, M. Wulder, B. St-Onge, and M. Flood. 2003. LiDAR Remote Sensing of Forest Structure. *Progress in Physical Geography*, Vol. 27, pp. 88–106.

McCombs, J. W., S. D. Roberts, and D. L. Evans. 2003. Influence of Fusing Lidar and Multispectral Imagery on Remotely Sensed Estimates of Stand Density and Mean Tree Height in a Managed Loblolly Pine Plantation. *Forest Science*, Vol. 49, pp. 457–466.

Means, J. E., S. A. Acker, B. J. Fitt, M. Renslow, L. Emerson, and C. J. Hendrix. 2000. Predicting Forest Stand Characteristics with Airborne Scanning Lidar. *Photogrammetric Engineering and Remote Sensing*, Vol. 66, pp. 1367–1371.

Means, J. E., S. A. Acker, D. J. Harding, J. B. Blair, M. A. Lefsky, W. B. Cohen, et al. 1999. Use of Large-Footprint Scanning Airborne Lidar to Estimate Forest Stand Characteristics in the Western Cascades of Oregon. *Remote Sensing of Environment*, Vol. 67, pp. 298–308.

Næsset, E. 2002. Determination of Mean Tree Height of Forest Stands by Digital Photogrammetry. *Scandinavian Journal of Forestry Research*, Vol. 17, pp. 446–459.

Næsset, E. 2002. Predicting Forest Stand Characteristics with Airborne Scanning Laser Using a Practical Two-Stage Procedure and Field Data. *Remote Sensing of Environment*, Vol. 80, pp. 88–99.

Næsset, E., and K. O. Bjerknes. 2001. Estimating Tree Heights and Number of Stems in Young Forest Stands Using Airborne Laser Scanner. *Remote Sensing of Environment*, Vol. 78, pp. 328–340.

Popescu, S. C. 2002. *Estimating Plot-Level Forest Biophysical Parameters Using Small-Footprint Airborne Lidar Measurements*. Blacksburg: Virginia Tech, College of Natural Resources, 155 pp.

Popescu, S. C., R. H. Wynne, and R. F. Nelson. 2003. Measuring Individual Tree Crown Diameter with Lidar and Assessing Its Influence on Estimating Forest Volume and Biomass. *Canadian Journal of Remote Sensing*, Vol. 29, pp. 564–577.

Popescu, S. C., R. H. Wynne, and J. A. Scrivani. 2004. Fusion of Small-Footprint Lidar and Multispectral Data to Estimate Plot-Level Volume and Biomass in Deciduous and Pine Forests in Virginia, USA. *Forest Science*, Vol. 50, pp. 551–565.

Popescu, S. C., R. H. Wynne, and R. F. Nelson. 2002. Estimating Plot-Level Tree Heights with Lidar: Local Filtering with a Canopy-Height Based Variable Window Size. *Computers and Electronics in Agriculture*. Vol. 37, 71–95.

Treuhaft, R. N., B. E. Law, and G. P. Asner. 2004. Forest Attributes from Radar Interferometric Structure and Its Fusion with Optical Remote Sensing. *BioScience*, Vol. 54, pp. 561–571.

Van Aardt, J. A. N., R. H. Wynne, and R. G. Oderwald. 2006. Object-Oriented Forest Volume and Biomass Estimation Using Small-Footprint Lidar Height and Intensity Distributions. *Forest Science*, Vol. 52, pp. 636–649.

van Aardt, J. A. N., R. H. Wynne, and R. G. Oderwald. 2006. Forest Volume and Biomass Estimation Using Small-footprint Lidar-Distributional Parameters on a Per-Segment Basis. *Forest Science*, Vol. 52, pp. 636–649.

Wynne, R. H., R. G. Oderwald, G. A. Reams, and J. A. Scrivani. 2000. Optical Remote Sensing for Forest Area Estimation. *Journal of Forestry*, Vol. 98, pp. 31–36.

Soil Brightness Line

Baret, F., and G. Guyot. 1991. Potentials and Limitations of Vegetation Indices for LAI and APAR Assessment. *Remote Sensing of Environment*, Vol. 35, pp. 161–173.

Baret, F., S. Jacquemond, and J. F. Hancoq. 1993. The Soil Line Concept in Remote Sensing. *Remote Sensing Reviews*, Vol. 7, pp. 65–82.

Ladoni, M., H. A. Bahrami, S. K. Alvaipanah, and A. A. Norouzi. 2009. Estimating Soil Organic Carbon from Soil Reflectance: A Review. *Precision Agriculture*, Vol. 11, pp. 82–99.

Nanni, M. R., and J. A. M. Dematte. 2006. Spectral Reflectance Methodology in Comparison to Traditional Soil Analysis. *Soil Science Society of America Journal*, Vol. 67, pp. 258–267.

Richardson, A. J., and C. L. Wiegand. 1977. Distinguishing Vegetation from Soil Background. *Photogrammetric Engineering and Remote Sensing*, Vol. 43, pp. 1541–1552.

Conservation Tillage

Arsenault, E., and F. Bonn. 2005. Evaluation of Soil Erosion Protective Cover by Crop Residues Using Vegetation Indices and Spectral Mixture Analysis of Multispectral and Hyperspectral Data. *Catena*, Vol. 62, pp. 157–172.

Daughtry, C. S. T., E. R. Hunt, P. C. Doraiswamy, and J. E. McMurtrey. 2005. Remote Sensing the Spatial Distribution of Crop Residues. *Agronomy Journal*, Vol. 97, pp. 864–871.

Serbin, G., C. S. T. Daughtry, E. R. Hunt, D. J. Brown, and G. W. McCarty. 2009. Effect of Soil Spectral Properties on Remote Sensing of Crop Residue Cover. *Soil Science Society of America Journal*, Vol. 73, pp. 1545–1558.

Sullivan, D. G., D. Lee, J. Beasley, S. Brown, and E. J. Williams. 2008. Evaluating a Crop Residue Cover Index for Determining Tillage Regime in a Cotton–Corn–Peanut Rotation. *Journal of Soil and Water Conservation*, Vol. 63, pp. 28–36.

Sullivan, D. G., T. C. Strickland, and M. H. Masters. 2008. Satellite Mapping of Conservation Tillage Adoption in the Little River Experimental Watershed, Georgia. *Journal of Soil and Water Conservation*, Vol. 63, pp. 112–119.

van Deventer, A. P., A. D. Ward, P. H. Gowda, and J. G. Lyon. 1997. Using Thematic Mapper Data to Identify Contrasting Soil Plains and Tillage Practices. *Photogrammtric Engineering and Remote Sensing*, Vol. 63, pp. 87–93.

CHAPTER EIGHTEEN

Earth Sciences

18.1. Introduction

This chapter addresses applications of remote sensing in the earth sciences, which are loosely defined here to include geology, geomorphology, and soil science. Despite their many differences, these disciplines share a common focus on the Earth's shape and structure and on the nature of the soils and sediments at its surface. Applications of remote sensing in the earth sciences are especially difficult. Often the subjects of investigation are geologic structures, soil horizons, and other features entirely or partially hidden beneath the Earth's surface. Seldom can remotely sensed images directly record such information. Instead, we must search for indirect evidence that may be visible at the surface. Thus interpretations of geologic, pedologic, and geomorphic information are often based on inference rather than direct sensing of the qualities to be studied. Furthermore, interpretations of geoscience information are frequently based on subtle distinctions in tone, texture, and spectral response. Even direct examination of many geologic and pedologic materials is subject to error and controversy, so it should be no surprise that applications of remote sensing in these fields can be equally difficult.

Finally, we must often depend on composite signatures formed as individual "pure" signatures of various types of soil, rock, and vegetation are mixed together into a single, composite spectral response that may be unlike any of its components (Chapter 10). Field observations are made by necessity at points such as outcrops, soil profiles, or boreholes. Because remote sensing instruments record radiation from ground *areas* rather than from ground *points*, it may be difficult to relate remotely sensed data to the field observations traditional for the earth sciences. Such problems present both practical and conceptual difficulties for applications of remote sensing in the earth sciences.

Remote sensing can, however, provide opportunities not normally available to geoscientists. The ability to observe reflectance and emittance over a range of wavelengths opens the door to the study of subjects that would not otherwise be possible. Also, the synoptic view of satellite images gives a broad-scale perspective of patterns not discernible within the confines of the large-scale, closeup view of ground observation. A single satellite image can record an integrated signature formed of many kinds of information, including shadow, soil, rock, and vegetation; although these mixtures may present problems, they also portray subtle differences in terrain that cannot be easily derived from other sources. Finally, although remotely sensed images cannot replace field and laboratory studies, they can form valuable supplements to more traditional methods and sometimes provide information and a perspective not otherwise available.

18.2. Photogeology

Geologists study many aspects of the Earth's surface in an effort to understand the Earth's structure, to guide the search for minerals and fuels, and to understand geologic hazards. Remote sensing contributes to several dimensions of the geological sciences by providing information concerning lithology, structure, and vegetation patterns. *Lithology* refers to fundamental physical and chemical properties of rocks, including, for example, the gross distinctions between sedimentary, igneous, and metamorphic rocks. *Structure* defines the kinds of deformation experienced by rocks, including folding, fracturing, and faulting. *Geobotanical studies* focus on relationships between plant cover at the Earth's surface and the lithology of underlying rocks.

Photogeology is the derivation of geological information from interpretation of aerial photographs. Photogeology originated early in the development of air photo interpretation. Many of its basic techniques were developed in the 1920s, then refined and applied into the 1950s and 1960s, when they seemed to approach the limits of their capabilities and were assimilated into newly developing research in geological remote sensing (Ray, 1960). Today photogeology is routinely applied to good effect. However, since aerial photographs now form only one of many forms of aerial images routinely available to the geologist, most research and innovation is now likely to occur in the broader context of geological remote sensing.

The practice of photogeology is based on direct application of the principles of image interpretation (Chapter 5) to geological problems. Image texture, size, shape, tone, shadow, and so on each have special significance in the realm of the geologist's view of the terrain. Likewise, the principles of photogrammetry (Chapter 3) have special applications in the context of photogeology for calculation of thicknesses of beds and determination of strike and dip from aerial photography. Such measurements permit derivation of structural information from aerial images.

Remotely sensed images can provide valuable information about structure and terrain, especially if stereoscopic views are available. Figures 18.1 and 18.2 show a stereo image of an arid landscape in New Mexico characterized by lava flows lying uncomfortably on gently dipping sedimentary rocks. This landscape is positioned just east of the Rio Grande River, between Albuquerque and Santa Fe. During the Quaternary and late Tertiary, this region experienced extensive volcanism.

At the upper edge of the photograph, upturned sandstone beds are visible as slightly curving arcs, dipping to the left and continuing toward the bottom of the photograph. In stereo, the vertical exaggeration creates distinctive relief, revealing the resistant character of the sandstone strata that form the ridges. These features are known as *hogbacks* (i.e., resistant strata upended to form steeped-sided ridges with jagged crests). Resistant sandstones form the ridges, with less resistant shales eroded to form intervening valleys. Figure 18.3 presents a ground view of the hogbacks, taken from a position near the upper right hand corner of Figure 18.2.

These photographs also show evidence of the volcanism that has shaped this landscape. Volcanic dikes cut across established topography as narrow, pronounced, steep-sided ridges. Notice that these dikes cut across the north–south oriented ridges, confirming the recent age of the volcanism relative to the other structures.

The lower edge of Figure 18.2 shows a portion of an extensive basaltic flow that forms the surface of a resistant, steep-sided mesa. Viewed in stereo, the resistant character of the basaltic surface is evident from its sharp relief and precipitous cliffs at its margins.

FIGURE 18.1. Stereo aerial photographs of the Galisteo Creek region, north central New Mexico. From U.S. Geological Survey and Army Map Service.

This region is a resistant bedrock surface covered with a thin cover of gravel, sparsely vegetated. As viewed on the aerial photograph, it has a distinctive appearance, even in tone, slightly rough in texture, and with a puffy appearance when viewed in stereo. Note that the basaltic plateau has covered preexisting sedimentary strata, especially where the flow has been superimposed over the rightmost hogback, creating a clear unconformity.

At the base of the cliff, the aerial photograph shows the contrasting textures of a thick sheet of stone, sand, and gravel, shaped by fluvial processes that transport geologic debris across the slope at the base of the cliff toward the stream flowing left to right across the photograph. This stream has maintained its course across the hogbacks as they were formed, carving a distinct V-shaped notch through the right-hand ridge. As the stream passes through the notch, it enters a terrain formed by erosion of shales exposed at the surface. Here surface erosion is much more effective than on the surfaces of the basaltic flows, for example, and the topography is much more disorganized and chaotic relative to the surface visible on the left-hand side of the image.

Although the discussion of this example is necessarily quite brief, it illustrates the power of aerial imagery to provide information and insight to the earth scientist, both to derive new information and to organize and interpret information already at hand. Subsequent sections explore some other dimensions of photogeology and the value of remotely sensed imagery for the earth sciences.

Galeisto
Creek

hogbacks

volcanic
dikes

stream

lava-topped
mesa

FIGURE 18.2. Annotated version of image shown in Figure 18.1

FIGURE 18.3. Ground photograph of the Galisteo Creek region, New Mexico, showing some of the same features represented in Figure 18.1 This photograph was taken from a location near the top right-hand corner of Figure 18.1, looking across the creek toward the hogbacks. Photograph by W.T. Lee, 1910. From U.S. Geological Survey Photographic Library, Denver, CO.

18.3. Drainage Patterns

Drainage patterns are often the most clearly visible features on remotely sensed imagery and also are among the most informative indicators of surface materials and processes. The character of drainage patterns permits geologists to infer valuable information about the surface materials and the geologic structure that prevail for a specific landscape (Figures 18.4 and 18.5). Drainage patterns as observed in nature may display characteristics of two or more of these patterns, so they may not match precisely to illustrations used to describe them.

A common drainage pattern is characterized by even branching of tributaries, similar to the patterns of veins on an oak or maple leaf—a *dendritic* drainage pattern. Dendritic patterns suggest uniformly resistant surface materials, gentle regional slopes, and the absence of major faults or structural systems. *Parallel* drainage is also found on uniform materials, but on landscapes that have pronounced regional slopes. Parallel drainage resembles the dendritic pattern but is distinctive by its elongated form derived from the increased typographic slope. In fact, there are many transitional forms that fall between these two forms.

If a landscape has been uplifted by tectonic forces or if base level has been lowered (e.g., by a decline in sea levels), then streams may become *entrenched*, or *incised*, into the landscape such that their channels are cut well below the surface of the surrounding landscape. The nature of the incision can reveal the character of the bedrock—sharp, deep, well-defined edges suggest the presence of strong, cohesive surface materials. More gently sloping terrain near the channel suggests the presence of less cohesive, weaker strata.

If the landscape is characterized by linear structural or lithologic features, drainage often develops to form a *trellis* pattern. Tributaries often follow the strike of the structure, whereas the main streams cut across the principal structures. Therefore, tributaries join main streams at right angles to form a system in which the main streams and the

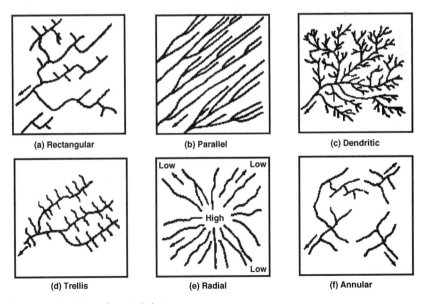

| (a) Rectangular | (b) Parallel | (c) Dendritic |
| (d) Trellis | (e) Radial | (f) Annular |

FIGURE 18.4. Sketches of varied drainage patterns.

FIGURE 18.5. Aerial photographs illustrate some of the drainage patterns described in the text and in Figure 18.4.

tributaries are oriented perpendicular to each other. When a landscape is dominated by a large central peak, the drainage system is organized to drain water away from the central upland, forming a *radial* pattern.

When landscapes are disturbed by broad-scale processes such as glaciation, faulting, surface mining, or volcanic deposition, the established drainage system is often buried or disturbed. The resulting landscape is characterized by incoherent drainage systems—lakes and marshes are connected by a confused system of streams. As time passes, an organized stream network will reestablish itself, but in the meantime the drainage system will be confused and disorganized.

Finally, *braided streams* have several channels that divide and rejoin, separated by ephemeral bars and islands. Braided streams are often found in locales that have variable stream discharge (such as those that are in arid regions or in glacial meltwater) and high sediment loads. Their appearance is usually quite distinctive—a wide, sparsely vegetated strip of open sand or gravel, with a network of ephemeral, anastomosing channels and elongated bars and islands.

Closer inspection of details of the drainage system can reveal specifics of the textures of the surface materials. Steep slopes are characteristic of coarse, loose surface materials. Noticeably flat, rounded slopes may indicate cohesive, plastic surface materials. The absence of surface drainage or a very simple drainage pattern often indicates well-drained pervious soil. Highly integrated drainage patterns often indicate impervious, plastic sediments that lose their strength when wet.

The shape of the cross-section of a drainageway is controlled largely by the cohesiveness of the surface materials. Abrupt changes in stream grade, channel direction, or its cross-section indicate changes in the underlying bedrock or the surface materials. Generally, short, V-shaped gullies with steep gradients are typical of well-drained, loosely structured sediments such as coarse sands. U-shaped stream profiles indicate the presence of deep, uniform silt deposits, especially the wind-deposited sediments known as *loess*. Poorly drained fine-textured surface materials usually form shallow drainageways with shallow, rounded saucer-like profiles.

On black-and-white aerial photography, the tones of surface materials can indicate the character of the drainage. Soft, light tones generally indicate pervious, well-drained soils. Large, flat areas of sand are frequently indicated by uniform light gray color tones, a flat appearance, and a lack of conformity; this indicates a natural surface drainage. Clays and organic soils frequently appear as dark-gray to black areas. In general, a sharp change in colors or tones signals a change in soil texture.

18.4. Lineaments

Lineaments is the name given by geologists to lines or edges, of presumed geologic origin, visible on remotely sensed images. Such features have also been referred to as *linears*, or *lineations*, although O'Leary et al. (1975) establish *lineaments* as the preferred term. Use of the term *lineament* in a geological context dates to 1904 and apparently has even earlier analogues in other languages. These early uses, prior to availability of aerial images, applied to specific geologic or geomorphic features, such as topographic features (ridgelines, drainage systems, or coastlines), lithologic contacts, or zones of fracture.

As early as the 1930s photogeologists studied fracture patterns visible on aerial photographs as a means of inferring geologic structure. These photo features apparently corresponded rather closely to faults and fractures defined in the field. More recently, in the context of geological remote sensing, the term has assumed a broader meaning. Any linear feature visible on an aerial image can be referred to as a *lineament*. A problem arises because it is difficult to establish a clear link between the features on images and corresponding features, if any, on the ground.

The strength of this link depends in part on the nature of the imagery. In the instance of interpretations from aerial photographs and photomosaics, scales are usually large, and ample detail is visible. Linear features visible on such images are likely to match to those that can be confirmed in the field. However, more extensive and subtle features cannot be easily detected on aerial photography. Each photograph shows only a small area; if mosaics are formed to show larger regions, variations in illumination and shadowing obscure more subtle lineaments that might extend over many photographs. Furthermore, in the era prior to routine availability of the CIR films now used for high-altitude photography, high-altitude photographs (which *can* show large areas under uniform illumination) were not of good visual quality due to effects of atmospheric scatter.

The advent of nonphotographic sensors and the availability of the broad-scale view of satellite images changed this situation, because it then became possible to view large areas with images of good visual clarity. Early radar images depicted large areas illuminated from a single direction at rather low illumination angles—conditions that tended to increase topographic shadowing in a manner that enhanced the visibility of linear features (Figure 18.6). Locations of these features did not always match previously known faults or fractures, so the term *lineament* was evoked to avoid an explicit statement of a geological origin. Later, similar features were again detected on Landsat MSS imagery. Initially, the detection of lineaments and the interpretation of their meaning generated some controversy; the broad scale and coarse resolution of the early satellite imagery caused lineaments to be subtly expressed and difficult to relate to previously known geological features. During the intervening years, investigation and debate have confirmed the value of lineaments as guides to understanding geologic phenomena.

Lineaments are, of course, "real" features if we consider them simply to be linear features detected on aerial imagery. The uncertainty arises when we attempt to judge their geological significance. There are sound reasons for assigning a geological meaning to some lineaments, even if they do not always correspond to clearly observable physical features at a specific point. In the simplest instances, a dip–slip fault may leave a subtle topographic feature that is visible as a linear shadow on aerial photography when illuminated obliquely from the elevated side (Figure 18.7). The axis of displacement for a

FIGURE 18.6. Radar mosaic of southern Venezuela. Here topography and drainage are clearly visible, as are the linear features in the upper right that illustrate the kinds of features important in lineament analysis. From Goodyear Aerospace Corporation.

FIGURE 18.7. Schematic illustration of a dip–slip fault. (a) Illumination from the elevated side produces a strip-like shadow on the image. (b) Illumination from the downthrust side may produce a bright strip as the sun is reflected from the linear facet that now faces the solar beam.

dip–slip fault is along the dip of the fault plane, to create a difference in relief, as depicted in Figure 18.7. In the alternative instance, a strike–slip fault, the axis of the fault displacement follows the strike of the fault (along the trace of the fault on the surface), creating fewer opportunities for surface expression. Of course, a fault that is oriented parallel to the direction of illumination may be indistinct or invisible. Not all faults are expressed topographically, but the fault plane may offer preferred avenues of movement for moisture and for the growth of plant roots (Figure 18.8). Therefore, the trace of the fault may be revealed by vegetation patterns, even though it has no obvious topographic expression. Faults of any form can alter drainage in a manner that creates linear drainage segments, which are then visible on aerial images. Such features, when clearly expressed, may have structural origins, and it may be possible to verify their existence through field observations. However, more subtle features may be observable only as broad-scale features, not easily verifiable at a given point. Such lineaments may be genuine structural features, but not easily confirmable as such.

Linear features that are clearly of structural origin are significant because they indicate zones of fracturing and faulting. It is often assumed that regions of intersection of lineaments of differing orientation are of special significance, as in theory they might identify zones of mineralization, stratigraphic traps, regions of abundant groundwater, or zones of structural compression. For this reason, much is often made of the orientation and angles of intersection of lineaments.

FIGURE 18.8. Hypothetical situation in which a fault with little or no surface expression forms a preferred zone of movement for moisture and for the growth of plant roots. Such a zone may create a favorable habitat for certain species or speed the growth of more prevalent plants located near the fault. The linear arrangement of such plants and their shadows may form a linear feature on an image.

Other lineaments may not have structural origins. It is conceivable that some may be artifacts of the imaging system or of preprocessing algorithms. In Figure 18.1, note the numerous man-made linear features, and in Figure 18.3 note the linear features caused by a jet contrail and its shadow, visible as the parallel white and dark streaks across the upper image. Some linear features may be purely surface features, such as wind-blown sediments, that do not reflect subsurface structure (Figure 18.9). Cultural patterns, including land-cover boundaries, edges of land ownership parcels, and political borders, can all have linear form and can be aligned in a manner that creates the linear features observed on aerial images (Figure 18.1). Because these features may not be readily distinguished from those of geologic origin, and because of the inconsistency of individual delineations of lineaments from the same image, lineament analysis has been regarded with skepticism by a significant proportion of the geological community (Wise, 1982).

Because the human visual system tends to generalize, manual interpretations may identify continuous linear features that are later found to be formed of separate segments of unrelated origin. Lineaments that are found to correspond to geologic features often are extensions of known fault systems rather than completely new systems. Results of manual interpretations vary greatly from interpreter to interpreter (Podwysocki et al., 1975). As a result, other research has attempted to automate the interpretation of lineaments. Because detection of edges and lines has long been an important task in the fields of image analysis and pattern recognition, the fields of image processing and machine vision were well equipped with techniques to apply to this inquiry. There is a rich literature that has investigated procedures for enhancing images to highlight linear features and then analyzed patterns to extract linear features that might be present.

Fracture Density Studies

Often images and field data can be analyzed to assess the abundance of lineaments by plotting their occurrence within cells used to partition the image. Analysis of the distribution and orientation of lineaments and fractures, in the context of field studies and other databases, can lead to assessment of groundwater or mineral potential.

Abrams et al. (1983) studied lineaments interpreted from a Landsat MSS color composite of a region in southwestern Arizona. Their analysis focused on the Helvetia–Rosemont area, known to be rich in porphyry copper. They attempted to determine whether their techniques would locate previously discovered mineralized zones. Paleozoic limestones, quartzites, shales, and dolomites cover Precambrian granites and shists in

FIGURE 18.9. Example of lineaments that do not reflect geologic structure. Wind-blown sand has been aligned in linear ridges that indicate the nature of the surface sediments only.

the study area. The Paleozoic deposits themselves were altered by Mesozoic uplift and erosion, then covered by further deposition. The Laramide orogeny during the late Cretaceous produced intrusion of granitic rocks with thrust faulting along a predominant northeast–southwest trend. This tectonic activity, together with later Paleocene intrusions, caused mineralization along some of the faults. Tectonic activity continued into the late Tertiary, producing extensive faulting and a complex pattern of lineaments.

Abrams et al. examined orientations of lineaments they interpreted from the Landsat image, then plotted their results in the form of a rose diagram (Figure 18.10). Each wedge in the circular pattern represents the numbers of lineaments oriented along specific compass azimuths. Because each lineament has two azimuths (e.g., a "north–south" line is oriented as much to the south as it is to the north), the diagram is symmetric, and some investigators prefer to show only half of the diagram.

Several features are noteworthy. First, the low numbers of lineaments oriented at (approximately) 110 (and 290), roughly east–southeast to west–northwest, correspond to the axis that parallels the orientation of the solar beam at the time that the Landsat data were acquired. The lineaments that might have this orientation will not be easily detected because shadowing is minimized. The northeast–southwest trend (B in Figure 18.10) is said to represent the predominant trend of Precambrian faulting that is observed throughout Arizona. Superimposed over this pattern is the northwest–southeast trend (A) arising from faulting in the Mesozoic rocks mentioned above. Finally, the east–west trend (C) is interpreted as the result of the Laramide faulting known to be associated with the mineralization that produced the copper deposits present in this region. Abrams and his colleagues counted the frequencies of the lineaments in each cell of a 10-km grid superimposed over the image and concluded that the highest frequencies, which they interpreted to form favorable locations for intrusion of magmas and for mineralization, correspond to zones of known mineral deposits.

18.5. Geobotany

Geobotany is the study of the relationships between plants and the underlying geologic substratum (Ustin et al., 1997). Such relationships include variations in the presence, abundance, vigor, and appearance of plants, indicator species, and distinctive phono-

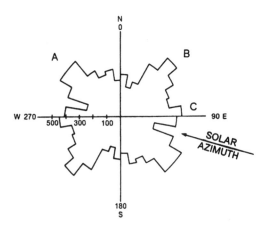

FIGURE 18.10. Strike-frequency diagram of lineaments for a region of southeastern Arizona. Features, as discussed in the text, include: (A) predominant northwest–southeast orientation; (B) predominant northeast–southwest orientation; and (C) low frequency along the axis that parallels the solar beam at the time of image acquisition. Based on Abrams et al. (1983, p. 593). Reproduced by permission.

logical behavior; often it can refer more specifically to examination of the relationships between plants and their response to unusual levels of nutrients or toxins released to the soil by weathering of geologic materials (Figure 18.11). *Regional geobotany* focuses on the study of vegetation patterns as indicators of lithologic variations. Specific plant species may have local significance as indicator species, as they tend to avoid or favor certain lithologic units. In other instances, variations in abundance or vigor may signal the occurrence of certain lithologies.

Geologic processes may concentrate trace elements in specific strata or regions. Geologic weathering may release trace elements in a form that can be absorbed by vegetation. Plants that absorb these trace elements at higher than normal levels may display abnormal spectral signatures, thereby signaling the existence and location of the anomalous concentration of elements.

Geobotany depends on knowledge of how geologic materials release elements into the nutrient pool, how these elements are absorbed by the soil and concentrated in plant tissues, and how they can alter the spectral signatures of plants. Because specific kinds of plants may thrive under certain unusual conditions of soil fertility, detailed knowledge of the spectral characteristics of such plants can be very valuable. Geobotanical studies are especially valuable in heavily vegetated regions where soil and rock are not exposed to direct view of the sensor, but they may also be useful in sparsely vegetated regions where it may be possible to identify individual plants or where vegetation patterns may be especially sensitive to subtle environmental variations.

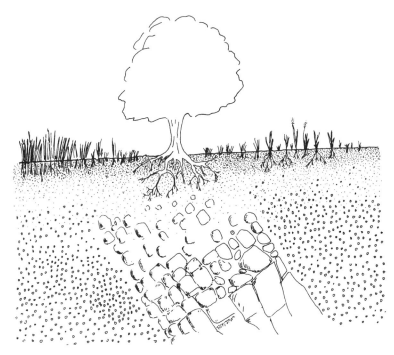

FIGURE 18.11. General model for geobotanical studies. Abundance, vigor, and patterns of plants are related to the geologic substratum. If geologists can isolate these geologic factors from others that influence vegetation patterns, botanical information can be used (with other data) to contribute to geological knowledge

The practice of geobotanical reconnaissance is restricted by several factors. First, geobotanical studies may depend on observations of subtle distinctions in vegetation vigor and pattern, so successful application may require data of very fine spatial, spectral, and radiometric resolution—resolution much finer than that of Landsat or SPOT data, for example. Thus some of the concepts of geobotanical exploration cannot be routinely applied with the imagery and data most commonly available. Hyperspectral remote sensing (Chapter 15) provides the necessary capabilities for developing geobotanical knowledge. This work, now conducted mainly in a research context, may provide the basis for development of sensors and analytical techniques applicable on an operational basis. Second, geology is only one of the many factors that influence plant growth. Site, aspect, and disturbance (both by mankind and by nature) influence vegetation distributions in very complex patterns. Effects of these several causes cannot always be clearly separated.

Timing of image acquisition is critical. Some geobotanical influences may be detectable only at specific seasons or may be observable as, for example, advances in the timing of otherwise normal seasonal changes in vegetation coloring. The issue of timing may be especially important because the remote sensing analyst may not have control over the timing of image acquisition or may have imagery only for a single date, whereas several may be necessary to observe the critical changes. Finally, many geobotanical anomalies may have a distinctively local character—specific indicators may have meaning only for restricted regions, so intimate knowledge of regional vegetation and geology may be necessary to fully exploit knowledge of geobotany.

Geobotanical knowledge can be considered first in the context of the individual plant and its response to the geological substratum. Growth of most plants is sensitive not only to the availability of the *major nutrients* (phosphorus [P], potassium [K], and nitrogen [N]), but also to the *micronutrients*, those elements (including barium [B], magnesium [Mg], sulphur [S], and calcium [Ca]) that are required in very small amounts. It is well established that growth and health of plants depends on the availability of elements present in very low concentrations and that sensitivity may be high in some plant species. In contrast, other elements are known to have toxic effects, even at very low concentrations, if present in the soil in soluble form. Heavy metals, including nickel (Ni), copper (Cu), chromium (Cr), and lead (Pb), may be present at sufficient levels to reveal their presence through stunted plant growth or by the localized absence of specific species that are especially sensitive to such elements. From such evidence, it may be possible to indirectly use the concentrations of these elements to reveal the locations of specific geologic formations or zones of mineralization worthy of further investigation.

Hydrocarbons may significantly influence plant growth if they are present in the root zone. Hydrocarbon gases may migrate from subsurface locations; at the surface, concentrations may be locally sufficient to influence plant growth. Hydrocarbons may be present in greater concentrations at the surface as petroleum seeps or coal seams. The presence of hydrocarbons in the soil may favor plant growth by increasing soil organic matter but may also alter soil atmosphere and soil biology in a manner that is toxic to some plants.

Specific geobotanical indicators include the following, which of course may have varied geologic interpretation depending on the local setting. Variations in biomass, either significantly higher or lower than expected, may indicate geobotanical anomalies. Or the coloring of vegetation may be significant. The term *chlorosis* refers to a general yellow discoloration of leaves due to a deficiency in the abundance of chlorophyll. Chlorosis can be caused by many of the geologic factors mentioned above, but it is not uniquely of geologic origin, so its interpretation requires knowledge of the local geologic and biologic setting.

Observation of these indicators may be facilitated by use of the vegetation indices and ratios mentioned in Chapter 17. Manual interpretation of CIR imagery often provides especially good information concerning the vigor and distribution of vegetation. Of special significance in the context of geobotany is the *blue shift* in the chlorophyll absorption spectra. The characteristic spectral response for living vegetation was discussed in Chapter 17. In the visible, peak reflectance is in the green region; the absorption spectra of chlorophyll decreases reflectance in both the blue and the red regions. However, in the near infrared, reflectance increases markedly (Chapter 17).

Collins et al. (1981) and others have observed that geochemical stress is most easily observed in the spectral region from about 0.55 to 0.75 μm (which includes portions of the green, red, and infrared regions). Most notably, the position and slope of the line that portrays the increase in reflectance at the edge of the visible region, although constant for healthy green plants, is especially sensitive to geochemical influences. As geochemical stress occurs, the position of this line shifts toward shorter wavelengths (i.e., toward the blue end of the spectrum; hence the term *blue shift*), and its slope becomes steeper (Figure 18.12).

It must be emphasized that, by everyday standards, the change is very subtle (0.007–0.010 μm) and has been observed only in data recorded by high-resolution sensors processed to filter out background reflectance. Such data are not acquired by remote sensing instruments available for routine use, although increased spectral, spatial, and radiometric resolution provided by hyperspectral remote sensing (Chapter 15) opens up opportunities to exploit this kind of knowledge. Nonetheless, Collins et al. (1983) report that the blue shift has been observed in all plant species studied thus far; they report that they believe this effect to be "universal for most green plants" (p. 739), so it is not confined to a few species.

The cause of the blue shift is not clearly understood at present. Unusual concentrations of heavy metals in the soil are absorbed by plants and apparently tend to be transported toward the actively growing portions of the plant, including the leaves. Concentra-

FIGURE 18.12. Blue shift in the edge of the chlorophyll absorption band. As heavy mineral concentration in plant tissue increases, the edge of the absorption band shifts toward shorter wavelengths (i.e., toward the blue end of the spectrum). The solid line shows the spectrum for a control plant. The dotted and dashed lines show how the magnitude of the shift increases as concentration of the heavy minerals increases. From Chang and Collins (1983, p. 727). Reproduced by permission.

tions of such metals can cause chlorosis, often observed by more conventional methods. The relationship between heavy metals and the blue shift is clear, but work by Chang and Collins (1983) indicates that the presence of heavy metals in the plant tissues does not alter the chlorophyll itself—the metallic ions apparently do not enter the structure of the chlorophyll. The heavy metals appear to retard the development of chlorophyll, thereby influencing the abundance, but not the quality, of the chlorophyll in the plant tissue.

Chang and Collins (1983) conducted laboratory and greenhouse experiments to confirm the results of spectral measurements in the field and to study the effects of different metals and their concentrations. Their experiments used oxides, sulfates, sulfides, carbonates, and chlorides of copper (Cu), zinc (Zn), iron (Fe), nickel (Ni), manganese (Mn), molybdenum (Mo), and vanadium (V). Selenium (Se), Ni, Cu, and Zn produced stress at concentrations as low as 100 ppm. Fe and Pb appeared to have beneficial effects on plant growth; Mo and V produced little effect at the concentrations studied. In combination, some elements counteract each other, and some vary in effect as concentrations change.

18.6. Direct Multispectral Observation of Rocks and Minerals

A second broad strategy for remote sensing of lithology depends on accurate observation of spectra from areas of soil and rock exposed to observation. Color and, by extension, the spectral characteristics of rocks and minerals can be distinctive identifying features in the direct examination of geologic samples. Geologists use color for identification of samples in the field, and in the laboratory sensitive instruments are used to measure spectral properties across a range of wavelengths in the ultraviolet, visible, and infrared regions. Spectral emittance and reflectance of rocks and minerals is an important property that is often closely related to their physical and chemical properties. In the laboratory, spectral characteristics can be observed in such sufficient detail that they can sometimes form diagnostic tests for the presence of specific minerals. In addition, small radiometers can be taken to the field to permit in situ observation of rocks and minerals in nonvisible portions of the spectrum. As a result, geologists have considerable experience in observing spectral properties of geologic materials and have developed an extensive body of knowledge concerning the spectral properties of rocks and minerals.

The application of this knowledge in the context of remote sensing can be very difficult. In the laboratory, spectra can be observed without the contributions of atmospheric attenuation, vegetation, or shadowing to the observed spectra. In the laboratory setting, scientists can hold secondary properties (such as moisture content) constant from one sample to the next or can compensate for their effects. In remotely sensed data, the effects of such variables often cannot be assessed. The usual remote sensing instruments do not have the fine spectral and radiometric resolutions required to make many of the subtle distinctions this strategy requires. Furthermore, the relatively coarse spatial resolutions of many remote sensing systems means that the analyst must consider composite signatures formed by the interaction of many landscape variables rather than the pure signatures that can be observed in the laboratory. Whereas photogeology distinguishes between units without attempting to make fine lithologic distinctions, much of the research in multispectral remote sensing in geology is devoted to more precise identification of specific minerals thought to be both spectrally distinctive and especially valuable in mineral exploration.

Locations of zones of hydrothermal alteration may be revealed by the presence of limonite at the surface. *Limonite* refers broadly to minerals bearing oxides and oxy-hydrites of ferric iron, including goethite and hematite; such minerals tend to exhibit broad absorption bands in the near infrared, visible, and ultraviolet regions. Typically their spectra decline below 0.5 µm, producing a decrease in the ultraviolet not normally observed in other minerals (Figure 18.13). A broad, shallow absorption region from 0.85 µm to 0.95 µm is observed in the near infrared. The presence of limonite, either as a primary mineral or secondarily as the product of geologic weathering, may identify the location of a zone of hydrothermal alteration, thereby suggesting the possibility of mineralized zones. However, the identification of limonite is not definitive evidence of hydrothermal activity, as limonite can be present without hydrothermal alteration and hydrothermal activity may occur without the presence of limonite. Thus the observation of these spectra, or any spectra, is not uniquely specific, and the analyst must always consider such evidence in the context of other information and his or her knowledge of the local geologic setting.

Clay minerals often decrease in reflectance beyond 1.6 µm; this behavior is common to enough minerals and occurs over a spectral region that is broad enough that it has been used to locate zones of hydrothermal alteration by means of the surface materials rich in clay minerals. At finer spectral resolution, these clay minerals and carbonates characterized by AlOH and MgOH structures display spectra with a narrow but distinctive absorption band from 2.1 to 2.4 µm (Figure 18.13).

Emittance in the spectral region from 8 to 12 µm (in the mid infrared) permits the identification of some silicate minerals and the separation of silicate minerals from nonsilicate minerals. Silicate minerals typically exhibit emittance minima in this region, whereas nonsilicate minerals have minima at longer wavelengths.

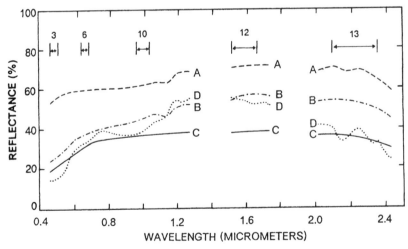

FIGURE 18.13. Spectra of some natural geologic surfaces: (A) White Joe Lott Tuff member of Mt. Belknap volcanics; (B) limonitic Joe Lott Tuff member of Mt. Belknap volcanics; (C) tan soil on rhyolite; (D) orange-layered laterite. Data were collected in the field near the surface using portable spectroradiometers. Limonitic minerals (B, C, D) show low reflectances at wavelengths below 0.5-m, relative to nonlimonitic materials (A). From Geotz et al. (1975), as given by Williams (1983, p. 1748).

18.7. Photoclinometry

Photoclinometry, loosely defined, is the understanding of the relationship between image brightness and the orientation of the surface that generated the brightness. If an irregular surface of uniform reflectance is illuminated at an angle, variations in image brightness carry information concerning the slopes of individual facets on the ground, and therefore the full image, composed of many such pixels, depicts the shape of the terrain (Eliason et al., 1981.)

The Earth's surface does not reflect uniformly, so the brightness caused by irregular terrain is mixed with the brightnesses caused by different surface materials. Remote sensing is usually concerned with the spectral differences that we observe at different places on the Earth's surfaces as clues to the local abundance of minerals and other resources. Therefore, the brightnesses caused by surface orientation are often regarded as complicating factors in our effort to extract and interpret spectral information.

As a result, the field of photoclinometry contributes to remote sensing by providing tools that permit analysis to separate the spectral information from the brightness that arises from the orientation of the surface. A diffuse reflector (Chapter 2) will have a brightness that is predictable from the angle of illumination (Figure 18.14). It is intuitively obvious that as the angle of illumination changes, so also will the brightness of the surface.

In the context of remote sensing, the surface and the source of illumination have a fixed geometric relationship, but the scene itself is composed of many individual facets, each with specific orientation with respect to the solar beam. That is, the solar beam illuminates the Earth at a fixed angle at any given instant, but topographic irregularities cause the image to be formed of many varied brightnesses. Thus, for a uniform surface of irregular topography, variations in image gray tone portray variations in surface slope and orientation.

FIGURE 18.14. Brightnesses of surfaces depend on the orientation of the surface in relation to the direction of illumination. If these surfaces all have the same surface material and behave in accordance with Lambert's cosine law, their brightnesses are related to the angles at which they are oriented.

Although few surfaces on the Earth meet the assumption of uniform reflectance, some extraterrestrial surfaces display only small variations in reflectivity, primarily because of the absence of vegetative cover and differences in color caused by weathering of geologic materials. Photogeologists who have studied the moon's surface and those of the planets have formalized the methods necessary to derive topographic information from images (Wildey and Pohn, 1964). Photoclinometry encompasses elements of remote sensing, photogrammetry, photometry, and radiometry. Although both the theory and practice of photoclinometry must still deal with many unsolved issues, it has important applications in remote sensing.

For the present discussion, let us note simply that it is possible to approximate the brightnesses of the terrain, measured in a specific pair of bands, by a ratio between them:

$$\frac{G_{band\ 1}}{G_{band\ 2}} = c\,\frac{R_1}{R_2} \qquad \text{(Eq. 18.1)}$$

where G represents the brightness of the gray tone on an image of spectral band selected from a multiband image, and c represents an arbitrary constant. This ratio image (R_1/R_2) has several useful properties.

First, the relationship holds both for shadowed pixels and for directly illuminated pixels. Therefore, a ratio image shows pure reflectance information without effects of topography. This result permits geologists to examine spectral properties of surfaces without the confusing effects of mixed brightnesses of topography and material reflectances. In the raw image, a difference in pixel brightness can be caused by a difference in slope, by shadowing, or by differences in the color of the surface material. In a ratio image, the geologist knows that differences in brightness are caused by differences in reflectance only.

This result is illustrated by the example in Figure 18.15. In the upper diagram, the brightness of the scene is caused by combined differences in topography and surface reflectance, and the analyst cannot resolve differences between the two effects. In the lower diagram, the ratios clearly separate the categories of surface reflectance, even though each is present on two different topographic surfaces.

In the context of photoclinometry, ratio images have a special application. An unsupervised classification of the several ratio images derived from a multispectral image permits identification of regions on the Earth's surface that have uniform spectral behavior. Remember that the ratio images convey only spectral, not topographic, information, so a classification performed on ratio images, not on the original digital values, will produce classes based on uniform spectral properties regardless of slope.

18.8. Band Ratios

Band ratios have further significance in geologic remote sensing. By removing effects of shadowing, which otherwise are mixed with the spectral information necessary to make lithologic discriminations, they can convey purely spectral information. Furthermore, ratios may minimize differences in brightness between lithologic units (i.e., ratios tend to

FIGURE 18.15. Example illustrating use of band ratios. In this contrived example, a topographic surface illuminated from the right side of the illustration consists of three contrasting surface materials, each partially illuminated by the solar beam, and on the reverse slope illuminated indirectly by diffuse light. The raw digital values (top) convey the mixed information of surface reflectance and shadowing. Because of shadowing, separate surface materials cannot be clearly separated into their distinctive classes. In the bottom illustration, ratios of the two bands isolate the effects of the different surface materials (removing the topographic effects) to permit recognition of distinct spectral classes. Based on Taranick (1978, p. 28).

emphasize color information and to deemphasize absolute brightness) and may facilitate comparisons of data collected on different dates, which will differ in solar angle.

Goetz et al. (1975) examined rock reflectances in Arizona using a portable spectrometer capable of collecting high-resolution spectra for a variety of natural surfaces. They found several ratios to be useful in discriminating between lithologic units. Because their bands are based on high-resolution data that do not match to familiar subdivisions of the spectrum, their ratios must be defined using their own designations, given in Figure 18.13. For example, a ratio between their bands 6 and 10 (roughly, red/near infrared) provided a good distinction between bare rock and vegetation-covered areas. Their 12/3 ratio (mid IR/blue) was successful in distinguishing between limonitic and nonlimonitic rocks. The spectrum for limonite decreases sharply in channel 3 (blue) because of the presence of ferric iron, which absorbs strongly in this region, whereas nonlimonitic rocks tend to be brighter in this region. Thus the 12/3 ratio tends to accentuate differences between limonitic and nonlimonitic rocks. Hydrothermally altered rocks containing clay minerals display absorption bands in channel 13, so the 12/13 ratio tended to identify hydrothermally altered regions.

Plate 22 shows a ratio image of the Cuprite region, a mining area in southwestern Nevada near the California border. This image depicts a sparsely vegetated region of known geology and proven mineral resources about 4.5 km × 7.0 km (about 2.8 mi. × 4.3 mi.) in size. The geology here is composed essentially of tufts and basalts within an area otherwise composed chiefly of Cambrian limestones and clastic sediments.

This image was produced for a study proposed by Vincent et al. (1984) in which data from two airborne multispectral scanners were obtained to evaluate geological uses of 12 different spectral regions. Plate 22 is formed from a selection of these data. In this image, the intensity of each of the three primaries is determined by the value of the ratio between two spectral channels. As the value of the ratio varies from pixel to pixel, so does the intensity of the color at each pixel. Because the image is formed from three primaries, and because each primary represents data from two bands, the image is derived from a total of six different bands.

Full description of the selection of spectral channels and their interpretation of the image would require an extensive summary of Vincent et al. (1984). However, a concise summary can be presented as follows:

1. The red band in the image is determined by the ratio (0.63–0.69 µm):(0.52–0.60 µm); greater intensity of red indicates high levels of ferric oxide.

2. The green band is a function of the ratio (1.55–1.75 m):(2.08–2.35 m); the intensity of green indicates areas in which there is high clay content in the surface soil.

3. The blue band is controlled by the value of the ratio (8.20–8.54 µm):(8.60–8.95 µm); the intensity of blue indicates levels of silica at the soil surface.

Thus a yellow color (both red and green) indicates the presence of both ferric oxide and clay, an indication of hydrothermal alteration, and a possible presence of a zone of mineralization. The blue region near the center of the image delineates a region of silification (high silica); it is surrounded by a zone of hydrothermal alternation, marked by yellow regions surrounding the blue center.

This example illustrates the power of using ratios between carefully selected spectral regions as aids for geological investigation. However, it is important to identify aspects of this study that differ from those that might be feasible on a routine basis. Although these data have 30-m resolution, they were collected at aircraft altitudes, so they are much less affected by atmospheric effects than would be the case for comparable satellite data. Also, this region has a sparsely vegetated and even topography, so it favors isolation of geological information that otherwise can be mixed with vegetation, shadowing, and topographic effects, as noted in the discussion of composite signatures in Chapter 15.

The use of ratio images carries certain risks, as mentioned in Chapter 17. In a geologic context, it is important that data be free of atmospheric effects or that such effects be removed by preprocessing. The significance of the atmosphere can be appreciated by reexamining the earlier discussion, in which the diffuse light values in the two band values were estimated to have predictable values that would cancel each other out. If severe atmospheric effects are present, they will differ from one band to the other, and the value of the ratio will no longer portray only spectral properties of the ground surface but instead will have values greatly altered by the varied atmospheric contributions to the separate bands.

18.9. Soil and Landscape Mapping

Soil scientists study the mantle of soil at the Earth's surface in an effort to understand soil formation and to map patterns of soil variation. *Soil* is a complex mixture of inorganic material weathered from the geologic substratum and mixed with decayed organic matter from tissues of plants and animals. Typically, a "soil" consists of three layers (known as *soil horizons*) of varying thickness and composition (Figure 18.16). Nearest the surface is the A horizon, usually dark in color and rich in decayed organic matter. The A horizon is sometimes known popularly as *topsoil*—the kind of soil one would like to have for a lawn or garden. In this layer plant roots are abundant, as are microorganisms, insects, and other animals. Below the A horizon is the second layer, the B horizon, usually lighter in color and more compact, where plant roots and biologic activity are less abundant. Sometimes the B horizon is known as the *subsoil*—the kind of hard infertile soil one would prefer *not* to have at the surface of a lawn or garden. At even greater depth is the C horizon, the deepest pedologic horizon, which consists of weathered geologic material,

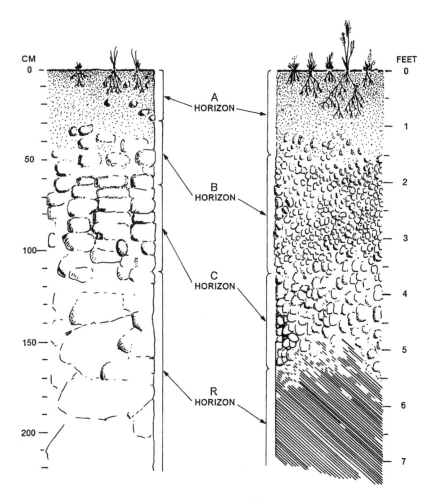

FIGURE 18.16. Sketches of two contrasting soil profiles.

decayed or fractured into material that is softer and looser than the unweathered geologic strata below. This is the material that usually forms the raw material for the A and B horizons. Finally, below the C horizon is the R horizon, consisting of unaltered bedrock.

The exact nature of the horizons at a given place is determined by the interaction between local climatic, topographic, geologic, and biologic elements as they act over time. Because varied combinations of climate, vegetation, and topography produce different soils from similar geologic materials, soil science is distinct from, although closely related to, geology and geomorphology.

The landscape is covered by a mosaic of patches of different kinds of soil, each distinct from its neighbors with respect to character and thickness (Figure 18.17). Soil surveyors outline on maps those areas covered by specific kinds of soil as a means of showing the variation of soils on the landscape. Each symbol represents a specific kind of soil, or, when the pattern is very complex, sometimes two or three kinds of soil that occur in an intimate pattern (Figure 18.18). Soil maps portray distributions of pedologic units, which, together with other maps and data, convey valuable information concerning topography, geology, geomorphology, hydrology, and other landscape elements. In the hands of a knowledgeable reader, they convey a comprehensive picture of the physical landscape. As a result, they can be considered to be among the most practical of all forms of landscape maps and are used by farmers, planners, and others who must judge the best locations for specific agricultural activities, for community facilities, or for construction of buildings and highways.

Soil mapping is conducted routinely by national soil surveys (or their equivalents) in most countries and by their regional counterparts at lower administrative levels. Details of mapping technique vary from one organization to another, but the main outlines of the procedure are common to most soil survey organizations. A soil map is formed by subdividing the landscape into a mosaic of discrete parcels. Each parcel is assigned to a class of soil—a *mapping unit*—that is characterized by specific kinds of soil horizons.

FIGURE 18.17. The soil landscape. From USDA Natural Resources Conservation Service.

FIGURE 18.18. Preparation of the soil map. In the mapping step, the soil surveyor (lower left image) delineates mapping units on aerial photographs (central image); eventually these mapping units are refined and combined to form the units presented on the final soil map (lower right). From Virginia Cooperative Soil Survey (lower left and central image) and USDA Natural Resources Conservation Service (lower right image).

Each mapping unit is represented on the map by a specific symbol—usually a color or an alphanumeric designation. In theory, mapping units are homogeneous with respect to pedologic properties, although experienced users of soil maps acknowledge the existence of internal variation, as well as the presence of foreign inclusions. Mapping units are usually defined with links to a broader system of soil classification defined by the soil survey organization, so that mapping units are consistently defined across, for example, county and state borders. For each region, a mapping unit can be evaluated with respect to the kind of agriculture that is important locally, so the map can serve as a guide to select the best uses for each soil. The following paragraphs outline an example of the soil survey process, although details may vary from one organization to the next.

The first step in the actual preparation of the draft map is the *mapping* process (Figure 18.18). The term *mapping* in this context means specifically to designate the delineation of landscape units on an aerial photograph or map base, rather than in the broader sense that refers to the entire set of activities necessary to generate the map. In the mapping step, the soil scientist examines aerial photographs, often using a stereoscope, to define the major boundaries between classes of soil. The soil scientist has already learned much about the region's soils through field observations, so he or she has knowledge of the kinds of soils present and their approximate locations. The soil scientist can use this field knowledge in interpretation of the photograph, using breaks in slope, boundaries between vegetation classes, and drainage patterns to define boundaries between soil mapping units. Seldom if ever can the soil classes actually be identified from the photographs—identification must be based on field observations—but in the mapping step, the scientist can subdivide the landscape into a mosaic of soil parcels, which are then each treated as independent units in the later steps.

Completion of the mapping step requires considerable field experience within the region to be mapped, as the surveyor must acquire knowledge of the numbers and kinds of soils present within a region, their properties and occurrence, and their uses and limitations. As a result, completion of this step requires a much longer time than that actually devoted to marking the aerial photographs if we include the time required to learn about the local landscape. Morphological descriptions of each mapping unit are made in the field, and mapping units are sampled for later analysis.

The second step is *characterization*. Samples are collected from each prospective mapping unit, then subjected to laboratory analysis for physical, chemical, and mineralogical properties.

These measurements form the basis for *classification*, the third step. In the United States, classification is the implementation of the classification criteria specified by Soil Taxonomy, the official classification system of the Soil Conservation Service, U.S. Department of Agriculture. In other countries, classification is conducted by applying the classification criteria established by each national soil survey organization.

Correlation, the fourth and possibly the most difficult step, matches mapping units within the mapped region to those in adjacent regions and to those in ecologically analogous areas. Whereas the other steps in a soil survey may be conducted largely on a local basis, correlation requires the participation of experienced scientists from adjacent regions, and even those from national or international levels in the organization, to provide the broader experience and perspective often required for successful correlation.

The fifth and final step is *interpretation*, in which each mapping unit is evaluated with respect to prospective agricultural and engineering uses. The interpretation step provides the user of the map with information concerning the likely suitability of each mapping unit to the land uses most common within the region.

Within this rather broad framework there are, of course, many variations in details of technique and in overall philosophy and strategy. The result is a map that shows the pattern of soils within a region and a report that describes the kind of soil that is encountered within each mapping unit. Each mapping unit is evaluated with regard to the kinds of uses that might be possible for the region, so that the map serves as a guide to wise use of soil resources.

18.10. Integrated Terrain Units

Definition of *integrated terrain units* is based on the concept that the varied and complex assemblages of soil, terrain, vegetation, and so on form distinctive spectral responses that can be recognized and mapped. This general approach has long been applied, under many different names, to the interpretation of resource information from aerial photographs and other forms of imagery. The best known, and oldest, of these methods is known as *land system mapping*.

Land system mapping subdivides a region into sets of recurring landscape elements based on comprehensive examination of distributions of soils, vegetation, hydrology, and physiography. The formalization of land system mapping procedures dates from the period just after World War II, although some of the fundamental concepts had been defined earlier. Today a family of similar systems are in use; all are based on similar principles and methods, although the terminology and details vary considerably. One of the oldest, best known, and most formally defined systems is that developed by the Australian CISRO Division of Land Research and Regional Survey (Christian, 1959); variations have been developed and applied throughout the world.

The method is based on a hierarchical subdivision of landscapes. *Land systems* are recurring contiguous associations of landforms, soils, and vegetation, composed of component *land units*. The basic units of the system (which are assigned varied definitions and names in alternative versions of this basic strategy) are areas of uniform lithology with relatively uniform soil and drainage. A characteristic feature of all versions of the land system method is a hierarchical spatial organization, so that subcategories are nested within the broader categories defined at higher levels. Designation of separate levels within the hierarchy differs among alternative versions. For example, Thomas (1969) defines the following sequence: *site* (the smallest units), *facet, unit landform, landform complex, landform system,* and *landform region.* Wendt et al. (1975) define a system using *landtypes* (the smallest regions), *landtype associations, landtype phases, subsections, sections,* and *provinces.* Although many versions imply that both land systems and their component elements recur in widely separated geographic regions (although under analogous ecological circumstances), closer examination of the method (Perrin and Mitchell, 1969) reveals that land systems are most effective when defined to be essentially local units.

Implicit, if not explicit, in most applications is the use of aerial photography as a primary tool for land system mapping, together with direct observation in the field. Aerial photography provides the broad overview and the map-like perspective favoring convenient definition and delineation of land systems and portrays the complex spatial patterns of topography, vegetation, and drainage in an integrated form that is compatible with the assumptions, methods, and objectives of this approach to terrain analysis. "There would be no point in defining any terrain class if its chances of being recognized from the air photographs and background information were small" (Webster and Beckett, 1970, p. 54).

An assumption of the procedure is that easily identified landscape features (e.g., vegetation and physiography) form surrogates for more subtle soil features not as easily defined from analysis of aerial photographs. Although the procedure has been applied at a range of scales from 1:1,000,000 to 1:25,000, most applications are probably at fairly broad scales, for reconnaissance surveys, or mapping of rather simple landscape divisions under circumstances in which the ultimate use will be rather extensive.

The method and its many variations have been criticized for their subjectivity and the variability of the results obtained by different analysts. Compared with most large-scale soil surveys, land system mapping presents a rather rough subdivision of the landscape, as the mapping units display much greater internal variability and are not as carefully defined and correlated as one would expect in more intensive surveys. As a result, the procedure may be best suited for application in reconnaissance mapping, for which broad-scale, low-resolution mapping is required as the basis for planning more detailed surveys.

The method of integrated terrain units has also been applied to analysis of digital data. At the coarse resolution of satellite sensors, individual spectral responses from many landscape components are combined to form composite responses. In this context, the usual methods of image classification for mapping of individual classes of soil, vegetation, or geology may be extraordinarily difficult. Therefore, many analysts have applied the integrated terrain unit strategy to digital classification in an effort to define image classes in a more realistic manner. In the ideal, we might prefer to define pure categories that each show only a single thematic class. But, given the complexity of the natural landscape, composite categories are often well suited to representation of the gradations and mixtures that characterize many environments (Green, 1986; Robinove, 1981).

18.11. Wetlands Inventory

Wetlands are areas of land characterized by saturated or inundated soil. In the United States, wetlands are legally defined by the presence of (1) hydrophytic vegetation (plants that occur predominantly in wetlands environments); (2) a wetlands hydrologic regimen, including evidence of inundation or saturation; and (3) diagnostic pedologic features caused by low oxygen levels, most notably gray or black colors, often with distinctive mottling indicative of prolonged saturation. Wetlands are not defined merely by proximity to bodies of open water but are encountered in a wide variety of ecological and physiographic settings, including uplands.

Understanding the occurrence of wetlands is significant for a broad spectrum of policy issues, including land use regulation, water quality, flood abatement, agriculture, and carbon sequestration. As a result, mapping their occurrence and extent is an important responsibility of local, state, and federal governments. For some purposes, field studies are necessary to identify and delineate the extent of wetlands.

However, for other purposes, remotely sensed imagery forms an important tool for inventory and monitoring of wetlands. At a national level in the United States, the National Wetlands Inventory (NWI) conducted by the U.S. Fish and Wildlife Service prepares a national inventory based on photointerpretation of NHAP imagery. Although wetlands vary greatly in their occurrence and distinctive properties (Lyon, 2001), optical, thermal, and SAR data all have distinctive capabilities that can detect some of the distinctive characteristics of wetlands. Because wetlands vary so greatly with respect to their distinct properties and vary seasonally, it may not be feasible to propose a single uniformly applicable approach to remote sensing of wetlands. It seems more likely that remotely sensed imagery may form the central resource for a suite of data that might be useful in delineating and monitoring wetlands.

18.12. Radar Imagery for Exploration

As a conclusion to this chapter, we note the application of SAR data to offshore petroleum exploration. Natural seepage of petroleum from the ocean floor causes a thin film of petroleum floating on the ocean surface (a "slick") that creates a locally smooth water surface that in turn induces conditions that promote specular reflection over that region of the ocean. Oil slicks occur when molecules of oil reach the sea surface to form a thin layer of petroleum that dampens the ocean-surface capillary waves. SAR technology is sensitive to differences in surface roughness, so it can easily discriminate between the smoother oil slick and the surrounding rough water. Thus oil slicks are regions of little to no backscatter, characterized by distinct areas of darkness on the radar image. Therefore, the dark returns assist in identifying regions where such slicks might originate. Because SAR has the ability to observe such regions day and night, and over a wide range of meteorological conditions, SAR data assist in identifying regions where such slicks form on a regular basis. Critical exploration maps can be produced by using a GIS system to overlay other information such as coastline, bathymetry, gravimetric data, shipping lanes, and existing oil rigs on the RADARSAT image. These maps allow decision makers to reduce magnetic field exploration risk and increase the cost-effectiveness of offshore drilling efforts.

18.13. Summary

Aerial photography and remote sensing have long been applied to problems in the earth sciences. Geology, topographic mapping, and related topics formed one of the earliest routine applications of aerial photography, and today remote sensing, used with other techniques, continues to form one of the most important tools for geologic mapping, exploration, and research.

Research for geologic remote sensing spans the full range of subjects within the field of remote sensing, including use of additional data in the thermal, near infrared, visible, and microwave regions of the spectrum. More than most other subjects, efforts such as geobotanical research are especially interesting because they bring knowledge of so many different disciplines together to bear on a single problem. In addition, geological investigations have formed important facets of research and applications programs for Landsat, SPOT, RADARSAT, ERS–1, and other systems. Geological studies are said to form a major economic component of the practice of remote sensing. Certainly, they form one of the most important elements in both theoretical and practical advances in remote sensing.

18.14. Some Teaching and Learning Resources

- Geologic Fieldwork with ASD: part 1 of 3
  www.youtube.com/watch?v=cjuitl_-_LU
- Geologic Fieldwork with ASD: part 2 of 3
  www.youtube.com/watch?v=xBknHsZdG2E&feature=related
- Geologic Fieldwork with ASD: part 3 of 3
  www.youtube.com/watch?v=2-AkC8jcL08&feature=related

Review Questions

1. Why is the timing of overpasses of Earth observation satellites such as Landsat or SPOT likely to be of special significance for geologic remote sensing?
2. Explain why the ability to monitor the presence of moisture (both open water bodies

and soil moisture at the ground surface) might be of special significance in geologic, geomorphic, and pedologic studies.

3. In the past there has been discussion of conflicts between earth scientists and plant scientists concerning the design of spectral sensitivity of sensors for the MSS and TM, based on the notion that the two fields have quite different requirements for spectral information. Based on information in this chapter, explain why such a distinct separation of information requirements may not be sensible.

4. Why should Landsat imagery have been such an important innovation in studies of geologic structure?

5. Summarize the significance of lineaments in geologic studies and the significance of remote sensing to the study of lineaments.

6. Compare relative advantages and disadvantages of photogeology and geologic remote sensing, as outlined in this chapter.

7. Some might consider lineament analysis, application of geobotany, and other elements of geologic remote sensing to be modern, state-of-the-art techniques. If so, why should photogeology, which is certainly not as technologically sophisticated, be so widely practiced today?

8. Full application of the principles and techniques of geologic remote sensing requires much finer spectral, radiometric, and spatial resolution than most operational sensors have at present. What difficulties can you envision in attempts to use such sensors for routine geologic studies?

9. Refer back to the first paragraphs of Section 12.1; compare and contrast the concept of image classification with that of integrated terrain units (Section 18.10). Write two or three paragraphs that summarize the major differences between the two strategies.

10. In what ways might radar imagery (Chapter 7) be especially useful for studies of geology and other earth sciences?

11. Write a short description of a design for a multispectral remote sensing system tailored specifically for recording geologic information. Disregard all other applications. Suggest how you would select the optimum spatial and radiometric resolution and the most useful spectral regions.

References

General

Abrams, M. J., D. Brown, L. Lepley, and R. Sadowski. 1983. Remote Sensing for Porphyry Copper Deposits in Southern Arizona. *Economic Geology*, Vol. 78, pp. 591–604.

Bishop, J. L., and E. Murad. 2005. The Visible and Infrared Spectral Properties of Jarosite and Alunite. *American Mineralogist*, Vol. 90, pp. 1100–1107.

Chang, S., and W. Collins. 1983. Confirmation of the Airborne Biogeophysical Mineral Exploration Technique Using Laboratory Methods. *Economic Geology*, Vol. 78, pp. 723–726.

Cole, M. M. 1982. Integrated Use of Remote Sensing Imagery in Mineral Exploration. *Geological Society of America Transactions*, Vol. 85, Part 1, pp. 13–28.

Collins, W., S. H. Chang, and J. T. Kuo. 1981. *Detection of Hidden Mineral Deposits by Airborne Spectral Analysis of Forest Canopies.* NASA Contract NSG-5222. Final Report. 61 pp.

Collins, W., S. Chang, G. Raines, F. Channey, and R. Ashley. 1983. Airborne Biogeochemical Mapping of Hidden Mineral Deposits. *Economic Geology,* Vol. 78, pp. 737–749.

Crippen, R. E., and R. G. Blom. 2001. Unveiling the Lithology of Vegetated Terrains in Remotely Sensed Imagery. *Photogrammetric Engineering and Remote Sensing,* Vol.67, pp. 935–943.

Goetz, A. F. H. 1984. High Spectral Resolution Remote Sensing of the Land. In *Remote Sensing* (P. N. Slater, ed.). Bellingham, WA: SPIE, pp. 56–68.

Goetz, A. F. H., F. C. Billingsley, A. R. Gillespie, M. J. Abrams, R. L. Squires, E. M. Shoemaker, I. Lucchitta, and D. P. Elston. 1975. *Application of ERTS Images and Image Processing to Regional Geological Problems and Geological Problems in Northern Arizona.* U.S. National Aeronautics and Space Administration, Jet Propulsion Laboratory, Technical Report 32-1597, 188pp.

Goetz, A. F. H., B. N. Rock, and L. C. Rowan. 1983. Remote Sensing for Exploration: An Overview. *Economic Geology,* Vol. 78, pp. 573–590.

Goetz, A. F. H., G. Vane, J. E. Solomon, and B. N. Rock. 1985. Imaging Spectrometry for Earth Remote Sensing. *Science,* Vol. 228, pp. 1147–1153.

Kaab, A. 2002. Monitoring High-Mountain Terrain Deformation from Repeated Air- and Spaceborne Optical Data: Examples Using Digital Aerial Imagery and ASTER Data. *ISPRS Journal of Photogrammetry and Remote Sensing,* Vol. 57, pp. 39–52.

King, R. B. 1970. A Parametric Approach to Land System Classification. *Geoderma,* Vol. 4, pp. 37–46.

Meyer, B. S., D. B. Anderson, R. H. Bohning, and D. G. Fratianne. 1973 *Introduction to Plant Physiology.* New York: Van Norstrand, 565 pp.

Perrin, R. M. S., and C. W. Mitchell. 1969. *An Appraisal of Physiographic Units for Predicting Site Conditions in Arid Areas.* Military Engineering Experimental Establishment (MEXE) Report 1111, Vols. I & II, Christchurch, Hants, U.K.

Prospero, J. M., P. Ginoux, O. Torres, S. E. Nicholson, and T. E. Gill. 2002. Environmental Characterization of Global Sources of Atmospheric Soil Dust Identified with the Nimbus 7 Total Ozone Mapping Spectrometer (TOMS) Absorbing Aerosol Product. *Reviews of Geophysics,* Vol. 40.

Rast, M., J. S. Hook, C. D. Elvidge, and R. E. Alley. 1991. An Evaluation of Techniques for the Extraction of Mineral Absorption Features from High Spectral Resolution Remote Sensing Data. *Photogrammetric Engineering and Remote Sensing,* Vol. 57, pp. 1303–1309.

Rowan, L. C., A. F. H. Goetz, and R. P. Ashley. 1977. Discrimination of Hydrothermally Altered and Unaltered Rocks in Visible and Near Infrared Multispectral Images. *Geophysics,* Vol. 42, pp. 522–535.

Vincent, R. K., P. K. Pleitner, and M. L. Wilson. 1984. Integration of Airborne Thematic Mapper and Thermal Infrared Multispectral Scanner Data for Lithologic and Hydrothermal Alteration Mapping. In *Proceedings, International Symposium on Remote Sensing of Environment, Third Thematic Conference, Remote Sensing for Exploration Geology.* Ann Arbor, MI: Environmental Research Institute of Michigan, pp. 219–226.

Williams, R. S. (ed.). 1983. Geological Applications. *Manual of Remote Sensing,* 2nd ed. (R. N. Colwell, ed.). Vol. II, Falls Church, VA: American Society for Photogrammetry, pp. 1667–1916.

Photoclinometry

Eliason, P. T., L. A. Soderblom, and P. S. Chavez. 1981. Extraction of Topographic and Spectral Albedo Information from Multispectral Images. *Photogrammetric Engineering and Remote Sensing*, Vol. 48, pp. 1571–1579.

Wildey, R. L., and H. A. Pohn. 1964. Detailed Photoelectric Photometry of the Moon. *Astronomical Journal*, Vol. 69, pp. 619–634.

Photogeology

Lueder, D. R. 1959. *Aerial Photographic Interpretation: Principles and Applications*. New York: McGraw-Hill, 462 pp.

Miller, V. C. 1961. *Photogeology*. New York: McGraw-Hill, 248 pp.

Ray, R. G. 1960. *Aerial Photographs in Geologic Interpretation and Mapping* (Geological Survey Professional Paper 373). Washington, DC: U.S. Government Printing Office, 230 pp.

Lineaments

Berlin, G. L., P. S. Chavez, T. E. Grow, and L. A. Soderblom. 1976. Preliminary Geologic Analysis of Southwest Jordan from Computer Enhanced Landsat I Image Data. In *Proceedings, American Society of Photogrammetry*. Bethesda, MD: American Society of Photogrammetry, pp. 543–563.

Casas, A. M., A. L. Cortés, A. Maestro, M. Asunción Soriano, A. Riaguas, and J. Bernal. 2000. LINDENS: A Program for Lineament Length and Density Analysis. *Computers and Geosciences*, Vol. 26, pp. 1011–1022.

Chavez, P. S., G. L., Berlin, and A. V. Costa. 1977. Computer Processing of Landsat MSS Digital Data for Linear Enhancements. In *Proceedings of the Second Annual William T. Pecora Memorial Symposium*. Falls Church, VA: American Society for Photogrammetry, pp. 235–251.

Clark, C. D., and C. Watson. 1994. Spatial Analysis of Lineaments. *Computers and Geosciences*, Vol. 20, pp. 1237–1258.

Ehrich, R. W. 1977. Detection of Global Edges in Textured Images. *IEEE Transactions on Computers*, Vol. C–26, pp. 589–603.

Ehrich, R. W. 1979. Detection of Global Lines and Edges in Heavily Textured Images. In *Proceedings, Second International Symposium on Basement Tectonics*. Denver, CO: Basement Tectonics Committee Conference held in Delaware. Publisher in Denver, pp. 508–513.

Moore, G. K., and F. A. Waltz. 1983. Objective Procedures for Lineament Enhancement and Extraction. *Photogrammetric Engineering and Remote Sensing*, Vol. 49, pp. 641–647.

O'Leary, D. W., J. D. Friedman, and H. A. Pohn. 1976. Lineament, Linear, Lineation: Some Proposed New Standards for Old Terms. *Geological Society of America Bulletin*, Vol. 87, pp. 1463–1469.

Podwysocki, M. H., J. G. Moik, and W. D. Shoup. 1975. Quantification of Geologic Lineaments by Manual and Machine Processing Techniques. In *Proceedings, NASA Earth Resources Survey Symposium*. Houston, TX: NASA, pp. 885–903.

Raghavan, V., K. Wadatsumi, and S. Masummoto. 1994. SMILES: A FORTRAN–77 Program for Sequential Machine Interpreted Lineament Extraction Using Digital Images. *Computers and Geosciences*, Vol. 20, pp. 121–159.

Ustin, S. L., M. D. Smith, S. Jacquemond, M. Verstrate, and Y. Govaerts. 1999. Geobotany Vegetation Mapping for the Earth Sciences. Chapter 3 in Manual of Remote Sensing

(Remote Sensing for the Earth Sciences) 3rd ed., Vol. 3. New York: John Wiley and Sons, pp. 198–248.

Vanderbrug, G. J. 1976. Line Detection in Satellite Imagery. *IEEE Transactions on Geoscience Electronics*, Vol. GE–14, pp. 37–44.

Vaughan, R. W. 1983. *A Topographic Approach for Lineament Recognition in Satellite Images*. Unpublished M. S. Thesis, Department of Computer Science, Virginia Polytechnic Institute, Blacksburg, 70 pp.

Wheeler, R. L., and D. U. Wise. 1983. Linesmanship and the Practice of Linear Geo-Art: Discussion and Reply. *Geological Society of America Bulletin*, Vol. 94, pp. 1377–1379.

Wise, D. U. 1982. Linesmanship and the Practice of Linear Geo-Art. *Geological Society of America Bulletin*, Vol. 93, pp. 886–888.

Yamaguchi, Y. 1985. Image-Scale and Look-Direction Effects on the Detectability of Lineaments in Radar Images. *Remote Sensing of Environment*, Vol. 17, pp. 117–127.

Integrated Terrain Units

Christian, C. S. 1959. The Eco-Complex and Its Importance for Agricultural Assessment. Chapter 36 in *Biogeography and Ecology in Australia. Monographiae Biologicae*, Vol. 8, pp. 587–605.

Green, G. M. 1986. Use of SIR-A and Landsat MSS Data in Mapping Shrub and Intershrub Vegetation at Koonamore, South Australia. *Photogrammetric Engineering and Remote Sensing*, Vol. 52, pp. 659–670.

Robinove, C. J. 1981. The Logic of Multispectral Classification and Mapping of Land. *Remote Sensing of Environment*, Vol. 11, pp. 231–244.

Thomas, M. F. 1969. Geomorphology and Land Classification in Tropical Africa. In *Environment and Land Use in Africa* (M. F. Thomas and G. W. Whittington, eds.). London: Methuen, pp. 103–105.

Wendt, G. E., R. A. Thompson, and K. N. Larson. 1975. *Land Systems Inventory: Boise National Forest, Idaho. A Basic Inventory for Planning and Management*. Ogden, UT: U. S. Forest Service Intermountain Region, 49 pp.

Soils and Geomorphology

Agbu, P. A., and E. Nizeyimana. 1991. Comparisons between Spectral Mapping Units Derived from SPOT Image Texture and Field Mapping Units. *Photogrammetric Engineering and Remote Sensing*, Vol. 57, pp. 397–405.

Franklin, S. E., M. B. Lavigne, E. R. Hunt Jr., B. A. Wilson, D. R. Peddle, G. J. McDermid, and P. T. Giles. 1995. Topographic Dependence of Synthetic Aperture Radar Imagery. *Computers and Geosciences*, Vol. 21, pp. 521–532.

McBratney, A. B., M. L. M. Santos, and B. Minasny. 2003. On Digital Soil Mapping. *Geoderma*, Vol. 117, pp. 3–52.

Metternicht, G. I., and J. A. Zinck. 2003. Remote Sensing of Soil Salinity: Potentials and Constraints. *Remote Sensing of Environment*, Vol. 85, pp. 1–20.

Post, D. F., E. H. Horvath, W. M. Lucas, S. A. White, M. J. Ehasz, and A. K. Batchily. 1994. Relations between Soil Color and Landsat Reflectance on Semiarid Rangelands. *Soil Science Society of America Journal*, Vol. 58, pp. 1809–1816.

Soil Survey Staff. 1966. *Aerial Photointerpretation in Classifying and Mapping Soils* (Agricultural Handbook 294). Washington, DC: U.S. Department of Agriculture, 89 pp.

Webster, R., and P. H. T. Beckett. 1970. Terrain Classification and Evaluation by Air Photography: A Review of Recent Work at Oxford. *Photogrammetria*, Vol. 26, pp. 51–75.

Weismiller, R. A., and S. A. Kaminsky. 1978. An Overview of Remote Sensing as Related to Soil Survey Research. *Journal of Soil and Water Conservation*, Vol. 33, pp. 287–289.

Wright, J. S., T. C. Vogel, A. R. Pearson, and J. A. Messmore. 1981. *Terrain Analysis Procedural Guide for Soil* (ETL-0254). Ft. Belvoir VA: U.S. Army Corps of Engineers, Geographic Services Laboratory, Engineering Topographic Laboratories, 89 pp.

Wetlands

Glenn, N. F., and J. R. Carr. 2003. The Use of Geostatistics in Relating Soil Moisture to RADARSAT-1 SAR Data Obtained over the Great Basin, Nevada, USA. *Computers and Geosciences*, Vol. 29, pp. 577–586.

Lyon, J. G. 2001. *Wetland Landscape Characterization: GIS, Remote Sensing, and Image Analysis*. Boca Raton, FL: Lewis.

Lyon, John G. and J. McCarthy. 1995. *Wetland and Environmental Applications of GIS*. Lewis Publishers.

Pantaleoni, E., R. H. Wynne, J. M. Galbraith, and J. B. Campbell. 2009. A Logit Model for Predicting Wetland Location Using ASTER and GIS. *International Journal of Remote Sensing*, Vol. 30, pp. 2215–2236.

Pantaleoni, E., R. H. Wynne, J. M. Galbraith, and J. B. Campbell. 2009. Mapping Wetlands Using ASTER Data: A Comparison between Classification Trees and Logistic Regression. *International Journal of Remote Sensing*, Vol. 30, pp. 3423–3440.

Interferometry

Franklin, S. E., M. B. Lavigne, E. R. Hunt, Jr., B. A. Wilson, D. R. Peddle, G. J. McDermid, et al. 1995. Topographic Dependence of Synthetic Aperture Radar Imagery. *Computers and Geosciences*, Vol. 21, No. 4, pp. 521–532.

Kervyn, François. 2001. Modeling Topography with SAR Interferometry: Illustrations of a Favorable and Less Favorable Environment. *Computers and Geosciences*, Vol. 27, pp. 1039–1050.

Massonnet, D., and K. L. Feigl. 1998. Radar Interferometry and Application to Changes in the Earth's Surface. *Reviews of Geophysics*, Vol. 36, pp. 441–500.

Strozzi, T., U. Wegmüller, L. Tosi, G. Bitelli, and V. Spreckels. 2001. Land Subsidence Monitoring with Differential SAR Interferometry. *Photogrammetric Engineering and Remote Sensing*, Vol. 67, pp. 1261–1270.

Glaciology

Jezek, K. C. 1999. Glaciological Properties of the Antarctic Ice Sheet from RADARSAT-1 Synthetic Aperture Radar Imagery. *Annals of Glaciology*, Vol. 29, pp. 286–290.

Rignot, E., J. L. Bamber, M. R. Van Den Broeke, C. Davis, Y. H. Li, W. J. Van De Berg, et al. 2008. Recent Antarctic Ice Mass Loss from Radar Interferometry and Regional Climate Modeling. *Nature Geoscience*, Vol. 1, pp. 106–110.

Rignot, E., and P. Kanagaratnam. 2006. Changes in the Velocity Structure of the Greenland Ice Sheet. *Science*, Vol. 311, pp. 986–990.

Rignot, E., and R. H. Thomas. 2002. Mass Balance of Polar Ice Sheets. *Science*, Vol. 297, pp. 1502–1506.

Zwally, H. J., B. Schutz, W. Abdalati, J. Abshire, C. Bentley, A. Brenner, et al. 2002. ICESat's Laser Measurements of Polar Ice, Atmosphere, Ocean, and Land. *Journal of Geodynamics*, Vol. 34, pp. 405–445.

CHAPTER NINETEEN

Hydrospheric Sciences

19.1. Introduction

Open water covers about 74% of our planet's surface. Oceans account for about 95% of the surface area of open water, but freshwater lakes and rivers (about 0.4%) have a significance for mankind that exceeds their small areal extent (Table 19.1).

In addition, soil and rock near the Earth's surface hold significant quantities of freshwater (but only about 0.01% by volume of the Earth's total water), as do the ice and snow of polar regions. Moisture in the form of "permanent" ice (about 5% of the Earth's surface) is largely beyond the reach of mankind, although the seasonal accumulation and melt of snowpack in temperate mountains is an important source of moisture for some agricultural regions in otherwise arid zones (such as the Great Basin of North America). Hydrologists and meteorologists monitor water as it occurs in all these forms, as it changes from liquid to vapor, condenses to rain and snow, and moves on and under the Earth's surface (Figure 19.1). In addition, studies of sea ice, movement of pollutants, and ocean currents are among the many other important subjects that attract the attention of the many scientists who study hydrology, oceanography, and related subjects.

Most traditional means of monitoring the Earth's water depend largely on measurements made at specific points or collections of samples from discrete locations. Oceans, lakes, and rivers can be studied by samples collected at the surface or by special devices that fill a collection bottle with water from a specified depth. Groundwater can be studied by collection of samples from wells or boreholes. Samples can be subjected to chemical and physical tests to measure levels of pollutants, to detect bacteria and other biological

TABLE 19.1. Water on Earth

	Percentage	
	By surface area[a]	By volume[b]
Oceans	94.90	97.1
Rivers and lakes	0.40	0.02
Groundwater	—	0.60
Permanent ice cap	4.69	2.20
Earth's atmosphere	—	0.001

Note. Calculated from values given by Nace (1967).
[a]Percentage by area of Earth's total water surface.
[b]Percentage by volume of Earth's water.

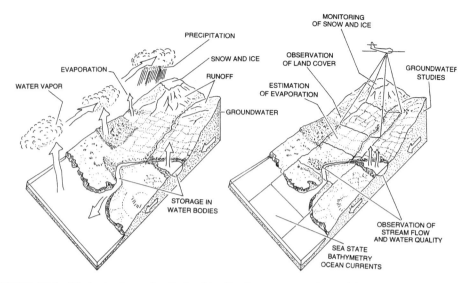

FIGURE 19.1. Hydrologic cycle. Left: idealized representation of key components of the hydrologic cycle, including evaporation, precipitation, snowfields, groundwater, runoff, and the like. Right: schematic view of the application of remotely sensed imagery to acquiring information about components of the hydrologic cycle.

phenomena, and to examine oxygen content, sediment content, and many other qualities of the water. Such measurements, or samples, of course, provide information about discrete points within the water body, whereas the analyst usually is interested in examining entire water bodies or regions within water bodies. Although measurements can be collected at several locations to build up a record of place-to-place variation within the water body, such efforts are at best a piecemeal approach to studying the complex and dynamic characteristics of water bodies that are of interest to hydrologists.

Therefore, remote sensing can provide a valuable perspective concerning broad-scale dynamic patterns that can be difficult to examine in detail using point measurements only. Careful coordination and placement of surface samples permits establishment of the relationship between the sample data and those collected by the remote sensor. Remotely sensed data can be especially valuable in studying phenomena over large areas. Satellite sensors provide the opportunity for regular observation of even very remote regions. Although remotely sensed images seldom replace the usual sources of information concerning water resources, they can provide valuable supplements to field data by revealing broad-scale patterns not recognizable at the surface, recording changes over time, and providing data for inaccessible regions.

19.2. Spectral Characteristics of Water Bodies

The spectral qualities of water bodies are determined by the interaction of several factors, including the radiation incident to the water surface, optical properties of the water, roughness of the surface, angles of observation and illumination, and, in some instances, reflection of light from the bottom (Figure 19.2). As incident light strikes the water surface, some is reflected back to the atmosphere; this reflected radiation carries little information about the water itself, although it may convey information about the roughness of the surface and therefore about wind and waves. Instead, the spectral properties (i.e., "color") of a water body are determined largely by energy that is scattered and reflected within the water body itself, known as *volume reflection* because it occurs over a range of depths rather than at the surface. Some of this energy is directed back toward the surface, where it again passes through the atmosphere and then to the sensor (Figure 19.2). This light, sometimes known as *underlight*, is the primary source of the color of a water body.

The light that enters a water body is influenced by (1) absorption and scattering by pure water and (2) scattering, reflection, and diffraction by particles that may be suspended in the water. For pure water, some of the same principles described previously for atmospheric scattering apply. Scattering by particles that are small relative to wavelength (*Rayleigh scattering*) causes shorter wavelengths to be scattered the most. Thus, for deep water bodies, we expect (in the absence of impurities) water to be blue or blue-green in color (Figure 19.3). Maximum transmittance of light by clear water occurs in the range 0.44 µm–0.54 µm, with peak transmittance at 0.48 µm. Because the color of water is determined by volume scattering, rather than by surface reflection, spectral properties of water bodies (unlike those of land features) are determined by transmittance rather than by surface characteristics alone. In the blue region, light penetration is not at its optimum, but at slightly longer wavelengths, in the blue-green region, penetration is greater. At these wavelengths, the opportunity for recording features on the bottom of the water body is greatest. At longer wavelengths, in the red region, absorption of sunlight is much greater, and only shallow features can be detected. Finally, in the near infrared region, absorption is so great that only land–water distinctions can be made.

As impurities are added to a water body, its spectral properties change. Sediments

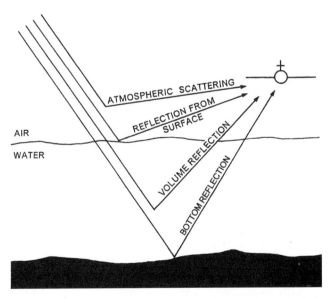

FIGURE 19.2. Diagram illustrating major factors influencing spectral characteristics of a water body. Modified from Alföldi (1982, p. 318).

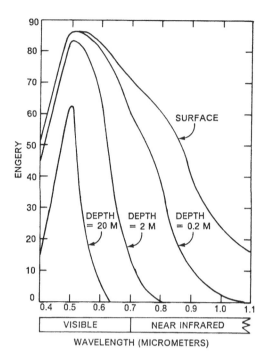

FIGURE 19.3. Light penetration within a clear water body. From Moore (1978, p. 458). Reproduced by permission of the Environmental Research Institute of Michigan.

are introduced both from natural sources and by mankind's activities. Such sediments consist of fine-textured silts and clays eroded from stream banks, or by water running off disturbed land, that are fine enough to be carried in suspension by moving water. As moving water erodes the land surface or the shoreline, it carries small particles as suspended sediment; faster flowing streams can erode and carry more and larger particles than can slower moving streams. As a stream enters a lake or ocean, the decrease in velocity causes coarser materials to be deposited. But even slow-moving rivers and currents can carry large amounts of fine-textured sediments, such as clays and silts, which can be found even in calm water bodies. Sediment-laden water is referred to as *turbid water*; scientists can measure *turbidity* by sampling the water body or by using devices that estimate turbidity by the transparency of the water. One such device is the *Secchi disk*, a white disk of specified diameter that can be lowered on a line from the side of a small boat. Because turbidity decreases the transparency of the water body, the depth at which the disk is no longer visible can be related to sediment content. Another indication of turbidity, nephelometric turbidity units (NTU), are measured by the intensity of light that passes through a water sample. A special instrument uses a light beam and a sensor to detect differences in light intensity. Water of high turbidity decreases the intensity of the light in a manner that can be related to sediment content.

Thus, as sediment concentration increases, the spectral properties of a water body change. First, its overall brightness in the visible region increases, so the water body ceases to act as a "dark" object and becomes more and more like a "bright" object as sediment content increases. Second, as sediment concentrations increase, the wavelength of peak reflectance shifts from a maximum in the blue region toward the green. The presence of larger particles means that the wavelength of maximum scattering shifts toward the blue-green and green regions (Figure 19.4). Therefore, as sediment content increases,

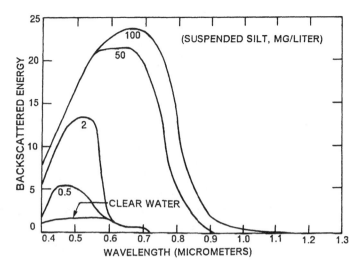

FIGURE 19.4. Effects of turbidity on spectral properties of water. From Moore (1978, p. 460). Reproduced by permission of the Environmental Research Institute of Michigan.

there tends to be a simultaneous increase in brightness and a shift in peak reflectance toward longer wavelengths, and the peak itself becomes broader, so that at high levels of turbidity the color becomes a less precise indicator of sediment content. As sediment content approaches very high levels, the color of the water begins to approach that of the sediment itself. Plate 23 illustrates the turbidity patterns that can be recorded as CIR images. The green reflectance of turbid water appears as light blue on the CIR image (Chapter 3), which provides a sharp contrast in color with the dark blue or black of clear water bodies.

19.3. Spectral Changes as Water Depth Increases

Figure 19.3 shows spectral characteristics of sunlight as it penetrates a clear water body. Near the surface, the overall shape of the curve resembles the spectrum of solar radiation, but the water body increasingly influences the spectral composition of the light as depth increases (Plate 23). At a depth of 20 m, little or no infrared radiation is present because the water body is an effective absorber of these longer wavelengths. At this depth, only blue-green wavelengths remain; these wavelengths are therefore available for scattering back to the surface, from the water itself, and from the bottom of the water body.

The attenuation coefficient (k) describes the rate at which light becomes dimmer as depth increases. If E_0 is the brightness at the surface, then the brightness at depth z (E_z) is given by

$$E_z = E_0 e^{-kz} \qquad \text{(Eq. 19.1)}$$

In hydrologic studies the influence of the atmosphere can be especially important. The atmosphere, of course, alters the spectral properties of incident radiation and also influences the characteristics of the reflected signal. Although these influences are also present in remote sensing of land surfaces, they assume special significance in hydrologic studies, in part because such studies often depend on subtle spectral differences (easily lost in atmospheric haze) and also perhaps because much of the hydrologic information is carried by the short wavelengths that are most easily scattered by the atmosphere.

Water bodies are typically dark, so the analyst must work with a rather restricted range of brightnesses relative to those available for study of land surfaces (Figure 19.5). As a result, analysts who specialize in remote sensing of water bodies must devote special attention to the radiometric qualities of the remotely sensed data. Typically, data to be analyzed for hydrologic information are examined carefully to assess their quality, the effects of the atmosphere, and sun angle. Geometric preprocessing is used with caution, to avoid unnecessary alteration of radiometric qualities of the data. In some instances, analysts calculate average brightness over blocks of contiguous pixels to reduce transient noisy effects of clouds and whitecaps. Sometimes several scenes of the same area, acquired at different times, can be used to isolate permanent features (e.g., shallows and shoals) from temporary features (including waves and atmospheric effects). Also, it is often advantageous to estimate original radiometric brightnesses from digital values as a means of accurately assessing differences in color and brightness.

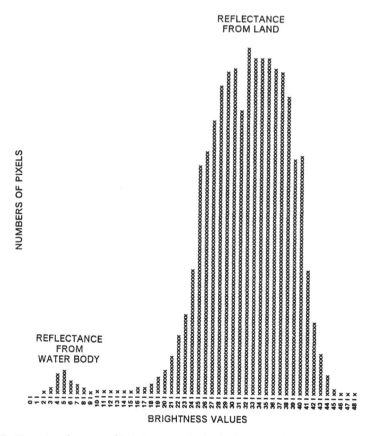

FIGURE 19.5. Restricted range of brightnesses in hydrologic setting.

19.4. Location and Extent of Water Bodies

Remote sensing provides a straightforward means to map the extent of water bodies, to inventory area occupied by open water, and to monitor changes in water bodies over time.

Although such tasks may seem very basic, there are numerous situations in which simple determination of the land–water boundary can be very important. Study of coastal erosion requires accurate determination of shoreline position and configuration at several dates. Likewise, comparisons of shoreline position before and after flooding permits measurement of areas flooded, as well as determination of locations of flooded areas. Such information can be difficult to acquire by conventional means.

Determination of the land–water body border is usually easiest in the near infrared region, where land, especially if vegetated, is bright, and open water dark. Usually it is possible to determine a sharp contact between the two (Figure 19.6). With Landsat 1–3 MSS data, the contrast is usually clear on bands 6 or 7. With photography, black-and-white infrared film is usually satisfactory. CIR film is suitable if the water is free of sediment. However, because CIR film is sensitive to green light, turbid water will appear bright, thereby preventing clear discrimination of the land–water boundary.

In the early 1970s Landsat MSS data were used to inventory impounded water throughout the United States. The failure of earth dams Buffalo Creek, West Virginia, in February 1972 and at Rapid City, South Dakota, in June 1972 produced catastrophic loss of life and property. In both instances, poorly constructed earth dams failed after periods of unusually heavy rainfall, which both increased the volume of impounded water and weakened the earthen dams. In August 1972 the U.S. Congress passed Public Law 92-367 requiring federal inspection of all dams impounding water bodies exceeding specified sizes. Because there existed no comprehensive record of such water bodies, NASA conducted a study to determine the feasibility of using Landsat MSS data as the means of locating water bodies impounded by dams requiring inspection under Public

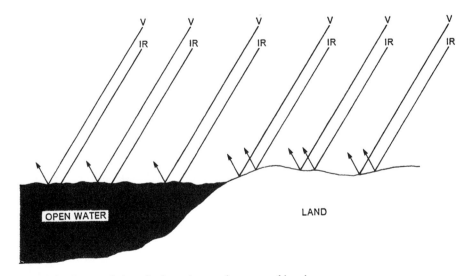

FIGURE 19.6. Determining the location and extent of land–water contact.

Law 92-367. NASA's method required selection of GCPs (Chapter 11) to register the data to accurate maps. A density slice technique separated water bodies from land areas; a printed output showing water pixels could then be matched to maps to locate impoundments not already delineated on the maps.

Philipson and Hafker (1981) studied application of Landsat MSS data to delineation of flooded areas. They concluded that simple visual interpretation of MSS band 4 images was as accurate as manual interpretation of multiband composite images and as accurate as digital analysis of MSS band 4 and of combinations of both bands 2 and 4. They experienced problems in applying the digital approach because of difficulties in registering the flooded and unflooded scenes. In addition, turbidity associated with flooding increased brightness in the green region (band 1) to the extent that there was insufficient contrast between the two dates to unambiguously separate flooded and unflooded areas. For their digital interpretation, they defined a ratio that compares the brightness values in near infrared (band 4) on the two dates:

$$\frac{(\text{MSS 1, unflooded}) - (\text{MSS 4, flooded or unflooded})}{(\text{MSS 1, unflooded})} \qquad (\text{Eq. 19.2})$$

Because open clear water is very dark in band 4, the ratio approaches 1 for flooded areas and 0 for unflooded areas. For example, if dry land has digital values near 35 in both bands, then for flooded areas the ratio is $(35 - 0)/35 = 1$, and for unflooded areas, $(35 - 35)/35 = 0$. Although Philipson and Hafker found that when water turbidity was high their procedure was not always successful, there are other procedures that can identify turbid water and separate such areas from unflooded land. In later studies, Frazier and Page (2000) found that density slicing of Landsat TM band 5 (mid infrared) was effective in identifying water bodies on a riverine flood plain.

Some laboratories have specifically dedicated their work to applications of multispectral data to observation of hydrologic events, notably coastal and riverine floods. For example, the widespread flooding of the upper midwestern United States in 1993 was examined in detail using remotely sensed imagery (Brakenridge et al., 1994). Remotely sensed data form the basis for research laboratories dedicated specifically to monitoring floods using imagery, in coordination with in situ data when available. Because of the routine widespread availability of satellite imagery throughout the world, such organizations can rapidly report extent and severity of events in both local and remote locations (Figure 19.7). The capabilities of the varied remote sensing systems discussed here complement each other, providing detail of fine-scale satellite systems, synoptic view of broad-scale systems, and the all-weather capabilities of imaging radars.

Examples of such laboratories include the Dartmouth Flood Observatory:

www.dartmouth.edu/~floods

and the USGS Rocky Mountain Mapping Center:

http://rmmcweb.cr.usgs.gov

http://nhss.cr.usgs.gov

FIGURE 19.7. Flood mapping using remotely sensed data. Flooded areas, New Orleans region in the aftermath of Hurricane Katrina, prepared from SPOT and Landsat data. From Brakenridge, Anderson, and Caquard (2005).

19.5. Roughness of the Water Surface

Figure 19.8 shows spectra for calm and wave-roughened surfaces in the visible and the near infrared. Wave-roughened surfaces are brighter than are smoother surfaces. Calm, smooth water surfaces direct only volume-reflected radiation to the sensor, but rough, wavy water surfaces direct a portion of the solar beam directly to the sensor. As a result, wavy surfaces are much brighter, especially in the visible portion of the spectrum. McKim et al. (1984) describe a procedure that uses polarizing filters to separate the surface brightness from that of the water body itself, using high-resolution data.

The most intensive studies of ocean roughness have used active microwave sensors (Chapter 7). *Sea state* refers to the roughness of the ocean surface, which is determined by wind speed and direction as they interact locally with currents and tides. Sea state is an important oceanographic and meteorological quality because, if studied over large areas and over time, it permits inference of wind speed and direction—valuable information for research and forecasts. Radars provide data concerning ocean roughness; the backscattering coefficient increases as wave height increases. Radars therefore provide a means of indirectly observing sea state over large areas and a basis for inferring wind

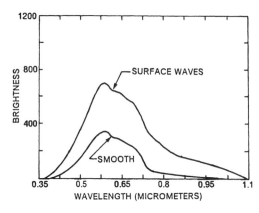

FIGURE 19.8. Spectra of calm and wind-roughened water surfaces (from laboratory experiments). Based on McKim et al. (1984, p. 358). Copyright 1984 by the American Society for Photogrammetry and Remote Sensing. Reproduced by permission.

speed and direction at the water's surface. If the imaging radar is carried by a satellite, as is the case with Seasat and the shuttle imaging radar (Chapter 7), then the analyst has the opportunity to observe sea state over very large areas at regular intervals. These observations permit oceanographic studies that were impractical or very difficult using conventional data. Furthermore, timely information regarding sea state has obvious benefits for marine navigation and can, in principle, form the basis for inferences of wind speed and direction that can contribute to meteorological data. Direct observation of sea state is relatively straightforward, but because ocean areas are so very large and conditions so rapidly changeable, the usual observations from ships in transit are much too sparse to provide reliable data.

Rigorous study of sea state by radar started soon after World War II and continued during the 1950s and 1960s, culminating with the radars carried by Seasat and the space shuttle. Although many experiments have been conducted using a variety of microwave instruments, one of the most important broad-scale sea state experiments was based on the Seasat A project, which used several instruments to monitor the Earth's oceans. The Seasat SAR and RADARSAT (Chapter 7) have provided an opportunity to observe large ocean areas on a repetitive basis. Calm ocean surfaces, with waves that are small relative to the radar wavelength, act as specular reflectors and appear as dark regions on the image, as energy is reflected away from the antenna. As wind speed increases, the ocean surface becomes rougher and acts more like a diffuse reflector, causing bright regions on the imagery (Figures 19.8 and 19.9). Because radar wavelength and system geometry are known, the received signal can form the basis for estimates of wave height and velocity.

In a test in the Gulf of Alaska, scientists found agreement between data derived from the Seasat SAR and those observed at the ocean surface. Wave direction, height, and wavelength were all estimated from the SAR imagery. Waves were most accurately observed when their direction of travel was within 25° of the look direction. Other studies have shown the SAR data to be useful for studying other kinds of ocean swells and currents.

19.6. Bathymetry

Bathymetry information concerning water depth and configuration of the ocean floor is one of the most basic forms of hydrographic data. Bathymetry is especially important

FIGURE 19.9. Seasat SAR image illustrating rough ocean surface, Nantucket Shoals, 27 August 1978. From Jet Propulsion Laboratory, Pasadena, California.

near coastlines, in harbors, and near shoals and banks, where shallow water can present hazards to navigation, and where changes can occur rapidly as sedimentation, erosion, and scouring of channels alters underwater topography.

Water depth can be measured by instruments carried on vessels, especially by acoustic (sonar) instruments that measure depths directly below the vessel. Ideally, bathymetric maps should be compiled from a more or less uniform network of depth measurements rather than from a limited set of data from traverses. Although modern side-scanning sonar can yield a continuous surface of depth information, aerial imagery is an important means of mapping subsurface topography in shallow water.

Photogrammetry can be applied to bathymetric measurements if high-quality, large-scale photography is available and if the water is clear. Filters have been used to separate radiation in the spectral region 0.44–0.54 µm, where solar energy is most easily transmitted by clear water. As Figure 19.3 indicates, in principle sunlight can penetrate to depths of about 20 m. Aerial photographs record information only from relatively shallow depths, although depths as deep as 16 m have been mapped using aerial photographs.

Special problems in applications of photogrammetry to bathymetry include estimation of differences in refraction between air and water (Figure 19.10) and the difficulty in acquiring a good set of underwater control points. Furthermore, mapping is difficult in regions far removed from the shoreline because of the difficulty in extending control across zones of deep water. Nonetheless, photogrammetric methods have been successfully used for mapping zones of shallow water.

Water depth can also be estimated by observing refraction of waves as they break

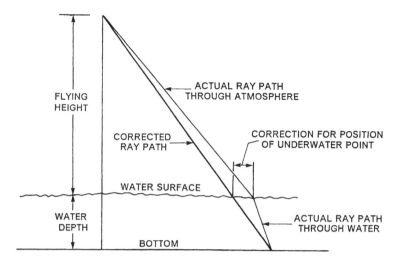

FIGURE 19.10. Bathymetry by photogrammetry.

in shallow water. As a wave approaches the shore, it begins to break at a depth that is determined by its velocity, wavelength, and period. If velocities and wavelengths can be measured on successive overlapping photographs, it is possible to calculate an estimate of depth. The method works well only when the water body is large and if the bottom drops slowly with increasing distance from shore. The wave refraction method is not as accurate as other procedures, but it does have the advantage of applicability in turbid water when other methods cannot be applied.

Multispectral Bathymetry

Depth of penetration of solar energy varies with wavelength. Longer wavelengths are absorbed most strongly; even modest depths—perhaps 2 m or so—of clear water will absorb all infrared radiation, so that longer wavelengths convey no information concerning depth of water or of the nature of the bottom. Visible radiation penetrates to greater depths but is still influenced by wavelength-dependent factors. For clear water, maximum penetration occurs in the blue-green region, which is approximated by Landsat MSS 1 (using the Landsat 4 and 5 band designation), although TM1 is likely to be more effective for bathymetric studies (Chapter 6). Polcyn and Lyzenga (1979) report that for their study region in the Bahamas under optimum conditions, maximum penetration for MSS 1 is as deep as 25 m and about 6 m for MSS 2. (For consistency in designation, Landsat 4 and 5 MSS numbering is employed here.)

Intensity of radiation decreases exponentially with depth (i.e., brightness decreases as depth increases). Thus MSS 4 (or TM1; both are positioned at or near the spectral region of maximum penetration) should be useful for estimating depth. Dark values suggest that the bottom is deep, beyond the 20-m range represented in Table 19.2. Bright values suggest that the bottom is near the surface (Figure 19.11). Many depth-extraction algorithms use a logarithmic transformation of MSS 4 brightness as a means of incorporating our knowledge that brightness decreases exponentially with depth.

Polcyn and Lyzenga (1979) defined a depth-finding algorithm:

TABLE 19.2. Logarithmic Transformation of Brightnesses

x	$\ln(x)$	Brightness	Depth
2	0.69	Dim	Deep
6	1.79		
10	2.30		
14	2.64		
18	2.89		
22	3.09		
26	3.26	Bright	Shallow

$$L_{\text{sen}} = L_{\text{atm}} + T_{\text{atm}} (L_{\text{surf}} + L_{\text{wat}}) \qquad \text{(Eq. 19.3)}$$

where L_{atm} is the radiance contributed by the atmosphere; T_{atm} is atmospheric transmission; L_{surf} is radiance from the water surface; and L_{wat} is radiance from beneath the water surface. L_{wat} for a specific band i is usually designated as L_i, and has been estimated as

$$L_i = L_{\text{w}(i)} + (L_{\text{b}(i)} \, L_{\text{w}(i)})e^{-2k_i z} \qquad \text{(Eq. 19.4)}$$

where i is a specific spectral band; $L_{\text{w}(i)}$ is the deep-water radiance for band i; $L_{\text{b}(i)}$ is the bottom reflectance for band i; and z is water depth.

L_{w} is defined as the energy reflected from a water body so deep that incident solar radiation is absorbed, and the energy observed by the sensor is so dim that it cannot

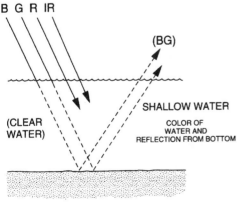

FIGURE 19.11. Multispectral bathymetry. Shallow water bodies filter radiation reflected from the bottom; the balance between spectral bands observed at the sensor reveals the approximate depth of the water body. See Figure 19.12.

be distinguished from sensor and environmental noise. Values for L_w vary depending on sensor sensitivity, spectral band, and sensor resolution; but for a given sensor, L_w is influenced also by atmospheric clarity, solar angle, and other factors. Therefore, for each specific study, L_w must be estimated, usually by selecting a set of pixels known to correspond to depths well beyond the range of penetration of solar energy.

Jupp et al. (1985) have summarized published values to produce the following estimates for maximum penetration of solar radiation within clear, calm ocean water under a clear sky:

- MSS band 1: 15–20 m (50–60 ft.)
- MSS band 2: 4–5 m (13–17 ft.)
- MSS band 3: 0.5 m (1.5 ft.)
- MSS band 4: 0

As expected from inspection of Figure 19.4, penetration is greatest within the blue-green region of MSS band 1 and least in MSS band 4, where solar radiation is totally absorbed. It is possible to use this information to designate each MSS pixel as a member of a specific depth class. If we find the numeric difference between the observed brightness in band i (L_i) and the deep-water signal for the same band ($L_{w(i)}$), then the quantity ($L_i - L_{w(i)}$) measures the degree to which a given pixel is brighter than the dark pixels of open water.

If the value for band 1 exceeds the band 1 deep-water limit but all other pixels are at deepwater values for their respective bands, then it is known that water depth is within the 15-m to 20-m range (Figure 19.12). If both band 1 and band 2 are brighter than their respective values for L_w but bands 3 and 4 are at their L_w values, then water depth is within the 4-m to 5-m range suggested above. If a pixel displays values in bands 1, 2, and 3 that all exceed their respective thresholds for L_w, then water depth is likely to fall within the 0.5-m range. Finally, if all four bands are at their threshold values, then the pixel corresponds to a depth beyond the 20-m limit. Because band 4 values for water pixels should always be very dim, bright values in this band suggest that water areas may have been delineated incorrectly, marsh vegetation rises above the water surface, or other factors have altered the spectral properties of the water.

This procedure assigns pixels to depth classes but does not provide quantitative estimates of depth, which, of course, would be more useful. Because of the exponential effect of the attenuation coefficient (Equation 19.1), brightnesses of water bodies are not linearly related to depth. That is, we cannot observe a given brightness in band i, then directly match that brightness to a specific depth. Lyzenga (1981) has attempted to solve this problem by devising a relationship that creates a new variable, x_i, for each band i; x_i is linearly related to water depth and therefore can be used to estimate depth within a given depth zone:

$$x_i = \ln (L_i - L_{w(i)})$$ (Eq. 19.5)

Thus x_i is defined as the natural logarithm of the difference between observed brightness in land i (L_i) and the deep-water reflectance ($L_{w(i)}$) in the same band. Table 19.2 shows that the logarithmic transformation enhances the significance of dim values from deep water. In Table 19.2, x represents a hypothetical set of digital brightnesses observed by a sensor; $\ln(x)$ does not itself measure depth but is linearly related to water depth, so that the investigator can define the local relationship between x_i and water depth.

Some methods have relied on estimates of k, which varies with wavelength and

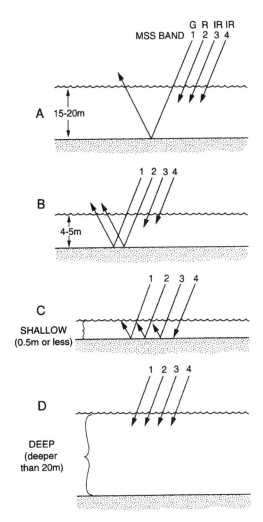

FIGURE 19.12. Multispectral bathymetry using Landsat MSS data. The relative brightnesses of the several spectral bands indicate depth classes, as discussed in the text.

depends on local effects that influence the clarity of the water. Polcyn and Lyzenga (1979) used $k = 0.075$ m^{-1} for MSS1 and $k = 0.325$ m^{-1} for MSS2 for their studies of water depths in the Bahamas. They were able to use this algorithm to achieve accuracies of 10% at depths of about 20 m (70 ft.). Under ideal conditions, they measured depths as deep as 40 m (130 ft.).

These procedures seem to be effective, provided that the water body is clear (otherwise turbidity contributes to brightness) and the bottom reflectance is uniform. When attenuation of the water and bottom reflectivity are known, accuracies as high as 2.5% have been achieved (although typical accuracies are lower). If the bottom reflectance is not uniform, then differences in brightness will be caused not only by differences in depth but also by differences in reflectivity from the bottom materials. Because such differences are commonly present, due to contrasting reflectances of different sediments and vegetation, it is often necessary to apply procedures to adjust for differing reflectances from the bottom of the water body. The effectiveness of this procedure varies with sun angle (due to variations in intensity of illumination) and is most effective when data of high radiometric resolution are available.

In some instances, ratios of two bands may remove differences in bottom reflectivity. Lyzenga (1979) defines an index that separates different reflectivities using different spectral bands. His index, calculated for each pixel, can be used to classify pixels into different bottom reflectivity classes, which can then each be examined, as classes, to estimate water depth.

19.7. Landsat Chromaticity Diagram

Colorimetry, the precise measurement of color, is of significance in several aspects of remote sensing but is especially significant in applications of remote sensing to studies of water quality. Even casual observation reveals that water bodies differ significantly in color, yet it can be very difficult to accurately measure these color differences and to relate color to the variables that we wish to study.

Specification of color is not as self-evident as it might seem. Scientists have established several alternative systems for describing color. Most establish a three-dimensional color space in which three primaries (red, green, and blue) form three orthogonal axes. Unique specification of a color requires identification of a location within this color space, giving hue (dominant wavelength), saturation (purity), and value (brightness) of a given color (Chapter 2).

These axes define *tristimulus space*, in which any color can be represented as a mixture of the three primaries—that is, as a location within the coordinate system represented in Figure 19.13. Such a system defines a color solid, which in principle permits specification of any color.

For many purposes, it is convenient to define a two-dimensional plane within the color solid by setting:

$$R = \frac{X}{X + Y + Z} \qquad G = \frac{Y}{X + Y + Z} \qquad B = \frac{Z}{X + Y + Z} \qquad \text{(Eq. 19.6)}$$

The values R, G, and B define locations within a two-dimensional surface that show variations in hue and saturation but do not portray differences in brightness. In effect, R, G, and B have been standardized by dividing by total brightness in each primary without conveying information concerning absolute brightness. Thus colors represented by this convention vary only with respect to hue and saturation, because brightness information has been removed by standardization. For example, R specifies "percent red" but does not convey information concerning whether it is a percentage of a very bright color or of a dark color.

In 1931 the Commission Internationale de l'Eclairage (CIE; International Commission for Illumination) met in England to define international standards for specifying color. The CIE system defines a tristimulus space as follows:

$$x = \frac{b}{r + g + b} \qquad y = \frac{g}{r + g + b} \qquad z = \frac{r}{r + g + b} \qquad \text{(Eq. 19.7)}$$

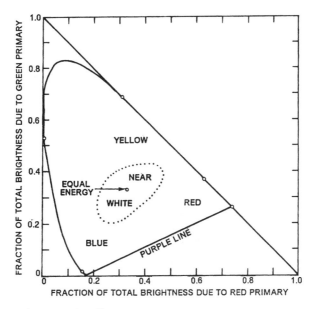

FIGURE 19.13. CIE chromaticity diagram.

in which x, y, and z correspond approximately to the red, green, and blue primaries and X, Y, and Z are the CIE chromaticity coordinates. Also, $x + y + z = 1$.

The three CIE primaries are "artificial" because, as a matter of convenience, they are defined to facilitate mathematical manipulation and do not correspond to physically real colors. Because x, y, and z sum to 1.0, only two are necessary to uniquely specify a given color, as the third can be determined from the other two. Therefore, the CIE chromaticity diagram has only two axes, using the x and y coordinates to locate a specific color in the diagram.

Figure 19.13 shows the CIE chromaticity diagram. The vertical axis defines the y (green) axis and the horizontal axis defines the x (red) primary, each scaled from 0 to 1.0 (to portray the proportion of total brightness in each band). At coordinates $x = 0.333$ and $y = 0.333$ (z must then equal 0.333) all three primaries have equal brightness, and the color is pure white. Colors to which the human visual system is sensitive are inside the horseshoe-shaped outline in Figure 19.13. The white colors near the center are "without color," as there is no dominant color; purity (*saturation*) increases toward the edges of the horseshoe, where colors are "fully saturated." The line at the base of the horseshoe, known as the *purple line*, locates *nonspectral* colors. (Nonspectral colors are those that do not appear in a spectrum of colors because they are formed by mixtures of colors from opposite ends of the spectrum.) Spectral colors are positioned in sequence around the curved edge of the horseshoe.

Because each of the x, y, and z primaries has been defined by dividing by the total brightness to yield the percentage brightness in each primary, they convey hue information without brightness information. That is, colors have been standardized to remove brightness, leaving only hue and saturation information. The CIE diagram is useful for precisely specifying color and for transferring color specifications from one context to another. Note, once again, that the CIE diagram conveys differences in color without portraying differences in brightnesses of colors—that is, it shows only hue and saturation.

Alföldi and Munday (1977) have applied the chromaticity concept to analysis of Landsat data. They defined a *Landsat chromaticity diagram*, analogous to the CIE chromaticity diagram, except that it defines a color space using MSS bands 4, 5, and 6 instead of the red, green, and blue primaries used for the CIE diagram. Thus infrared data from band 3 are substituted for the missing primary. Because the raw MSS digital counts do not accurately portray absolute brightnesses, it is necessary to use the appropriate calibration data to estimate the brightnesses in each band as shown in Chapter 11.

The Landsat chromaticity coordinates can be used to assess the turbidity of water bodies imaged by the MSS. Figure 19.14 shows the Landsat chromaticity diagram, which is identical to the CIE chromaticity diagram except for the use of MSS bands 4, 5, and 6 as substitutes for the CIE primaries. "E" signifies the *equal radiance point*, where radiances in MSS bands 4, 5, and 6 are equal. (Note that the labels in Figure 19.13 do not apply to the Landsat chromaticity diagram, as it does not use the CIE primaries.) The curved line extending to the right of the equal radiance point is the experimentally defined locus that defines the positions of water pixels as their radiances are plotted on the diagram. Pixels with positions near the right end of this line represent water bodies with clear water. As chromaticity coordinates shift closer to the other end of the line, near the equal radiance point, the spectral properties of the pixels change, indicating increases in turbidity as the pixel position approaches the left-hand end of the line. This line was defined experimentally by Alföldi and Munday (1977) for their study area, using sediment samples collected at the time the imagery was acquired. (For detailed studies, it is important to define a specific locus for each individual study region, but for illustrative purposes it is sufficient to use their line as a general-purpose locus.)

Thus, in Figure 19.15, the position of pixel A reveals a clear water body, and pixel B represents a turbid water body. As pixels known to represent water bodies drift away from the line (pixel C is an example), we suspect that atmospheric haze or thin clouds have altered their radiances. Although our inspection of the chromaticity diagram is essentially a qualitative analysis of pixel positions on the chromaticity diagram, Alföldi

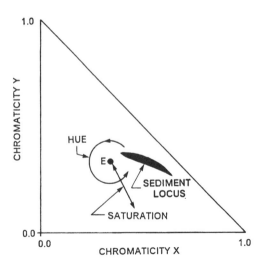

FIGURE 19.14. Landsat MSS chromaticity space. The sediment locus is experimentally derived and may vary slightly from region to region or over time within a given region. The radial dimension (outward from E) presents spectral purity information, and radial shifts in the position of the locus are attributed to atmospheric scattering, thin clouds, air pollution, or whitecaps. Diagram based on Alföldi and Munday (1977).

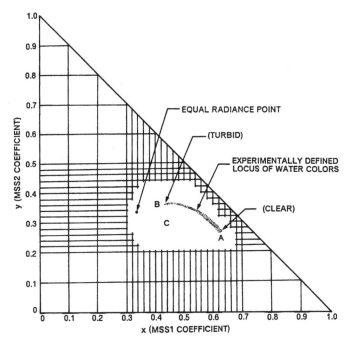

FIGURE 19.15. Landsat MSS chromaticity diagram. From Alföldi and Munday (1977, p. 336). Reproduced by permission of the Canadian Aeronautics and Space Institute.

and Munday have conducted more quantitative studies, using observed values of suspended sediment to calibrate the technique.

As an example of a practical application, Table 19.3 presents Landsat MSS data selected from sequential imagery of the same water body as observed over the course of a year. In Table 19.3, these MSS digital counts (introduced in Chapter 4) for MSS bands 1, 2, 3, and 4 are tabulated at the left. To plot these values in chromaticity space, it is necessary first to covert these values to radiances (introduced in Chapter 11), as shown in the next columns. (Table 19.3 shows, as examples, calculated radiances for the first two rows.) Next, chromaticity coordinates can be calculated from radiances using Eq. 19.7, and then plotted on the Landsat chromaticity diagram (Figures 19.15 and 19.16). Chromaticity coordinates for each month in the sequence are plotted in Figure 19.16, using the first letters of months to indicate seasonal variations. Although the results plotted in Figure 19.16 are approximate, they illustrate seasonal variations in turbidity—the clearest water is observed in November, and highest turbidity with the spring runoff, when streamflow is highest and vegetation cover is low. The values that have plotted to the lower left of the sediment locus (and therefore, off the locus representing sediment content) have been influenced by atmospheric haze, chiefly in the summer months.

19.8. Drainage Basin Hydrology

A less obvious component of water resources concerns the management of land areas to promote orderly retention and flow of moisture over the land surface and then within the

TABLE 19.3. Data for Chromaticity Diagram

Date	MSS Digital Counts				Radiances			Sum	Chromaticity coordinates	
	MSS1	MSS2	MSS3	MSS4	MSS1	MSS2	MSS3		X	Y
May	36	33	24	7	0.703	0.520	0.333	1.555	0.452	0.334
May	36	36	27	8	0.703	0.567	0.374	1.644	0.426	0.345
Feb	30	32	19	3					0.433	0.372
Feb	29	32	17	3					0.434	0.366
Feb	21	16	22	12					0.424	0.261
Apr	28	21	14	4					0.510	0.309
Apr	28	21	14	5					0.510	0.309
Apr	31	22	15	3					0.522	0.299
May	40	24	24	5					0.524	0.253
May	40	25	25	6					0.513	0.259
May	42	39	22	5					0.472	0.353
Jun	37	28	26	6					0.474	0.269
Jun	37	30	31	11					0.445	0.291
Jun	37	30	29	12					0.452	0.296
Sep	26	25	16	3					0.452	0.351
Sep	28	25	16	3					0.470	0.339
Sep	27	25	16	2					0.461	0.345
Nov	18	9	5	1					0.625	0.252
Nov	16	11	6	2					0.549	0.305
Nov	18	11	7	1					0.565	0.279

Note. Blank cells are left open for use as a student exercise.

stream system, so that flooding and erosion are minimized and an even, steady flow of moisture is maximized. The nature and distribution of land use and land cover within drainage basins influences runoff from the land surface and eventually also influences the flow of water within stream channels. Although there is no set relationship between these elements that can be said to form a universally "normal" balance, each climatic and ecological situation tends to be characterized by an established pattern of rainfall, evapotranspiration, and streamflow. Extensive changes to natural land-cover patterns can disrupt this local balance and cause accelerated soil erosion, flooding, and other undesirable consequences. Hydrologists attempt to study such processes by constructing *hydrologic models*: mathematical simulations that use measurements of local rainfall, together with information concerning patterns of land use, soil, topography, and drainage, to predict the nature of the streamflow that might be caused by storms of a given magnitude. Such models provide a means of assessing effects of proposed changes in land use (e.g., new construction, harvesting of timber) or in channel configuration because they can estimate the effects of such changes before they actually occur. Often such models can be evaluated and refined using data from watersheds that have very accurate records of rainfall and streamflow.

Hydrologic models require large amounts of detailed and accurate data, so many of the costs involved in applying a given model to a specific region arise from the effort required to collect accurate information concerning soil and land-cover patterns—two of the important variables required for such models. Some of the information concerning land use can be derived from remotely sensed data. Manual interpretation of land use and land cover from aerial photographs (Chapter 20) is a standard procedure, and, of course, aerial photographs are a source of much of the information regarding topography and soils. Many studies have evaluated the use of image classification of digital remote sensing data (Chapter 12) as a source of land-cover information for such models. Advantages include reduced costs and the ability to quickly prepare new information as changes occur.

Jackson et al. (1977) applied this kind of approach in a study of a small watershed in northern Virginia. Their study, using both manual interpretations of aerial photography and computer classifications of Landsat MSS data, was able to compare results based on MSS data with those of conventional estimates. Their results indicated that the use of Landsat data was much cheaper than the manual method but tended to produce results of comparable reliability.

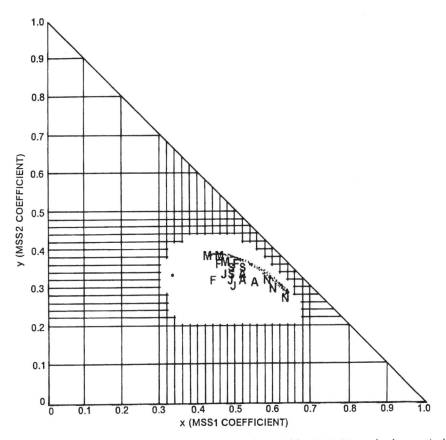

FIGURE 19.16. Chromaticity plot of data tabulated in Table 19.3. Here the letters indicate the chromaticities of pixels from a single water body in Eastern Virginia as assessed using MSS imagery acquired in February, March, June, August, September, and November of the same year. These results illustrate the ability of multispectral remote sensing to monitor seasonal variations in water quality.

19.9. Evapotranspiration

Evapotranspiration is defined as combined *evaporation*, moisture lost to the atmosphere from soil and open water, and *transpiration*, moisture lost to the atmosphere from plants. Evapotranspiration measures the total moisture lost to the atmosphere from the land surface, and as such it (together with other variables) forms an important variable in understanding the local operation of the hydrologic cycle.

However, accurate measurement of evapotranspiration is difficult. Reliable data are usually available only for well-equipped meteorological stations and agricultural research facilities that are regularly attended by a trained staff. The vast majority of the Earth's land area is not near such installations, so reliable measurements of evapotranspiration are not available. Evapotranspiration can, of course, be estimated using measurements of air temperature, wind speed, plant cover, cloudiness, and other variables, but such estimates often lack the detail and accuracy required for good environmental studies and also require data that are themselves difficult to obtain.

Remotely sensed data can form the basis for *spatially referenced estimates* of evapotranspiration—that is, estimates that apply to specific areas of small areal extent (Jackson, 1985). In contrast, estimates based on meteorological measurements can usually provide only a single estimate for a large region in the vicinity of a climate station. In effect, spatially referenced estimates provide detailed, independent estimates of evapotranspiration for each pixel.

Direct sensing of the soil surface using remote sensing instruments (especially passive) can provide estimates of soil moisture, or calculation of such estimates can be based on digital elevation data that represent topographic elevation as a matrix of values, similar in concept to digital remote sensing images. For the elevation data, each pixel value represents the topographic elevation at the center of each pixel rather than the brightness of the energy reflected from the land at that pixel. From such data a computer program can calculate the topographic slope at each pixel and can also determine the *aspect*, the compass direction at which the slope faces.

With the elevation data in hand, it is possible to overlay Landsat data and data from meteorological satellites, such as MODIS and AVHRR (Chapters 6, 17, and 21), so that all three forms of data register to one another. A classification of the Landsat data (as described in Chapter 12) can provide a map of the major vegetation classes (forest, pasture, bare ground, etc.) present. The data gathered by meteorological satellites can be used to estimate temperature and total solar radiation received during a 24-hr. interval. These vary depending on the slope and aspect of each pixel, as derived from the elevation data. From these separate forms of data, it is then possible to estimate evapotranspiration for each pixel. With independent estimates of evapotranspiration for each pixel within a study area, the analyst can then prepare much more detailed hydrologic data.

Jackson (1985) describes the estimation of evapotranspiration from remotely sensed data and observes that some difficulties still remain. Atmospheric degradation of the remotely sensed data can cause errors in the estimates, and the usual procedures cannot consider effects of factors such as wind speed or surface roughness. Also, estimates apply only to a specific time (when the remotely sensed data were acquired), although usually the requirement is to estimate evapotranspiration over intervals of a day, several days, or even longer. Therefore, it is necessary to extend the estimates from a specific time to a longer interval—a process that is likely to introduce error. Nonetheless, it should be emphasized that further development of such procedures is one of the most interesting and practical applications of remote sensing, as they can make major contributions to studies of hydrology, climatology, and agriculture in many of the Earth's remote and inaccessible regions.

19.10. Manual Interpretation

The following examples illustrate how techniques of manual image interpretation (Chapter 5), applied with knowledge of band combinations (Chapter 3) and of the spectral behavior of water bodies as outlined earlier in this chapter, can be used to derive information concerning water bodies and coastal features. Figures 19.17 and 19.18 show the mouth of a tributary at the east bank of the Patuxent River, a tidal estuary in Maryland, south of Washington, DC. The two images show the same scene as recorded by different film/filter combinations. The panchromatic image (Figure 19.17) shows the land area near the river as a pattern of cropped agricultural fields bordered by hedgerows and groves of trees. Here the banks of the Patuxent are fairly steep and the land–water contact is well defined, but near the center of Figure 19.17 it is possible to see a sparsely vegetated patch of land that is periodically exposed and submerged as river and tidal influences change the water level of the river. The tributary carries a high sediment load, visible as the darker gray color that contrasts with the bright tone of water in the main channel.

Figure 19.18 shows the same scene as recorded by black-and-white infrared imagery (Figure 3.10b). Here the land–water distinction is especially clear: there is a sharp contrast between the dark of the open water and the bright infrared reflectance of the vegetated land areas. Some of the variations in vegetative cover are visible, but the variations in tone within the water bodies (previously visible in Figure 19.17) are no longer visible.

FIGURE 19.17. Black-and-white panchromatic photograph of the same area shown in Figure 19.18. From NASA.

FIGURE 19.18. Black-and-white infrared photograph of the same area depicted in Figure 19.17. From NASA.

We have in general a clearer view of the turbidity patterns that were at least partially visible in Figure 19.17.

The Patuxent River itself carries a modest load of sediment, but the tributary has much higher levels, possibly due to runoff from the cultivated land that occupies most of its drainage basin. As it discharges into the Patuxent, the current carries the turbid water downstream in a plume that parallels the shoreline. Within the wetlands at the mouth of the tributary, the black-and-white infrared film portrays details of the differences in kinds and densities of wetland vegetation that are not clearly visible on the other image. Plate 24 shows the mouth of the Chesapeake Bay as photographed 8 June 1991 from the space shuttle. Land areas appear in silhouette due to underexposure to record hydrographic features. As a result, the land–water interface is very sharply delineated. The Chesapeake Bay occupies the left-hand portion of the image. Virginia's coastal waters are visually distinct from those of the bay along the curved front visible at center right of the image, due to currents, surface oils, temperature, and wind and wave patterns.

Figures 19.19 through 19.22 represent this same region from a different perspective. Thermal images of the Atlantic coast were acquired just after noon on 23 July 1972, using an infrared scanner sensitive in the thermal region. The setting is depicted in a broader context in Plate 24 and Figures 19.21 and 19.22; the coastline is protected by a series of barrier islands formed by sediment transported from the north and periodically reshaped by currents and storms.

The islands themselves are long, low strips of sand shaped by the action of both water and wind (Figure 19.20). At the edge of the ocean, a long beach slopes toward the water; the lowest sections of the beach are influenced each day by the effects of tides and

FIGURE 19.19. Eastern shore of Virginia: overview showing areas imaged in Figures 19.21 and 19.22.

currents. The upper, higher sections are affected only by the highest tides and strongest storms, so these regions may experience major changes only once or twice a year or even less often. Inland from the upper beach ridges is windblown sand formed into dunes that are generally above the reach of waves, although very strong storms may alter these regions. The dunes are reshaped by wind, but their general configuration is often stabilized by grasses and small shrubs that cover much of this zone. Inland from the dunes, elevations decrease, and water again assumes the dominant influence in shaping the eco-

FIGURE 19.20. Eastern shore of Virginia: diagram of barrier islands and tidal lagoons.

FIGURE 19.21. Thermal infrared image of Eastern shore of Virginia, 23 July 1972, approximately 12:50 P.M.: Cobb Island and Wreck Island. From NASA.

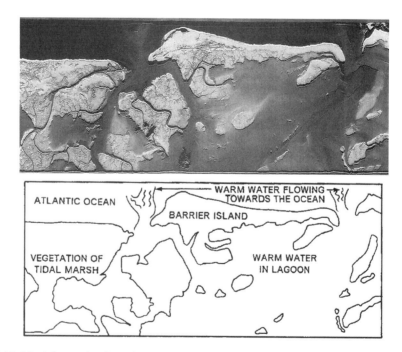

FIGURE 19.22. Thermal infrared image of Eastern shore of Virginia, 23 July 1972, approximately 12:50 P.M.: Paramore Island. From NASA.

systems. Water from the tidal marshes between the islands and the mainland rises and falls with the tides but is without the strong waves that characterize the seaward side of the islands. These tidal flats are covered with vegetation adapted to the brackish water; they experience tidal fluctuations and have poorly drained soils.

Many of these features are clearly visible on the infrared images of this region (Figures 19.21 and 19.22). The bright white strip along the seaward side of the island contrasts sharply with the cool dark of the open ocean. This bright strip is, of course, the hot surface of the open sandy beach, which has received the direct energy of the solar beam for several hours and now, at midday, is very hot. Inland from the beach the dunes are visible as a slightly darker region on the image with darker areas caused by shadowing. The topographic structure of the beach ridges are clearly visible in several regions. In the tidal marshes, the image appearance is controlled largely by vegetative cover, which gives a clear delineation of the edges of the open channel not usually visible on other images. The open water within the tidal lagoons has a lighter tone than that of the open ocean, indicating warmer temperatures. Here the shallow water and the restricted circulation have enabled the water in these areas to absorb the solar radiation it has been receiving now for several hours. Nonetheless, this water is still cooler than water flowing from the land surface; it is possible to see the bright (warm) plumes of water from the streams that flow into the tidal marshes from the mainland. Finally, tidal currents are clearly visible as the bright (warm) stream of water passing between the barrier islands to enter the darker (cooler) water of the open ocean. This image captures the tide flowing outward towards the open ocean (ebb tide).

Plate 25 illustrates another example of the use of remotely sensed data to study currents and tidal flow. These images show a portion of the Belgian port of Zeebrugge as observed using 12 multispectral channels encompassing portions of the visible, near infrared, and thermal regions. With the use of onsite observations, the image has been analyzed to reveal suspended sediment within the upper 1 m of the water column. Reds and yellows indicate high sediment content; blues and greens indicate low sediment content at low tide (top), and at high tide (bottom). From such images it is possible to estimate the broad-scale patterns of sediment transport and deposition and to plan efficient dredging and construction activities.

19.11. Sea Surface Temperature

Remote sensing has become one of the most effective means of assessing *sea surface temperature* (SST), the average temperature of the ocean surface, indicative of temperatures within the upper meter or so of the ocean surface. SST is an important variable for oceanographic studies, including characteristics and dynamics of currents and distributions and movements of marine life. Because SST is major influence for fish and their sources of food, SST maps are important for marine fishing fleets. Commercial services have been successful in providing timely interpretations of satellite data and images for fishing fleets at sea. Because of the significance of oceans in absorbing, radiating, and transporting thermal energy, SST has profound meteorological significance not only for marine weather and climate but also for understanding general atmospheric processes. Therefore, satellite sensing of SST has formed an important component of attempts to model global climate.

The radiation detected by sensors is determined by the temperature of the upper 1 mm or so of the ocean surface. Although there is mixing of water in the upper portion of the water column, these values may not accurately represent true temperature of water volume at the upper meter of the ocean. Nonetheless, estimates of SST temperature based on observations from AVHRR and related satellite sensors provide some of the most valuable and widely used sources of SST data.

To reduce or eliminate effects of cloud cover, images with differing patterns of cloud cover or cover acquired at different times (but within a short interval) are used to create composites that show cloud-free views of the ocean surface. To correct for contributions of atmospheric moisture on SST estimates, two spectral channels in different atmospheric windows are used—the two-channel split-window method (McClain et al., 1985). Wavelengths selected for this purpose are positioned close to the peak of the blackbody radiation expected from the Earth (roughly, 10 µm) and lie within well-defined atmospheric windows, so the radiation will be transmitted to the atmosphere.

Plate 26 shows the Gulf Stream as it meanders along the east coast of North America, imaged by the MODIS spectroradiometer. Red pixels in this image show the warmer areas, greens are intermediate values, and blues are relatively low values; note the detail in the swirls and gyres of the current patterns in the Gulf Stream. This image was produced using MODIS detectors that are sensitive to wavelengths near 11 µm, which record the "brightness temperature" observed at the top of the atmosphere, which represents thermal radiation from a combination of the sea surface and the heat released from the ocean into the overlying atmosphere, increasing the humidity. Thus this image depicts the complex interaction of the sea with the atmosphere. Satellites such as AVHRR and MODIS provide synoptic views of the ocean and a high frequency of repeat views, allowing the examination of basin-wide upper ocean dynamics. A given area may be viewed several times a day, collecting data at an intensity that would be infeasible using other methods.

19.12. Lidar Applications for Hydrospheric Studies

Lidar contributes to several dimensions of hydrologic and hydrographic studies, especially to monitoring of coastal features, to bathymetry, and to examination of subsurface topography. White and Wang (2003) interpreted sequential digital elevation data compiled from Lidar imagery to examine changes in the form of North Carolina beaches over the interval 1995–2005 (Figure 19.23). These images show a small section of beach, with the open water of the Atlantic at the right as the uniformly dark region, and the white polygons representing beach houses constructed immediately adjacent to the left edge of the open water. The gray areas that form the background for the left-hand regions represent the slightly elevated beaches and coastal topography. These sequential images show the varied impacts of beach erosion and beach nourishment, with the final image in the sequence showing the decline of the beach's ability to protect the structures. The economy, detail, and precision provided by Lidar imagery offer advantages over previous methods, especially for dynamic landforms, such as coastlines. As Lidar data become more widely established, such analyses will become more common.

Under favorable circumstances, Lidar has the ability to assess water depth and to separate among varied classes of subsurface vegetation. Lidar returns from water bodies

FIGURE 19.23. Lidar DEMs depicting beach erosion, North Carolina, 1997–2005. The bright, rectangular shapes at the left represent beachfront homes; the black area at the right is open water of the Atlantic Ocean. Lidar point data were collected in the period 1997–2005 by NASA/NOAA coastal mapping programs and are hosted by the Coastal Service Center (CSC) of NOAA at Charleston, SC. The data were retrieved online from the CSC web download site and rasterized and processed at East Carolina University. Ground reference data were collected between 2006 and 2008. See White and Wang (2003) and Allen and Wang (2010).

have characteristic form, if the water surface is calm, the water clear, and depths are within limits of penetration of the signal. Lidar penetration in clear water is often described as within two to three Secchi depths—perhaps as deep as 30 m in some instances. Wang and Philpot (2007), found that their analysis could correct for scan angle effects, assess slope of subsurface topography, and separate among classes of bottom vegetation. Figure 19.24 depicts an idealized Lidar waveform (i.e., the accumulated history of the returned signal) from a water body. The base of the waveform is formed by backscattering from the water column, with distinct peaks formed by returns from the water surface and from the subsurface topography.

19.13. Summary

Hydrologic studies in general cover a very broad range of subject matter, from the movement of currents in bodies of open water to the evaporation of moisture from a soil surface. Such studies can be very difficult, especially if the goal is to examine the changes in hydrologic variables as they occur over time and space, because the usual methods of surface observation gather data at isolated points or at specific times. Two of the great advantages of remote sensing in this context are the synoptic view of the aerial perspective and the opportunity to examine dynamic patterns at frequent intervals.

Yet remote sensing encounters difficulties when applied to hydrologic studies. Many of the standard sensors and analytical techniques have been developed for study of land areas and are not easily applied to the special problems of studying water bodies. Analyses often depend on detection of rather subtle differences in color, which are easily

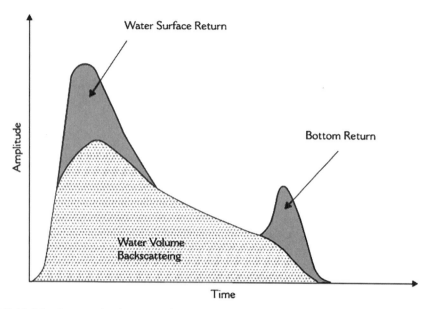

FIGURE 19.24. Generic bathymetric lidar waveform, illustrating contributions of the Lidar return contributed by the water surface, water column, and shallow subsurface features. Reprinted from Wang and Philpot (2007), Fig. 1, p. 12. Copyright 2007, with permission from Elsevier.

lost by effects of atmospheric degradation of the remotely sensed energy. Currents and other dynamic features may change rapidly, requiring frequent observation to record the characteristics of significance to the analyst. Important hydrologic variables, such as groundwater, are not usually directly visible, and others, such as evapotranspiration, are not at all visible but must be estimated through other quantities that may themselves be difficult to observe. These problems and others mean that further development of hydrologic remote sensing is likely to be one of the most challenging research areas in remote sensing.

19.14. Some Teaching and Learning Resources

- NASA/USGS|Landsat: A Space Age Water Gauge
  www.youtube.com/watch?v=Go45F1QviZA

- Physical Geography: The Hydrologic Cycle
  www.youtube.com/watch?v=dabPFNxo844

- microBRIAN (1987)
  www.youtube.com/watch?v=7pEFhusz_IQ

- Marine Algal Blooms as Renewable Fuel Sources
  www.youtube.com/watch?v=U6uSW_XPeAk

- Observing the Ocean
  www.youtube.com/watch?v=WlbVhsLzE5E

- Observing the Biosphere from Space
  www.youtube.com/watch?v=Hn_ffF_KvIU

- NASA Lockheed ER-2 Aids Deepwater Horizon Oil Spill Response
  www.youtube.com/watch?v=TKrmlRNc2pU

- Gulf of Mexico Oil Spill Satellite Animation
  www.youtube.com/watch?v=kVlUQ7znFIo

- NASA|Satellites View Growing Gulf Oil Spill
  www.youtube.com/watch?v=mCWW5xt3Hc8&feature=related

- Tracking the Oil Spill off Grand Isle Day 65
  www.youtube.com/watch?v=a6BQY9EUPtk

- Water Quality: Turbidity
  www.youtube.com/watch?v=kJ6fZrUyWj4

Review Questions

1. List qualities of water bodies that present difficulties for those who study them using only surface observations collected from a ship. Identify those difficulties that are at least partially alleviated by use of some form of remote sensing.

2. Review Chapter 7 to refresh your memory of the qualities of Seasat and Chapter 6 to find corresponding information for Landsat. List differences between the two systems (orbit, frequency of repeat observation, spectral bands, etc.), identifying those qualities that are particularly well or especially poorly suited for hydrographic applications.

3. It is probably best to compile bathymetric information using directly observed surface data because of their significance for navigation. Yet there are some special situations in which use of remotely sensed data may be especially advantageous; can you identify at least two such circumstances?

4. Can you think of important applications of accurate delineation of edges of water bodies (i.e., simple separation of land vs. open water)? Be sure to consider observations over time, as well as use of images from a single date.

5. Contrast SPOT (Chapter 6) and Seasat (Chapter 7) with respect to their usefulness for oceanographic studies. (Chapter 7 discusses only some of Seasat sensors, but it provides enough information to conduct a partial comparison.)

6. Using information in Sections 11.6 and 19.7, calculate the radiances for data tabulated in Table 19.3, then plot the chromaticity coordinates on the chromaticity diagram shown on the following page. Assess both the degree of turbidity (high, moderate, low) and the atmospheric clarity for each entry. Compare your results with those presented in Figure 19.16.

7. Write a short description of a design for a multispectral remote sensing system tailored specifically for collecting hydrologic information and accurate location of the

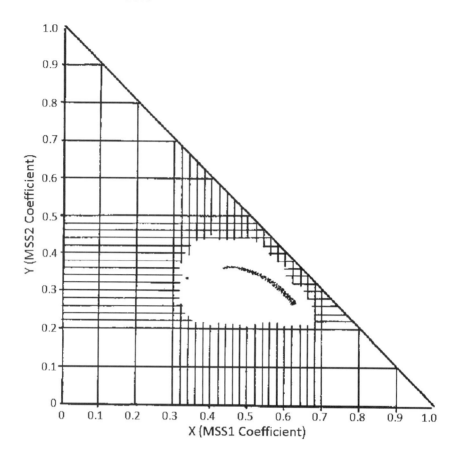

edges of water bodies. Disregard all other applications. Suggest some of the factors that might be considered in choosing the optimum time of day for using such a system.

8. Outline the dilemma faced by scientists who wish to use preprocessing in the application of SPOT or TM data for bathymetric information. Describe the reasons why a scientist would very much like to use preprocessing in some instances, as well as the counterbalancing reasons why he or she would prefer to avoid preprocessing.

9. Outline some of the ways that the methods of image classification (Chapter 12) might be useful for hydrologic studies. Outline also problems and difficulties that limit the usefulness of these methods for studies of water bodies.

References

General

Alföldi, T. T. 1982. Remote Sensing for Water Quality Monitoring. Chapter 27 in *Remote Sensing for Resource Management* (C. J. Johannsen and J. L. Sanders, eds.). Ankeny, IA: Soil Conservation Society of America, pp. 317–328.

Bhargava, D. S., and D. W. Mariam. 1991. Effects of Suspended Particle Sizes and Concentrations on Reflectance Measurements. *Photogrammetric and Remote Sensing*, Vol. 57, pp. 519–529.

Brakenridge, G. R., E. Anderson, and S. Caquard. 2005. *Flood Inundation Map DFO 2003–282*. Hanover, NH: Dartmouth Flood Observatory, Hanover, USA, digital media, *www.dartmouth.edu/~floods/2003282.html*.

Brakenridge, G. R., J. C. Knox, F. J. Magilligan, and E. Paylor. 1994. Radar Remote Sensing of the 1993 Mississippi Valley Flood. *EOS: Transactions of the American Geophysical Union*, Vol. 75, pp. 521–527.

Curran, P. J., and E. M. M. Novo. 1988. The Effects of Suspended Particle Size and Concentration on Reflectance Measurements: A Review. *Journal of Coastal Research*, Vol. 4, pp. 351–368.

Fishes, L. T., F. Scarpace, and R. Thomson. 1979. Multidate Landsat Lake Quality Monitoring Program. *Photogrammetric Engineering and Remote Sensing*, Vol. 45, pp. 623–633.

Frazier, P. S., and K. L. Page. 2000. Water Body Detection and Delineation with Landsat TM Data. *Photogrammetric Engineering and Remote Sensing*, Vol. 66, pp. 1461–1465.

Han, L. 1997. Spectral Reflectance with Varying Suspended Sediment Concentrations in Clear and Algae-Laden Waters. *Photogrammetric Engineering and Remote Sensing*, Vol. 63, pp. 701–705.

Han, L., and D. C. Rundquest. 1997. Comparison of NIR/RED Ratio and First Derivative of Reflectance in Estimating Algal-Chlorophyll Concentration: A Case Study in a Turbid Reservoir. *Remote Sensing of Environment*, Vol. 62, pp. 253–261.

Han, L., and D. C. Rundquest. 2003. The Spectral Response of *Ceratophyllum demersum* at Varying Depths in an Experimental Tank. *International Journal of Remote Sensing*, Vol. 24, pp. 859–864.

Holden, H., and E. Ledrow. 1999. Hyperspectral Identification of Coral Reef Features. *International Journal of Remote Sensing*, Vol. 20, pp. 2545–2563.

Jackson, R. D. 1985. Evaluating Evapotranspiration at Local and Regional Scales. *Proceedings of the IEEE*, Vol. 73, pp. 1086–1096.

Jackson, T. J., R. M. Ragan, and W. N. Fitch. 1977. Test of Landsat-Based Urban Hydrologic Modeling. *Journal of Water Resources Planning and Management Division, American Society of Civil Engineers*. Vol. 103, No. WRI, pp. 141–158.

Khorram, S. 1980. Water Quality Mapping from Landsat Digital Data. *International Journal of Remote Sensing*, Vol. 2, pp. 143–153.

Klemas, V., R. Sicna, W. Treasure, and M. Otley. 1973. Applicability of ERTS-1 Imagery to the Study of Suspended Sediment and Aquatic Forms. In *Symposium on Significant Results Obtained from Earth Resources Technology Satellite-1*. Greenbelt, MD: Goddard Space Flight Center, pp. 1275–1290.

Lee, Z. P., K. L. Carder, and R. A. Arnone. 2002. Deriving Inherent Optical Properties from Water Color: A Multiband Quasi-Analytical Algorithm for Optically Deep Waters. *Applied Optics*, Vol. 41, pp. 5755–5772.

Liedtke, T. A., A. Roberts, and J. Lutenauer. 1995. Practical Remote Sensing of Suspended Sediment Concentration. *Photogrammetric Engineering and Remote Sensing*, Vol. 61, pp. 167–175.

Lillesand, T. M., E. L. Johnson, R. L. Deuell, O. M. Lindstrom, and D. E. Miesner. 1983. Use of Landsat Data to Predict the Trophic State of Minnesota Lakes. *Photogrammetric Engineering and Remote Sensing*, Vol. 49, pp. 219–229.

McKim, H. L., C. J. Merry, and R. W. Layman. 1984. Water Quality Monitoring Using an

Airborne Spectroradiometer. *Photogrammetric Engineering and Remote Sensing*, Vol. 50, pp. 353–360.

Moore, G. K. 1978. Satellite Surveillance of Physical Water Quality Characteristics. In *Proceedings of the Twelfth International Symposium on Remote Sensing of Environment*. Ann Arbor: Environmental Research Institute of Michigan, pp. 445–462.

Philipson, W. R., and W. R. Hafker. 1981. Manual versus Digital Landsat Analysis for Delineating River Flooding. *Photogrammetric Engineering and Remote Sensing*, Vol. 47, pp. 1351–1356.

Ritchie, J. C., P. V. Zimba, and J. H. Everitt. 2003. Remote Sensing Techniques to Assess Water Quality. *Photogrammetric Engineering and Remote Sensing*, Vol. 69, pp. 695–704.

Schmugge, T. J., W. P. Kustas, J. C. Ritchie, T. J. Jackson, and A. Rango. 2002. Remote Sensing in Hydrology. *Advances in Water Resources*, Vol. 25, pp. 1367–1385.

Schwab, D. J., G. A. Leshkevich, and G. C. Muhr. 1992. Satellite Measurements of Surface Water Temperatures in the Great Lakes. *Journal of Great Lakes Research*, Vol. 18, pp. 247–258.

Strandberg, C. 1966. Water Quality Analysis. *Photogrammetric Engineering*, Vol. 32, pp. 234–249.

Stumpf, R. P. 1992. Remote Sensing of Water Clarity and Suspended Sediment in Coastal Waters. In *Proceedings of the First Thematic Conference on Remote Sensing for Marine and Coastal Environments. Proceedings of the International Society of Optical Engineering*, Vol. 1930, pp. 1436–1473.

Tolk, B. L., L. Han, and D. C. Rundquist. 2000. The Impact of Bottom Brightness on Spectral Reflectance of Suspended Sediments. *International Journal of Remote Sensing*, Vol. 21, pp. 2259–2268.

Westawy, R., M. S. N. Lane, and D. M. Hicks. 2001. Remote Sensing of Clear-Water, Sallow, Gravel-Bed Rivers Using Digital Photogrammetry. *Photogrammetric Engineering and Remote Sensing*, Vol. 67, pp. 1271–1282.

Chromaticity Analysis

Alföldi, T. T., and J. C. Munday. 1977. Progress toward a Landsat Water Quality Monitoring System. In *Proceedings, Fourth Canadian Symposium on Remote Sensing*. Montreal, Quebec, Canada: Canadian Remote Sensing Society of the Canadian Aeronautics and Space Institute, pp. 325–340.

Markham, B. L., and J. L. Barker. 1986. Landsat MSS and TM Post-Calibration Dynamic Ranges, Exoatmospheric Reflectance, and At-Satellite Temperatures. *Landsat Technical Notes*, No. 1, pp. 3–8.

Munday, J. C., and T. T Alföldi. 1975. Chromaticity Changes from Isoluminous Techniques Used to Enhance Multispectral Remote Sensing Data. *Remote Sensing of Environment*, Vol. 4, pp. 221–236.

Munday, J. C., T. T. Alföldi, and C. L. Amos. 1979. Bay of Fundy Verification of a System for Multidate Landsat Measurement of Suspended Sediment. In *Satellite Hydrology* (M. Deutsch, D. R. Wiesner, and A. Rango, eds.). Minneapolis, MN: American Water Resource Association, pp. 622–640.

Nace, R. L. 1967. *Are We Running Out of Water?* (U.S. Geological Circular 586). Reston, VA: U.S. Geological Survey, 7 pp.

Nelson, R. 1985. Reducing Landsat MSS Scene Variability. *Photogrammetric Engineering and Remote Sensing*, Vol. 51, pp. 583–593.

Bathymetric Mapping

Hallada, W. A. 1984. Mapping Bathymetry with Landsat 4 Thematic Mapper: Preliminary Findings. In *Proceedings, Ninth Canadian Symposium on Remote Sensing.* Halifax, Nova Scotia, Quebec, Canada: Canadian Aeronautics and Space Institute, pp. 629–643.

Han, L. 1997. Spectral Reflectance with Varying Suspended Sediment Concentrations in Clear and Algae-Laden Waters. *Photogrammetric Engineering and Remote Sensing,* Vol. 63, pp. 701–705.

Jupp, D. L. B., K. K. Mayo, D. A. Kucker, D. Van, R. Classen, R. A. Kenchinton, et al. 1985. Remote Sensing for Planning and Managing the Great Barrier Reef of Australia. *Photogrammetria,* Vol. 40, pp. 21–42.

Lyzenga, D. R. 1979. Shallow-Water Reflectance Modeling with Applications to Remote Sensing of the Ocean Floor. In *Proceedings, Thirteenth International Symposium on Remote Sensing of Environment.* Ann Arbor: Environmental Research Institute of Michigan, pp. 583–602.

Lyzenga, D. R., N. M. Malinas, and F. J. Tarris. 2006. Multispectral Bathymetry Using a Single Physically Based Algorithm. *IEEE Transactions on Geoscience and Remote Sensing,* Vol. 44, pp. 2251–2259.

Lyzenga, D. R., R. A. Shuchman, and R. A. Arnone. 1979. Evaluation of an Algorithm for Mapping Bottom Features under a Variable Depth of Water. In *Proceedings, Thirteenth International Symposium on Remote Sensing of Environment.* Ann Arbor: Environmental Research Institute of Michigan, pp. 1767–1780.

Mishra, D., S. Narumalani, M. Lawson, and D. Rundquist. 2004. Bathymetric mapping using IKONOS multispectral data. *GIScience and Remote Sensing,* Vol. 41, pp. 301–321.

Polcyn, F. C., and D. R. Lyzenga. 1979. Landsat Bathymetric Mapping by Multispectral Processing. In *Proceedings, Thirteenth International Symposium on Remote Sensing of Environment.* Ann Arbor: Environmental Research Institute of Michigan, pp. 1269–1276.

Satzman, B. (ed.). 1985. *Satellite Oceanic Remote Sensing: Advances in Geophysics.* New York: Academic Press, 511 pp.

Tanis, F. J., and W. A. Hallada. 1984. Evaluation of Landsat Thematic Mapper Data for Shallow Water Bathymetry. In *Proceedings, Eighteenth International Symposium on Remote Sensing of Environment.* Ann Arbor: Environmental Research Institute of Michigan, pp. 629–643.

Sea Surface Temperature

McClain, E. P., W. G. Pichel, and C. C Walton. 1985. Comparative Performance of AVHRR-Based Multichannel Sea Surface Temperatures. *Journal of Geophysical Research,* Vol. 90, p. 11587–11601.

Reynolds, R. W. 1988. A Real-Time Global Sea Surface Temperature Analysis. *Journal of Climate,* Vol. 1, pp. 75–86.

Reynolds, R. W., C. K. Folland, and D. E. Parker. 1989. Biases in Satellite-Derived Sea Surface Temperature Data. *Nature,* Vol. 341, pp. 728–731.

Reynolds, R. W., and D. C. Marsico. 1993. An Improved Real-Time Global Sea Surface Temperature Analysis. *Journal of Climate,* Vol. 6, pp. 114–119.

Reynolds, R. W., and T. M. Smith. 1994. Improved Global Sea Surface Temperature Analyses Using Optimum Interpolation. *Journal of Climate,* Vol. 7, pp. 929–948.

Lidar

Allen, T. R., and Y. Wang. 2010. Selected Scientific Analyses and Practical Applications of Remote Sensing: Examples from the Coast. Chapter 24 in *Manual of Geospatial Sciences and Technology* (J. D. Bossler, ed.). London: Taylor & Francis, pp. 467–485.

Gao, J. 2009. Bathymetric Mapping by Means of Remote Sensing: Methods, Accuracy, and Limitations. *Progress in Physical Geography*, Vol. 33, pp. 103–116.

Houma, F., R. Belkessa, and N. Bachari. 2006. Contribution of Multispectral Satellite Imagery to the Bathymetric Analysis of Coastal Sea Bottom. *Revue des Energies Renouvelables*, Vol. 9, pp. 165–172.

Wang, C.-K., and Philpot, W. D. 2007. Using Airborne Bathymetric Lidar to Detect Bottom Type Variation in Shallow Waters. *Remote Sensing of Environment*, Vol. 106, pp. 123–135.

White, S. A., and Y. Wang. 2003. Utilizing DEMs Derived from LIDAR Data to Analyze Morphologic Change in the North Carolina Coastline. *Remote Sensing of Environment*, Vol. 85, pp. 39–47.

CHAPTER TWENTY

Land Use and Land Cover

20.1. Introduction

Land use describes use of the land surface by humans. Normally, use of land is defined in an economic context, so we think of land as it is used for agricultural, residential, commercial, and other uses. However, strictly speaking, we can seldom really see the use of the land, except on the very closest inspection, so we consider also *land cover*—the visible features of the Earth's surface—included in the vegetative cover, natural and as modified by humans, its structures, transportation and communications, and so on (Figure 20.1). As a practical matter, we must consider land use and land cover together, while also recognizing the distinction between the two.

Modern society depends on accurate land use data for both scientific and administrative purposes. They form an essential component of local and regional economic planning to ensure that various activities are positioned on the landscape in a rational manner. For example, accurate knowledge of land use patterns permits planning to avoid placing residential housing adjacent to heavy industry or in floodplains. Accurate land use information can ensure that residential neighborhoods are logically placed with respect to commercial centers and access to transportation services. In another context, land use is an important component of climatic and hydrologic modeling to estimate the runoff of rainfall from varied surfaces into stream systems. And accurate land use data are important for transportation planning, so traffic engineers can estimate the flow of vehicles from one region to another and design highways with appropriate capacities.

Land use patterns also reflect the character of a society's intimate interaction with its physical environment. Although land use patterns within our own society seem self-evident and natural, other societies often organize themselves in contrasting patterns. This fact becomes obvious when it is possible to observe different economic and social systems occupying similar environments. Plate 27 shows a Landsat TM quarter scene depicting Santa Rosa del Palmar, Bolivia (northwest of Santa Bruise) acquired in July 1992. This image illustrates the close relationship between land use patterns and the social and cultural histories of the communities occupying regions within the image. In the southeastern portion of the scene, the land use is dominated by broad-scale mechanized agriculture practiced by a Mennonite community. In contrast, upper right (northeastern) region of the image shows a mountainous area occupied by a diminishing population of Indians, who practice a form of slash-and-burn agriculture, visible as dispersed patches of light green. Throughout the northwestern and central portions of the image, the Bolivian government has encouraged broad-scale clearing of forest for agriculture.

LAND USE = ECONOMIC FUNCTION

LAND COVER = SURFACE FEATURES

FIGURE 20.1. Land use and land cover. *Land use* refers to the principal economic enterprises that characterize an area of land: agriculture, manufacturing, commerce, and residential. *Land cover* indicates the physical features that occupy the surface of the Earth, such as water, forest, and urban structures.

These regions are visible in the upper left region of the image as light green spots aligned northwest to southeast. At the center of each patch of cleared land is a central facility supporting colonists with fertilizers, pesticides, and staples. In the southwestern corner of the image, the complex field pattern reflects an established landscape occupied by Japanese immigrants. This clearly depicts in a single image the unique patterns that each cultural perspective has created as humans have occupied the Earth's surface.

20.2. Aerial Imagery for Land Use Information

Although some forms of land use data (local zoning or tax data) must be acquired through direct observation by officials who visit each site in person, the vast majority of land use and land cover data are acquired by interpretation of aerial photography and similar imagery. Aerial photography provides the overhead view, the spatial perspective, and the comprehensive detail that permits accurate, systematic, and effective study of land use patterns. Early proponents of the use of aerial photography in civil society envisioned that aerial photographs would provide a rich source of information regarding the structure of cities and landscapes. By the 1930s, the staff of the Tennessee Valley Authority was among the first to pioneer methods for using aerial photography for land use survey. Later, the USGS developed classification systems specifically tailored for use with aerial photography. These techniques form the foundations for today's methods of applying aerial imagery to land use survey.

Remotely sensed images lend themselves to accurate land cover and land use mapping in part because land cover information can be visually interpreted more or less directly from evidence visible on aerial images and because objects can be seen in the context of the neighboring features.

Land use maps are routinely prepared at a wide variety of scales, typically ranging from 1:12,500 and larger to 1:250,000 and smaller. At one end of this spectrum (the larger scale maps), remotely sensed imagery may itself contribute relatively little information to the survey; its main role may be to form a highly detailed base for recording data gathered by other means (Figure 20.2). At such large scales the land use map may actually serve as a kind of reference map, having little cartographic generalization. Such products are often used at the lowest levels of local governments, perhaps mainly in urban areas, that have requirements for such detailed information and the financial resources to acquire it.

As the scale of the survey becomes broader (i.e., as map scale becomes smaller), the contribution of the image to the informational content of the map becomes greater, although even at the smallest scales there must always be some contribution from collateral information. Differing scales and levels of detail serve different purposes and different users. For the regional planner, the loss of resolution and detail at smaller scales may actually form an advantage in the sense that often analysts sometimes prefer an integration and a simplification of the information that must be examined. For the medium- and small-scale land use surveys that are most often compiled by use of remotely sensed images, the product is a thematic map that depicts the predominant land cover within relatively homogeneous areas, delineated subject to limitations of scale, resolution, generalization, and other constraints accepted (or at least recognized) by both compilers and users of thematic maps.

20.3. Land Use Classification

Preparation of a land use–land cover map requires that the mapped area be subdivided in the discrete parcels, each labeled with a distinct, mutually exclusive, nominal label. That is, each parcel must be identified using a single, distinct label in a classification system. "Nominal" simply means that the labels are names rather than values. So, for example, when we use a symbol such as "22" to designate a category, we do not mean that it

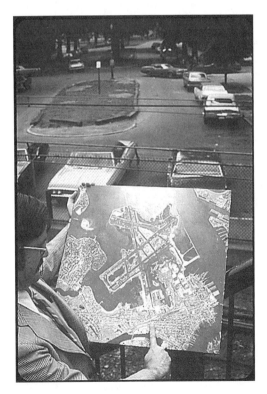

FIGURE 20.2. A community planning official uses an aerial photograph to discuss land use policy relating to planning land use near Boston's Logan Airport, 1973. From National Archives and Records Administration, ARC548443.

has twice the value of "11," but simply that the separate designations convey different qualitative meanings, such as "urban" and "agricultural" land uses. Usually, we also use colors, or colors and symbols, to designate land use classes on maps.

Perhaps the most widely used classification system for land use and land cover data derived from aerial photography is the U.S. Geological Survey's Land Use and Land Cover classification system, developed during the 1970s (Table 20.1).

This classification system has many advantages over previous systems. Whereas previous systems did not consider the unique advantages of aerial imagery as a source of land use and land cover data, the USGS system was specifically prepared for use with aerial photography and related imagery. Its categories are tailored for use with information derived from aerial images. It has a hierarchical design that lends itself to use with images of varied scales and resolutions. Level I categories, for example, are designed for broad-scale, coarse-resolution imagery (e.g., broad-scale satellite imagery or high-altitude aerial photography).

Although the USGS system specifies Level I and II categories, more detailed classes at Level III and below must be defined by the analyst to meet the specific requirements of a particular study and a specific region. When an interpreter defines Level III categories, the Level I and II categories should be used as a framework for the more detailed Level III classes.

20.4. Visual Interpretation of Land Use and Land Cover

Image interpreters apply the elements of image interpretation to delineate separate land cover classes, applied in an organized and systematic manner. Land use interpretation proceeds by marking boundaries between categories as they occur on the imagery. Often interpreters mark transparent overlays that register to the image, or digitize directly as the image is displayed on a computer screen. As each land use parcel is outlined, its identity is marked with a symbol (usually one to three numerals) matching to the classification system (Figure 20.3).

A land use map is then prepared by delineating regions of consistent land use and assigning them to appropriate classes in land use classification. An example can show an application of land use mapping to an aerial photograph, using principles outlined in Chapter 5.

Steps can be seen in Figure 20.4 (1) the original aerial photograph (A); (2) then, the land use boundaries, as defined by a photointerpreter, to separate land use parcels at Anderson Level II (B and C); (3) the separate parcels are then assigned to classes and symbolized with colors or symbols (C); and finally (4) a map is displayed with a legend (E). Land use–land cover information must be presented to the user in a planimetrically correct form, so the imagery that is used for the interpretation must be either corrected beforehand (e.g., as in the case of a DOQ), or the boundaries must later be rectified.

Identification of land cover parcels is based on the elements of image interpretation, as presented in Chapter 5. Although sometimes land use may be characterized by specific objects, usually the primary task is one of consistent delineation of uniform parcels that match the classification system. As an example, *cropped agricultural land* (Anderson symbol 21) (as it occurs in midlatitude landscapes) is usually easily recognizable by the systematic division of fields and the smooth, even textures. Tone varies with the crop and the growth stage (Figure 20.5).

TABLE 20.1. USGS Anderson et al. (1976) Land Use and Land Cover Classification

1 Urban or Built-Up Land
    11   Residential
    12   Commercial Services
    13   Industrial
    14   Transportation, Communications
    15   Industrial and Commercial
    16   Mixed Urban or Built-Up Land
    17   Other Urban or Built-Up Land
2 Agricultural Land
    21   Cropland and Pasture
    22   Orchards, Groves, Vineyards, Nurseries
    23   Confined Feeding Operations
    24   Other Agricultural Land
3 Rangeland
    31   Herbaceous Rangeland
    32   Shrub and Brush Rangeland
    33   Mixed Rangeland
4 Forest Land
    41   Deciduous Forest Land
    42   Evergreen Forest Land
    43   Mixed Forest Land
5 Water
    51   Streams and Canals
    52   Lakes
    53   Reservoirs
    54   Bays and Estuaries
6 Wetland
    61   Forested Wetlands
    62   Nonforested Wetlands
7 Barren Land
    71   Dry Salt Flats
    72   Beaches
    73   Sandy Areas Other than Beaches
    74   Bare Exposed Rock
    75   Strip Mines, Quarries, and Gravel Pits
    76   Transitional Areas
    77   Mixed Barren Land
8 Tundra
    81   Shrub and Brush Tundra
    82   Herbaceous Tundra
    83   Bare Ground
    84   Wet Tundra
    85   Mixed Tundra
9 Perennial Snow and Ice
    91   Perennial Snowfields
    92   Glaciers

Source: Anderson, J. R., Hardy, E. E., Roach, J. T. and Witmer, R. E., 1976, *A Land Use and Land Cover Classification System for Use with Remote Sensor Data*. U.S. Geological Survey, Professional Paper 964, p. 28, Reston, VA.

Pasture (Anderson symbol 21) usually has more irregular boundaries and a mottled texture with medium tones and is often characterized by isolated trees or small groves both at edges and in the interiors of parcels (Figure 20.6). *Deciduous forest* (Anderson symbol 41) is characterized by rough textures and medium dark tones, occurring usually in relatively large parcels with irregular edges (Figure 20.7). Roads and clearings are common.

FIGURE 20.3. Land use and land cover maps. (a) Large scale, fine detail; (b) small scale, coarse detail.

FIGURE 20.4. Representation of key steps in interpretation of land use from aerial photography. (A) Aerial photograph only; (B) delineation of boundaries; (C) symbolization of principal classes; (D) boundaries without the backdrop of the areal photograph; (E) map with symbols and key. From Jessica Dorr.

FIGURE 20.5. Visual interpretation of aerial imagery for land use–land cover information: cropped agricultural land (Anderson symbol 21). Distinctive characteristics include smooth texture and the contrasting tones that separate varied crops and growth stages. Irrigated cropland, western Kansas, principally wheat, alfalfa, maize, and sorghum. From USDA.

FIGURE 20.6. Visual interpretation of aerial imagery for land use–land cover information: pasture (Anderson symbol 21). Pastureland is often recognizable by its medium gray tones, often without sharp boundaries, and the trees that occur in small groves or isolated individuals. Fields at top center are cropped agricultural land. Copyright Virginia Department of Transportation. Used by permission.

FIGURE 20.7. Visual interpretation of aerial imagery for land use–land cover information: deciduous forest (Anderson symbol 41). In the growing season, deciduous forests are usually characterized by a coarse texture, dark tones, and large, uneven parcels. Copyright Virginia Department of Transportation. Used by permission.

Land in transition (Anderson symbol 76) is usually recognizable by the bright tones characteristic of bare soil exposed during construction and irregular outlines. Sometimes, the outlines of roadways, foundations for buildings, and partially completed structures are visible (Figure 20.8).

Transportation (Anderson symbol 14) is often recognizable by linear patterns that cut across the predominant land use pattern (Figure 20.9) and by the distinctive loops of interchanges. *Residential* (Anderson symbol 11) land uses, especially in suburban developments, are often distinctive because of the even placement of structures, the curving street patterns, and the even background of lawns and ornamental trees (Figure 20.10). *Commercial and services* (Anderson symbol 12) is characterized by bright tones (because of the predominance of pavement, parking lots, and rooftops), uneven textures (because of shadowing of buildings), and close proximity to transportation and other high-density land use classes (Figure 20.11).

As the map is prepared, remember that some land use maps have legal standing and that some maps may be intended to match data collected previously or by a neighboring

FIGURE 20.8. Visual interpretation of aerial imagery for land use–land cover information: land in transition (Anderson symbol 76). Land in transition is often characterized by bright tones, uneven texture, and irregular parcels. Copyright Virginia Department of Transportation. Used by permission.

FIGURE 20.9. Visual interpretation of aerial imagery for land use–land cover information: transportation (Anderson symbol 14). Divided highways display the linear forms and the distinctive loops at interchanges shown in this photograph. Copyright Virginia Department of Transportation. Used by permission.

jurisdiction. Therefore, the interpreter should not make off-the-cuff adjustments in procedures to solve needs of local interpretation problems.

Interpreters must take care to define delineations that are consistent, clear, and legible. Each parcel is completely enclosed by a boundary and is labeled with a symbol keyed

FIGURE 20.10. Visual interpretation of aerial imagery for land use–land cover information: residential land (Anderson symbol 11). Suburban residential land, such as that shown here, is often characterized by evenly spaced rooftops and systematic patterns of paved streets. Copyright Virginia Department of Transportation. Used by permission.

FIGURE 20.11. Visual interpretation of aerial imagery for land use–land cover information: commercial and services (Anderson symbol 12). In this instance, the identifying characteristics are the large commercial buildings positioned in a linear form, with large parking lots nearby. Copyright Virginia Department of Transportation. Used by permission.

to the classification system. The interpreter delineates only those features that occupy areas at the scale of the final map, so usually linear features (such as streams and rivers, railways, and power lines) or point data are not mapped.

Some parcels will encompass categories other than those named by the parcel label. Such inclusions are permissible but must be clearly described in an accompanying report and must be consistent throughout the map. When several interpreters work on the same project, consistency is especially important. Interpreters must coordinate their work with those assigned to neighboring areas, so that detail is uniform throughout the mapped area.

The entire area devoted to a specific use is delineated on the overlay. Thus the delineation of an airfield normally includes not only the runway but also the hangar, passenger terminals, parking areas, access roads, and, in general, all features inside the limits of the perimeter fence (i.e., the outline of the parcel encompasses areas occupied by all of these features, even though they are not shown individually on the map). In a similar manner, the delineation of an interstate highway includes not only the two paved roadways but also the median strip and the right of way.

The example of an aerial image of a quarry can also illustrate this practice (Figure

20.12). The quarry itself consists of several features, each with a distinctive appearance on the aerial photograph. Even though it would be possible to individually delineate the transportation and loading structures, the milling and administration buildings, the quarries, and the spoil, these features constitute a single land use, so all the features are outlined and labeled as a single parcel. Only at the very largest scales, possibly required for planning and management of the facility itself, would the individual components be delineated as separate features.

The issue of *multiple use* occurs because we assign a single label to each parcel, even though we know that there may in fact be several uses. For example, a forested area may

FIGURE 20.12. Delineation of the entire area devoted to a given use. This quarry consists of a complex of individual activities that together constitute a single enterprise, so the individual components need not be delineated separately. Copyright Virginia Department of Transportation. Used by permission.

simultaneously serve as a source of timber, as a recreational area for hunters and hikers, and as a source of runoff that supplies water for an urban region.

The interpreter must select an appropriate *minimum size* for the smallest parcels to be represented on the final map. Areas smaller than this size will be omitted. Even though they are visible to the interpreter, they are too small to depict legibly.

Usually the label of each category identifies the *predominant category* present within each parcel. Although at small scales there may be unavoidable inclusions of other categories, the interpreter should aspire to define categories that include consistent mixtures of such inclusions.

20.5. Land Use Change by Visual Interpretation

Land use patterns change over time in response to economic, social, and environmental forces. The practical significance of such changes is obvious. For planners and administrators, they reveal the areas that require the greatest attention if communities are to develop in a harmonious and orderly manner. From a conceptual perspective, study of land-use changes permits identification of long-term trends in time and space and the formation of policy in anticipation of the problems that accompany changes in land use (Anderson, 1977; Estes and Senger, 1972; Jensen and Toll, 1982).

Remote sensing and photointerpretation provide the primary vehicles for compiling *land use change maps*—maps that represent changes in land use from one date to another. Such maps form important tools in planning land use policy, in hydrologic management of watersheds, in transportation planning (to understand changes in the population patterns that generate traffic flow), and in environmental studies.

Land use and land cover change is very simple in concept: Two maps representing the same region, prepared to depict land use patterns at different dates, are compared, point by point, to summarize differences between the two dates (Figure 20.13). In practice, compiling land use and land cover change requires mastery of several practical procedures that often reveal previously hidden difficulties.

First, the two maps must use the same classification system, or at least two classification systems that are compatible. Compatible classifications mean that the classes can be clearly matched to one another without omission or ambiguity. For example, if one map represents forested land as a single category, "forest," and a second uses three classes—"coniferous," "deciduous"—and "mixed coniferous/deciduous," then the two maps are compatible in the sense that the three categories of the second maps can be combined into

FIGURE 20.13. Schematic representation of compilation of land use change using sequential aerial imagery.

a single forested class that is logically compatible with the "forested" class on the first map. As long as the classes are matched at that level of generalization, one can be confident that the comparisons would reveal changes in the extent of forested land.

However, if the second map employs classes such as "densely forested," "partially forested," and "sparsely forested," we could not reasonably match these classes to the "forested class" on this first map—the two sets of classes are not logically compatible. The meanings conveyed by these classes, as represented on the two maps, are not equivalent, so they cannot be compared to determine whether the differences represent true change or simply differences in the way the maps were prepared.

Second, the two maps must be compatible with respect to scale, geometry, and level of detail. Can we match one point on one map with the corresponding point on the second and be sure that both points refer to the same place on the Earth's surface? If not, we cannot conduct a change analysis using the two maps because we cannot be sure that differences in the two maps reflect genuine differences in land use and not simply differences in the projection, scale, or geometric properties of the two maps.

These two basic conditions must be satisfied before change data can be complied using any two maps or images: compatibility of the information portrayed and geometric compatibility.

Often, land use maps are prepared to meet the very specific objectives of the sponsoring organization. Therefore, it may be very difficult to reuse that map in a land use change study many years later. Objectives may have changed, or the level of detail, the classification system, and so forth, reflect the needs of a previous era. It is for this reason that it is common in land use change studies for maps to be prepared at the time of the study, using consistent materials, methods, and procedures, even if earlier studies already exist and are available.

20.6. Historical Land Cover Interpretation for Environmental Analysis

A special application of land use change analysis has been developed to address environmental hazards arising from land use change in regions where hazardous industrial materials have been abandoned. Aerial imagery is especially valuable because it provides a spatial perspective and has the ability both to provide a detailed historical record and to observe outside the visible spectrum. These capabilities allow analysts to examine problems over time, to detect the existence of specific environmental problems, and to see them in their geographical context. Image analysis can provide a historical record of conditions and actions at specific sites to assess the nature and severity of conditions, often when other sources of information are not available.

Over the decades, in industrial regions of the United States, hazardous industrial wastes have been deposited in varied disposal sites, including ponds, lagoons, and landfills that were not designed to safely store dangerous substances. Although many of these areas were once positioned in rural or sparsely populated regions, urban growth has brought many of these sites close to residential neighborhoods. As a result, many such sites are now near populated regions and in some instances have been converted to other uses, including residential housing, schoolyards, and recreational areas. Deterioration of containers has released these materials to contaminate nearby soils and to enter groundwater systems to migrate beyond the original site. Often, the use of the site for waste disposal was not properly recorded or the nature of the wastes was never documented, so remediation is limited by ignorance of specifics of the situation.

Because of the difficulty of establishing the historical pattern of use at such sites and the absence or inaccuracy of official records, aerial photography has formed one of the most valuable resources for investigating such sites. For many areas in the United States, the archives of aerial photography extend as far back as the 1930s. Thus it may be possible to assemble photographs to form a sequence of snapshots that record the use of the site over time. Such images, when combined with field data, official records, and health surveys, can permit development of an understanding of the sequence of events and assessment of the risks to the environment and to nearby populations. In some instances use of photography permits estimation of the amount of waste present and even specification of the class of materials by identification of the kinds of containers used. Aerial photography permits development of an understanding of the relationship of each site with respect to nearly drainage, population, economic activities, and other waste disposal sites.

A sequence of photographs (Figure 20.14) illustrates the EPA's ability to track the history of waste disposal sites. The most recent photograph shows how current land use can mask evidence that reveals the history of the site's use for waste disposal. These photographs show the Elmore Waste Disposal site and the Sunnyside Dump, Spartanburg County, South Carolina. These sites are outlined on each photograph. The Elmore Waste Disposal site is about one-half acre in area, situated near a residential neighborhood near Greer. The Elmore site was designated as a superfund site, so it has been monitored by the Environmental Protection Agency (EPA).

1955

1981

DETAIL OF ACTIVE
LANDFILL. 1955

DETAIL OF RESIDENTIAL
CONSTRUCTION. 1965

FIGURE 20.14. Aerial photographs depicting land use changes at the Elmore/Sunnyside waste disposal area in Spartanburg County, South Carolina. These photographs, selected from a longer sequence, show the active landfill (1955, left), newly constructed residential housing at the same site (1965, center), and the mature residential neighborhood (1981, right). Historical aerial photography permits unraveling of the historical sequence that is no longer visible. From Environmental Protection Agency, Environmental Photographic Interpretation Center.

It is now known that a large number of drums containing liquid wastes were placed at this site between 1975 and 1977. In 1977, the property owner signed a consent order with the State of South Carolina for cleanup of the site. However, state health and environmental officials judged the actions taken to be inadequate for the situation and ordered the owner to stop use of the site. Between 1981 and 1984, state and federal EPA officials investigated the site and found arsenic, chromium, and other heavy metals, as well as a number of volatile organic compounds (VOCs), in site soils.

At various times during this interval, there were between 150 and 300 drums present onsite, as well as a 6,000-gallon tank, partially buried, containing contaminated waste. When the owner died in 1983, ownership passed to his heirs, one of whom continued to operate the site and accept waste drums. After many failed efforts to compel this owner to clean the site, in June 1986 state officials completed a state-funded removal of 5,500 tons of contaminated soil and 16,800 pounds of liquid wastes, which were taken to an appropriate hazardous waste facility.

After this removal, groundwater monitoring wells and EPA's remedial investigation in 1991–1992 identified a contaminated groundwater plume extending some 700 feet north from the site. The estimated area underlain by this plume (north of the Elmore property) is 6–10 acres. Although no private water wells are located near this plume, the groundwater discharges to a creek. Additionally, surface soil in a one-quarter acre area at one end of the Elmore property was found to be contaminated by lead and arsenic at levels exceeding health-based residential standards for those metals.

Aerial photography provides documentation of these events and an overview that places the disposal site, the residential areas, and the drainage system in their spatial context.

20.7. Other Land Use Classification Systems

A land use classification system is not simply a list of categories. It must organize its classes according to an underlying logic that defines the relationships between classes. The organizing logic must impose a consistent organization on the classes. Classes must be mutually exclusive, and subclasses must nest within each other to form the hierarchical structure that we encountered in the Anderson system, for example. Further, the distinctions between the classes must be consistent from class to class. For example, if some classes are defined on the basis of economic function, then the entire classification structure must reflect that logic.

General-Purpose Land Use Classification

The Andersen (USGS) land use classification system (Table 20.1) is designed specifically for use with aerial photography and related imagery. It is a good example of a *general-purpose land use classification system*, intended to provide a comprehensive classification of land use. General-purpose classifications are among the most widely used and are probably the classifications most likely to be encountered by photointerpreters. General-purpose land use classification attempts to provide a classification that serves many purposes, although it is not specifically tailored for any specific application.

There were several predecessors to the Andersen system, with respect to scope and purpose.

• L. Dudley Stamp, a British economist, organized a national survey of the land use of Britain during the interval 1931–1938 (Table 20.2). He gathered large-scale land use information for Britain using volunteers, who recorded land use near their residences by annotating large-scale topographic maps, which were then mailed to a central office where information was edited and posted to maps at 1:63,360. Although completion of this ambitious project was delayed by World War II, the information was still valuable (despite the delay) when it was used in planning Britain's economic recovery after the war.

• In the United States, during the same interval, photointerpreters at the *Tennessee Valley Authority* (TVA) were defining procedures to interpret land use from aerial photography. TVA photointerpreters devised a system (not now in use) that classified each parcel according to both its economic use and its physical properties.

• During the mid-1960s scientists at Cornell University designed New York's Land Use and Natural Resources (LUNR) Survey, a database to record land use and related information for the entire state of New York. Much of the information was interpreted from aerial photography (1:24,000 panchromatic aerial photography). Photointerpreters coded information according to predominant land use within 1-km cells, using 100 classes defined specifically by the LUNR project (Table 20.3). This project was one of the first computerized land data inventories, forming an early precedent for the geographic information systems now in use today.

Special-Purpose Land Use Classification

Another approach to land use classification—*special-purpose classifications*—is also significant. Special-purpose land use classifications are designed to address a specific classification issue, with no attempt to provide comprehensive scope.

A good example of a special-purpose land use classification system, and one that is of practical significance to photointerpreters, is the wetlands classification developed by Cowardin (1979), devoted specifically to classification of wetlands (Table 20.4). Its logic therefore reflects the concerns of scientists who wish to portray the distribution of wetlands in detail and to indicate specific characteristics of wetlands that are of significance for the hydrologist and the ecologist. For example, in the Anderson system, wetlands at levels I and II are subdivided simply as the distinction between forested and nonforested wetlands, whereas in Cowardin's system, wetlands are distinguished first by two hydrologic criteria ("Marine," "Riverine," etc., and "Tidal," "Subtidal," etc.), and then by geo-

---

TABLE 20.2. Land Utilization Survey of Britain (Stamp, 1951)

Forest and woodland
  High Forest (specified as Coniferous, Deciduous, or Mixed)
  Coppice
  Scrub
  Forest Cut and Not Replanted
Meadowland and permanent grass
Arable or Tilled land
Fallow Land
Heathlands, Moorland, Commons, Rough Hill Pasture
Gardens
Land Agriculturally Unproductive, Buildings, Yards, and Mines
Ponds

TABLE 20.3. New York's Land Use and Natural Resources (LUNR) Inventory Classification (Hardy et al., 1971)

Area Land Use
A	Agriculture (10 subclasses)
F	Forestland (3 subclasses)
W	Water (3 subclasses)
W	Wetlands (3 subclasses)
N	Non-productive Land (2 subclasses)
P	Public
R	Residential (7 subclasses)
S	Shoreline
C	Commercial (4 subclasses)
I	Industrial (2 subclasses)
OR	Outdoor recreation
E	Extractive (7 subclasses)
T	Transportation (7 subclasses)

Point Data
   Water (3 classes)
   Livestock (3 subclasses)
   Water (5 subclasses)
   Specialty Agriculture (4 subclasses)
   Residential Public land
   Airport
   Outdoor Recreation (10 subclasses)
   Communication (5 subclasses)
   Underground extractive

Linear Data Civil Divisions
   Civil Division Boundaries (7 subclasses)
   Streams, Rivers
   Lake, Reservoir, Shoreline,
   Highways (7 subclasses)
   Barge Canal
   Public Land
   Railways (5 subclasses)

morphic and ecologic criteria ("Reef," "Streambed," etc.). Special-purpose systems such as Cowardin's can be seen as specialized alternatives to the general purpose strategy, or, in some instances, they might serve as Level III or IV within a hierarchical system such as the Anderson system.

It is important to select or design a system that is tailored to the needs of the client who will use the data. It is especially important to consider compatibility with previous classification systems if the results are to be compared with data from earlier dates and to consider compatibility with neighboring jurisdictions or with higher or lower units (such as state or city systems, in the case of a county survey), as it may be important for the classes to match with these other datasets. A careful selection of classification that meets immediate needs may not be ideal if does not permit comparison with data from these other units.

20.8. Land Cover Mapping by Image Classification

Land cover can be mapped by applying image classification techniques discussed in Chapter 12 to digital remote sensing images. In principle, the process is straightforward;

TABLE 20.4. Wetland Classification (Cowardin et al., 1979)

Systems
   1. Marine
   2. Esturaine
   3. Riverine
   4. Lacustrine
   5. Palustrine
Subsystems
   1. Subtidal
   2. Intertidal
   3. Tidal
   4. Lower Perennial
   5. Upper Perennial
   6. Intermittent
   7. Limnetic
   8. Littoral
Classes
   1. Rock Bottom
   2. Unconsolidated bottom
   3. Aquatic bed
   4. Reef
   5. Rocky shore
   6. Unconsolidated Shore
   7. Streambed
   8. Emergent Wetland
   9. Scrub-Shrub Wetland
 10. Forested Wetland
 11. Moss-Lichen Wetland

in practice, many of the most significant factors are concealed among apparently routine considerations:

1. *Selection of images.* Success of classification for land cover analysis depends on the astute selection of images with respect to season and date. Therefore, the earlier discussion (Section 6.9) of the design and interpretation of searches of image archives, although ostensibly mundane in nature, assumes vital significance for the success of a project. What season will provide the optimum contrasts between the classes to be mapped? Possibly two or more dates might be required to separate all the classes of significance.

2. *Preprocessing.* Accurate registration of images and correction for atmospheric and system errors (Chapter 11) are required preliminary steps for successful classification. Subsetting of the region to be examined requires careful thought, as discussed in Section 11.7.

3. *Selection of classification algorithm.* The discussion in Chapter 12 reviewed many of the classification algorithms available for land cover analysis. Selection of the classification procedure should also be made on the basis of local experience. AMOEBA, for example, tends to be accurate in landscapes dominated by large homogeneous patches, such as the agricultural landscapes of the midwestern United States, and less satisfactory in landscape composed of many smaller heterogeneous parcels, such as those found in mountainous regions. Local experience and expertise are likely to be more reliable guides for selection of classification procedures than

are universal declarations about their performance. Even when there is comparative information on classification effectiveness, it is difficult to anticipate the balance between effects of the choice of classifier, selection of image date, characteristics of the landscape, and other factors.

4. *Selection of training data.* Accurate selection of training data is one process that is universally significant for image classification, as we emphasized in Chapter 12. Training data for each class must be carefully examined to be sure that it is represented by an appropriate selection of spectral subclasses to account for variations in spectral appearance due to shadowing, composition, and the like. Many individual laboratories and image analysis software packages have applied unsupervised classification in various forms to define homogeneous regions from which to select training fields for supervised classification (e.g., Chuvieco and Congalton, 1988). Another approach to the same question is an algorithm that permits the analyst to select a pixel or group of pixels that forms the focal point for a region that grows outward until a sharp discontinuity is encountered. This process identifies a region of homogenous pixels from which the analyst may select training fields for that class.

5. *Assignment of spectral classes to informational classes.* Because of the many subclasses that must be defined to accurately map an area by digital classification, a key process is the aggregation of spectral classes and their assignment to informational classes. For example, accurate classification of the informational class *deciduous forest* may require several spectral subclasses, such as *north-facing forest*, *south-facing forest*, *shadowed forest*, and the like. When the classification is complete, these subclasses should be assigned a common symbol to represent a single informational class.

6. *Display and symbolization.* The wide range of colors that can be presented on color displays and the flexibility in their assignment provides unprecedented opportunity for effective display of land cover information. Although unconventional choices of colors can sometimes be effective, it is probably sensible to seek some consistency in symbolization of land cover information to permit users to quickly grasp the meaning of a specific map or image without detailed examination of the legend. Therefore, the color symbols recommended by Anderson et al. (1976) may be useful guides. Another strategy for assignment of colors to classes is to mimic the colors used for USGS 7.5-minute quadrangles.

Within such general strategies, it is usually effective to assign related colors to related classes to symbolize Level II and Level III categories. For example, subclasses of agricultural land can be represented in shades of brown (using Anderson et al.'s strategy as a starting point), water and wetlands in shades of blue, subclasses of forest in shades of green, and so forth.

20.9. Broad-Scale Land Cover Studies

The availability of multispectral AVHRR data (and data from similar meteorological satellites) on a regular basis has provided the capability to directly compile broad-scale land cover maps and data. In this context, *broad scale* refers to images that represent entire continents, or even entire hemispheres, based on data collected over a short period of time, perhaps about 10 days to 2 weeks. Previously, data for such large regions could be acquired only by generalizing more detailed information—a task that was difficult and inaccurate because of the incompleteness of coverage and the inconsistencies of the many detailed maps required to prepare small-scale maps. The finer resolution data from the Landsat and SPOT systems provide information of local and regional interest but are not really suitable for compilation of data at continental scales because of the effects of cloud cover, differences in sun angle, and other factors that prevent convenient comparisons and mosaicking of many scenes into a single dataset representing a large region.

AVHRR data, described in Section 17.9, provide coverage of entire continents over relatively short time periods. Accumulation of data over a period of a week to 10 days usually permits each pixel to be observed at least once under cloud-free conditions. Although the scan angle varies greatly, data are acquired at such frequent intervals that it is often possible to select coverage of the region of interest from the central section of each scene to reduce the effects of the extreme perspective at the edges of each scene.

Tucker et al. (1985) examined AVHRR data for Africa, using images acquired over a 19 month time interval. They examined changes in the vegetation index (the "normalized difference" described in Chapter 12) at intervals of 21 days; their results clearly illustrate the climatic and ecological differences between the major biomes, as observed using the vegetation index and seasonal variations in the vegetation index. They conducted a second analysis using eight of the 21-day segments selected from the 19-month interval mentioned above. They calculated principal components (Chapter 11) as a means of condensing the many variables into a concise yet potent dataset that describes both seasonal and place-to-place variations in the vegetation index. Their land cover map (based on the first three principal components) is an extraordinary representation of key environmental conditions over an entire continent. It shows what are clearly major climate and ecological zones, as defined by conventional criteria, although they have been derived from data that are completely independent of the usual climatic data. The map has a uniformity, a level of detail, and (apparently) an accuracy not obtainable using conventional vegetation and climatic analysis. Although the authors have not fully evaluated their map by systematic comparison with field observations, it seems clear that products of this sort will provide an opportunity to examine broad ecological zones at continental scales and, furthermore, to examine seasonal and year-to-year changes in such zones.

More recently, AVHRR data have been used to compile other kinds of broad-scale land cover maps, to be discussed in Chapter 21.

20.10. Sources of Compiled Land Use Data

During the past decades, remote sensing has contributed greatly to the ability of organizations to prepare comprehensive land cover surveys of large regions. Satellite imagery in particular has provided near-simultaneous acquisition, broad areal coverage, uniform detail, and other qualities not available from alternative sources. In the United States, several governmental initiatives have exploited this capability to prepare broad-scale land cover surveys.

Land Use and Land Cover

In the 1970s the USGS developed a national land cover mapping program (described in sections 20.4 and 20.5 and in Anderson et al. [1976] and Campbell [1983]) based on manual interpretation of high-altitude aerial photography at scales of 1:60,000 and smaller. The resulting USGS Land Use and Land Cover (LULC) data files describe the vegetation, water, natural surfaces, and cultural features for large regions of the United States. The USGS National Mapping Program provides these datasets and associated maps using map bases at 1:250,000. (In some cases, such as Hawaii, 1:100,000 scale maps are also used as the base.) For more information, see

http://edc.usgs.gov/glis/hyper/guide/1_250_lulc#lulc10

Multiresolution Land Characteristics Consortium

To economize in the acquisition of image data, in 1992 several agencies of the U.S. government formed a consortium designed to acquire satellite-based remote sensor data for their separate environmental monitoring programs. The MultiResolution Land Characteristics (MRLC) Consortium was sponsored originally by the USGS, the U.S. EPA, the NOAA, and the U.S. Forest Service (USFS). MRLC was designed as an effort to pool financial resources to acquire satellite remote sensing imagery data in a cost-effective manner. Mapping efforts are based on TM data, and future plans are directed at acquisition of Landsat 7 ETM data for all 50 states, to be used as the basis for land cover mapping. For more information, see

www.epa.gov/mrlc/nlcd.html

During the 1990s, MRLC supported land cover mapping programs such as the Gap Analysis project and the National Land Cover Data (NLCD) project, both mentioned below.

Gap Analysis

In the United States, a national program to assess biodiversity and threats to biodiversity has been based on assessment of land cover derived from multispectral Landsat TM data. The objective of *gap analysis* is to identify "gaps" in biodiversity by surveying land cover patterns to assess cover types. For example, regions with high species diversity are often associated with high ranges of elevations, contrasts in soil characteristics, and the presence of varied vegetation classes. Gap analysis permits identification of high priorities for conservation and for development and land-use policies (Scott and Jennings, 1998).

Each state conducts its own survey, using its own classification, mapping, and accuracy assessment strategy, developed in collaboration with neighboring states. The national project is coordinated by the National Gap Analysis Program within the USGS Biological Resources Division (USGS-BRD). For more information, see

www.gap.uidaho.edu

National Land Cover Dataset 2001 (NCLD 2001)

One of the projects sponsored by the MRLC Consortium was production of land cover data for the conterminous United States. Land cover was mapped using general land cover classes. For example, forest is classified as either deciduous, evergreen, or mixed. Land cover classification was based on the Landsat 5 TM data archive and a host of ancillary sources. Once completed, NLCD will be the first national land cover dataset produced since the early 1970s, effectively replacing the land cover data known to many as LUDA or GIRAS. Available data have been used for many environmental applications, including land use planning, hydrological analyses, habitat assessments, and many others. Assessments of the accuracy of the data are undertaken after completion of large regions.

The MRLC produced the National Land Cover Dataset (NLCD) based on Landsat imagery circa 1992, based loosely on Anderson et al. (1976) Level II classes. In 2001 MRLC was expanded (MRLC 2001) to include additional institutions, including the Bureau of Land Management (BLM), NASA, the National Park Service (NPS), and the U.S. Fish and Wildlife Service (USFWS).

MRLC 2001 used multitemporal imagery from three dates to prepare the land cover dataset called NLCD 2001. NLCD 2001 is intended to be flexible enough to meet the needs of a spectrum of users. It is based on a database designed to provide intermediate products tailored for local applications (users can access materials to prepare their own classifications). It is designed to apply simple, efficient, and consistent methods and to maintain compatibility with earlier products, such as the NLCD.

MRLC 2001 is based on normalized TM imagery for three seasons (spring, summer, and fall) and ancillary data, especially DEMs (used to prepare slope, aspect, and slope position). Imagery and ancillary DEM products were used to prepare intermediate products for each pixel, including (1) normalized tasseled cap (TC) transformations for each season, (2) DEM data and related slope and aspect, (3) per-pixel estimates of percent imperviousness and percent tree canopy (Figure 20.15), (4) 29 classes of land cover data based on the imagery, and (5) metadata and classification rules

These intermediate products were then used to form a database that could be employed to derive land use and land cover products tailored to meet the specific needs of individual institutions. This approach differs from the classical classification approach strategies described in Chapter 12 in which a single classification was intended to serve the needs of diverse users. The database approach permits each institution to tailor its classification in detail to suit its specific needs within the broad framework of the more general classes.

This approach is applied to the U.S. states and Puerto Rico by defining 66 mapping zones (Figure 20.16) according to physiography and overall landscape characteristics. Studies have shown that land cover mapping accuracies can be improved by stratifying by physiographic provinces. Therefore, the 66 regions are designed to provide a degree of consistency in implementing the classification system.

Further details are presented by Homer et al. (2004), and at *www.mrlc.gov.*

20.11. Summary

Study of land use and land cover reveals the overall pattern of mankind's occupation of the Earth's surface and the geographic organization of its activities. At broad scales,

Legend

percent/pixel

1 - 10
11 - 20
21 - 30
31 - 40
41 - 50
51 - 60
61 - 70
71 - 80
81 - 90
91-100

Legend

percent/pixel

1 - 20
21- 40
41 - 60
61 - 80
81-100

TREE CROWN COVER   IMPERVIOUSNESS

FIGURE 20.15. Sample images illustrating tree canopy and imperviousness data used for NLCD 2001. Left: percent tree canopy cover for each pixel (for a mountainous area in Utah). Right: Imperviousness, derived from TM data (for a built-up region in Virginia). From USGS, Homer et al. (2004).

the land cover map provides a delineation of the broad patterns of climate and vegetation that form the environmental context for our activities. At local and regional scales, knowledge of land use and land cover forms a basic dimension of resources available to any political unit; both the citizens and the leaders of any community must understand the land resources available to them and the constraints that limit uses of land and environmental resources.

Although the formal study of land use and land cover dates from the early 1800s, systematic mapping at large scale was not attempted until the 1920s, and aerial photography and remote sensing were not routinely applied until the 1960s. Thus effective land

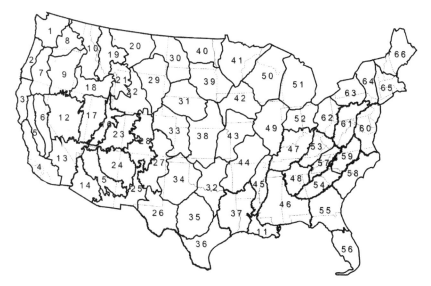

FIGURE 20.16. The 66 regions within the United States defined to provide consistent terrain and land use conditions for the NLCD 2001. From USGS, Homer et al. (2004).

use mapping is a relatively recent capability. We have yet to fully assemble and evaluate all of the data that are available and to develop the techniques for acquiring and interpreting imagery.

Without the aerial images acquired by remote sensing, there can be no really practical method of observing the pattern of land cover or of monitoring changes. Systems such as MSS, TM, SPOT, and AVHRR have provided a capability for observing land cover at broad scales and at intervals that previously were not practical. Images from such systems have not only provided vital information but have also presented data at new scales that have changed the intellectual perspective with which we consider the environment by recording broad-scale patterns and relationships that otherwise could not be accurately perceived or analyzed.

20.12. Some Teaching and Learning Resources

- Growth in Las Vegas 1984–2009
  www.youtube.com/watch?v=-xwquYkf43o&NR=1

- Spot Image—Shanghai: Evolution of the City from 1994 to 2010
  www.youtube.com/watch?v=bJNlNwAQib4&feature=channel

- NASA|Science for a Hungry World: Part 1
  www.youtube.com/watch?v=1RJ6AqWAOEg&feature=channel

- NASA|Science for a Hungry World: Part 2
  www.youtube.com/watch?v=3rucWGuj0Lk&feature=channel

- NASA|Science for a Hungry World: Part 3
  www.youtube.com/watch?v=GXuS9gX7CRM

- NASA|Science for a Hungry World: Part 4
  www.youtube.com/watch?v=HDQ7XIOIAfY

- NASA|Science for a Hungry World: Part 5
  www.youtube.com/watch?v=o1QsCa7RmmU&feature=channel

- Landsat Land Use Change Deforestation ASTER
  www.youtube.com/watch?v=uHz3J-iehq0&feature=related

Review Questions

1. Using aerial photographs and other information provided by your instructor, design Level III categories (compatible with the USGS classification) for a nearby region.

2. Review Chapters 6 and 10 to refresh your memory of resolutions of satellite sensors and the effects of mixed pixels. A typical city block is said to be about 300 ft. × 800 ft. in size. Make rough assessments of the effectiveness of the MSS, TM, SPOT, and systems for depicting land use and land cover within urban regions. List factors other than the sizes of objects that would be important in making such assessments.

3. Outline some of the difficulties that would be encountered in compiling land use and land cover maps if aerial photography and remotely sensed data were not available.

4. Compare relative advantages and disadvantages of alternative kinds of imagery, including photography, thermal imagery, and radar imagery (at comparable scales and resolutions), for compiling land use and land cover maps. List advantages and some problems that might be encountered in using all three kinds of images in combination.

5. Review Section 20.5, and then prepare a diagram or flow chart that illustrates manual preparation of a land use change map.

6. The following issues all require, directly or indirectly, use of accurate land use and land cover information. For each, identify, in a few sentences or in short paragraphs, the role of accurate land use maps and data.

   a. Solid waste disposal
   b. Selecting a location for a new electrical power plant
   c. Establishing boundaries of a state park or wildlife preserve
   d. Zoning decisions in a suburban region near a large city
   e. Abandoned toxic waste dumps

7. Section 20.8 is closely related to material in Chapters 16 (Section 16.8) and 20 (Sections 20.3 and 20.4). Review these sections and prepare a summary of links between the topics, as well as some differences.

8. Some scientists have advocated development of a classification system with categories based on the appearance of features on specific kinds of remotely sensed images. In contrast, the approach used by Anderson et al. (Table 20.1) is based on the idea that remotely sensed data should be categorized using classes that remain the same for all forms of remotely sensed images and that match those used by planners. Compare advantages and disadvantages of both strategies, considering both the ease of application to the imagery and the ease of use by those who must actually apply the information.

9. About 18% of urban land is devoted to streets. About 20% of the land area of large cities is said to be undeveloped. Assess the ability of remotely sensed images to contribute to assessing the amount and patterns of these two kinds of land use; consider TM and SPOT data, as well as aerial photography at 1:10,000.

References

General

Anderson, J. R. 1977. Land Use and Land Cover Changes: A Framework for Monitoring. *Journal of Research, U.S. Geological Survey*, Vol. 5, No. 3, pp. 143–153.

Anderson, J. R., E. E. Hardy, J. T. Roach, and R. E. Witmer. 1976. *A Land Use and Land-Cover Classification for Use with Remote Sensor Data* (USGS Professional Paper 964). Washington, DC: U.S. Government Printing Office, 18 pp.

Baker, R. D., J. E. deSteiger, D. E. Grant, and M. J. Newton. 1979. Land-Use/Land-Cover Mapping from Aerial Photographs. *Photogrammetric Engineering and Remote Sensing*, Vol. 45, pp. 661–668.

Budreski, K. A., R. H. Wynne, J. O. Browder, and J. B. Campbell. 2007. Comparison of Segment and Pixel-Based Nonparametric Land Cover Classification in the Brazilian Amazon Using Multitemporal Landsat TM/ETM+ Imagery. *Photogrammetric Engineering and Remote Sensing*, Vol. 73, pp. 813–827.

Campbell, J. B. 1983. *Mapping the Land: Aerial Imagery for Land Use Information* (Resource Publications in Geography). Washington, DC: Association of American Geographers, 96 pp.

Campbell, J. B. 2001. Land Use and Land Cover Inventory. Chapter 9 in *Manual of Photographic Interpretation* (2nd ed.; Warren R. Philipson, ed.). Bethesda, MD: American Society of Photogrammetry and Remote Sensing, pp. 335–364.

Christian, C. S. 1959. The Eco-Complex and Its Importance for Agricultural Assessment. Chapter 36 in *Biogeography and Ecology in Australia*. Monographiae Biologicae, Vol. 8, pp. 587–605.

Chuvieco, E., and R. G. Congalton. 1988. Using Cluster Analysis to Improve the Selection of Training Statistics in Classifying Remotely Sensed Data. *Photogrammetric Engineering and Remote Sensing*, Vol. 54, pp. 1275–1281.

Cicone, R. C., and M. D. Metzler. 1984. Comparison of Landsat MSS, Nimbus–7 CZCS, and NOAA–7 AVHRR Feature for Land-Cover Analysis. *Remote Sensing of Environment*, Vol. 15, pp. 257–265.

Cihlar, J. 2000. Land Cover Mapping of Large Areas from Satellites: Status and Research Priorities. *International Journal of Remote Sensing*, Vol. 21, pp. 1093–1114.

Cowardin, L. M., V. Carter, F. C. Golet, and E. T. LaRoe. 1979. *Classification of Wetlands and Deepwater Habitat of the United States*. Washington, DC: U.S. Department of the Interior, Fish and Wildlife Service.

DeFries, R. S., R. A. Houghton, M. C. Hansen, C. B. Field, D. Skole, et al. 2002. Carbon Emissions from Tropical Deforestation and Regrowth Based on Satellite Observations for the 1980s and 1990s. *Proceedings of the National Academy of Sciences of the United States of America*, Vol. 99, pp. 14256–14261.

Dobson, J. E., E. A. Bright, P. R. Coleman, R. C. Durfee, and B. A. Worley. 2000. A Global Population Database for Estimating Population at Risk. *Photogrammetric Engineering and Remote Sensing*, Vol. 66, pp. 849–857.

Drake, B. 1977. Necessity to Adapt Land Use and Land Cover Classification Systems to Readily Accept Radar. In *Proceedings, 11th International Symposium on Remote Sensing of Environment*. Ann Arbor: Environmental Research Institute of Michigan, pp. 993–1000.

Friedl, M. A., D. K. McIver, J. C. F. Hodges, X. Y. Zhang, D. Muchoney, A. H. Strahler, et al. 2002. Global Land Cover Mapping from MODIS: Algorithms and Early Results. *Remote Sensing of Environment*, Vol. 83, pp. 287–302.

Grainger, A. 2008. Difficulties in Tracking the Long-Term Global Trend in Tropical Forest Area. *Proceedings of the National Academy of Sciences of the U.S.A.*, Vol. 105, pp. 818–823.

Herold, M., P. Mayaux, C. E. Woodcock, A. Baccini, and C. Schmullius. 2008. Some Challenges in Global Land Cover Mapping: An Assessment of Agreement and Accuracy in Existing 1-km Datasets. *Remote Sensing of Environment*, Vol. 112, pp. 2538–2556.

Hoeschele, W. 2000. Geographic Information Systems and Social Ground Truth in Attappadi, Kerala State, India. *Annals, Association of American Geographers*, Vol. 90, pp. 293–321.

Homer, C., C. Q. Huang, L. M. Yang, B. Wylie, and M. Coan. 2004. Development of a 2001 National Land-Cover Database for the United States. *Photogrammetric Engineering and Remote Sensing*, Vol. 70, pp. 829–840.

Janssen, L. F., and F. J. M. van der Wel. 1994. Accuracy Assessment of Satellite Derived

Lane-Cover Data: A Review. *Photogrammetric Engineering and Remote Sensing*, Vol. 94, pp. 419–426.

Jensen, J. R. 1978. Digital Land Cover Mapping Using Layered Classification Logic and Physical Composition Attributes. *American Cartographer*, Vol. 5, pp. 121–132.

Kalnay, E., and M. Cai. 2003. Impact of Urbanization and Land-Use Change on Climate. *Nature*, Vol. 423, pp. 528–531.

Kasetkasem, T., M. K. Arora, and P. K. Varshney. 2005. Super-Resolution Land Cover Mapping Using a Markov Random Field-Based Approach. *Remote Sensing of Environment*, Vol. 96, pp. 302–314.

King, R. B. 2002. Land Cover Mapping Principles: A Return to Interpretation Fundamentals. *International Journal of Remote Sensing*, Vol. 23, pp. 3525–3545.

Latifovic, R., Z. L. Zhu, J. Cihlar, C. Giri, and I. Olthof. 2004. Land Cover Mapping of North and Central America: Global Land Cover 2000. *Remote Sensing of Environment*, Vol. 89, pp. 116–127.

Liverman, D., E. F. Moran, R. R. Rindfuss, and P. C. Stern. 1998. *People and Pixels: Linking Remote Sensing and Social Science*. Washington, DC: National Academy Press, 244 pp.

Lo, C. P., and D. A. Quattrochi. 2003. Land-Use and Land-Cover Change, Urban Heat Island Phenomenon, and Health Implications: A Remote Sensing Approach. *Photogrammetric Engineering and Remote Sensing*, Vol. 69, pp. 1053–1063.

Loveland, T. R., B. C. Reed, J. F. Brown, D. O. Ohlen, Z. Zhu, L. Yang, et al. 2000. Development of a Global Land Cover Characteristics Database and IGBP DISCover from 1 km AVHRR data. *International Journal of Remote Sensing*, Vol. 21, pp. 1303–1330.

Loelkes, G. L. 1977. *Specifications for Land Cover and Associated Maps* (U.S. Geological Survey Open File Report 77–555). Reston, VA: U.S. Geological Survey.

Meyer, W. B., and B. L. Turner (eds.). 1994. *Changes in Land Use and Land Cover: A Global Perspective*. New York: Cambridge University Press, 537 pp.

Morton, D. C., R. S. DeFries, Y. E. Shimabukuro, L. O. Anderson, E. Arai, F. D. Espirito-Santo, et al. 2006. Cropland Expansion Changes Deforestation Dynamics in the Southern Brazilian Amazon. *Proceedings of the National Academy of Sciences of the U.S.A.*, Vol. 103, pp. 14637–14641.

Nunnally, N. R., and R. E. Witmer. 1970. Remote Sensing for Land Use Studies. *Photogrammetric Engineering and Remote Sensing*, Vol. 36, pp. 449–453.

Olthof, I., R. Latifovic, and D. Pouliot. 2009. Development of a Circa 2000 Land Cover Map of Northern Canada at 30 m Resolution from Landsat. *Canadian Journal of Remote Sensing*, Vol. 35, pp. 152–165.

Ostrom, E., and H. Nagendra. 2006. Insights on Linking Forests, Trees, and People from the Air, on the Ground, and in the Laboratory. *Proceedings of the National Academy of Sciences of the U.S.A.*, Vol. 103, pp. 19224–19231.

Parmenter, A. W., A. Hansen, R. E. Kennedy, W. Cohen, U. Langner, R. Lawrence, et al. 2003. Land Use and Land Cover Change in the Greater Yellowstone Ecosystem: 1975–1995. *Ecological Applications*, Vol. 13, pp. 687–703.

Petit, C., T. Scudder, and E. Lambin. 2001. Quantifying Processes of Land-Cover Change by Remote Sensing: Resettlement and Rapid Land-Cover Changes in Southeastern Zambia. *International Journal of Remote Sensing*, Vol. 22, pp. 3435–3456.

Rindfuss, R. R., S. J. Walsh, B. L. Turner, J. Fox, and V. Mishra. 2004. Developing a Science of Land Change: Challenges and Methodological Issues. *Proceedings of the National Academy of Sciences of the U.S.A.*, Vol. 101, pp. 13976–13981.

Roberts, D. A., M. Keller, and J. V. Soares. 2003. Studies of Land-Cover, Land-Use, and Biophysical Properties of Vegetation in the Large-Scale Biosphere Atmosphere Experiment in Amazonia. *Remote Sensing of Environment*, Vol. 87, pp. 377–388.

Robinove, C. J. 1981. The Logic of Multispectral Classification and Mapping of Land. *Remote Sensing of Environment*, Vol. 11, pp. 231–244.

Scott, J. M., and M. D. Jennings. 1998. Large Area Mapping of Biodiversity. *Annals of the Missouri Botanical Garden*, Vol. 85, pp. 34–47.

Smith, J. H., J. D. Wickham, S. V. Stehman, and L. M. Yang. 2002. Impacts of Patch Size and Land-Cover Heterogeneity on Thematic Image Classification Accuracy. *Photogrammetric Engineering and Remote Sensing*, Vol. 68, pp. 65–70.

Sobrino, J. A., and N. Raissouni. 2000. Toward Remote Sensing Methods for Land Cover Dynamic Monitoring: Application to Morocco. *International Journal of Remote Sensing*, Vol. 21, pp. 353–366.

Sponseller, R. A., E. F. Benfield, and H. M. Valett. 2001. Relationships between Land Use, Spatial Scale, and Stream Macroinvertebrate Communities. *Freshwater Biology*, Vol. 46, pp. 1409–1424.

Stefanov, W. L., M. S. Ramsey, and P. R. Christensen. 2001. Monitoring Urban Land Cover Change: An Expert System Approach to Land Cover Classification of Semiarid to Arid Urban Centers. *Remote Sensing of Environment*, Vol. 77, pp. 173–185.

Turner, B. L., W. C. Clark, R. W. Kates, J. F. Richards, J. T. Mathews, and W. B. Meyer (eds.). 1990. *The Earth as Transformed by Human Action: Global and Regional Changes in the Biosphere over the Past 300 Years*. New York: Cambridge University Press.

Vogelmann, J. E., T. Sohl, and S. M. Howard. 1998. Regional Characterization of Land Cover Using Multiple Sources of Data. *Photogrammetric Engineering and Remote Sensing*, Vol. 64, pp. 45–57.

Change Detection

Congalton, R. G., M. Balough, C. Bell, K. Green, J. Milliken, and R. Ottman. 1998. Mapping and Monitoring Agricultural Crops and Other Land Cover in the Lower Colorado River Basin. *Photogrammetric Engineering and Remote Sensing*, Vol. 64, pp. 1107–1113.

Estes, J. E., and L. Senger. 1972. Remote Sensing in the Detection of Regional Change. In *Proceedings, 8th International Symposium on Remote Sensing of Environment*. Ann Arbor: Environmental Research Institute of Michigan, pp. 317–324.

Jensen, J. R. 1979. Spectral and Textural Features to Classify Elusive Land Cover at the Urban Fringe. *Professional Geographer*, Vol. 31, pp. 400–409.

Jensen, J. R. 1981. Urban Change Detection Mapping Using Landsat Digital Data. *American Cartographer*, Vol. 8, pp. 127–147.

Jensen, J. R. 1996. *Introductory Digital Image Processing: A Remote Sensing Perspective*. Upper Saddle River, NJ: Prentice-Hall, 316 pp.

Jensen, J. R., and D. L. Toll. 1982. Detecting Residential Land-Use Development at the Urban Fringe. *Photogrammetric Engineering and Remote Sensing*, Vol. 48, pp. 629–643.

Lunetta, R. S., and S. D. Elvidge. 1998. *Remote Sensing Change Detection: Environmental Monitoring Methods and Applications*. Ann Arbor, MI: Ann Arbor Press, 318 pp.

Merteng, B., and E. F. Lambin. 2000. Land-Cover-Change Trajectories in Southern Cameroon. *Annals, Association of American Geographers*, Vol. 90, pp. 467–494.

AVHRR for Land-Cover Studies

Defries, R. S., and J. R. G. Townshend. 1994. NDVI-Derived Land Cover Classifications at a Global Scale. *International Journal of Remote Sensing*, Vol. 15, pp. 3567–3586.

Erb, T. L., et al. 1981. Analysis of Landfills with Historic Airphotos. *Photogrammetric Engineering and Remote Sensing*, Vol. 47, pp. 1363–1369

Homer, C., C. Huang, L. Yang, B. Wylie, and M. Coan. 2004. Development of a 2001 National Land-Cover Database for the United States. *Photogrammetric Engineering and Remote Sensing*, Vol. 70, pp. 829–840.

Livingston, G. P., C. A. Clark, and M. H. Trenchard. 1984. *An Investigation of NOAA/AVHRR Data in the Classification and Mapping of Landcover* (L. B. Johnson Space Center Technical Report 18918). Houston, TX: Johnson Space Center, 40 pp.

Loveland, T. R., T. L. Shol, S. V. Stehman, A. L. Gallant, K. L. Sayler, and D. E. Napton. 2002. A Strategy for Estimating the Rates of Recent United States Land Cover Changes. *Photogrammetric Engineering and Remote Sensing*, Vol. 68, pp. 1091–1100.

Lyon, J. G. 1987. Use of Maps, Aerial Photographs, and Other Remote Sensor Data for Practical Evaluations of Hazardous Waste Sites. *Photogrammetric Engineering and Remote Sensing*, Vol. 53, pp. 515–519.

Stehman, S. V., and R. L. Czaplewski. 1998. Design and Analysis for Thematic Map Accuracy Assessment: Fundamental Principles. *Remote Sensing of Environment*, Vol. 64, pp. 331–344.

Townshend, J. R. G. 1994. Global Data Sets for Land Applications from the Advanced Very High Resolution Radiometer: An Introduction. *International Journal of Remote Sensing*, Vol. 15, pp. 3319–3332.

Tucker, C. J., J. R. G. Townshend, and T. E. Goff. 1985. African Land Cover Classification Using Satellite Data. *Science*, Vol. 227, pp. 369–375.

Vogelmann, J. E., D. Helder, R. Morfit, M. J. Choate, J. W. Merchant, and H. Bulley. 2001. Effects of Landsat 5 Thematic Mapper and Landsat 7 Enhanced Thematic Mapper Plus Radiometric and Calibrations and Corrections on Landscape Characterization. *Remote Sensing of Environment*, Vol. 78, pp. 55–70.

Vogelman, J. E., S. M. Howard, L. Yang, C. R. Larson, B. K. Wylie, and N. Van Driel. 2001. Completion of the 1990s National Land Cover Data Set for the Coterminous United States from the Landsat Thematic Data and Ancillary Data Sources. *Photogrammetric Engineering and Remote Sensing*, Vol. 70, pp. 650–662.

Wu, S.-S., X. Qiu, E. L. Usery, and L. Wang. 2009. Using Geometrical, Textural, and Contextual Information of Land Parcels for Classification of Detailed Urban Land Use. *Annals of the Association of American Geographers*, Vol. 99, pp. 76–98.

Wulder, M. A., J. C. White, S. N. Goward, J. G. Masek, J. R. Irons, M. Herold, et al. 2008. Landsat Continuity: Issues and Opportunities for Land Cover Monitoring. *Remote Sensing of Environment*, Vol. 112, pp. 955–969.

Wynne, R. H., K. A. Joseph, J. O. Browder, and P. M. Summers. 2007. Comparing Farmer-Based and Satellite-Derived Deforestation Estimates in the Amazon Basin Using a Hybrid Classifier. *International Journal of Remote Sensing*, Vol. 28, pp. 1299–1315.

CHAPTER TWENTY-ONE

Global Remote Sensing

21.1. Introduction

The history of remote sensing can be visualized as one of increasing scope of vision, with respect both to space and time (Figure 21.1). Before the availability of aerial photography, human observation of the landscape was limited in scope to the area visible from a high hill or building. As aerial photography became available, it became possible to view larger areas over longer intervals.

However, even as aerial photographs came to represent larger and larger areas of the Earth's surface, the ability to represent broader regions was limited by the effects of atmospheric scattering as photography was attempted at higher altitudes and by the uneven illumination within individual frames (Figure 21.2). Later, as the routine availability of infrared emulsions permitted high-altitude aerial photography, larger areas could be represented within a single frame. Still later, the larger areas viewed by satellite images permitted analysts to consider even broader perspectives. Finally, by merging images from several dates to remove effects of cloud cover, it became possible to compile images representing continents, hemispheres, and the entire globe (Plate 28).

Such images have made it possible to directly examine broad-scale environmental patterns and their changes over years, decades, and even longer intervals. This capability has had profound impact on perspectives that scientists and the general public apply to their consideration of broad-scale environmental issues. Issues that once could be only matters of speculation now can be considered in the light of data and information based on systematic observation.

21.2. Biogeochemical Cycles

Such capabilities open new opportunities for understanding basic processes that underlie the Earth's climatic and biologic patterns. The elements and nutrients that make life on Earth possible reside differentially in biologic, geologic, and atmospheric components of the Earth. Over years, decades, and centuries, atmospheric and geologic processes cycle these nutrients and chemicals through the varied components of the biosphere. The Earth's atmosphere and oceans move at rates measured in minutes, hours, or days. At the other extreme, minerals of the Earth's crust can be considered mobile only when viewed over intervals of many millions of years.

Understanding the Earth's environment requires examination of movements and

614

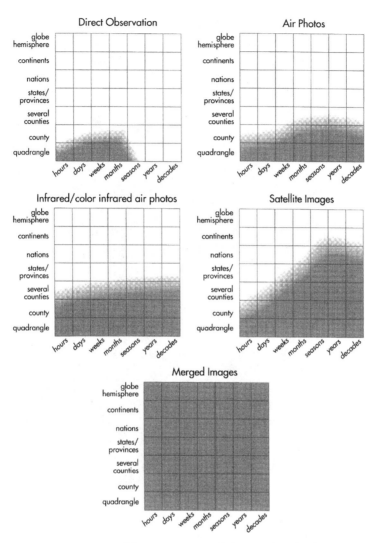

FIGURE 21.1. Increasing scope of human vision of the Earth. The horizontal axis in each graph represents time (in an arbitrary scale of increasing intervals from hours and days at one extreme to years and decades at the other). The vertical axis represents space (in arbitrary units, from quadrangle-sized areas at one extreme to continents and hemispheres at the other). From MacEachren et al. (1992). Copyright 1992 by Rutgers University Press. Reproduced by permission.

interchanges between its varied components. For example, minerals weather from geologic material, enter the soil, are absorbed by plant roots, then are incorporated into plant tissues as plants grow. As these plants die and decompose, nutrients are released into groundwater or surface runoff, then again assume geologic form as sedimentological and lithological processes concentrate and mineralize materials. Such are the biogeochemical processes that govern the movement of carbon, sulfur, nitrogen, oxygen, and other elements.

FIGURE 21.2. Mosaic of black-and-white photographs. Uneven illumination prevents rigorous assessment of image detail across the individual photographs. From Virginia Department of Transportation. Reproduced by permission.

The following sections outline major components of the most important biogeochemical cycles; these are not intended to provide complete descriptions but to illustrate the kinds of issues that require global perspectives.

Hydrologic Cycle

The hydrologic cycle is the best known and most central of the many processes that cycle materials through the Earth's biosphere (Figure 21.3). Because of the central role of moisture in biologic, climatic, and geologic processes, the hydrologic cycle (discussed earlier in Chapter 19) has obvious significance for all other processes that cycle materials through the Earth's environment. Hydrologic processes at all scales are intimately connected with land cover (Chapters 17 and 20), so the broad-scale land cover data provided by AVHRR data form a central component of efforts to examine the hydrologic cycle at continental and global scales. Further, aerial photography and the satellite data outlined in previous chapters provide, at least in some regions, a record of land cover changes extending several decades into the past, which offer the opportunity to develop broad-scale analyses of land cover changes and their impact on key components of the hydrologic cycle.

Carbon Cycle

Carbon (C) in its varied forms is one of the essential components of living organisms and is of key interest in any consideration of global cycling within the biosphere (Figure 21.4). Carbon is available for living organisms as carbon dioxide (CO_2) and as organic carbon (plant and animal tissues), sometimes known as fixed carbon, because it is much less mobile than CO_2. Photosynthesis reduces CO_2 to organic carbon, also much less mobile than CO_2; within organisms respiration can convert organic carbon to its gaseous form.

Chapter 2 described the significance of atmospheric CO_2 in the Earth's energy balance; although the atmosphere holds only a very small proportion of the Earth's carbon, atmospheric CO_2 is very effective in absorbing long-wave radiation emitted by the Earth. Larger amounts of carbon are stored in oceans.

The largest reservoir of carbon is formed by carbonate rocks (such as limestone). Very slowly, carbon from the atmosphere, oceans, and living organisms enters this reservoir as sedimentary rocks are formed. And equally slowly, carbon is released through weathering of these rocks. An exception, of course, is carbon residing as fossil fuels, which is rapidly released as it is burned for heat, to generate electrical power, and to propel automobiles and aircraft. Estimation of the amounts of carbon released by these activities and of their effects on the Earth's climate, is a high priority for those who hope to gauge effects of human activities on Earth's climates.

Key components in the carbon cycle are illustrated in Figure 21.4. Atmospheric CO_2 is incorporated into the structure of living organisms by photosynthesis. The decay of

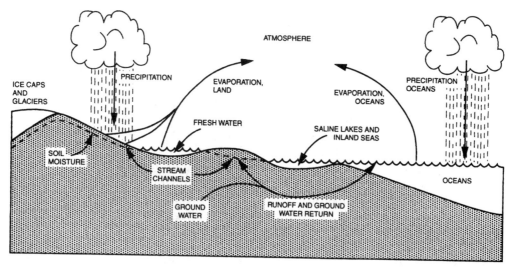

FIGURE 21.3. Hydrologic cycle. Adapted from Strahler and Strahler (1973). Copyright 1973 by J. Wiley & Sons, Inc. Reproduced by permission.

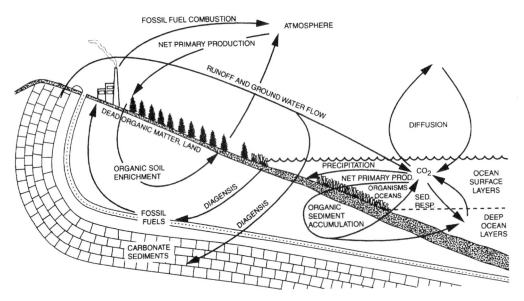

FIGURE 21.4. Carbon cycle. Adapted from Strahler and Strahler (1973). Copyright 1973 by J. Wiley & Sons, Inc. Reproduced by permission.

organic matter releases CO_2 into the atmosphere. CO_2 is also released by the burning of fossil fuels.

Remote sensing instruments assist scientists in understanding the carbon cycle by providing estimates of the areas covered by plants, identifying the kinds of plants, and estimating the time period for which they are photosynthetically active.

Nitrogen Cycle

Nitrogen (N) occurs in large amounts on Earth and is of great significance in the food chain, because it occurs as a principal element in the atmosphere and in tissues of plants and animals (Figure 21.5). It occurs in several forms, both organic and inorganic, and participates in important biological processes. Very large amounts of the Earth's nitrogen are considered inactive; it is held in sedimentary and crustal rocks and does not participate in the short-term cycling of nitrogen. Within the active pool, the largest amount of nitrogen resides in the atmosphere, where it forms one of the major constituents of the atmosphere. Atmospheric nitrogen, however, is not available for biological use.

Nitrogen becomes available for biological use through *fixation*, the chemical reaction in which nitrogen combines with hydrogen (H) to create ammonia (NH_3). Although nitrogen fixation can occur within the atmosphere (through lightening strikes), in the burning of fossil fuels, and industrially (in the manufacture of artificial fertilizers), the most significant source by far is biologically based. *Biological fixation* occurs by the action of living plants that change atmospheric nitrogen (N_2) to a fixed form (NH_3) that can be used by plants and animals. Microorganisms, both free-living and those symbiotic with specific plants (legumes), are the only animals that can accomplish this process. Amino acids from plant or animal tissue can be broken down by the process of *ammonification*, which releases energy, H_2O, and CO_2. Ammonification is accomplished by decomposing microorganisms.

By this process, nitrogen enters the food chain and the tissues of plants and animals. *Denitrification*, the process by which nitrites (NO_{2-}) or nitrates (NO_{-3}) are converted to molecular nitrogen or nitrous oxide, is accomplished by soil bacteria under anaerobic conditions. This occurs in a two-step process during which ammonia is oxidized to nitrate. The first step, completed by the bacteria *Nitrosommonas*, yields nitrate ions and water. The second step, accomplished by the bacteria *Nitrobacter*, consumes nitrite ions to yield nitrate. Industrial fixation accomplishes the same process using energy to transform atmospheric nitrogen to ammonia, which can be used in other industrial processes to manufacture fertilizers and explosives.

One of the largest reservoirs of nitrogen is decaying organic matter. Remote sensing assists in understanding the nitrogen cycle by providing accurate estimates of areas in which organic matter accumulates for decay, such as swamps and marshes.

Sulfur Cycle

In comparison with other biogeochemical cycles discussed here, the sulfur (S) cycle (Figure 21.6) is somewhat less significant with respect to total amounts in active circulation and with respect to its role in the food chain. But it has great significance if we examine ways in which humans influence the sulfur cycle. Sulfur resembles nitrogen in the sense that it occurs in several forms within the environment and participates in several organic and inorganic reactions that alter its forms and behavior. By far the largest amount of sulfur is in geologic form, in sedimentary deposits that are removed from short-term par-

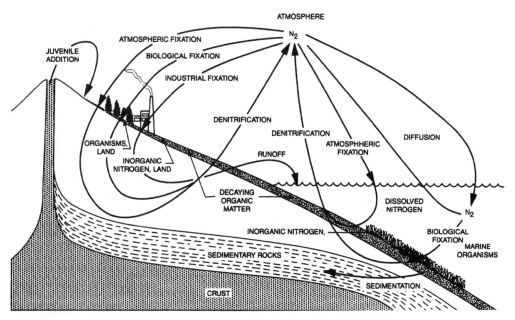

FIGURE 21.5. Nitrogen cycle. Adapted from Strahler and Strahler (1973). Copyright 1973 by J. Wiley & Sons, Inc. Reproduced by permission.

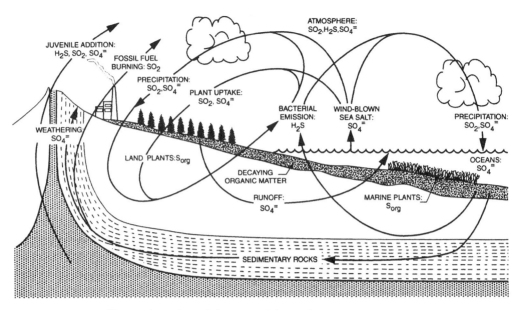

FIGURE 21.6. Sulfur cycle. Adapted from Strahler and Strahler (1973). Copyright 1973 by J. Wiley & Sons, Inc. Reproduced by permission.

ticipation in the sulfur cycle. Weathering releases small amounts of geologic sulfur to the atmosphere. Of greater interest, however, is the release of sulfur into the atmosphere by the removal and burning of fossil fuels.

Oceans form the main reservoir for sulfur in active circulation. Only small amounts reside in the atmosphere. On land, some is incorporated in the tissues of plants and animals, and some is retained as decaying organic matter (e.g., in swamps and marshes). Sulfur enters the atmosphere from four sources: emissions from bacteria as organic matter decays, combustion of fossil fuels, airborne dust from evaporating sea spray, and volcanic sources.

Accelerated release of geologic sulfur by burning of fossil fuels has contributed to reactions within the atmosphere that have increased the acidity of rainfall near industrialized regions of northern latitudes, which in turn is believed to have decreased the pH of lakes, rivers, and even soils in some regions. Such changes in water and soil acidity, if sustained, would have profound impact on wildlife, vegetation patterns, and soil fertility. Because such impacts have already been observed, governments have instituted efforts to reduce sulfur emissions and to encourage industries to switch to low-sulfur fuels.

Significance of Remote Sensing

Remote sensing is the primary means by which we can observe the dynamic character of the Earth's biosphere. Although basic outlines of biogeochemical cycles are well known, specific quantities, rates of movement, and regional variations are not well understood, nor is the true impact of human activities on their functioning. Such detail is necessary for scientists to assess the significance of human influences on biogeochemical processes and their significance in relation to natural variations. Therefore, scientists seek not only to understand the general structure of such cycles but also to develop a more precise knowledge of quantities and rates of movement within these cycles. Preparation of such estimates is very difficult, because the figures commonly given are necessarily very rough estimates, and conventional data are not adequate to provide the required information. For example, the existing network of climate stations consists of a series of point observations at locations not representative of specific environments. This network can provide only the very crudest estimates of temperature patterns and is clearly inadequate for answering the critical questions concerning the possible warming of the Earth's atmosphere during recent history.

Mather and Sdasyuk (1991, pp. 96–113) review existing sources of environmental data suitable for assessing broad-scale environmental trends, identifying the many defects in the present sources of data. They, like many others, conclude that broad-scale remote sensing provides the only practical means of acquiring data of sufficient scope to address these issues.

21.3. Advanced Very-High-Resolution Radiometer

Previously introduced in Chapters 6 and 17, the advanced very-high-resolution radiometer (AVHRR) is a multispectral radiometer carried by a series of meteorological satellites operated by NOAA in near-polar, sun-synchronous orbits. They can acquire imagery over a swath width of approximately 2,400 km. AVHRR acquires global coverage on a daily basis; the frequent global coverage is possible because of the wide angular field of view, but areas recorded near the edges of images suffer from severe geometric and angular effects. As a result, AVHRR data selected from the regions near the nadir provide the most accurate information. Although designed initially for much narrower purposes, AVHRR is the first sensor to provide the basis for collection of worldwide datasets that permit assessment of environmental issues.

AVHRR pixels are about 1.1 km on each side at nadir, with fine radiometric resolution at 10 bits. Details of spectral coverage vary with specific mission, but in general AVHRR sensors have recorded radiation in the visible and infrared regions of the spectrum. AVHRR spectral regions have been selected to attain meteorological objectives, including spectral data to distinguish clouds, snow, ice, and open water. Nonetheless, AVHRR has been employed on an ad hoc basis for land resource studies. The AVHRR sensor includes one channel in the visible spectrum, one in the near infrared, and three in the thermal infrared. It is possible to use the data to calculate vegetation indices of the type described in Chapter 17.

From these AVHRR data, the global vegetation index (GVI) dataset has been prepared for the period 1982–1985. GVI data are presented using an arbitrary grid system, with cells that vary in size with latitude, from about 13 km at the equator to 26 km at the poles. Data collected within a 7-day period are examined for the brightest ("greenest") NDVI value in each cell for the reporting interval. These data form the AVHRR composites described in Chapter 17 and referred to indirectly in Chapter 6. GVI images remove effects of clouds, except for instances in which clouds are present in all seven images. For more information, see

http://edcdaac.usgs.gov/1KM/1kmhomepage.html

Such data provided one of the first tools that permitted study of the Earth's major biomes. Previously, the areas and extents of major regions of forest, savanna, tundra, deserts, and the like could be estimated only roughly from climate and vegetation patterns as observed at ground level. Although satellite data show that these earlier approximations were quite accurate, the use of satellite data permits closer studies of detail of the patterns and examination of seasonal variations and long-term trends with accuracy and detail that were not previously possible. Specifically, AVHRR permits closer examination of effects of droughts and other broad-scale climatic changes, as well as changes related to human activities.

Pathfinder Data

In 1990 NASA and NOAA began a program to compile *Pathfinder datasets*, designed to provide data for analysis of long-term environmental processes at continental and global scales. Pathfinder data are compiled from systems placed in service prior to the Earth Observing System (EOS) to provide experience with global-scale data systems and to extend the period of analysis to as early a date as possible. These objectives require use of systems with accurate calibration and consistent qualities to permit analysis over time and from one sensor system to another. The first Pathfinder datasets have been prepared from AVHRR data, other meteorological satellite data, and selected Landsat data. Pathfinder data are sampled over time (in part to select cloud-free data and to reduce other atmospheric effects) and over space (to create consistent levels of detail and data sets of manageable sizes) and then processed to permit convenient use by a wide variety of users. Separate Pathfinder initiatives are under way to compile data describing ocean surface temperature, atmospheric conditions, and land surfaces.

Pathfinder AVHRR land data are global in scope but sampled over time and space. The *daily dataset* consists of AVHRR terrestrial data, mapped to an equal area projection on an 8-km grid. Pixels near the edges of the AVHRR swath are excluded. The daily data can be used directly for analysis or as source data for composite datasets. The *composite dataset* is derived from the daily dataset; the compositing process filters out effects of clouds and selects data for each date from the scene with the highest NDVI. The *climate dataset* is compiled to a 1° × 1° resolution for 8- to 11-day composite periods. The result forms a set of cloud-free, averaged NDVI data. Thirty-six such images are produced each year to provide data for examination by scientists studying broad-scale climatic and biosphere changes. *Browse images* are prepared from the daily dataset and the composite dataset to provide a reduced dataset to assist in data selection and evaluation. Browse images are not designed for analysis. For more information, see

AVHRR Pathfinder Program: *http://xtreme.gsfc.nasa.gov/pathfinder*

21.4. Earth Observing System

Although existing satellite data such as AVHRR and those described in previous chapters might seem at first consideration to provide a means of studying global environmental problems, in fact these images provide piecemeal rather than systematic long-term coverage. Likewise, much of the existing archive of imagery records arbitrary regions of the spectrum (rather than regions targeted on those required for specific tasks) and are not coordinated with related data-collection efforts. Attacking these broad-scale, long-term data-collection requirements takes instruments tailored for specific tasks at hand. This is the context in which NASA, in the late 1980s and early 1990s, developed a program to acquire the environmental data needed to address questions posed by concerns over global environmental change.

Earth Science Enterprise (formerly *Mission to Planet Earth* [MTPE]) is one of NASA's four strategic enterprises designed to establish a long-term, coordinated research effort to study the total Earth system and the effects of natural and human-induced changes on the global environment. Earth Science Enterprise is a long-term effort to observe the Earth's environments and how they change over time, as an essential component to developing an understanding of the nature, rates, and consequences of broad-scale environmental change. The early activities of Earth Science Enterprise (in the 1990s, when it was still named MTPE) focused on the design of numerous satellites that require specialized orbits and instruments.

The second, long-term, component of the Earth Science Enterprise is a system of satellites that form an integrated system of Earth-orbiting satellites designed to provide a continuous stream of data for a period of 15 years or more. The Earth Observing System (EOS) is to consist of several satellites carrying numerous sensors designed to monitor environmental variables on a worldwide basis. EOS efforts have been coordinated nationally through various agencies of the U.S. government and internationally through agreements with other nations. The design of EOS has changed in response to scientific, operational, and budgetary considerations and will no doubt continue to change. Thus the following sections outline some of the instruments likely to be of interest to readers of this volume but do not attempt an exhaustive description of the complete system.

- NASA Earth Observing System: *http://eospso.gsfc.nasa.gov*
- NASA Science Mission Directorate: *http://science.hq.nasa.gov/directorate/index. html*
- Destination Earth: *www.earth.nasa.gov*

21.5. Earth Observing System Instruments

EOS is designed to collect data to provide a comprehensive overview of the dynamic components of the Earth's atmosphere and land and water surfaces (Figure 21.7). To accomplish this objective, the EOS plan has included 30 or more instruments designed to monitor physical and biological components of the biosphere. However, as budget priorities have changed, the program design has evolved. Therefore, individual sensors have been redesigned or dropped from the plan. Here we briefly describe only a few of the many instruments proposed for EOS that are most closely related to those already discussed in this volume.

Moderate-Resolution Imaging Spectroradiometer

The moderate-resolution imaging spectroradiometer (MODIS) (Figure 21.8), previously introduced in Chapter 6, was launched on the Terra satellite in December 1999:

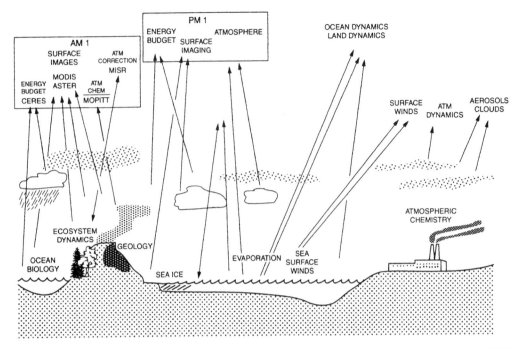

FIGURE 21.7. EOS overview. This diagram represents the planned configuration for EOS missions. Based on a NASA diagram given in Assrar and Dokken (1993).

http://modis.gsfc.nasa.gov

MODIS is designed to acquire imagery that will permit compilation of global datasets at frequent intervals. It uses a cross-track scanning mirror and CCD detectors in 36 spectral bands, covering spectral regions from the visible to the thermal infrared. Spatial resolutions vary from 500 m to 1 km; the satellite provides coverage every 2 days. MODIS instruments are designed to extend the kinds of data collected by AVHRR, TM, and some of the meteorological satellites (Figure 21.9).

MODIS is designed to monitor the Earth's land areas; in essence, it is an instrument similar to AVHRR, but tailored to monitoring land resources (in contrast to the meteorological mission of AVHRR). *Moderate resolution* refers to its 500-m pixels, which may seem coarse relative to SPOT or Landsat but are fine relative to the 1-km pixels of AVHRR and achieve global coverage to meet EOS objectives. MODIS data permit analysts to monitor biological and physical processes at the Earth's land surfaces and water bodies and to collect data describing cloud cover and atmospheric qualities. MODIS collects data describing range and forest fires, snow cover, chlorophyll concentrations, surface temperature, and other variables.

Earth Observing System Ocean Color Experiment

The EOS Ocean Color Experiment (EOS-Color) is an optical–mechanical imaging radiometer provided for EOS service by Japan's Ministry of International Trade and Industry.

It is designed to operate from a sun-synchronous orbit, with a noon equatorial crossing. It collects data in 8 spectral channels over a swath width of 2,800 km. Spatial resolution at nadir is 1.1 km and as much as 4.5 km off-nadir. The instrument is designed to acquire data to study marine phytoplankton, the oceanic role in the carbon cycle, mixing of oceanic and coastal waters, and the nature of heat exchange within the upper layers of the ocean.

Multi-Angle Imaging Spectroradiometer

The multi-angle imaging spectroradiometer (MISR) employs nine CCDs to observe the Earth's surface from nine angles (Figure 21.9). One looks to the nadir, and four each look to the fore and aft along the ground track of the satellite in each of four spectral regions, centered on 0.443 m, 0.555 m, 0.670 m, and 0.865 m. These channels have resolutions of 240 m, 480 m, 960 m, and 1.92 km, respectively.

The instrument is designed to observe the atmosphere at different angles and to observe features through differing path lengths and angles of observation. Data from MISR will permit the study of atmospheric properties, including effects of pollutants, and the derivation of information to correct data from other sensors for atmospheric effects. The instrument will also contribute to the development of models for the optical behavior of surfaces (e.g., forest canopies) by developing accurate BRDFs. The configura-

FIGURE 21.8. Sample MODIS image. From NASA.

FIGURE 21.9. This computer-generated image shows the Terra spacecraft, with the MISR instrument on board, orbiting Earth. Direction of flight is toward the lower left. The actual locations imaged by the nine cameras, each with four color bands, along Earth's surface are illustrated here with translucent surfaces. From JPL, image P-49081.

tion of the sensors provides the ability to acquire stereo imagery of both ground surfaces and clouds.

Advanced Spaceborne Thermal Emission and Reflection Radiometer

The Advanced Spaceborne Thermal Emission and Reflection Radiometer (ASTER) is an imaging radiometer designed to collect data in 14 spectral channels (Figure 21.10). Three channels in the visible and near infrared provide 15-m resolution in the region 0.5 m–0.9 m; six channels in the short-wave infrared provide 30-m resolution in the interval 1.6 m–2.5 m; and five channels in the thermal infrared provide 90-m resolution in the interval 8 μm–12 μm. Imagery can be acquired in a 60-km swath centered on a pointable field of view ±8.5° in the mid infrared and thermal infrared and ±2.4° in the visible and near infrared (Figure 21.10). The instrument acquires a stereo capability through use of an aft-pointing telescope that acquires imagery in channel 3. The ASTER can provide 16-day repeat coverage in all 14 bands and 5-day coverage in the visible and near infrared channels. The instrument is designed to provide data to describe land use, cloud coverage and kinds and heights of clouds, land and ocean temperature, topographic data, and information on glacial activity and snow cover. The ASTER archive has been used to compile a worldwide digital elevation model (ASTER Global Digital Elevation Model). This DEM, compiled by stereo analysis of imagery from the ASTER archive, represents topography of the Earth's land surface between 83°N and 83°S latitudes (Figure 21.11). Details of the complete DEM about 30 meters are available at

http://asterweb.jpl.nasa.gov/gdem.asp

21.6. Earth Observing System Bus

EOS sensors are to be placed on a series of satellites designed to provide standard support functions, including communications, thermal control, power supply, and related functions. The standardized components are known as the *EOS bus* (Figure 21.12).

The first of the EOS satellites, EOS-AM, known as Terra, was launched in December 1999 and began service in February 2000, carrying CERES, MISR, MODIS, ASTER, and MOPITT sensors. Terra was placed in an orbit designed for morning observation (local sun time), which optimizes observation of terrestrial surfaces by minimizing land areas obscured by cloud cover. A second satellite, EOS-PM (Aqua), launched in May 2002, is in an orbit designed to observe the Earth at a later time of day, designed for optimum observation of ocean and atmospheric features, as well as selected terrestrial phenomena. Aqua carries six sensors: the atmospheric infrared sounder (AIRS), the advanced microwave sounding unit (AMSU-A), the humidity sounder for Brazil (HSB), the advanced microwave scanning radiometer for EOS (AMSR-E), the moderate resolution imaging spectroradiometer (MODIS), and clouds and the Earth's radiant energy system (CERES). Aqua carries some of the same sensors (e.g., MODIS) carried on Terra to permit features to be imaged by the same instrument at different times of day, providing an opportunity to examine diurnal changes such as those suggested in Chapter 9. For more information, see

FIGURE 21.10. Sample ASTER image. From NASA.

FIGURE 21.11. ASTER global digital elevation model. This DEM, compiled by stereo analysis of imagery from the ASTER archive, represents topography of the Earth's land surface between 83°N and 83°S latitudes. From Japan's Ministry of Economy, Trade and Industry (METI) and NASA.

FIGURE 21.12. EOS bus. This illustration of Terra depicts fields of view for some of the sensors carried by the satellite. From NASA.

http://terra.nasa.gov

http://aqua.nasa.gov/about

21.7. Earth Observing System Data and Information System

The EOS Data and Information System (EOSDIS) was established to handle the unprec-edented volume of data to be acquired by the system of satellite sensors. Operation of EOS requires a system for acquiring, processing, and distributing data to scientists for interpretation and analysis. Key components within EOSDIS are the eight distributed active archive centers (DAACs), established to process and distribute data. Each DAAC specializes in handling specific kinds of data (e.g., "land processes," "sea ice and polar processes," "hydrology"), with the objective of providing sustained access to the data for a broad community of users (Figure 21.13 and Table 21.1). Components of EOSDIS are connected electronically, so users can quickly access archives, indices, and data from their own institutions.

http://nasadaacs.eos.nasa.gov

http://edcdaac.usgs.gov/main.asp

Related global-scale environmental datasets include:

- Hydrology: *http://edcdaac.usgs.gov/gtopo30/hydro/index.html*
- Elevation: *http://edcdaac.usgs.gov/gtopo30/gtopo30.html*

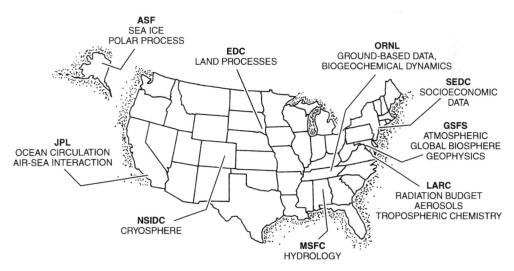

FIGURE 21.13. EOS distributed active archive centers (see Table 21.1). Based on NASA dia-gram in Assrar and Dokken (1993).

TABLE 21.1. EOS Distributed Active Archive Centers

ASF: Alaska SAR Facility: *www.asf.alaska.edu*	Acquire, process, archive, and distribute satellite synthetic aperture radar (SAR) University of Alaska, Fairbanks
EDC: EROS Data Center: *http://lpdaac.usgs.gov/*	Process, archive, and distribute land-related data main. asp collected by EOS sensors Sioux Falls, SD
GSFC: Goddard Space Flight Center: *http://daac.gsfc.nasa.gov*	Atmospheric chemistry; atmospheric dynamics; field experiments; hydrology; global land, ocean, and atmospheric data; land biosphere; ocean color Greenbelt, MD
JPL: Jet Propulsion Laboratory: *http://podaac-www.jpl.nasa.gov*	Physical oceanography Pasadena, CA
LARC: Langley Research Center: *http://eosweb.larc.nasa.gov/chemistry*	Radiation budget, clouds, aerosols, tropospheric Langley, VA
NSIDC: National Snow and Ice Data Center: *http://nsidc.org/NASA/GUIDE/index.html*	Snow and ice data Boulder, CO
ORL: Oak Ridge National Laboratories: *www-daac.ornl.gov*	Biogeochemical and ecological data Oak Ridge, TN
SEDAC: Socioeconomic Data and Applications Center: *http://sedac.ciesin.columbia.edu*	Socioeconomic data Columbia University, New York, NY

21.8. Long-Term Environmental Research Sites

In this context, information provided by field data assumes a new meaning. The usual field observations, collected to match with specific images on a one-time basis at isolated times and places, have less significance when the objectives extend over very large areas and very long time intervals. To support the search for long-term, broad-scale patterns of change, field data must exhibit continuity over time and be positioned to represent ecosystems at continental scales. To address these needs, a network of long-term ecological research sites has been established within many of the Earth's major ecological zones. At each site long-term studies are under way to document ecological behavior and changes over time and to aid in understanding relationships between natural changes and those that might be caused or accelerated by human activities.

Research conducted at some sites records local patterns of nutrient cycling as a component in understanding changes in the global cycles. Some long-term ecological research sites (LTER; Table 21.2) were already established as reserves or research sites, so some have 30 or 40 years' worth of records of climate and ecological conditions. Figure 21.14 and Table 21.2 show a network of LTER stations supported in part by the U.S. National Science Foundation. Most of these sites represent principal ecological zones of the western hemisphere and a broad sample of global environments. Similar efforts are under way throughout the world to establish a network of international sites representing a full range of biomes.

www.lternet.edu

21.9. Earth Explorer

Earth Explorer, developed by the USGS EROS Data Center (Sioux Falls, SD, 57198-0001; Tel: 800-252-4547), is an interactive computer system designed to permit scientists to identify information pertaining to the Earth's land surfaces. Earth Explorer consists of metadata (Chapter 6) describing dates, coverages, cloud cover, quality ratings, and other characteristics of broad-scale datasets of the Earth's land areas. Included are summaries of land cover, soil, geologic, cultural, topographic, and remotely sensed data. Users can connect to Earth Explorer computers remotely to review data for their specific regions of interest. Earth Explorer permits examination of the suitability of data for specific projects and inspection of sample data projects and provides a vehicle for placing orders to acquire data.

Earth Explorer descriptions are organized to permit users to search for coverage of specific regions by date to ascertain availability of data from specific years or by class of data (aerial photography, digital line graphs, maps, elevation data, or satellite imagery). For more information, see

http://edcsns17.cr.usgs.gov/EarthExplorer

Another data distribution effort, NASA's Visible Earth, is intended to provide imagery to the press and public, chiefly though a website at

TABLE 21.2. Long-Term Ecological Research Sites

1. **Andrews LTER:** Temperate coniferous forest: J. J. Andrews Experimental Forest, Blue River, OR
2. **Arctic Tundra LTER:** Arctic lakes and tundra, Brooks Range, AK
3. **Baltimore Ecosystem Study:** Eastern deciduous forest/suburban agriculture fringe, Baltimore, MD
4. **Bonanza Creek LTER:** Tiaga, Fairbanks, AK
5. **Central Arizona-Phoenix:** Sonoran Desert scrub, Phoenix, AZ
6. **California Current Ecosystem:** Upwelling current biome, San Diego, CA
7. **Cedar Creek LTER:** Eastern deciduous forest and tall grass prairie, Cedar Creek Natural History Area, Minneapolis, MN
8. **Coweeta LTER:** Eastern deciduous forest, Coweeta Hydrological Laboratory, Otto, NC
9. **Florida Coastal Everglades:** Marsh ecosystems, Miami, FL
10. **Georgia Coastal Ecosystems LTER:** Coastal barrier island/marsh complex, Sapelo Island, GA
11. **Harvard Forest:** Eastern deciduous forest, Petersham, MA
12. **Hubbard Brook LTER:** Eastern deciduous forest, West Thornton, NH
13. **Jordana Basin:** Hot desert, Jordana Experimental Range, Las Cruces, NM
14. **Kellogg Biological Station:** Row-crop agriculture, Hickory Corners, MI
15. **Konza LTER:** Tall-grass prairie, Manhattan, KS
16. **Luquillo LTER:** Tropical rain forest, San Juan, PR
17. **McMurdo Dry Valleys:** Antarctic valleys, Antarctica
18. **Moorea Coral Reef:** Tahiti, French Polynesia
19. **Niwot Ridge LTER:** Alpine tundra, Boulder, CO
20. **North Temperate Lakes:** Madison, WI
21. **Palmer Station:** Polar marine, Antarctica
22. **Plum Island Ecosystem:** Coastal estuary, Plum Island, MA
23. **Santa Barbara Coastal LTER:** Semi-arid coastal zone, Santa Barbara, CA
24. **Sevilleta LTER:** Arid mountains/hot desert/cold desert, Albuquerque, NM
25. **Short-grass Steppe:** Nunn, CO
26. **Virginia Coastal Reserve:** Coastal barrier island, Oyster, VA
27. **LTER Network Office:** Albuquerque, NM: *lternet.edu/sites*

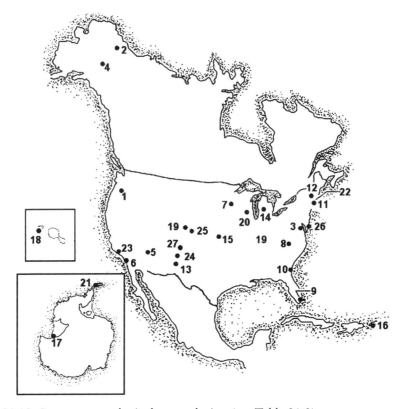

FIGURE 21.14. Long-term ecological research sites (see Table 21.2).

http://visibleearth.nasa.gov/about.html

21.10. Global Monitoring for Environment and Security

In 1998 the European Commission (the executive agency of the European Union) and the European Space Agency (ESA) defined the Global Monitoring for Environment and Security (GMES) program to acquire and assess data to better understand broad-scale environmental issues, to understand and mitigate the effects of climate change, and to support civil security programs. GMES focuses principally on collecting and interpreting satellite imagery. ESA is developing and managing a series of five satellite missions, known as "Sentinels," specifically for GMES, scheduled for launch in 2013.

The satellite component of GMES includes several systems developed under mandates from the European Parliament to address priorities relating to natural and technological crises, humanitarian aid, monitoring compliance with international treaties, peacekeeping missions, and surveillance of EU borders. Most of these systems are in conceptual and planning stages, based on policy studies and technical evaluations. Some systems will build on existing programs such as SPOT and ENVISAT (Chapter 6). GMES is expected to include a commercial component to provide data to communities of users who can process value-added data products tailored for specific markets of users.

- *Sentinel-1.* The Sentinel-1 series of satellites will address the issue of data continuity for SAR imagery, especially for C-band data. Therefore, it is designed to maintain continuity with capabilities of ERS, Radarsat-2, and Envisat (Chapters 6 and 7). Sentinel-1 (described below) is scheduled for launch in 2011.

- *Sentinel-2.* Sentinel-2 is designed to provide multispectral data along the lines of the SPOT and Landsat systems to maintain continuity with these systems to ensure availability of data to service providers and users.

- *Sentinel-3.* Sentinel-3 is dedicated to providing operational oceanographic services, continuing and building on capabilities of ERS and ENVISAT (Chapter 6) to provide both regional and worldwide coverage.

- *Sentinels-4 and -5.* The Sentinel 4 and 5 systems are dedicated to monitoring atmospheric chemistry on a real-time basis, including atmospheric chemistry, pollution, ozone, and aerosols. Sentinels 4 and 5 would be space-based systems operating from geostationary (Sentinel-4), and low Earth orbits (Sentinel-5).

21.11. Gridded Global Population Data

In an effort to better understand the local and regional impact of human population distributions, several organizations have examined worldwide demographic data to approximate global population distributions. This task, although ostensibly straightforward, is in fact quite difficult, due to varying census definitions and procedures, differing census dates, wide variations in sizes of demographic units, and many other obstacles. As a result, efforts to compile worldwide demographic data must apply a variety of approximations and improvisations to estimate population patterns.

LandScan

LandScan is a worldwide population database with 1-km spatial detail. It was developed as part of the U.S. Department of Energy's (USDOE) Oak Ridge National Laboratory's (ORNL) Global Population Project to provide data for demographic cartography and for population estimates to support disaster response, among other applications. LandScan provides spatially explicit estimates of population data, based on a likelihood estimator applied to national census estimates, topography, transportation, urban infrastructure, remotely sensed imagery, and GIS. The LandScan dataset is compiled on a 30-in. × 30-in. latitude–longitude grid.

The LandScan dataset files are available via the Internet in ESRI grid format by continent and for the world, and in ESRI raster binary format for the world. LandScan dataset licenses are available free of charge for the U.S. federal government, for United Nations humanitarian efforts, for and educational research use. (Educational research is considered within the United States to include K–12 schools, colleges, and universities.) See

www.ornl.gov/sci/landscan/index.html

http://gcmd.nasa.gov/records/GCMD_Landscan.html

Gridded Population of the World

Gridded Population of the World (GPW) is prepared by the Socioeconomic Data and Applications Center (SEDAC; *sedac.ciesin.columbia.edu/gpw*; CIESIN; Center for International Earth Science Information Network). SEDAC's mission is to support the integration of socioeconomic and earth science data and interactions between the social and earth sciences. The GPW dataset estimates residential population density based on where people reside and are likely to be at nighttime. GPW pixels are 2.5 arc-minutes, five times coarser than LandScan pixels.

CIESIN seeks to apply the most detailed census data available. Within nations using detailed census reporting, the maps have higher resolution because the census data have greater detail. As a result, population distributions of these nations are represented with higher detail than is likely for comparable LandScan data. GPW methods account for land use effects, such as water bodies, and placement of urban population is within delineated urban regions. In rural regions, they distributed population evenly within administrative units. In 2008 Ellis and Ramankutty examined GPW data to define 21 anthropogenic biomes based on population density, land use, biota, climate, terrain, and geology. They subdivided their anthropogenic biomes into six major categories: dense settlements, villages, croplands, rangeland, forested land, and wildlands.

21.12. Summary

The instruments and programs outlined in this chapter are designed to gather data that provide the geographic coverage and consistency required to compile global datasets and to examine fundamental questions concerning place-to-place variations in climate, vegetation patterns, and biophysical processes and their changes over time. Although existing data have permitted humans to define the general framework for understanding these phenomena, it is only by collection of consistent data over large areas over long intervals of time that we will be able to develop an intimate understanding that will permit us to assess effects of human actions in the context of inherent geographic and historical variations.

21.13. Some Teaching and Learning Resources

- NASA|Destination Earth
  www.youtube.com/watch?v=1V2J0oBj42Y&feature=related

- Earth System Science
  www.youtube.com/watch?v=ciV6Uaeobxk&feature=related

- Long-Term Ecological Research site in Michigan
  www.youtube.com/watch?v=ggtiYfTfbdQ

- A Hike in HJA LTER
  www.youtube.com/watch?v=0BeoW0bx2zI&feature=related

- NASA: The Carbon Cycle
  www.youtube.com/watch?v=OByqdUhWERk&feature=related

- The Water Cycle
  www.youtube.com/watch?v=0_c0ZzZfC8c&feature=fvw

Review Questions

1. Enumerate some of the obstacles important in limiting, until recently, the practice of global remote sensing.

2. How might the practice of global remote sensing differ from the practice of other, more conventional, remote sensing?

3. Discuss problems in collecting and evaluating field data to support global remote sensing.

4. How might global remote sensing assist in global change studies?

5. In what ways might hyperspectral data assist in global change studies?

6. Global remote sensing developed initially from applications of data collected by meteorologic satellites for purposes other than their intended function. Can you identify problems (either practical or conceptual) that arise from these origins?

7. What are some of the special practical and theoretical problems inherent in the practice of global remote sensing?

8. Enumerate some of the special problems or research questions that can be addressed by global remote sensing but not by more conventional remote sensing.

9. The question of continuity of data (i.e., maintaining a continuous and consistent data archive) is a crucial issue for global change research. Discuss some of the problems this presents. For example, what is a sensible response when new, improved technology becomes available? Is it better to improve instruments or to maintain continuity of the older instruments?

10. Discuss some of the problems you might expect to encounter in maintaining an archive for global change data and in distribution of data to scientific analysts throughout the world.

References

General

Claussen, E. (ed.). 2001. *Climate Change: Science, Strategies, and Solutions.* Leiden, Netherlands: Brill Academic, 400 pp.

Edwards, P. N. 2010. *Computer Models, Climate Data, and the Politics of Global Warming.* Boston: MIT Press, 518 pp.

Foody, G., and P. Curran. 1994. *Environmental Remote Sensing from Global to Regional Scales.* New York: Wiley, 238 pp.

Goward, S. N., D. Dye, A. Kerber, and V. Kalb. 1987. Comparison of North and South American Biomes from AVHRR Observations. *Geocarto International*, Vol. 1, pp. 27–39.

MacEachren, A. M., et al. 1992. Visualization. Chapter 6 in *Geography's Inner Worlds* (R. F. Abler, M. G. Marcus, and J. M. Olson, eds.). New Brunswick, NJ: Rutgers University Press, pp. 99–137.

Mather, J. R., and G. V. Sdasyuk. 1991. *Global Change: Geographical Approaches*. Tucson: University of Arizona Press, 289 pp.

National Research Council. 2010. *Understanding the Changing Planet: Strategic Directions for the Geographical Sciences*. National Research Council Committee on Strategic Directions for the Geographical Sciences in the Next Decade. Washington, DC: National Academies Press. 156 pp.

Schneider, S., A. Rosencranz, and M. Mastranda. 2009. *Climate Change Science and Policy*. Washington, DC: Island Press. 544 pp.

Sensors and Systems

Assrar, G., and D. J. Dokken. 1993. *EOS Reference Handbook*. Washington, DC: NASA, 140 pp.

EOS Science Steering Committee. (n.d.). *From Pattern to Process: The Strategy of the Earth Observing System* (EOS Science Steering Committee Report, Vol. 3). Washington, DC: NASA, 140 pp.

Gabrynowicz, J. I. 1995. The Global Land 1-km AVHRR Project: An Emerging Model for Earth Observations Institutions. *Photogrammetric Engineering and Remote Sensing*, Vol. 61, pp. 153–160.

Harries, J. E., D. T. Llewellyn-Jones, C. T. Mutlow, M. J. Murray, I. J. Barton, and A. J. Prata. 1995. The ASTER Programme: Instruments, Data and Science. Chapter 1 in *TERRA 2: Understanding the Terrestrial Environment: Remote Sensing Data Systems and Networks* (P. M. Mather, ed.). New York: Wiley, pp. 19–28.

Hastings, D. A., and W. J. Emery. 1992. The Advanced Very High Resolution Radiometer (AVHRR): A Brief Reference Guide. *Photogrammetric Engineering and Remote Sensing*, Vol. 58, pp. 1183–1188.

Iverson, L. R., E. A. Cook, and R. L. Graham. 1989. A Technique for Extrapolating and Validating Forest Cover across Large Regions: Calibrating AVHRR Data with TM Data. *International Journal of Remote Sensing*, Vol. 10, pp. 1085–1812.

Loveland, T. R., J. W. Merchant, J. F. Brown, D. O. Ohlen, B. K. Reed, P. Olson, et al. 1995. Seasonal Land Cover Regions of the United States. *Annals of the Association of American Geographers*, Vol. 85, pp. 339–335 (with map at 1:7,500,000).

Loveland, T., B. Reed, J. Brown, D. Ohlen, Z. Zhu, L. Yang, et al. 2000. Development of a Global Land Cover Characteristics Database and IGBP DISCover from 1-km AVHRR data. *International Journal of Remote Sensing*, Vol. 21, pp. 1303–1330.

Roller, N. E. G., and J. E. Colwell. 1986. Coarse-Resolution Satellite Data for Ecological Surveys. *BioScience*, Vol. 36, pp. 468–475.

Biogeochemical Cycles

Delwiche, C. C. 1970. The Nitrogen Cycle. *Scientific American*, Vol. 223, pp. 136–146.

Jenkinson, D. S. 1990. An Introduction to the Global Nitrogen Cycle. *Soil Use and Management*, Vol. 6, No. 2, pp. 56–61.

Powlson, D. S. 1993. Understanding the Soil Nitrogen Cycle. *Soil Use and Management*, Vol. 9, No. 3, pp. 86–94.

Strahler, A. N., and A. H. Strahler. 1973. *Environmental Geoscience: Interaction between Natural Systems and Man*. New York: Wiley, 511 pp.

Global Population Patterns

Bhaduri, B., E. Bright, P. Coleman, and J. Dobson. 2002. Landscan: Locating People Is What Matters. *Geo Informatics*, April/May.

Dobson, J. E., E. A. Bright, P. R. Coleman, R. C. Durfee, and B. A. Worley. 2000. A Global Population Database for Estimating Population at Risk. *Photogrammetric Engineering and Remote Sensing*, Vol. 66, pp. 849–857.

Ellis, E. C., and N. Ramankutty. 2008. Putting People in the Map: Anthropogenic Biomes of the World. *Frontiers in Ecology and the Environment*, Vol. 6, pp. 439–447.

Schneider, A., M. A. Friedl, and D. Potere. 2010. Mapping Global Urban Areas Using MODIS 500-m Data: New Methods and Datasets based on "Urban Ecoregions." *Remote Sensing of Environment*, Vol. 114, pp. 1733–1746.

Tatum, A. J., S. J. Goetz, and S. I. Hay. 2004. Terra and Aqua: New Data for Epidemiology and Public Health. *Journal of Applied Earth Observation and Geoinformation*, Vol. 6, pp. 33–46.

Global Land Use

Defries, R. S., and J. R. G. Townshend. 1994. NDVI-Derived Land Cover Classifications at Global Scale. *International Journal of Remote Sensing*, Vol. 15, pp. 3567–3586.

Feddema, J. J., K. W. Oleson, G. B. Bonan, L. O. Mearns, L. E. Buja, G. A. Meehl, et al. 2005. The Importance of Land-Cover Change in Simulating Future Climates. *Science*, Vol. 310, pp. 1674–1678.

Foley, J. A., R. DeFries, G. P. Asner, C. Barford, G. Bonan, S. R. Carpenter, et al. 2005. Global Consequences of Land Use. *Science*, Vol. 309, pp. 570–574.

Herold, M., P. Mayaux, C. E. Woodcock, A. Baccini, and C. Schmullius. 2008. Some Challenges in Global Land Cover Mapping: An Assessment of Agreement and Accuracy in Existing 1-km Datasets. *Remote Sensing of Environment*, Vol. 112, pp. 2538–2556.

McCallum, I., M. Obersteiner, S. Nilsson, and A. Shvidenko. 2006. A Spatial Comparison of Four Satellite Derived 1-km Global Land Cover Datasets. *International Journal of Applied Earth Observation and Geoinformation*, Vol. 8, pp. 246–255.

Moody, E. G., M. D. King, S. Platnick, C. B. Schaaf, and F. Gao. 2005. Spatially Complete Global Spectral Surface Albedos: Value-Added Datasets Derived from Terra MODIS Land Products. *IEEE Transactions on Geoscience and Remote Sensing*, Vol. 43, pp. 144–158.

Siebert, S., P. Doll, J. Hoogeveen, J. M. Faures, K. Frenken, and S. Feick. 2005. Development and Validation of the Global Map of Irrigation Areas. *Hydrology and Earth System Sciences*, Vol. 9, pp. 535–547.

Sterling, S., and A. Ducharne. 2008. Comprehensive Data Set of Global Land Cover Change for Land Surface Model Applications. *Global Biogeochemical Cycles*, Vol. 22, GB 3017, 20 pp.

Tucker, C. J., W. W. Newcomb, and H. E. Dregne. 1994. AVHRR Data Sets for Determination of Desert Spatial Extent. *International Journal of Remote Sensing*, Vol. 15, pp. 3547–3565.

Global Carbon Monitoring

Bala, G., K. Caldeira, M. Wickett, T. J. Phillips, D. B. Lobell, C. Delire, et al. 2007. Combined Climate and Carbon-Cycle Effects of Large-Scale Deforestation. *Proceedings of the National Academy of Sciences of the U.S.A.*, Vol. 104, pp. 6550–6555.

Hansen, M. C., S. V. Stehman, P. V. Potapov, T. R. Loveland, J. R. G. Townshend, R. S.

DeFries, et al. 2008. Humid Tropical Forest Clearing from 2000 to 2005 Quantified by using Multitemporal and Multiresolution Remotely Sensed Data. *Proceedings of the National Academy of Sciences of the United States of America*, Vol. 105, pp. 9439–9444.

Mayaux, P., P. Holmgren, F. Achard, H. Eva, H. Stibig, and A. Branthomme. 2005. Tropical Forest Cover Change in the 1990s and Options for Future Monitoring. *Philosophical Transactions of the Royal Society B—Biological Sciences*, Vol. 360, pp. 373–384.

McMillan, W. W., C. Barnet, L. Strow, M. T. Chahine, M. L. McCourt, J. X. Warner, et al. 2005. Daily Global Maps of Carbon Monoxide from NASA's Atmospheric Infrared Sounder. *Geophysical Research Letters*, Vol. 32, L11801.

van der Werf, G. R., J. T. Randerson, L. Giglio, G. J. Collatz, P. S. Kasibhatla, and A. F. Arellano. 2006. Interannual Variability in Global Biomass Burning Emissions from 1997 to 2004. *Atmospheric Chemistry and Physics*, Vol. 6, pp. 3423–3441.

Marine

Alvain, S., C. Moulin, Y. Dandonneau, and F. M. Breon. 2005. Remote Sensing of Phytoplankton Groups in Case 1 Waters from Global SeaWiFS imagery. *Deep-Sea Research: Part I. Oceanographic Research Papers*, Vol. 52, pp. 1989–2004.

Behrenfeld, M. J., R. T. O'Malley, D. A. Siegel, C. R. McClain, J. L. Sarmiento, G. C. Feldman, et al. 2006. Climate-Driven Trends in Contemporary Ocean Productivity. *Nature*, Vol. 444, pp. 752–755.

Chiu, L. S., R. Chokngamwong, Y. K. Xing, R. X. Yang, and C. L. Shie. 2008. "Trends" and Variations of Global Oceanic Evaporation Data Sets from Remote Sensing. *Acta Oceanologica Sinica*, Vol. 27, pp. 124–135.

Gregg, W. W., and N. W. Casey. 2004. Global and Regional Evaluation of the SeaWiFS Chlorophyll Data Set. *Remote Sensing of Environment*, Vol. 93, pp. 463–479.

The Outlook for the Field of Remote Sensing

The View from 2011

This section presents a concluding look at the prospects for the field of remote sensing during the next decade or so, revised and updated from the previous version (Campbell, 2007). It is pertinent to note that Chapter 1 identified several other prospective views looking forward from earlier eras, including Lee's (1922) review of the capabilities of aerial photography and the National Academy of Sciences (1970) report *Remote Sensing with Special Reference to Agriculture and Forestry*, which looked forward to the beginning of multispectral remote sensing within civil society. And, in 1973, the *Report of the Federal Mapping Task Force on Mapping, Charting, Geodesy, and Surveying* (Executive Office of the President, 1973), identified significant issues with interagency coordination and computerization that were implemented within the following decades. Such earlier reviews illustrate the value of the forward-looking spirit that has characterized the field, which has sought to identify the technological and social implications of present-day activities.

Each topic discussed here identifies a continuation of a process now already well under way, so it requires no acute insight to identify them, but presenting them as a list may highlight their significance. Here they are not presented in any particular order, and we do not imply that they form a comprehensive inventory of significant developments.

Society and Institutions

1. At the time of this writing, present developments indicate continued uncertainty as to the role of the U.S. government in supporting flagship remote sensing enterprises such as Landsat (Chapter 6) and EOS (Chapter 21). Since 2007, the Landsat Data Continuity Mission (LDCM) has been approved, funded, and, at present, appears to be on schedule for launch in 2012. In the interim, the aging Landsat sensors have continued their stalwart service. However, both Landsats 5 and 7 are experiencing large technical difficulties that now make a data gap probable. Landsat 7, in addition to the SLC failure in 2003, is low on fuel and has substantially reduced solid state recording capacity. The most significant issue facing Landsat 5 is, at present, the unlikely ability of the primary traveling wave tube amplifier (TWTA) to last until the launch of LDCM because of rising helix current.

Even if the Landsat sensors remain in service till the LDCM program begins operation, the broader issue of instrument and data continuity continues (Jensen, 2006), as the future of other successful and valuable systems remains in doubt. For example, the replacements for many of the EOS sensors, scientifically validated and of proven value, remain in question. Although U.S. federal support for high-visibility remote sensing programs may well continue, we can anticipate reduced levels of support and greater difficulty in marshalling political and public support in the face of competing priorities for public financing. In the United States, consolidation of federal governmental infrastructures supporting remote sensing across agencies and across application areas will provide a more focused, less diverse spectrum of activities.

2. We can expect continued internationalization of broad-scale remote sensing enterprises, in part because of the inherent significance of such activities and in part because other nations will fill the gap created by the contraction of U.S. government programs. Other players will likely include present participants, such as the European Union, China, Japan, India, Indonesia, Malaysia, and Canada, as well as others. Although there are many advantages in nations operating multiple systems with similar capabilities, these advantages and economies can be fully exploited when standards and capabilities are coordinated internationally.

3. Corporate remote sensing enterprises will continue to grow and play a significant role in the practice of remote sensing, often with a strong international component. Although the most visible organizations may focus on collection of imagery, those enterprises that process and distribute value-added products will increase with respect to the number and diversity of their specializations. Many will specialize less in the traditional remote sensing-based products than in unconventional products that will combine a remote sensing focus with unexpected application areas—such as real estate, banking, public health, and law enforcement, for example. Sophistication of remote sensing applications in disaster management at the local, state, and federal levels will increase.

4. Availability of imagery to the public, and the availability of tools to use it, will increase. Society will accelerate its rate of establishing new applications that directly influence day-to-day concerns of individual citizens who are not specialists in the field. The number and range of applications will increase, extending to include not only institutions with specialized expertise in remote sensing but also individuals, households, businesses, local governments, and community enterprises that previously made little use of imagery or that required products produced by specialists. This has the potential to empower individuals and organizations that previously had little practical access to the information extractable from remotely sensed data. Nonprofit organizations, such as SkyTruth and Amnesty International's Science for Human Rights program, will increasingly bring imagery to bear on both environmental and human rights issues (Plate 29).

5. During the interval since the release of the previous edition, remotely sensed imagery has increased its significance in the public arena to an extent that could not have been anticipated, reaching audiences far beyond the rather focused community of specialists that was typical of the discipline in past decades. Remotely sensed imagery reaches the public through media that did not exist even a decade before. There are

many more widely available applications in which imagery plays a prominent role, such as Google Earth, mobile mapping, location-based services, and image backdrops for city and county-level community data. Such innovations influence the practice of remote sensing by increasing public awareness of the nature and value of accurate and timely imagery at the local level and of its significance for government, commerce, and science. We expect such innovations to continue.

Such developments will create societal benefits, economies, and efficiencies but will also create controversy, including disputes about issues such as privacy, use by law enforcement, ownership of imagery (and the information derived from it), and issues that arise from merging aerial imagery with other forms of data. Such developments will create a context for the discipline that differs greatly from the recent past, in which the field was seen as the province of specialists largely detached from such issues.

Technology and Analysis

1. As this is written, the digital format has displaced the film format for most aerial survey applications and is rapidly displacing those remaining applications of film.

2. The opening of the full U.S. Landsat archive for public use, via Internet delivery and without charge to customers, represents a landmark achievement that is likely to transform several important dimensions of remote sensing science. The availability of such geographic and temporal depth and consistent calibration will support an already growing number of change detection investigations. Such studies will lead to improved understanding of long-term ecological processes and to innovation in the development of change detection procedures and algorithms. Perhaps as important, the U.S. policy has spurred the European Space Agency to adopt a comparable policy for Sentinel-2.

3. Although basic photointerpretation skills have been deemphasized in many, if not most, curricula, their significance remains even in the realm of the digital format and advanced instruments. The availability of increased image detail often leads to increase in the demand for extraction of increasingly detailed information. Thus far, advances in machine-assisted interpretations are not yet adequate for the needs of many fields that require spatially detailed, discipline-specific information. This development will require a cadre of image analysts—skilled interpreters, with discipline-specific knowledge, proficient both in image interpretation and manipulation of digital imagery.

4. Accelerated interest in analysis of sets of data and assemblages and sequences of data (per the discussion of HDFs and NetCDFs in Chapter 4) that integrate very large arrays of multispectral or hyperspectral data, in combination with other large datasets such as meteorological or environmental data, is the explicit goal of the Global Earth Observation System of Systems (GEOSS). Remote sensing will play a key role for such analyses. Image fusion may well assume greater significance with the development of more sophisticated algorithms for combining information for many spectral channels.

5. We can expect the convergence of remote sensing with allied technologies, espe-

cially GPS and GIS, and with related systems such as wireless communications, and miniaturization of imaging systems. Remote sensing will be influenced by progress in advanced technologies, such as "formation flying" of remote sensing satellites and innovations in UAVs (Chapter 13).

6. We can expect expansion of applications of remote sensing into fields with immediate significance for the public, including biosecurity, plant pathology, epidemiology, support of disaster relief, natural hazards, and homeland security. Although remote sensing will often contribute an essential component of such efforts, other geospatial technologies will form significant components of such systems.

7. One consequence of combined effects of such developments will be recognition of the significance of human–computer interaction (HCI) with respect to remote sensing imagery and its interpretation. Advanced display systems and related software dedicated to visualization of complex data will be required to achieve many of the advances listed above. Although some of these advances will be based on improvements in display hardware, many of the critical advances may well be based on display software that will permit presentation of complex data in ways that enhance the ability of humans to understand its meaning, especially when it forms a part of a dynamic series or must be evaluated with respect to related datasets.

References

Campbell, J. B. 2007. The Outlook for the Field of Remote Sensing. Conclusion in *Introduction to Remote Sensing*. New York: Guilford Press, pp. 608–610.

Campbell, J. B., and V. V. Salomonson. 2010. Remote Sensing: A Look to the Future. Chapter 25 in *Manual of Geospatial Sciences and Technology*, 2nd ed. (J. D. Bossler, ed.). London: Taylor & Francis, pp. 487–509.

Executive Office of the President. 1973. *Report of the Federal Mapping Task Force on Mapping, Charting, Geodesy, and Surveying*, Washington, DC: U.S. Government Printing Office.

Jensen, J. R. 2002. Remote Sensing: Future Considerations. Chapter 23 in *Manual of Geospatial Sciences and Technology* (J. D. Bossler, ed.). London: Taylor & Francis, pp. 389–398.

Lee, W. T. 1922. *The Face of the Earth as Seen from the Air* (American Geographical Society Special Publication No. 4). New York: American Geographical Society, 110 pp.

National Academy of Sciences. 1970. *Remote Sensing with Special Reference to Agriculture and Forestry*. Washington, DC: National Academy of Sciences, 424 pp.

National Research Council. 2008. *Satellite Observations to Benefit Science and Society: Recommended Missions for the Next Decade, Committee on Earth Science and Applications from Space: A Community Assessment and Strategy for the Future* (R. Henson, ed.). Washington, DC: National Research Council.

Index

(*f* indicates a figure; *t* indicates a table)

About the Authors

James B. Campbell, PhD, is Professor of Geography at Virginia Tech, where he teaches remote sensing, quantitative methods, and geomorphology. He has worked closely with students and faculty in related fields such as forestry, geology, agronomy, environmental sciences, and planning. Since 1997 he has served as codirector of Virginia Tech's Center for Environmental Applications of Remote Sensing (CEARS). The author of numerous technical articles and several books, Dr. Campbell received the Outstanding Service Award of the American Society for Photogrammetry and Remote Sensing in 1994 and its Fellow Award in 1996. In 1997 he received the Outstanding Service Medal awarded by the Remote Sensing Specialty Group of the Association of American Geographers. Dr. Campbell's research has been sponsored by numerous academic, governmental, and private organizations, including NASA, the National Science Foundation, and the United States Geological Survey. He has been active in the AmericaView Program, including service as a principal investigator for the VirginiaView consortium, as a member of the AmericaView Board of Directors (2003–present), and as chair of the Board of Directors (2006–2007).

Randolph H. Wynne, PhD, is Professor in the Forest Resources and Environmental Conservation Department at Virginia Tech. He also serves as Remote Sensing Team Leader for the Forest Productivity Cooperative, Associate Director of the Conservation Management Institute, and Co-Director of the Center for Environmental Applications of Remote Sensing. He teaches courses in forest photogrammetry and spatial data processing and remote sensing of natural resources. Dr. Wynne's research interests are in the applications of remote sensing to forestry, natural resource management, ecology, ecosystem services, and earth system science.